Modern Solder Technology for Competitive Electronics Manufacturing

Electronic Packaging and Interconnection Series

Charles M. Harper, Series Advisor

Alvino PLASTICS FOR ELECTRONICS
Classon SURFACE MOUNT TECHNOLOGY FOR CONCURRENT ENGINEERING AND MANUFACTURING
Ginsberg and Schnoor MULTICHIP MODULE AND RELATED TECHNOLOGIES
Harper ELECTRONIC PACKAGING AND INTERCONNECTION HANDBOOK
Harper and Miller ELECTRONIC PACKAGING, MICROELECTRONICS, AND INTERCONNECTION DICTIONARY
Harper and Sampson ELECTRONIC MATERIALS AND PROCESSES HANDBOOK, 2/e
Lau BALL GRID ARRAY TECHNOLOGY
Licari MULTICHIP MODULE DESIGN, FABRICATION, AND TESTING

Related Books of Interest

Boswell SUBCONTRACTING ELECTRONICS
Boswell and Wickam SURFACE MOUNT GUIDELINES FOR PROCESS CONTROL, QUALITY, AND RELIABILITY
Byers PRINTED CIRCUIT BOARD DESIGN WITH MICROCOMPUTERS
Capillo SURFACE MOUNT TECHNOLOGY
Chen COMPUTER ENGINEERING HANDBOOK
Coombs ELECTRONIC INSTRUMENT HANDBOOK, 2/e
Coombs PRINTED CIRCUITS HANDBOOK, 4/e
Di Giacomo DIGITAL BUS HANDBOOK
Di Giacomo VLSI HANDBOOK
Fink and Christiansen ELECTRONICS ENGINEERS' HANDBOOK, 3/e
Ginsberg PRINTED CIRCUITS DESIGN
Juran and Gryna JURAN'S QUALITY CONTROL HANDBOOK
Jurgen AUTOMOTIVE ELECTRONICS HANDBOOK
Manko SOLDERS AND SOLDERING, 3/e
Rao MULTILEVEL INTERCONNECT TECHNOLOGY
Sze VLSI TECHNOLOGY
Van Zant MICROCHIP FABRICATION

Modern Solder Technology for Competitive Electronics Manufacturing

Jennie S. Hwang, Ph.D.

McGraw-Hill
New York San Francisco Washington, D.C. Auckland Bogotá
Caracas Lisbon London Madrid Mexico City Milan
Montreal New Delhi San Juan Singapore
Sydney Tokyo Toronto

Library of Congress Cataloging-in-Publication Data

Hwang, Jennie S.
Modern solder technology for competitive electronics manufacturing / Jennie S. Hwang.
p. cm. (Electronic packaging and interconnection series)
Includes bibliographical references and index.
ISBN 0-07-031749-6 (hc)
1. Electronic packaging. 2. Solder and soldering. I. Title. II. Series.
TK7870.15.H93 1996
621.381'046—dc20 95-44641
CIP

McGraw-Hill

*A Division of The **McGraw-Hill** Companies*

1 2 3 4 5 6 7 8 9 0 DOC/DOC 9 0 1 0 9 8 7 6

ISBN 0-07-031749-6

The sponsoring editor for this book was Stephen S. Chapman, the editing supervisor was Peggy Lamb, and the production supervisor was Suzanne W. B. Rapcavage. It was set in Century Schoolbook by Terry Leaden of McGraw-Hill's Professional Book Group composition unit.

Printed and bound by R. R. Donnelley & Sons Company.

This book is printed on recycled, acid-free paper containing a minimum of 50% recycled, de-inked fiber.

to the memory of my mother
and grandfather

to Leo for his assistance, endurance,
and encouragement

to Raymond and Rosalind
for their ever-bright future

Contents

List of Figures

List of Tables

Preface

Advanced technologies are transforming manufacturing, and the information highway is getting packed.

In this ever-changing, fast-paced industry, new developments in semiconductors, integrated circuit (IC) packages, and circuit board assemblies have been introduced through various formats. Innovative technologies are unveiled through press releases, technical conferences, panel discussions, dedicated seminars, and journal publications. Yet an integrated and systematic source is largely lacking. It is hoped that this book will fill the void in one of the most important fields in electronics manufacturing—modern solder technology.

It is the author's belief that value-added industrial applications stem from the synergistic result of general understanding of the big picture of technologies and applications and an in-depth knowledge of a targeted application area. That belief prompted the writing of this book in an effort to facilitate the information flow and enhance the knowledge base, thus ultimately contributing to the manufacturing of globally competitive electronics products in performance, in reliability, and in cost.

With the expected wide spectrum of readers in terms of expertise and level of experience, the author has sought to provide a middle ground in the level of treatment of topics. The material is intended as neither too theoretical for the reader with only practical applications in mind nor overly descriptive for the reader who has extensive background in the subject. It is however the author's intention to offer a comprehensive and complete book which covers all essential elements that are associated with the applications of solder material and the soldering process for electronics and microelectronics packaging and assembly.

The book is written to serve as working knowledge and as a guide necessary for implementing a manufacturing system as well as to provide insights into future technology and process advancement. The book can also serve as a text for college and graduate studies to gain a quick grasp of the state of the art of manufacturing technology for the electronics and microelectronics industry which is one of the largest job offering sectors within the United States and other industrialized nations.

Solders have been successfully used as an interconnecting material for electronic packaging and assembly, providing electrical, thermal, and mechanical functions. With continued implementation of surface mount technology, solder paste and soldering will continue to be the most practical material and process system in mass production, playing a crucial role in the overall performance and reliability of electronic circuits.

Of the 20 chapters, the first two provide highlights of market trends, technological demands, and new developments of semiconductors and IC packages. Fundamental material properties and characteristics including flux chemistry, solderability factors, solder metallurgy, solder paste technology, and microstructure of solder joints are covered in Chapters 3 through 7. Chapters 8 and 9 discuss manufacturing processes in water-clean and no-clean as well as controlled atmosphere soldering. Chapters 10 and 11 are dedicated to two primary bonding architectures of IC packages—peripheral fine pitch (QFP) and grid array bonding. Soldering methodologies are discussed in Chapter 12. Commonly encountered production-related problems and reliability issues are aggregated in Chapter 13. With respect to new materials, approaches to strengthen solder materials and the development of lead-free solders are separately treated in Chapters 14 and 15. Solder joint failure modes and factors affecting solder joint reliability are illustrated in Chapter 16. Chapters 17 and 18 summarize the U.S. and International industry standards and specifications related to solders and application of solders. The industry has made significant progress in reducing defect rate and the extent of rework and repair. Nonetheless, inspection, rework, and repair continue to be a part of the manufacturing operation. Chapter 19 covers some basic concepts and techniques in this area. The book concludes with a summary of emerging technologies which directly or indirectly affect the needs and requirements of solder technology and with some thoughts about future efforts and studies warranted in the field.

It is hoped that the integrated treatment of information coupled with the author's assessment and analysis will be of value to many in manufacturing electronic packages and assemblies. It is also hoped that the text will simulate further innovations and developments of end-use products. The products are featured with faster speed, smaller size, lighter weight, increased ruggedness, and lower cost in the broad spectrum of applications ranging from communication, computing equipment, such as personal computers and cellular phones, to daily tools, such as automobiles and dishwashers, and to critical electronics, such as weaponry and life support systems.

As always, the author wishes to acknowledge her debt of gratitude to the considerable number of people (the enumeration of whom would not be practical here), who have contributed in direct or indirect ways to the knowledge, insights, and inspiration that have enhanced the author's career and nurtured her as a person. Special mention is given to Rev. William P. Cooke for his constant commendation of the author's efforts and progress which has been a lively encouragement for nearly three decades. Special thanks should also be given to the publisher's staff for their support and effort, particularly Mr. Steve Chapman and Ms. Peggy Lamb, and to Mr. Charles M. Harper for coordinating the Electronic Packaging and Interconnection Series. The author gratefully acknowledges her indebtedness to Mrs. Martha Payne and Mrs. Vivian Sobieski for their outstanding work in typing the entire manuscript in a skillful and patient manner. Finally, the author gives deepest gratitude to her family, Leo, Raymond, and Rosalind, for their all-time support, encouragement, and understanding in the face of lost evenings, weekends, and holidays. In contributing to the impressive growth of the industry over the past 15 dynamic years, the author has been passion ately embarking on technologies, products, and manufacturing know-how, and information dissemination. During this period, Raymond has been brought up to be a well-rounded and bright young man, now a sophomore at MIT; and Rosalind, a sensible and talented young lady, now a junior at Hawken School. Having done both, nothing can be more fulfilling or rewarding.

Jennie S. Hwang, Ph.D.
January 1996
Cleveland, Ohio U.S.A.

Acknowledgments

My grateful thanks are tendered to the following companies and individuals who have provided photographs, testing results, information, and/or advice:

AMTX

CEERIS International, Inc.

Compix Inc.

Four Pi Systems

Glenbrook Technologies, Inc.

H-Technologies Group, Inc.

IBM—Assembly Process Design

IBM—Microelectronics, Ceramic Packaging Business

IHS Publishing Group

Motorola—Semiconductor Products Sector

Photo Stencil, Inc.

Ricon Electronic Systems

Tessera, Inc.

The Panda Project—Archistrat Division

Kathleen Beaulieu

Bruce Bollinger

John Carey

Thomas Caulfield, Ph.D.
Richard S. Clothier
William E. Coleman, Ph.D.
Mary Cretekos
Thomas Di Stefano. Ph.D.
Gary Freeman
Steve Hall
Charles M. Harper
Lisa Heck
Stanley Jasne, Ph.D.
Charles-Henri Mangin
Stephen M. Miller
Giana Phelan
Maria M. Portuondo
Jim Walcutt
Kathryn Weaver
Gil Zweig

And collective thanks to all those individuals and publications cited in the references of each chapter.

About the Author

Dr. Jennie S. Hwang is a recognized top leader in the field of electronics packaging and interconnection. She is a popular lecturer worldwide and a consultant to OEMs, U.S. military, and SMT manufacturers.

Dr. Hwang received her Ph.D. in Materials and Metallurgical Engineering from the Engineering School of Case Western Reserve University, an M.S. in Liquid Crystals from the Liquid Crystal Institute of Kent State University, and a second M.S. in Physical Chemistry from Columbia University. She also has a bachelor's in chemistry. In conjunction with multidisciplinary training, her many years of experience as a technology and business manager covering various industries has brought her a unique breadth and depth of knowledge in materials and processes.

In addition to many publications, she is the sole author of two text books: *Solder Paste in Electronics Packaging* and *BGA & Fine Pitch Peripheral Interconnection*. She is a contributing author to three edited books: *Solder Joint Reliability, Electronic Packaging and Interconnection Handbook*, and *Fine Pitch Technology Handbook*. She writes a monthly column for SMT magazine. She holds several patents in the area of electronic solders.

Dr. Hwang serves on the International Materials Review Committee of the British Institute of Metals and American Society of Metals. She has served as the president and a board director of Surface Mount Technology Association. She is on the advisory board of *Surface Mount Technology Magazine*. She serves on the advisory group to IPC and IEC working on Surface Mount International Standards and on the Advisory Board of NEPCON Conferences/Exhibitions. She is a member of The Mic roelectronic Society (ISHM), American Ceramic Society,

American Chemical Society, American Society of Metals, and Surface Mount Technology Association, and an expert member of the American National Standards Institute.

She is inducted into the Executive Order of Ohio Commodore and selected as a most interesting person in the Greater Cleveland area. She is appointed as a Commerce Department's District Export Council—Northern Ohio.

She serves on a number of professional, trade, and civic committees and boards, including IPC Public Policy Council and Tax & Competitiveness Chair.

Dr. Hwang is currently CEO of H-Technologies Group, Inc., specializing in providing business and manufacturing solutions to the electronics/microelectronics industry. Prior to this position, she has held senior managerial and research positions with Martin Marietta Corp., SCM Corp., Sherwin Williams Co., and IEM Corp.

1 Introduction

1.1 Purpose and Utility of the Book

Advanced technologies are transforming manufacturing, and the information highway is in sight. In this exciting and changing time, the electronics industry has responded and will continue to respond to the need for competitive products in the global marketplace. In our industry the general features of the desired end-use products are faster speed, lighter weight, and smaller size, in addition to lower cost and increased ruggedness.

What has transpired from these market demands is continued technology innovation. The current and future technology is not only derived from the conventional material theory of circuits, which is based on the mobility and conductivity of electrons and phonons, but also from that of photons. Looking at the hierarchy of electronics, it is seen that semiconductor devices, following Moore's Law, double their complexity every year. For example, the one-megabit (Mbit) dynamic random access memory (DRAM) of 1990 has progressed to a 16-Mbit generation this year and is projected to approach a 1-Gigabit (Gbit) generation at the turn of the century. At the time of this writing, microprocessors have reached a speed of 133 megahertz (MHz) (Intel Pentium) using an 8-inch (in) [200-millimeter (mm)] wafer and a 0.35-micrometer (μm) process. The dimension of integrated circuits (IC) continues to shrink down to a deeper submicrometer level. The submicrometer geometries of silicon circuitry naturally demand newer and better technologies in wafer processing, die packaging, and subsequent assembly. Chip scale (size) packages and other advanced surface mount devices such as flip chip or controlled collapse chip connection (C-4), ball grid array (BGA) or pad array carrier (PAC), tape automated bonding (TAB), fine-pitch quad flat pack (QFP) in single-chip packages as well as in multichip modules (MCM) are taking shape. Multichip modules are also adding to the electronics hierarchy another level of packaging which fits between the chip and the board.

As electronic packaging and assembly continues to increase in circuitry density and in functionality per unit area or space, the interconnection technology in materials, equipment, and process becomes increasingly important. A thorough examination of technologies and techniques at all levels is warranted in an effort to deliver quality and reliable interconnections.

Solders have been commonly used as interconnecting material for electronic packaging and assembly, providing electrical, thermal, and mechanical functions. With the implementation of surface mount technology, solder paste and soldering, which are the most practical and viable material and process system, respectively, in mass production, continue to play a crucial role to the overall performance and reliability of electronic circuit assemblies. Figure 1.1 depicts the distribution of defects contributing to the overall faults on surface mount manufacturing. Solder-related defects constitute the largest fault, although

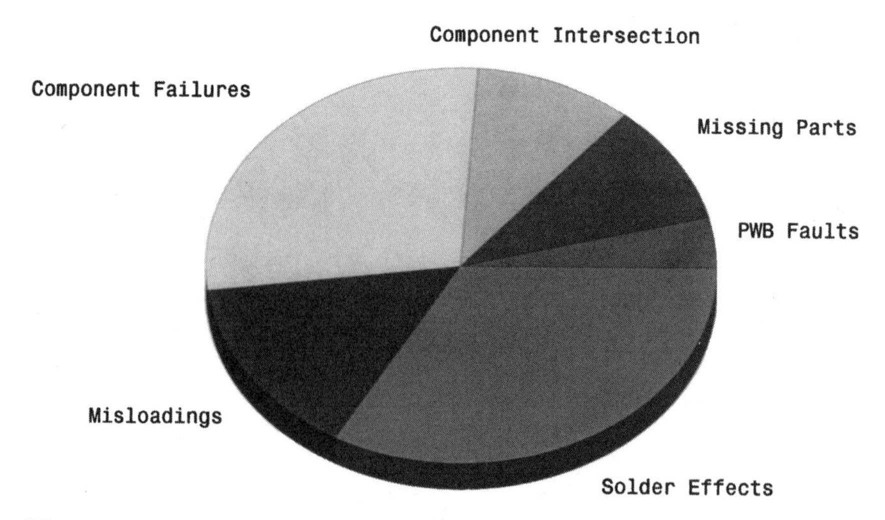

Figure 1-1. Distribution of surface mount manufacturing defects.

some solder defects may originate from the faults of boards, components, and the process employed. Understanding the fundamentals underlying solders and soldering and gaining current knowledge of techniques and applications are the means to alleviate the fault of solder defects, thus achieving an increasingly competitive manufacturing.

This book is intended to provide complete coverage on all relevant and related technologies and techniques which are associated with the applications of solder and soldering for electronic/microelectronic packaging and assembly. The book also includes a proper level of underlying fundamentals that are important to the practical applications.

It is hoped that the book will serve as a working knowledge and a guide necessary for implementing a manufacturing system as well as providing insights to future technology and process advancement.

The book consists of 20 chapters. This chapter (Chap. 1) provides an introduction and highlights the market trends and technological demands. Chapter 2 provides an overview of newly developed IC packages for surface mount manufacturing with an emphasis on the design, materials, and process involving solders and soldering. The fundamental material properties are discussed in Chap. 3 followed by soldering chemistry and solderability, which are covered in Chaps. 4 and 5, respectively. The microstructure, which is one of the most important characteristics of solder alloys, has not been adequately studied in the literature. Thus Chap. 6 is dedicated to the discussion of all aspects of the microstructure of solders suitable for the manufacturing sector, including the examples of micrograms of various alloy compositions and the relationship among the soldering process, the microstructure, and the resulting solder joint properties. Chapter 7 covers the essentials of solder paste technology. Chapter 8 is devoted to two manufacturing solder chemistries, water-clean and no-clean, which are deemed to be the viable solutions to eliminate the use of chlorofluorocarbons (CFCs). The subject of protective and reactive atmospheric soldering, because of its scope and demand, is treated separately in Chap. 9.

The advances in the development of IC packages are expected to direct the board-level surface mount manufacturing into two paths based on the two primary bonding architectures in order to accommodate increased input/output (I/O) counts:

1. *Peripheral package.* The I/O pin counts with narrow pitches between pins distributed along the perimeter only, such as fine-pitch QFP.

2. *Array package.* The I/O interconnecting sites distributed over the whole substrate area, such as BGA or PAC.

In the foreseeable future, it is anticipated that QFP and BGA IC packages will share the surface mount manufacturing shelves for their respective merits and individual production fitness. Quad flat pack applications for surface mount fine-pitch technology are covered in Chap. 10, and BGA applications are covered in Chap. 11.

Chapter 12 discusses the soldering methodologies and associated equipment. Chapter 13 covers the prevalent concerns and questions about soldering-related issues, particularly the role of intermetallics, gold-plated substrate contamination effects, plastic package cracking, solder balling, and solder voids.

In view of increased demands in the performance of solder material and the need for lead (Pb)-free solders, research and development of improved solder alloys have been conducted through various efforts. Improvements in solder materials are covered in Chap. 14 and those for lead-free solders in Chap. 15. Chapter 16 outlines the basic solder joint failure modes and the factors affecting solder joint reliability including case studies that exemplify solder joint or assembly failures. The quality issue and various industry and military specifications and standards from US and international efforts are covered in Chap. 17. The commonly accepted solder joint fillet criteria for the standard component configurations are presented in Chap. 18. Chapter 19 covers solder joint inspection, rework, and repair.

The last chapter, Chap. 20, provides a summary of thoughts about future effort and studies warranted in this field, including pending issues. The chapter also outlines the emerging technologies which directly or indirectly affect the solder and soldering area, and thus the electronics industry.

1.2 Driving Forces

There are mainly three immediate driving forces which direct the electronic/microelectronic market and industry: (1) increasing circuit density in semiconductor level, (2) addressing environmental/health issues in all levels of manufacturing, and (3) meeting global competition in quality, reliability, and cost. Looking from semiconductor level down to the end-use products, the schematic of Fig. 1.2 illustrates the hierarchy of the electronics/microelectronics industry in terms of key technologies in electronic packaging and assembly.

It is generally recognized that DRAMs and microprocessors are two

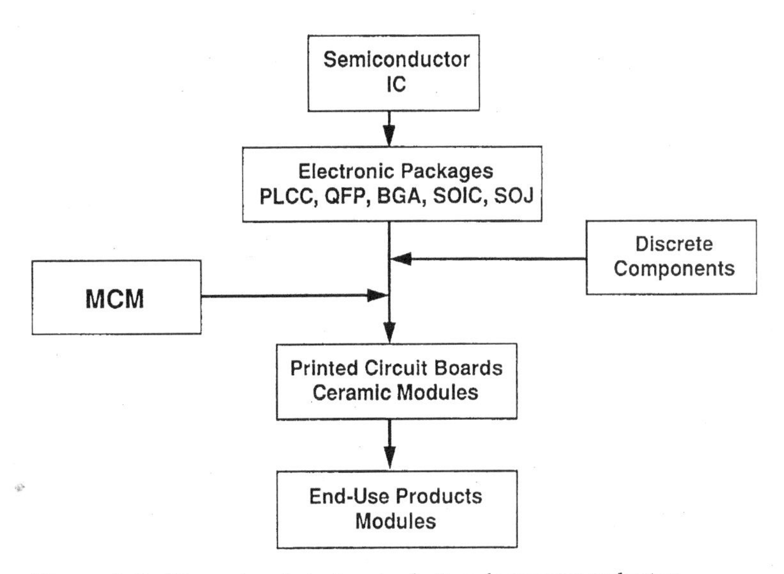

Figure 1-2. Hierarchy of electronics/microelectronics industry.

primary technology drivers,[1] although DRAM and microprocessors along with application specific integrated circuit (ASIC) are considered major application technologies. As depicted in Fig. 1.3, the use of the design rule for the current 16-Mbit generation of DRAM will continue to decrease and a 1-Gbit generation will dominate at the turn of the century in order to meet the market pressure. The developmental mile-

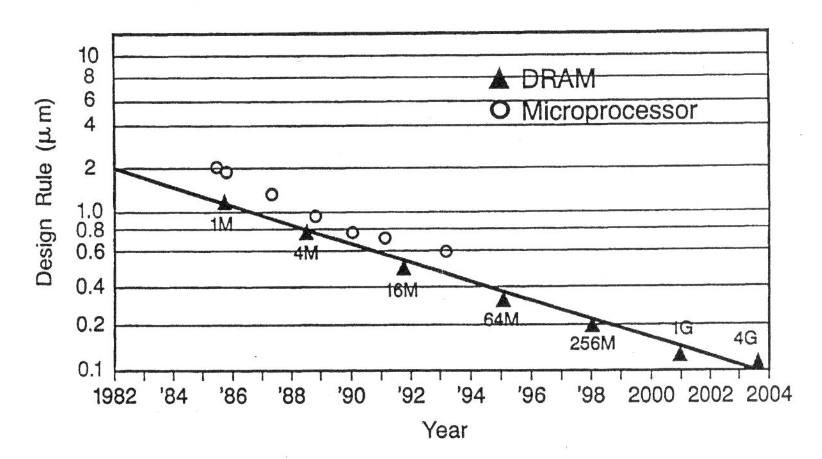

Figure 1-3. Design rule of DRAM.

Table 1-1. DRAM Milestones, Worldwide Averages

DRAM generation	Minimum geometry, μm	New tool development	Customer qualification	Peak DRAM production in bits
4 Mbits	0.80	1983/84–1988	1988/89	1994
16 Mbits	0.50	1985/86–1990/91	1991	1997
64 Mbits	0.35	1987/88–1993	1993/94	2000
256 Mbits	0.25	1989/90–1995/96	1996	2003
1 Gbits	0.15	1991/92–1998	1998/99	2006

SOURCE: Kopp Semiconductor Engineering.

stones of DRAM are outlined in Table 1.1.[2] The design rules for ASICs, however, tend to lag behind the memory chips by about 1½ to 2 years. It is further anticipated that IC complexity continues to follow Moore's Law, i.e., roughly doubles every year.[3] This applies to key parameters such as feature size reduction, die/wafer size increase, and component density increase. In addition, design innovation and reliability improvement will also follow suit. As the silicon geometries go to deeper submicrometer, the design, manufacturing, and reliability of the IC packages and their subsequent interconnections will become the major tasks for the industry. In order to accommodate the requirement of I/O sites, new and renewed packaging designs and interconnecting technologies were introduced. These lead to the conventional surface mount packages in finer pitches of lead configurations and the advanced array surface mount packages. These packages can house not only a single chip but also multiple chips. Section 1.3 outlines the commonly used IC packages and emerging IC packages, and Chap. 2 compares various packaging and assembly technologies.

1.3 IC Packages

Since surface mount technology was implemented in the printed circuit industry in the early 1980s, surface mount manufacturing has continued to grow and to gain market share at the expense of through-hole technology. It is estimated that 50 to 60 percent of total IC packages are in a surface mount configuration at the time of this writing and that surface mount packages will surpass through-hole packages by 1996 as shown in Fig. 1.4. This trend is largely attributed to the inherent merits of surface mount technology:

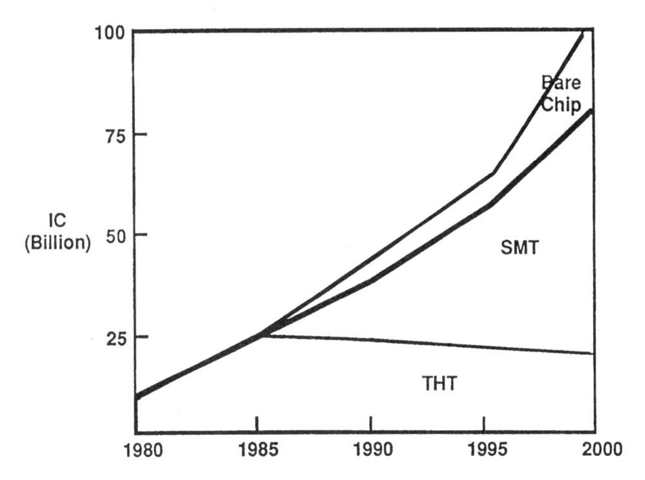

Figure 1-4. IC packaging evolution.

Increased circuit density

Decreased component size

Decreased board size

Reduced weight

Shorter leads

Shorter interconnections

Improved electrical performance

Facilitation of automation

Lower cost in volume production

1.3.1 Conventional surface mount IC packages

The selected surface mount components under the Joint Electronic Device Engineering Council (JEDEC) Committee or the Institute for Interconnecting and Packaging Electronic Circuits (IPC) standards are listed as follows, and their configurations are illustrated in Fig. 1.5.

- Small outline integrated circuits (SOICs) with gull-wing leads
- SOICs with J leads
- Plastic leaded chip carriers (PLCCs)
- Thin small outline packages (TSOPs)

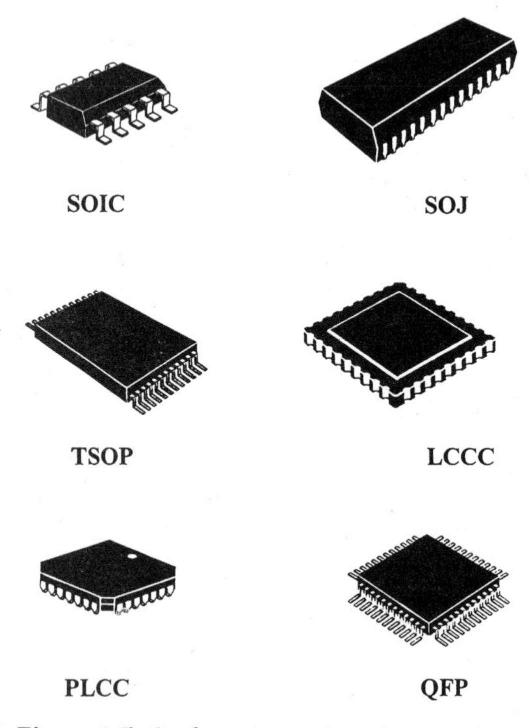

Figure 1-5. Surface mount IC package configurations.

- QFPs with lead pitch larger than 0.020 in (0.50 mm)
- Leadless ceramic chip carriers (LCCCs)

1.3.2 Advanced surface mount packages and chip-attach technologies

- QFPs with lead pitch equal to or less than 0.020 in (0.5 mm)
- Flip chip
- Tape automated bonding (TAB)
- Chip-on-board (COB)
- Ball grid array (BGA)
- Pad array carrier (PAC)

- Multichip module (MCM)
- Chip scale (size) packages and assembly

1.4 Environment and Health Perspectives

In addition to technology advances, another strong driving force affecting the industry is the environmental and health/safety issues. The public perception on these issues is well reflected by the result of a poll conducted by the *Wall Street Journal* and NBC.[4] The three questions asked and the corresponding responses are as follows:

- Overall, do you feel the environment has gotten better, gotten worse, or stayed about the same over the past 20 years?

Better	20%
Worse	66%
Same	13%

- Are you willing to make sacrifices for a better environment?

Yes	80%
No	20%

- In the past 6 months, have you purchased any product specifically because the product or the manufacturer has a good reputation for protecting the environment?

Yes	46%
No	45%

The responses to these questions clearly indicate that the condition of the environment that we live in is of concern and that the effort from the manufacturing community to retain and/or improve its quality is warranted. Although environment-friendly products did not govern the market share at the time of the poll, it is anticipated that manufacturers that produce products which are not environmentally safe will not be competitive in the reasonably near future.

Within the electronic packaging and assembly industry, the top four issues related to global environment and public health are

- Ozone-depleting CFC elimination
- Wastewater handling

- Lead control
- Volatile organic compound (VOC) control

1.4.1 CFC elimination

Chlorofluorocarbons have been one of the most useful chemical groups serving several major industries effectively for many years. They are used as a coolant in refrigeration and air-conditioning, a foaming agent for plastic-foam metals, a propellant in aerosols, and a cleaning solvent for the electronics and metal industry. The worldwide output of CFCs in 1991 was 680,000 metric tons[5] with application shares as outlined in Fig. 1.6. It is believed that CFCs are released mostly in the northern hemisphere and are then quickly dispersed throughout the troposphere, from where they slowly enter the stratosphere causing the depletion of the ozone layer as illustrated in Fig. 1.7. The loss of ozone increases the amount of ultraviolet (UV) radiation reaching the earth's surface. Scientific studies have shown that the enhanced UV radiation is linked to skin cancer and agricultural damages.

Both government and industry have recognized the threat of ozone depletion and its relation to CFCs since the 1970s. This resulted in the ban of CFC use in aerosol applications in 1978.

It was reported that global ozone levels for the second half of 1992 and for early 1993 were the lowest ever observed in the 14 years dur-

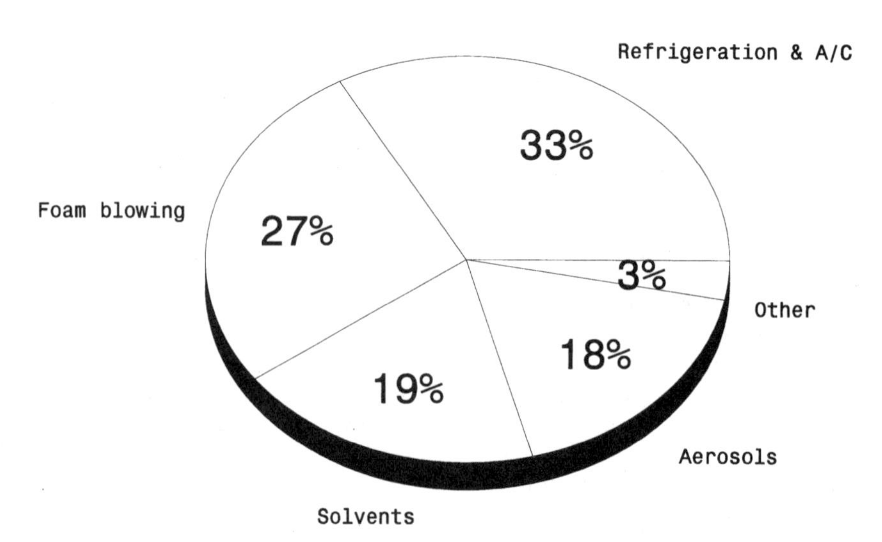

Figure 1-6. Worldwide CFC applications.

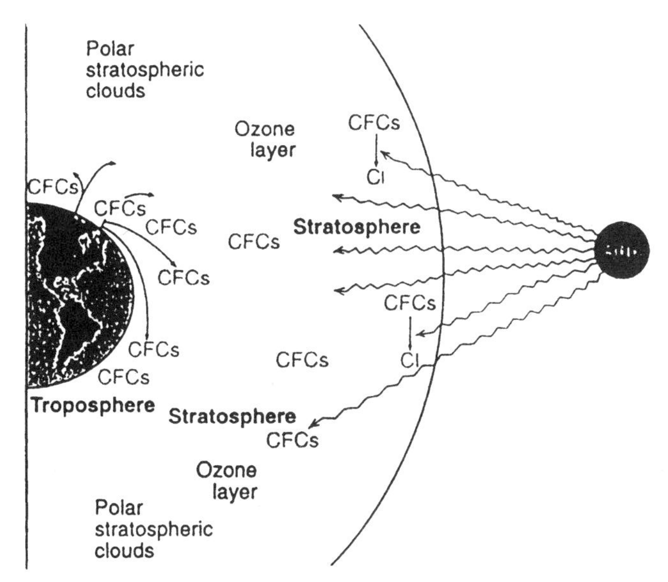

Figure 1-7. Ozone layer shielding earth from solar UV radiation.

ing which the National Aeronautics and Space Administration (NASA) has been monitoring the stratosphere from space. This observation further heightened the concern by all.

In 1987, governments from around the world formed the Montreal Protocol which called for a 50 percent reduction in CFC use by 1998. In the United States the government has taken action to speed up the phase out of ozone-depleting chemicals and has urged other nations to do the same. As a result, U.S. production of CFCs will be eliminated by December 31, 1995, and beginning May 15, 1993, all products manufactured with a process that uses CFCs must display a warning label stating that "the product was manufactured using a substance that depletes ozone." Ozone researchers estimate that the amount of ozone-depleting chemicals in the atmosphere will peak around the year 2000 before slowly beginning to decline, assuming that the nations of the world comply with the Montreal Protocol.

Although there are opposite opinions about the role of CFCs in the depletion of ozone in the stratosphere and about the validity of ozone depletion reports, significant scientific results conclude the existence of ozone depletions and the linkage of CFCs to ozone depletion. According to Professor James G. Anderson of Harvard University in a lecture: "We know unequivocally that CFCs are responsible for ozone destruction in the Antarctic....Chlorine and bromine top the list of suspects." Ozone researchers believe that chlorine and bromine catalyze the ozone destruction process. The constant motion of the atmosphere, which makes it difficult to assess the changes in ozone, is the result of atmospheric dynamic fluctuation or of the chemical destruction. Yet the concluded ozone depletion is the result after the seasonal variations and solar cycle have been factored in.

The basic chemistry involved in the balance of ozone dynamics can be expressed as follows:

Formation of ozone in the upper stratosphere

$$O_2 + \text{UV radiation} \rightarrow O\cdot + O\cdot$$

$$O\cdot + O_2 \rightarrow O_3$$

Consumption of ozone

$$O_3 + \text{UV radiation} \rightarrow O_2 + O\cdot$$

Thus ozone is rapidly formed in the stratosphere and that is consumed when it absorbs UV light since the process produces atomic oxygen that reacts with oxygen molecules to produce another ozone molecule. Ozone can also react with other radicals such as chlorine and nitrogen. The fully halogenated compounds as CFCs and halons, which are essentially inert in the troposphere, can rise above the bulk of the ozone layer and be photolyzed by UV light. This produces halogen atoms which capture and destroy ozone molecules. The corresponding chemical reactions are expressed as follows:

$$CCl_3F + \text{UV radiation} \rightarrow CCl_2F + Cl\cdot$$

or

$$CCl_2F_2 + \text{UV radiation} \rightarrow CCl_2F_2 + Cl\cdot$$

$$Cl\cdot + O_3 \rightarrow ClO + O_2$$

$$ClO + O\cdot \rightarrow Cl\cdot + O_2$$

Lately, some major skepticisms countering the finding that CFCs deplete ozone have been confronted with the following reasonings.[6]

- *Skepticism.* CFCs are heavier than air and should not rise up to the stratosphere.
- *Reasoning.* Heavy molecules can rise when they are mixed with light molecules and are evenly dispersed by winds.
- *Skepticism.* There are more abundant natural sources of chlorine than CFCs, such as saltwater and volcanoes.

- *Reasoning.* Since the chlorine source is in NaCl form which is readily water-soluble, rain removes it rapidly from the troposphere. Chlorofluorocarbons, by contrast, are not removed by rainfall because of their insolubility in water.

1.4.2 Wastewater handling

The wastewater from the cleaning step of the electronic packaging and assembly process needs attention for not only the environmental effect but also for economic reasons.

The concerns about contamination can be grouped into several areas:

- pH
- Temperature
- Heavy metal
- Biological oxygen demand (BOD) and chemical oxygen demand (COD)
- Suspended solids
- Toxic chemicals or hazardous substances
- Lead content

The alkalinity or acidity, as represented by pH, and the temperature of water can affect the activity of biological organisms and the integrity of structural treatment facilities. Heavy metals such as copper, lead, nickel, chromium, silver, and iron in significantly increased concentrations in water can be hazardous. The large concentration of BOD/COD can cause the deficiency of oxygen, suffocating plant and animal life. This usually can be remedied by chemical oxidation, biological oxidation, or carbon adsorption. Excessive amounts of suspended solids contribute to sludge building. Thus whether the wastewater from cleaning equipment can be discharged to sewage depends on the contamination level of the wastewater and the local and state regulations in water treatment. Table 1.2 lists the results of a survey about pretreatment regulations covering various regions in the United States.[7,8]

The availability of a closed-loop water recycling system for cleaning process makes in-process wastewater which needs disposal handling to a minimum. Most commercially available systems are capable of treating the bulk of in-process water for use as rinse supply and of removing waste removal without facing routine wastewater discharge problems. More details are covered in Sec. 7.7.

Table 1-2. Wastewater Pretreatment Regulation in the United States

	Average limitations						
Category	Northeast	Mid-Atlantic	Southeast	Midwest	South-central	Southwest	Northwest
Temperature, °F	160	147	145	127	132	143	147
Grease, oil, fat	100	84	100	125	133	283	100
pH	5.5–9.5	5.3–10.0	5.5–10.0	5.7–9 .5	5.8–9.8	5.5–11.2	5.5–10.5
BOD	275	317	383	250	275	300	300
COD	700	450	580	—	—	—	900
Suspended solids	300	333	250	258	288	300	350
Copper	3.1	3.1	1.4	6.0	2.5	6.9	2.5
Lead	1.9	0.9	0.4	4.8	2.2	2.5	1.4
Nickel	2.3	3.3	1.2	5.3	2.8	8.2	2.7
Chromium	3.3	2.9	1.4	7.7	3.7	4.3	3.7
Silver	2.5	1.8	0.2	0.8	2.1	3.6	0.5
Cadmium	0.3	0.6	0.1	0.4	0.4	5.6	1.4
Zinc	2.9	4.2	1.5	6.4	4.0	11.8	3.4
Fluorides	—	—	14.0	—	—	10.0	15.0
Total toxic compounds	5	1.42	2.1	3.1	2.1	1.0	1.4

1.4.3 VOC control

The control of volatile organic compound (VOC) emissions has been a major task in the coating industry for a couple of decades. The first international legislation to reduce VOC emissions was introduced in 1991.[9] A VOC is generally defined as an organic compound that participates in atmospheric photochemical reactions. The regulatory definition of VOCs has evolved into the following statement by the EPA:[10–12]

> Any organic compound that participates in atmospheric photochemical reactions; that is, any organic compound other than those which the administrator designates as having negligible photochemical reactivity. VOC may be measured by a reference method, an equivalent method, an alternative method, or by procedures specified under any subpart. A reference method, an equivalent method, or an alternative method, however, may also measure nonreactive organic compounds. In such cases, an owner or operator may exclude the nonreactive organic compounds when determining compliance with a standard. The administrator has designated the following organic compounds as negligibly reactive: methane, ethane, 1,1,1-trichloroethane(methyl chloroform), methylene chloride (dichloromethane), dichlorotrifluoroethane (HCF-123), tetrafluoroethane (HCC-134a), dichlorofluoroethane (HCFC-14lb) and chlorodifluoroethane (HCFC-142B), trichlorofluoromethane (CFC-11), dichlorodifluoromethane (CFC-12), chlorodifluoromethane (CFC-22), trifluoromethane (CFC-23), trichlorotrifluoroethane (CFC-113), dichlorotetrafluoroethane (CFC-114), and chloropentafluoroethane (CFC-15).

According to ASTM-D-3960-92, the VOC content is expressed as a function of the coating volume less water and exempt solvent or the volume of coating solids. Thus the general expression for VOC content is:

$$\text{VOC} = \frac{(\text{weight \% of total volatiles less water less exempt solvent})}{(100\% - \text{vol\% of water} - \text{vol\% of exempt solvent}) \times \text{density of material}}$$

For the systems that do not contain water or exempt solvent:

$$\text{VOC} = \frac{\text{weight \% of total volatiles}}{100\% \times \text{density of the system}}$$

A specific dry condition to determine the total volatiles is 110 ±5°C for one hour in the aluminum dish by:

Weight percent of total volatiles =

$$\frac{\text{weight before drying} - \text{weight after drying} \times 100\%}{\text{total weight of the system}}$$

The effort is to minimize or eliminate the content of VOCs in the development of solder chemistry and in the CFC alternatives for cleaning. Specific regulations for the allowable VOC emissions vary with state and city.

1.4.4 Lead and its regulations and legislation

Lead has been used for applications in glass, glazes, and enamels since ancient times. The oldest artifacts that were made of lead reportedly dated back to 1600 to 2000 BC.[13]

Today the largest use of lead, which accounts for 60 percent of all lead product, is in the manufacture of storage batteries. The US primary lead production,[14] which constitutes about 20 percent of world production, exhibited a trend of decline during the period of 1980 to 1991, as shown in Fig. 1.8. The lead consumption in electronic solders is reported to be less than 7 percent of the total.[15]

The use of lead in plumbing supplies, gasoline, and the paint industry has been heavily regulated. For example, the use of lead in con-

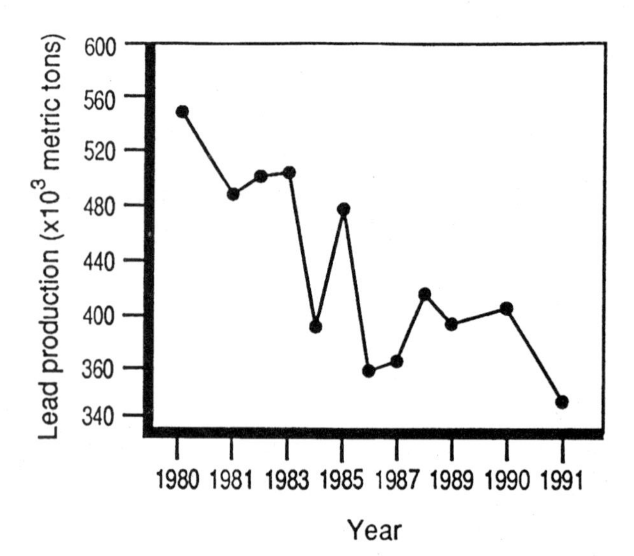

Figure 1-8. US primary lead production.

sumer paints has been banned since 1978. The present campaign against lead poisoning from housings with lead-based paints could put a high price tag on the paint industry. A bill at the congressional level is expected to be introduced this year which would dictate that lead be removed from these housings with a raised fund of $10 billion over a period of 10 years. The Department of Housing and Urban Development estimated that as many as 57 million private sector homes still contain lead-based paint, and the cost to remove those paints would exceed $500 billion.[16]

The paint industry has responded to this finding: "There is a great chance that the source of lead poisoning comes from diverse sources as plumbing pipe, solder, smelting, industrial emissions and lead-based gasoline.[16] Recently, public-health officials have classified lead as the number one environmental threat to children.[17] The top seventeen hazardous chemicals that impose the greatest threat to human health[18] are

Benzene
Cadmium and cadmium compounds
Carbon tetrachloride
Chloroform
Chromium and chromium compounds
Cyanides
Dichloromethane
Lead and lead compounds
Methyl ethyl ketone
Mercury and mercury compounds
Methyl isobutyl ketone
Nickel and nickel compounds
Tetrachloroethylene
Toluene
1,1,1-trichloroethane
Trichloroethylene
Xylene

It is reported that the US Food and Drug Administration is about ready to recommend that the United States join Europe in banning the use of tin-coated lead foil capsules for wine bottles.[19]

On another front, the assertion that lead poisoning is responsible for damaging the intelligence of children, which prompted US environmental and health officials to tighten the guidelines on acceptable levels of lead, is being challenged by a different group of scientists.[20] Separately, the Lead Industry Association, Inc., reported that: "All current uses of lead combined have an insignificant health effect on most sensitive population-young children, as concluded in a recent study by Environ Corporation."[21]

It is evident that recent findings have provided the basis for the formation of two opposite positions regarding the hazards and regulations of lead. Nevertheless, the Center for Disease Control has lowered the threshold definition of dangerous levels of lead in the blood stream by 60 percent, from 25 micrograms per deciliter (μg/dL) to 10 μdL.

The public mood and political climate appear to exert increasing pressure on the use of lead regardless of industry and end-use products. As to current US legislation and regulations regarding the use of lead, the Lead Exposure Reduction Act (S. 729) was introduced in April of 1993. In addition to banning lead solders for plumbing, this Act requires that the Environmental Protection Agency (EPA) must inventory all lead-containing products and then develop a *concern list* of all products that may reasonably be anticipated to present an unreasonable risk of injury to human health or the environment. Any person can petition the EPA at any time to add a product to the concern list. In addition, under this act, anyone who manufactures or imparts a lead-containing product that is not on the inventory list must submit a notification to the EPA. Products on the concern list must be labeled as such. On May 25, 1994, the Lead Exposure Reduction Act passed on the US Senate floor. Furthermore, the Lead Tax Act (Representative Ben Cardin, HR 2479; Senator Bill Bradley, S. 1347) was also introduced in June of 1993. That would place a 45¢ per pound tax on all lead smelted in the United States and on the lead content of all imported products. Tactics and technologies to replace lead are covered in Chap. 15.

1.5 References

1. Peter Singer, "DRAMs Microprocessors to Drive Technology in the 90's," *Semiconductor International,* February 1993, p. 17.
2. Bob Kopp, "Forecast 1993—Fitting the Pieces Together," *Semiconductor International,* June 1993, p. 24.
3. *The Wall Street Journal,* August 2, 1991, section A, column 1, page A.
4. In-Stat, Inc., Scottsdale, Arizona.
5. "United National Environment Program," *1991 Report of the Technology & Economic Assessment Panel.*
6. *ASHRAE Journal,* December 1992, American Society of Heating and Refrigeration and Air Conditioning Engineers.
7. James J. Andrews, "PWB Aqueous and Semiaqueous Cleaning System—Approaches and Tradeoffs," *NEPCON West Proceedings,* 1991, p. 283.
8. Davins Samsami, "Cleaning Without CFCs: The Semi-Aqueous Solutions," *Electronic Packaging and Production,* June 1991, p. 62.
9. "Solvent and the Environment," *Chemistry International,* vol. 14, no. 6, November, 1992.
10. Superintendent of Documents, U.S. Government Printing Office, Washington D.C., 20402.

11. *Federal Register,* January 1989, vol. 54, no. 11, 1990.
12. "Manual on Determination of Volatile Organic Compounds in Paints, Inks and Related Products," ASTM Manual Series, MNL 4, ASTM, May 1989.
13. J. S. Nordyhe, "Lead Products," *Ceramic Bulletin* vol. 70, no. 5, 1991, p. 872.
14. U.S. Bureau of Mines.
15. William B. Hampshire, "The Search for Lead-Free Solders," *SMI Annual Conference,* 1993, p. 729.
16. Kenneth Silverstein, "Lead Issues Gain Steam at State and Federal Levels," *Modern Paint and Coatings,* April 1993, p. 30.
17. Steven Waldman, "Lead and Your Kids," *Newsweek,* July 15, 1991, p. 42.
18. EPA, Washington, D.C.
19. *Packaging Digest,* February 1993, p. 2.
20. Gary Putka, "Research in Lead Poisoning is Questioned," *The Wall Street Journal,* March 6, 1992, B8.
21. J. Miller, "Senator Reid's Lead Use Reduction Bill—S. 729," Lead Industries Association, Inc., April 21, 1993.

2
Advanced Surface Mount and Die Attach Technologies

Various packaging and assembly technologies have been introduced in an effort to connect between ICs and to connect ICs onto the mother board for improved performance and reliability. These include flip-chip, chip-on-board, tape automated bonding, ball grid array/pad array carrier and multichip module technologies. Each technology differs in its design of circuitry and interconnections, and each demands a different set of requirements on process and equipment in order to implement its application. Yet they all fall into the realm of surface mount technology in terms of how they mount on the mother board. The following sections provide a brief introduction for each technology and its merits and drawbacks. Information is specifically tailored for solder and soldering applications.

2.1 Flip-chip Technology[1-6]

Flip-chip technology is considered to be an advanced form of surface mount technology, where a bare semiconductor chip processed with solder bumps on its surface is turned upside down and bonded directly to the substrate without requiring any intermediate chip packaging or lead frame mounting. The technology is applicable to either single-chip packages or multiple-chip modules.

The technology originally invented by IBM in 1962 focused on solid logic technology with silicon transistors mounted on thick-film substrate. Several years later, the concept of limiting the solder-wettable area on the mating pads of the components was introduced, that is, controlled collapse chip connection (C-4) technology. Until recently C-4 has been primarily a technology for mainframe computers and some niche applications in automotive sensors and ignition modules. Today, the application of the concept of flip chip is extended to house high I/O count semiconductor chips.

2.1.1 Merits and limitations

The major merits and features that the flip chip can offer are

- High package density without the need of a lead frame.
- No intermediate chip packaging required.
- Achieves a high number of interconnections; bonding sites can be configured in a peripheral or array pattern.
- Enhanced electrical performance as a result of the short distance between the chip and substrate; includes minimal propagation delay, minimal transmission losses at high frequency, and low parasitic capacitance and inductance values.
- Low thermal resistance; solder bonds primarily serve as heat-dissipating paths and are thus suitable for high-speed electronic devices.
- No leads-related problems (bending, misalignment).
- Self-alignment during reflow.
- Better fit for making small electronic modules, due to the inherent low profile and reduced weight and space required.
- Better fit to automotive under-the-hood applications where erosion of wire bondings, which is catalyzed by chlorine, often occurs because no wire bonding is required.
- Suitable for high-volume applications.
- Repairable.

The main disadvantages and hurdles are

- Substantial capital investment required

- Limited availability of bumped ICs
- Potential thermal stress-induced failure at the bump interconnections particularly for large ICs
- Involves bare chip handling and testing
- Lack of known-good die

2.1.2 System and process

On the chip side, the silicon IC wafer is fabricated according to the conventional chip process. The aluminum contact pads of the IC wafer then receive glass passivation, such as silicon nitride to define the opening of aluminum metallization. The solderable metallizations are prepared by depositing a single-metal or multiple-metal system on the aluminum contact pads. Generally, there are two techniques employed: physical vapor deposition (PVD) or lithography. The commonly used metal system consists of chromium followed by a chromium-copper alloy, then copper layer, followed by gold surface layer. In this case, the metallization of chromium on aluminum is to provide good adhesion and to serve as a diffusion barrier to copper. The chromium-copper alloy provides continuity to the copper layer which gives a solderable surface. The gold on copper prevents copper oxidation. Other less elaborate metallization systems are also successfully used, these include

1. Chromium (Cr), nickel (Ni), gold (Au)
2. Titanium-tungsten (Ti-W), sputtered copper, electroplated copper
3. Aluminum-nickel (Al-Ni), nickel (Ni), copper (Cu)

Once the IC chip and a proper metallization system are in place, the IC chip is ready for solder bumping for which PVD and electroplating can be used. Both of the two techniques have their merits and drawbacks. Physical vapor deposition does not need electrode connections; however, it involves an expensive vacuum system. The process is normally limited to a thin layer of deposition within a reasonable process time. In comparison, electroplating does not require need expensive equipment, although it requires a strike layer wafer surface for electrode connections. Conventional solder bump alloys are high-temperature solder alloys such as 5 tin (Sn)/95 Pb or Pb-indium (In) compositions. The solder-bumped chips at this time undergo the solder reflow step to make the well-defined contour of solder bumps. The size of solder bumps generally ranges from 90 to 125 μm (0.0036 to 0.005 in) in diameter and 80 to 100 μm (0.0032 to 0.004 in) in height. The bumped chips can be manually or machine pick-and-placed onto the substrate for connection which can be accomplished by means of the solder reflow process. The substrate side obviously requires solderable pads. For ceramic substrates, thick-film conductors serve as connecting pads. In some cases, solder paste is used as additional solder material. Figure 2.1 is the schematic of the process; and Fig. 2.2 illustrates the cross section of a flip-chip package.

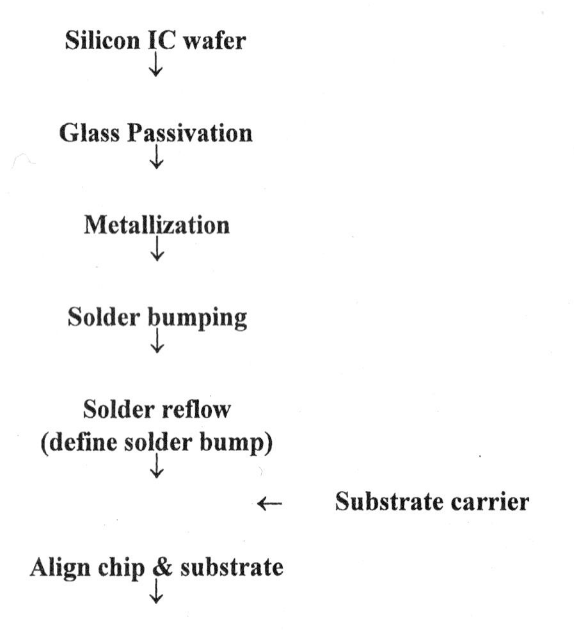

Figure 2-1. Flip-chip process.

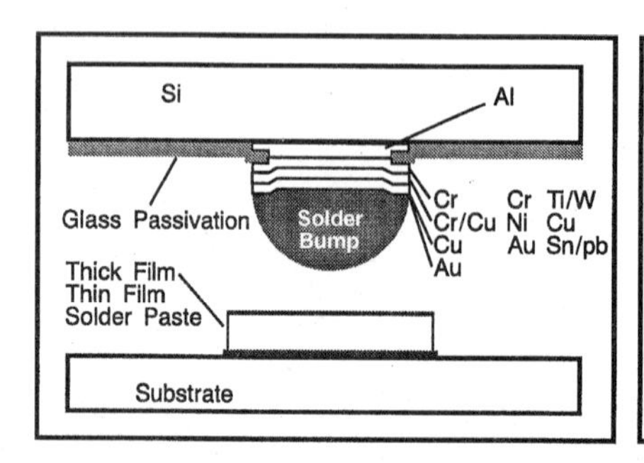

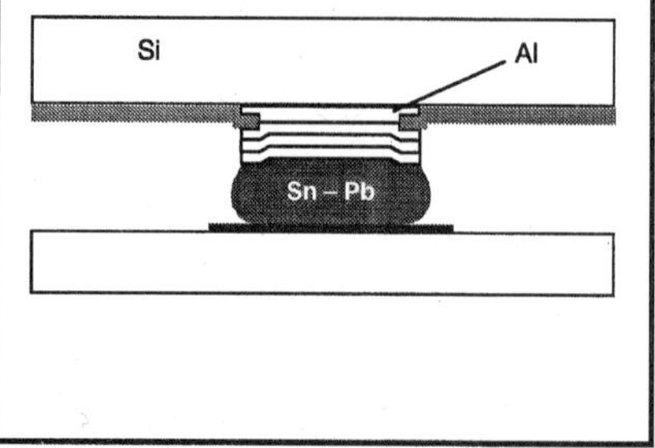

Figure 2-2. Cross-sectional view of flip-chip package.

2.1.3 Flip chip on printed circuit boards

In contrast to the assembly onto ceramic substrates, the use of flip chips on printed circuit boards (PCB) as shown in Fig. 2.3 generates a couple of additional requirements. These are the need to lower solder bump reflow temperature and the need to overcome the large mismatch in the coefficient of thermal expansion (CTE) between the chip and the PCB substrate as listed in Table 2.1. The process as represented in Fig. 2.4 is therefore modified to accommodate the above two requirements. The process may involve the redeposit of solder bumps on the substrate by using 63 Sn/37 Pb solder or by using solder paste containing 63 Sn/37 Pb during the chip-substrate interconnection. The chip-substrate interconnection can be accomplished by the conventional reflow process. The space between the chip and the substrate is then filled with epoxy resin or an equivalent by means of capillary action so that the chip and resin are firmly bonded upon the complete cure of the resin as shown in Fig. 2.5. The additional encapsulation is considered necessary in order to minimize the effect of a large mismatch in CTE among the materials involved in the assembly.[7]

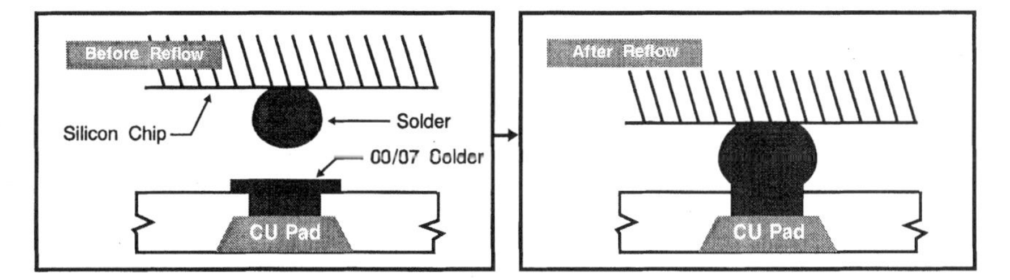

Figure 2-3. Cross section of mounting flip chip on PCB.

Table 2-1. Coefficient of Thermal Expansion

Component	TEC, $10^{-6}/°C$
Chip	2.5–3.5
PCB	14.0 (xy)
63 Sn/37 Pb	24.0
5 Sn/95 Pb	29.0
Encapsulant	26–60 (0–100°C)

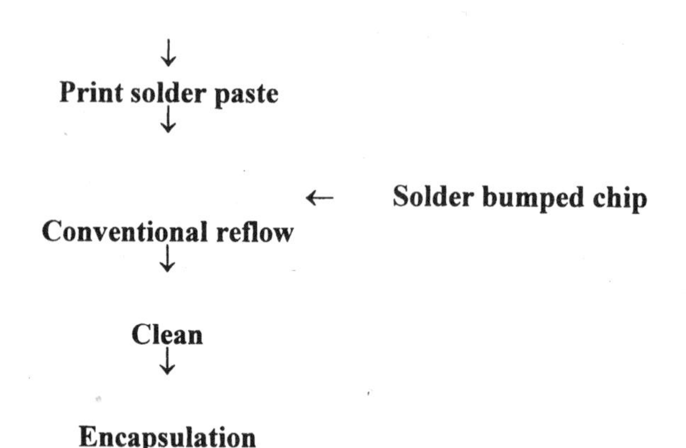

Figure 2-4. Process of mounting flip chip on PCB.

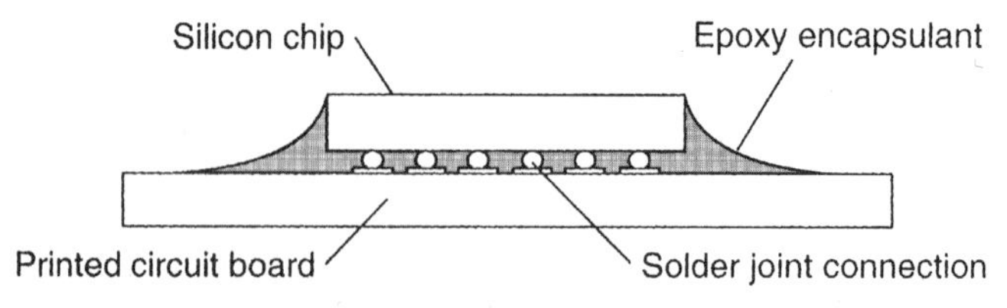

Figure 2-5. Schematic of flip chip on PCB.

In evaluating the performance of an encapsulant, the strain development as measured at the bump farthest from the center of the chip is 4.5 percent for FR-4 substrate and 1.5 percent for CTE-controlled PCB. With added encapsulant, the strain of 4.5 percent for FR-4 board is dropped to 1.0 percent.[7] The encapsulant is believed to be capable of converting the strain on the solder bumps into the deformation of the whole system of chip and substrate, thus reducing the strain on the bump.

2.1.4 Repair and replacement

It has been a general practice that unsatisfactory single-chip IC packages be discarded because the cost required for repair usually does not warrant the rework. However, repair and replacement of one or more chips for multichip products becomes necessary, particularly in the cases where multiple chips are involved, since the cost prohibits discarding the MCMs which still contain good chips.

Three main processes are required in order to carry the repair and replacement:[8]

- Chip removal
- Solder cleanup
- New solder joint formation

The chip can be removed either by mechanical means without the aid of heating, such as torque or ultrasound, or by means of thermally induced techniques which preheat the assembly in conjunction with the application of heat to the back side of the chip (the top side of the assembly). The heating sources can be conduction heating, focused infrared with shield, laser, and inert gas heating.

To clean up the residual solder after the chip is removed, three techniques are outlined:[8]

- Mechanical shaving
- Heated N_2 flow with critical velocity to propel molten solder away from the site into the gas stream
- Use of a sintered solderable metal slug to absorb molten solder by capillary action

Once the residual solder is properly removed, a regular reflow process can be used to make new solder connections.

2.2 Tape Automated Bonding Technology

Tape automated bonding (TAB) is not a brand new concept as it was patented in the late 1960s. As the I/O count increases, the wire bonding becomes unfavorable due to its high cost. Thus, TAB becomes an alternative interconnecting technique, particularly for high-volume and high-I/O-count devices. Tape automated bonding (TAB) has been widely used in consumer products, and its applications have been dominated by Japan, who is the worldwide leader in the application of TAB technology. It has also been utilized in very large scale integrated (VLSI) circuits.

With respect to circuit density, wire bonding is generally capable of handling a pad centerline space of 100 to 150 μm (0.004 to 0.006 in), and TAB is able to reach the capability of 50 to 100 μm (0.002 to 0.004

QFP

MOLDING
WIRE
DIE
SUBSTRATE

TAB

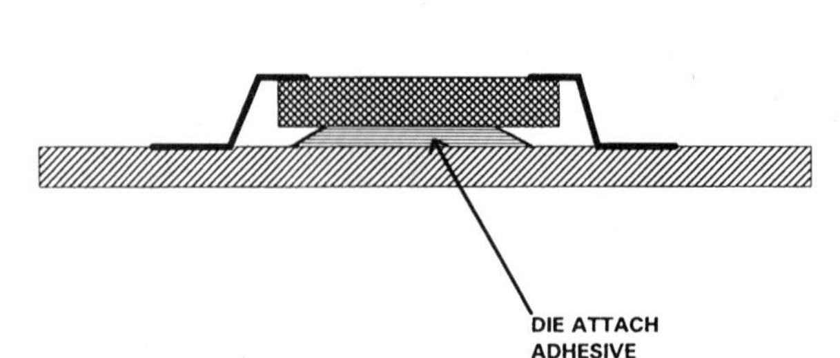

Figure 2-6. Schematic comparison between TAB and QFP.

in). The schematic comparison between TAB and the standard surface mount QFP package with wire bonding is shown in Fig. 2.6.

2.2.1 Merits and limitations

Three main attributes of TAB technology are

- Assurance of good die prior to final assembly
- Capability of high yield
- Higher bond strength than wire bonding or flip-chip bonding

Major drawbacks are

- High tooling cost
- Dedicated tooling for each package and assembly design

- Long lead time in development
- Potential heat-dissipation deficiency
- Difficulty in rework
- Scarce availability of bumped wafers
- Limited availability of TAB bonder equipment

2.2.2 System and process

The system consists of three major portions: tape manufacturing, die, and tape-die assembly. The tape resembles a miniature flexible printed circuit which is supplied in a continuous sprocked film format. The tape is constructed with copper conductors supported on a dielectric substrate, commonly polyimide, in two layers (copper/polyimide) or three layers (copper/adhesive/polyimide). The die is fabricated with bonding pads which are sputtered with a metallization barrier, such as titanium/tungsten. The bonding area is formed through a series of steps: photoresist, exposure, development. Gold is then plated onto the bonding area.

During assembly, a reel of tape is mounted on an inner lead bonder (ILB) together with a mounted and sawed wafer. The tape is unwound and fed through a tape track to the bonding location. A good die is selected and fed to the bonding site where heat and pressure are applied to make the bond between the tape and the die. Actual bonding conditions can vary; for example, common parameters are temperature of 400 to 500°C, pressure of 50 to 60 pounds per square inch (lb/in^2), time of 0.8 to 1.6 second (s), force per bump of 30 to 35 grams (g). Following ILB, the tape and die are rewound on a reel which then goes to an outer lead bonder (OLB). Between the OLB and ILB processes, the chips on tape can be tested so that only good dies are bonded to the final package. During the OLB process, the chip is excised from the tape frame and the leads are formed (if the package required), which are ready to be bonded to the package. A schematic of the process is shown in Fig. 2.7.

2.3 Chip-on-Board Technology

In this text, chip-on-board (COB) refers to the traditional definition of the direct attachment of bare die to PCBs. The detailed discussion of COB technology is outside the scope of this book. Since it rarely involves solders and soldering, this section is intended to cover only

Bump chip
↓
Inner lead bond (ILB)
tape to chip
↓
Excise chip from tape &
form lead
↓
Place package on PCB & outer lead bond (OLB)
↓

Inspection
↓
encapsulate
↓
Cure

Figure 2-7. Process of TAB.

certain aspects of COB technology in relation to other high-density interconnection technologies for comparison purposes.

2.3.1 Merits and limitations

Some inherent advantages of COB technology are

- Low tooling cost
- No special die processing (in contrast to flip chip and TAB)
- Choice of substrate
- Reduced area, height, and weight
- No solder bumping needed
- Capability of handling high I/O count
- Good for very small assemblies

Disadvantages include

- Demands high-quality PCB, such as thinner multilayer board with 0.003-in (0.075-mm) lines and spacings and less than 0.010-in (0.25-mm) vias
- Requires boards with high glass transition temperature (T_g) when gold wire bonding is used
- Requires gold plating for bondability
- Has a high profile when using aluminum wire bonding
- Limited power dissipation
- Limited use for high-power bipolar circuits
- Lack of hermeticity
- Low yield
- Requires a relatively clean room (class 1000-10,000) compared with other board-level assembly technologies
- Cost uncertainty

2.3.2 System and process

Individual silicon chips are aligned to bond sites on the substrate surface and secured with epoxy and with the curing process. Fine wire (gold or aluminum) is then welded first to the bonding pads on the perimeters of the silicon die and then to a corresponding circuit pattern on the substrate. Figure 2.8 illustrates the basic process, and Fig. 2.9 exhibits the cross section of the COB package. The die-attach adhesive is typically epoxy-based, possessing curing temperature and time compatible with the properties of PCB, particularly the glass transition temperature (T_g). For example, for the PCB with T_g in the range of 125 to 150°C, the required curing temperature must be below this range with a safe margin. A silver-filled epoxy system is commonly used, and for high-power devices solder is the choice. The encapsulant material is commonly epoxy-based or a room temperature vulcanizing (RTV) material. Its coefficient of thermal expansion is critical to the reliability of the assembly.

The wire bonding in the COB process essentially follows the basic principles of wire bonding technology for conventional IC packages, yet the wire bonder (equipment) needs a higher level of flexibility to handle various dies and boards. Two techniques, wedge bonding and ball bonding, are commonly adopted. Wedge bonding uses ultrasonic energy to weld an aluminum wire to a printed circuit trace without

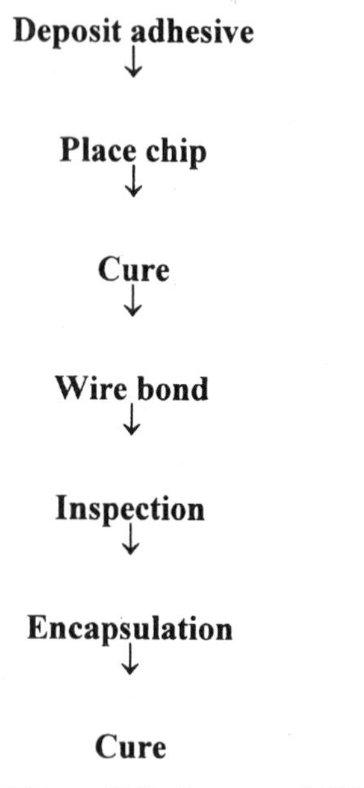

Figure 2-8. Process of COB.

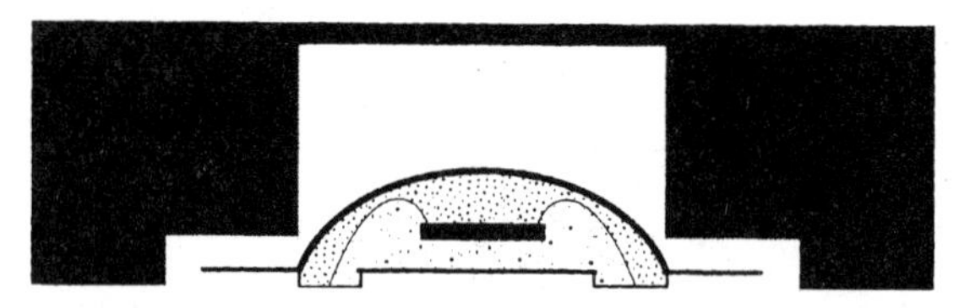

Figure 2-9. Cross-sectional view of COB.

heating the substrate; and ball bonding uses thermocompression to join gold wires to the traces usually at above 180°C. The pros and cons of these two processes are summarized in Table 2.2. For detailed information about wire bonding refer to Ref. 8.

Although ball bonding traditionally has used gold or aluminum, copper ball bonding has also been developed. The COB technology appears to have a natural fit to MCM-L application which will be discussed in Sec. 2.6.

Table 2-2. Comparison between Wedge Bonding and Ball Bonding

Characteristic	Wedge bonding	Ball bonding
Process temperature	Cold	>180°C (356°F)
Pad on substrate	Cu, Ni, Au	Au
Substrate quality	More tolerant	Less tolerant
Wire	Al	Au
Material	Less corrosion resistant	More corrosion resistant
Bonding susceptible to contamination	Less susceptible	More susceptible
Loop profile	Lower loop height	Higher loop height
Throughput	Slower	Faster

2.4 Fine-Pitch QFP Technology

As the I/O count increases in the die level, the conventional surface mount package in the QFP will have to shrink in its lead pitches to maintain the functional density per unit area, leading to fine-pitch technology which will be covered in Chap. 10.

2.4.1 Merits and limitations

Major advantages of fine-pitch QFP in comparison with other packaging approaches are

- Pretested ICs
- Easier to rework
- Known yield and cost
- Established SMT manufacturing base in process and equipment

Its limitations are

- All leads-related problems
- Still limited I/O counts to be accommodated

2.4.2 System and process

The processes involved include conventional die attach, wire bonding on both chip and lead frame, molding, and lead trimming and forming. The resulting IC packages are completely sealed, and the leads are formed along the perimeters serving as the interconnecting sites. When comparing with BGA/PAC packages, QFPs are referred to as peripheral IC packages.

2.5 BGA/PAC Technology

The bonding sites of the completed packages are in an array pattern over the whole substrate area in contrast to QFP; thus the BGA/PAC is often referred to as an array IC package.

2.5.1 Merits and limitations

The major advantages of array packages in comparison with peripheral packages are

- No lead compliance problem
- High I/O density
- Larger pitch, particularly relieving the stringent requirements of pick-and place equipment and printing performance
- Can adapt to existing production equipment in the surface mount technology industry
- Registered as JEDEC standard

However, its negative side includes

- Difficulty in inspection and repair
- Needs special rework technique
- Unknown cost

2.5.2 Process

The process may involve the same steps as for QFP except that no lead frame is needed; instead a bumping process is required to form the array bonding sites on the bottom of the substrate. The packages, materials, and processes, and reliability aspect will be discussed in Chap. 11.

2.6 MCM Technology

2.6.1 Definition and advantages

The multichip module (MCM) can be generally described as bare chips mounted and interconnected on a multilayer substrate which is then packaged as a single functional unit with a defined interconnect density. This is in contrast to the IC packages in which one single chip is mounted on a substrate and then packaged. The functional module is ready to be interconnected onto the board (mother board) in the subsequent assembly. The die-level substrate can be silicon-based; ceramic-based, such as Al_2O_3, AlN, SiC, BeO; and polymer-based, such as FR-4, bismaleimide-triazine (BT) epoxy, polyimide, cyanate ester. The die-to-substrate interconnecting technique may involve conventional wire bonding, tape automated bonding and solder bumped flip-chip methods.[9–12] Both flip-chip and TAB methods require less space than wire bonding, and TAB has the further advantage of being easy to inspect.

In addition to decreased size and weight and enhanced reliability as the common desired feature for electronics, increased electrical performance is constantly being sought after. The features of electrical performance include faster clock rate, reduced signal noise, lower inductance and capacitance values, and lower dielectric constant. In order to achieve the desired level of performance, both materials used in the packaging and the packaging design technology are important. For example, every additional interconnection imposes a signal propagation delay which is in turn directly proportional to the square root of the dielectric constant of the substrate material. Table 2.3 lists the dielectric constants of common die-level substrates. In addition to the dielectric constant, other material properties, including the coefficient of thermal expansion, thermal conductivity, insulation resistance, and bending strength are all important to the overall electrical performance of the package.

In the case of single-chip packages the delay of signal propagation via the transport of electrons can be attributed mainly to two sources: (1) the delay within the IC die and (2) the delay between IC dies. As the I/O counts or lead counts increase, the conventional IC packages are generally larger than the die itself. Therefore, under this condition, even the single-chip packages are mounted as close together as possible, and the propagation delay increases with the increasing size of the package. In order to minimize the propagation delay, it is advisable to have the size of the single-chip package be as close as possible to the size of the die itself or to have multiple chips in one package. In addition, obtaining the substrate possessing the lowest dielectric constant as possible is also a means to reduce the propagation delay.

Table 2-3. Dielectric Constants of Common Substrate Materials

Material	Dielectric constant, room temperature,1 MHz
FR-4	4.8
Polyimide	3.5–4.5
BT Epoxy	4.2
Cyanate ester	4.5
Al_2O_3 (99%)	9.2
Al_2O_3 (96%)	9.0
BeO	6.7
AlN	6.0–8.5
SiC	40.0–45.0
Si	11.7
GaAs	12.8
Diamond [chemical vapor deposition (CVD)]	5.5–5.9

In comparison with single-chip packages, the use of multiple chips within one module offers several advantages:

- *Eliminates packaging of individual chips.* This potentially increases package density and shortens the distance between IC chips which is the most effective way of reducing parasitic capacitance and thus reducing signal noise. The shorter length between chips lowers signal propagation time, resulting in faster speed of the system. The combination of shorter signal length and the lower dielectric constant of substrate provides the fast clock rate.
- *Increases packaging efficiency.* Packaging efficiency is defined as the ratio of active chip area to the total package area. For example, typical single-chip packaging represents 1 to 15 percent of total package area and MCM may achieve 60 percent. Table 2.4 depicts the comparison of packaging efficiency of various technologies.
- *Reduced number of interconnections.* It is reported that approximately 90 percent of electronic failures may be attributed to package and board interconnections. Multichip modules can reduce the number of interconnections significantly, which potentially may translate into higher reliability.

Table 2-4. Packaging Efficiency of Various Technologies[13]

Packaging technology	Packaging efficiency, %
IC chip	100
Through-hole	1–3
SMT	6–14
Thick-film hybrid	10–30
COB	15–30
MCM-D	30–60

- *Flexibility to mix digital, analog, complementary metal oxide semiconductor (CMOS), and bipolar-chip circuit technologies.* Ability to mix the inherent advantages of each chip circuit technology as listed in Table 2.5 fulfills the requirement of design features. For example, both digital and analog circuits can be made within one module.

Therefore, MCM becomes a superior packaging idea when one or all of the following conditions is met:

- One circuitry function cannot be integrated on a single chip.
- Multiple ICs in one package are required on a PCB substrate.
- Circuitry function requires two or more chip technologies (CMOS, bipolar, analog, digital).
- The cost of MCMs is less than the sum of costs of single-chip packages.
- Specific functional performance, e.g., clock rate or data transfer rate, is needed.

Table 2-5. Characteristics of Chip Circuit Technology

Chip circuit technology	Characteristics
CMOS	Low cost, low speed, lower power
Bipolar	High cost, high speed, high power
Analog	Resolution advantage
Digital	Integration advantage

In contrast to conventional COB, MCM today is more accurately defined by the interconnection density than by the description as multiple chips mounted and interconnected on a single substrate. The accepted interconnection density includes

- Ratio of active device area and substrate area is larger than 30 percent.
- Substrate has more than four layers and is thinner than 1.5 mm (0.06 in).
- Package contains more than 100 leads (or I/O count).
- Fabrication processes involve the capability of making 0.1 mm (0.004 in) or smaller lines and 0.15 mm (0.006 in) or smaller spacings.

In the electronics hierarchy, MCMs offer another level of packaging which fills in between the chip and board. Based on the differentiation of substrate material and the fabrication process involved in constructing the module, three types of MCM are classified: MCM-D where D denotes the *d*eposited dielectric, MCM-C where C refers to *c*eramics, and MCM-L, where L represents *l*aminates.

2.6.2 Hurdles

The main hurdles in the use of MCMs include

- Lack of known-good die
- Handling bare die
- Lack of bare die testing equipment
- Difficult to repair
- Unknown or high cost

Figure 2.10 illustrates the distribution of the category of hurdles.[14]

2.6.3 MCM-D

Multichip module type D is normally composed of dielectrics and conductors deposited on a base substrate by means of thin-film technology to form sputtered, thin-film multilayers.[15,16] The common substrate material used is silicon, and the insulator is polyimide. Copper, gold, and aluminum are viable signal-layer metallization materials. Chips can be gold-tin eutectic bonded and interconnected by conventional

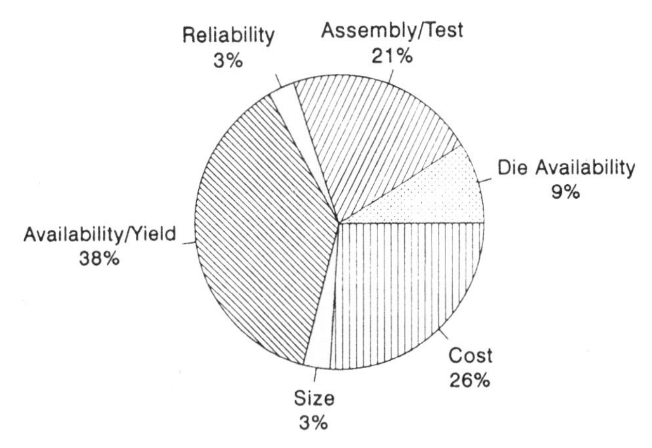

Figure 2-10. Acceptance hurdles for MCM use.

wire bonding (gold) onto the pads on the die and substrate. Tape automated bonding or flip-chip techniques can also be used.

In order to maintain the function of metallization, the metal selected must possess good conductivity, long-term resistance to corrosion, and good thin-film deposition characteristics. Generally, aluminum has good adhesion to polymers and good conductivity; gold imparts low resistivity and good evaporation characteristics facilitating the formation of an intact thin film; copper possesses excellent conductivity, but its proneness to oxidation which may result in poor adhesion constitutes its weakness. Ideal insulator polymers are expected to have a low dielectric constant,[17] low H_2O absorption, good adhesion, and a CTE match with the rest of the materials. Thermal stability and good mechanical properties are additional virtues required. The general fabrication technology, essentially borrowed from IC fabrication technology, imparts superior resolution in line width and via size. The achievable range in line width reaches 10 to 25 μm and that for vias is 10 to 20 μm when formed by plasma or laser etch technique.

The general characteristic feature of MCM-D is its ability to achieve the speed for high-performance application, i.e., higher than 100-MHz clock rate. Thus the circuitry density over 200 pads per square inch is considered to be a good match with MCM-D. It is reported that a new development of MCM employs silicon wafer as its substrate, is processed with standard IC fabrication, and is capable of operating up to 300 MHz. In this case, solder bumps spaced at 1200 μm (0.048 in) are used for the chip-to-module connection rather than wire bonding in order to reduce inductance (noise). Silicon substrate in conjunction with the flip-chip connection are key features of this technique.[18] The primary hurdle to implementing MCM-D is associated with its cost and manufacturability outside the thin-film industry.[19]

2.6.4 MCM-C

Multichip module type C consists of a cofired ceramic substrate with a metallization of either plated thin film or screen-printed thick film. Integrated circuit chips are attached on the substrate with epoxy or Au-Sn eutectic solder. Aluminum wire bonding, TAB, and flip chip are the optional techniques for chip-to-chip interconnections. The fabrication process uses traditional thick-film hybrid circuit technology which is well established in the hybrid industry. In fact, the availability of facilities and resources in hybrid manufacturing is one of the primary driving forces to adopt MCM-C.

The high I/O MCM-C mounting to printed circuit board has been successfully assembled using BGA interconnections.[20] In this assembly, MCM-C is composed of a multilayer ceramic substrate carrier constructed with an array of 10 Sn/90 Pb solder sites in a 1.25-mm (0.050 in) grid. The solder sites serve as module-to-board interconnections. The solder sites are in either ball or column form as shown in Fig. 2.11. Solder balls are 0.88 mm (0.35 in) in diameter, and columns are 2.2 mm (0.087 in) in height and 0.50 mm (0.020 in) in diameter. The MCM-C is attached to the board with 63 Sn/37 Pb solder using the standard surface mount process. Because of the high melting temperature of 10 Sn/90 Pb solder, the ball or column does not melt during module-to-board soldering under 63 Sn/37 Pb soldering conditions. Thus a gap or standoff is created between the module and the board. The gap and standoff are designed to be beneficial to the integrity of solder joints by better absorbing the cyclic strain generated during power cycling and in the meantime to provide clearance for post-solder cleaning.

Comparing the performance between solder ball and solder column, test results showed that the solder column as specified above withstood a higher number of power/temperature cycling.[20] Based on this study, a solder column design is recommended for the applications where the following conditions exist:

- Large temperature gradient within a cycle
- Higher number of cycles required
- I/O count greater than 500

(a)

(b)

Figure 2-11. Ceramic BGA package: (*a*) solder ball, and (*b*) solder column. (*Courtesy of IBM Microelectronics.*)

Between the two package architectures, QFP and BGA, MCM-C in QFP is considered to be a low-cost approach when the applications require less than 200 I/Os and do not require a ground plane. When applications require more than 200 I/Os and a ground plane, MCM-C in BGA has a cost advantage because of the demand in thermal dissipation.

2.6.5 MCM-L[21]

The substrate material for MCM-L is multilayer PCB. Base chips are attached on the board by polymeric adhesive. Fine wire is welded first to the bonding pads on the die and then to the substrate board by the conventional wire bonding process. The type of polymer used for the substrate board dictates the wire material and bonding technique to be used due to the temperature required for bonding as indicated in Table 2.2. For instance, when FR-4 board is the substrate board, aluminum wire with ultrasonic bonding is the suitable choice. To use thermosonic gold wire bonding which requires a stage temperature higher than 150°C, higher-temperature-resistant polymers [polyimide, BT (see Table 2.6 for their glass transition temperatures)] are needed.

For chip-to-substrate interconnections, MCM-L largely borrows COB technology in high-density interconnections as specified in Sec. 2.3. Flip-chip and TAB methods are also alternatives for chip attachments.

With respect to the substrate board, several features set MCM-L apart from conventional PCBs. These include the overall thickness of the board being less than 1.55 mm (0.062 in) and via hole diameters going down to 0.15 mm (0.006 in). In addition, the board usually contains five to nine layers, and the surface is coated with a soft gold finish.

Among the three types of MCM, one of the main advantages of MCM-L is its cost. The cost comparison among the three MCM types is listed in Table 2.7.[19]

Table 2-6. Glass Transition Temperature of Substrate Polymers

Polymer type	T_g, °C
FR-4	125
BT epoxy	180
Polyimide	250–350
Cyanate ester	230

Table 2-7. Cost Comparison of MCMs

MCM type	Cost ratio
MCM-L	1
MCM-C	1.5–2.0
MCM-D (thin film on Al_2O_3)	2–3
MCM-D (silicon)	7–10

When comparing MCM-L with the conventional single-chip packaging, except functional equivalency, the practical factors such as manufacturing facility, die handling, test equipment and techniques, and yield target should also be considered. Test and repair and their effects on cost are still an uncertain area. A newly developed technology using a contact array of bare die on a flexible carrier for both testing and permanent bonding[22] and a membrane probe technology for bare die testing[23] have been introduced. Because of the inherent poor conductivity and the large CTE of PCB substrate, two technological areas call for attention. These are thermal management to facilitate heat dissipation and minimization of CTE mismatch between the board and chip.

One approach to minimize the CTE mismatch is to tailor the CTE of the PCB substrate so that it is close to that of the chip. Kevlar-based PCB with its high tensile strength, low CTE, and low dielectric constant is one promising candidate.[24]

Although the infrastructure for bare PCB manufacturing and surface mount assembly have been well established, the bare chip handling required in MCM-L demands a better controlled manufacturing environment. The cleanliness of 1000 to 10,000 class is generally recommended for manufacturing of MCM-L.

2.7 Module-level Packaging

Multichip module type L can be packaged in standard surface mount IC packages such as QFP or in pad array or ball grid array carriers as well as in a flip-chip package. The packaged modules are ready to be connected to the subsequent assembly (mother board) by using surface mount manufacturing. Figures 2.12 through 2.14 illustrate examples of the module packages.[25]

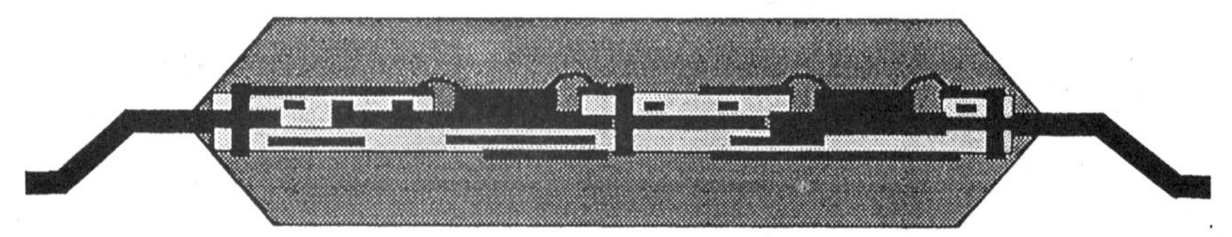

Figure 2-12. Plastic molded quad flat pack MCM-L.

Figure 2-13. Plastic molded pad array MCM.

Figure 2-14. Flip-chip MCM-L.

2.7.1 QFP MCM-L

One example as shown in Fig. 2.12 uses a seven-layer substrate and employs conventional wire bonding to integrate the lead frame and the substrate.[25] After attachment of the die and wire bonding, the combined substrate and chip set is molded with plastics similar to a single-chip plastic package. The signal entering or leaving a plastic QFP module goes through the lead frame into the substrate and through a wire bond into the IC. The lead frame also helps dissipate heat generated from the powered circuitry.

Quad flat pack MCM-L covers a selection of standard lead counts and lead pitches as shown in Table 2.8.[26] Since the series of packages are designed around the existing industry standard for single-chip QFPs, they can readily fit to the surface mount manufacturing process.

2.7.2 PAC MCM-L

The I/O contacts are designed in a grid array across the bottom surface of the module. This design addresses the requirement of high I/O

Table 2-8. QFP MCM-L Packages

Body size, mm	Total leads	Lead pitch	
		mm	in
28	128	0.8	0.032
28	160	0.65	0.025
28	208	0.5	0.020
40	232	0.65	0.025
40	304	0.5	0.020

counts without encountering fine-pitch problems as with QFPs. The signal is routed through a low-inductance solder ball/pad and through the thin substrate and a wire bond into the IC chip as shown in Fig. 2.13. For a high-heat-generated assembly, heat dissipation can be enhanced by using thermal vias.

2.7.3 Flip-chip MCM-L

As shown in Fig. 2.14, the bare chips are attached by flip-chip technology with an array of bumps on a 0.25 to 0.30 mm (0.010 to 0.015 in) pitch across the surface of the chip. The signal goes directly from the IC to the substrate through a solder ball. This transmission route minimizes lead inductance which is associated with wire bond or TAB lead frame, thus providing better electrical performance. The heat dissipation problem can be readily remedied by using a metal module lid. Polymeric thermal gel can be added between the metal lid and the back face of the chip to further enhance heat dissipation. In this design, the CTE mismatch between the interconnected parts: silicon die, flip-chip solder ball, laminate substrate, substrate solder ball, and metal lid should be considered.

2.7.4 Pending issues

In summary, to implement MCMs, several manufacturing issues need to be addressed:

- Die availability
- Known-good die
- Die testing methodology

- Die testing equipment
- Substrate testing methodology
- Substrate testing equipment
- Yield versus cost
- Industry standard
- Reliability
- System cost relative to single-chip package
- Investment versus customer demand
- Confidence level and comfort factor
- Demand versus supply dilemma

2.8 Hybrid of Array and Peripheral IC Package

The concept of combining the merits of both array and peripheral interconnections is being implemented. One example is the very small peripheral array package (VSPA).[27] The VSPA has a prefabricated semiconductor package housing multiple rows of leads (pins) which are laid out around the periphery of the package as shown in Fig. 2.15. Its footprint for interconnections is illustrated in Fig. 2.16. The leads can be made of phosphor bronze, Alloy 42, or other metals and arranged in three, four, or five tiers through the holes the plastic package wall as exhibited in Fig. 2.17 for a conventional die mounting and in Fig. 2.18 for a cavity-down mounting. The package also distinguishes itself from the conventional plastic QFP process as a prefabricated package structure ready for die mounting, as Fig. 2.19 so illustrates. The package is designed for easy inspection of peripheral solder joints and to achieve a high I/O density by taking up multiple rows of leads.

Figure 2.20 compares the packaging density among ceramic ball grid array (CBGA), plastic ball grid array (PBGA), QFP, and the peripheral array package in terms of I/O per unit density over a range of I/O counts. The physical size of this peripheral array package is significantly reduced as compared with QFPs and PGAs containing an equivalent number of I/O pins as shown in Fig. 2.21. The concept is applicable to single-chip modules as well as to multichip modules as exhibited in Fig. 2.22.

Figure 2-15. Example of array-peripheral package—VSPA. (*Courtesy of The Panda Project.*)

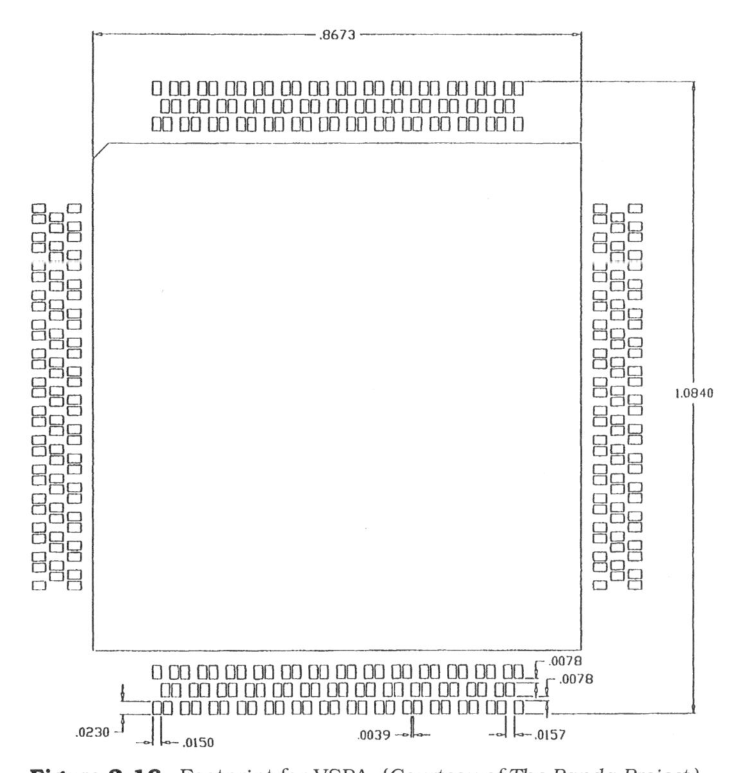

Figure 2-16. Footprint for VSPA. (*Courtesy of The Panda Project.*)

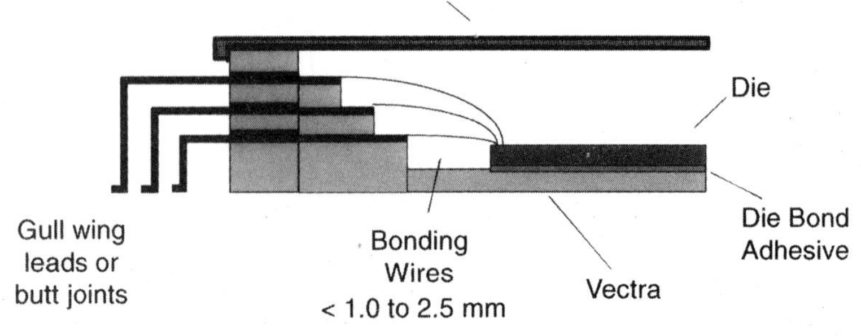

Figure 2-17. Lead configuration of VSPA for conventional die mounting. (*Courtesy of The Panda Project.*)

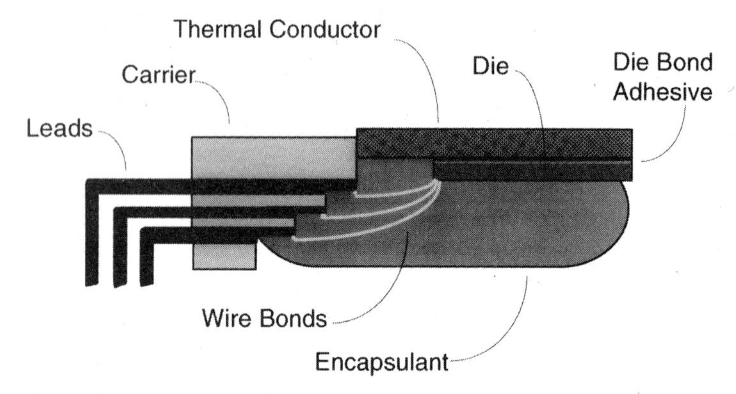

Figure 2-18. Lead configuration of VSPA for cavity-down mounting. (*Courtesy of The Panda Project.*)

2.9 PCMCIA

2.9.1 Features

The Personal Computer Memory Card International Association (PCMCIA) formed a standard committee in 1989 with the objective to develop an industrywide standard for credit card–like memory cards. The standard then extended to nonmemory applications such as I/Os. The PCMCIA card brings memory expansions to the portable class of computers. It enhances the portability of notebook and sub-notebook computers. To date, three PCB card types are defined:[28–30]

Type I 3.3-mm-thick package, mostly for memory applications, such as flash and static random access memory

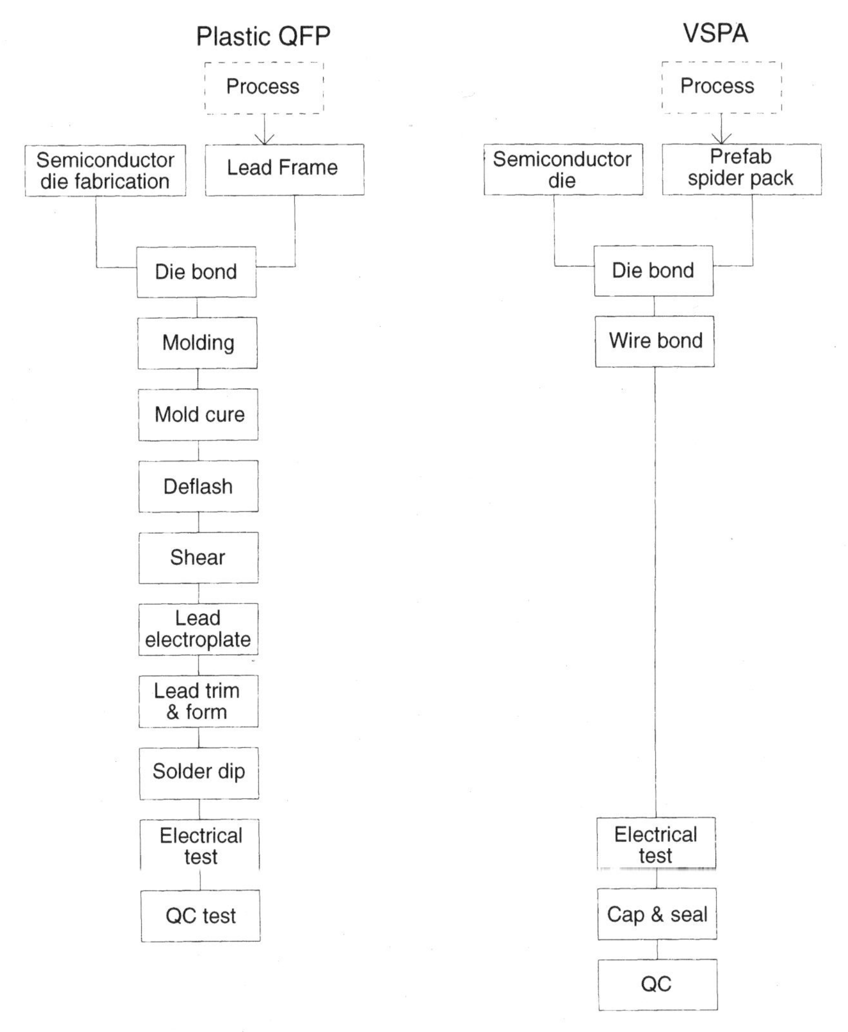

Figure 2-19. Process comparison between QFP and VSPA.

Type II	5.0-mm-thick package, for simple I/O type applications such as local area network (LAN), modem, and fax
Type III	10.5-mm-thick package, primarily for rotating disk applications and originally to accommodate 1.8-mm hard disk drive.

Computers are configured with either single- or dual-slot connectors to accommodate the personal computer cards. The dual-slot configuration can accept two Type I, two Type II, one Type I and one Type II, or one Type III card. PCMCIA connector slots that can accept a Type III

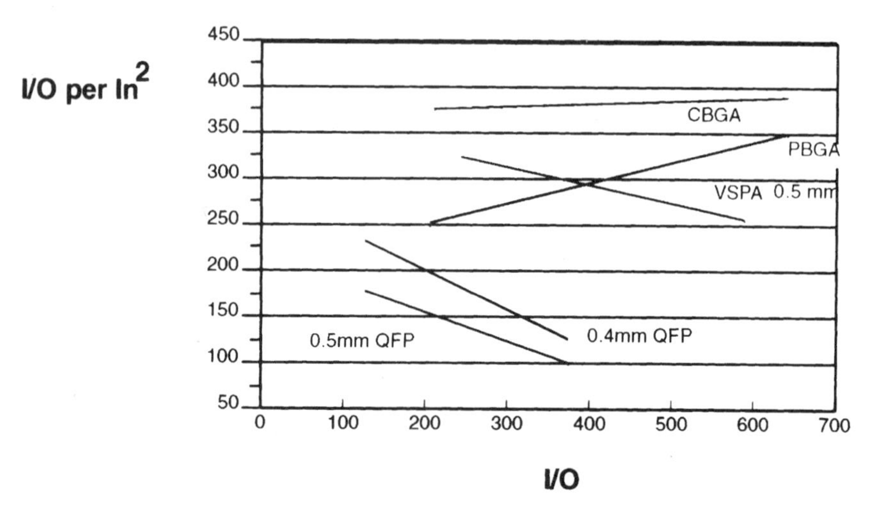

Figure 2-20. Relative package density of CBGA; PBGA, QFP, and VSPA. (*Courtesy of The Panda Project.*)

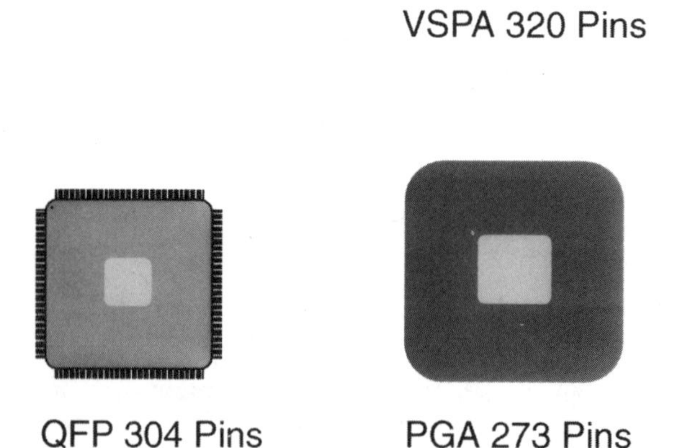

Figure 2-21. Physical size comparison of VSPA-320, QFP-304, and PGA-273. (*Courtesy of The Panda Project.*)

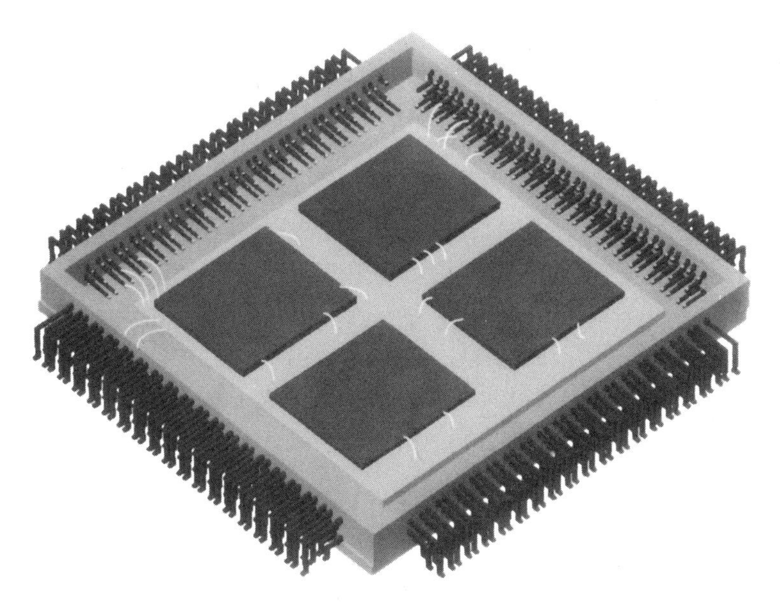

Figure 2-22. MCM in VSPA package. (*Courtesy of The Panda Project.*)

card are gaining increasing popularity. Computers featuring card slots for the convenience of a removable card-mounted hard drive or the flexibility of an easily upgradable hard drive are also becoming a popular choice. Electromagnetic interference (EMI) protection is more important to a PCMCIA system. Reduced power from 5.0 to 3.3 volts (V) is another trend.

To fit Type I, the card itself must be less than 0.5-mm thick and the components less than 1.20-mm thick, such as TSOP with very short leads and low stand-off. Other packages which are sufficiently thin, such as tape quad flat pack (TQFP), tape automated bonding (TAB), wire bonding, and flip chip, are also suitable. Each package has its merits and limitations. The TSOP is commonly available and suitable for memory applications. Its solder joints may have a shorter fatigue life than the TQFP. However, the TQFP has lower electrical and thermal performance, limited extendability to high I/O, and less real estate efficiency. TAB provides better electrical performance than TQFP, yet its cost is higher and rework more difficult. A flip-chip attachment is characterized by best electrical performance and high I/O capability. An extra process step of encapsulation may be needed to enhance the reliability of flip-chip solder joints. Flip-chip rework is also more difficult. Wire bonding connections experience a limitation for high I/Os and require encapsulation. However, wire bonding has the merits of known establishment and flexibility.

For discrete components, capacitor and resistors in 0603 (0.06×0.030 in) and 0402 (0.040 × 0.020 in) sizes are available to offer miniaturization. Other devices, for example, switches and oscillators, must be repackaged into a low-profile form. The PCMCIA standard specifies that connectors must be capable of withstanding 10,000 insertions and be designed to provide 68 or more SMT connections on a limited card edge length.[29]

To compare card-level solder joint density, PCMCIA cards generally contain 8 to 12 solder joints per square centimeter area in contrast to 3 to 6 solder joints for a double-sided board. Flip-chip technology, MCMs and array packages are likely to become an integral part of PCMCIA cards.

2.9.2 Process

Assembling PCMCIA cards generally involves a double-sided SMT process. One example,[29]

Print solder paste on the card
Pick and place components
Mass reflow (nitrogen)
Invert card
Print solder paste
Pick and place components
Mass reflow
Separately attach TAB or flip-chip packages
Attach connectors by dual thermode soldering
Clean with deionized (DI) water (or no-clean)
Inspect
Encapsulate
Mount plastic frame and battery contacts
Package into stainless-steel covers and weld the covers
Insert battery holder into the assembly
Test and do final inspection

2.10 Chip Scale Packaging and Assembly

2.10.1 Definition

It is generally accepted that a chip scale assembly covers both types of ICs—bare die and chip scale package whose area is equal to or smaller than 1.2 times the bare die. These chip scale packages are connected to the outside world through surface mount or modified surface mount processes onto the board substrate. The substrate may be either dedicated board or those containing other surface mount or through-hole components. Interconnecting techniques essentially utilize direct chip attach (DCA) through solder bumps or a conventional wire bonding process.

Direct die attachment, largely borrowing from the flip-chip concept, involves mounting chips or chip scale packages that have electrical interconnections facing down onto the substrate. Thus the chip's electrical interconnect pads are in direct contact with the electrical interconnect pads of the substrate, resulting in the shortest electrical path. The interconnections between the chip and substrate are made by reflow soldering using solder paste or other forms of material. In contrast, the well-established wire bonding process connects the chip's electrical interconnect pads that are facing up with the substrate electrical interconnect pads by means of gold or aluminum wire; and the chip underside is attached to the substrate by adhesive.

2.10.2 Chip scale package

New chip scale packages continue to emerge, such as slightly larger than IC carrier (SLICC), micro-ball grid array (μBGA), mini-ball grid array (mBGA), micro-SMT (μSMT), and others. These packages are discussed in Chap. 11.

2.10.3 Technological challenges

In order to utilize chip scale assemblies reliably and economically, several technological factors need to be addressed. These include:

Reduced CTE substrate materisal

Known-good die

Controlled CTE underfill material

Reworkable underfill polymer

High-density, low-warpage board with acceptable cost

Enhanced reliability of solder interconnections

Thermal management

2.11 References

1. Terry Cestlow, "IBM Reveals Flip-Chip Technology Details," *Electronic Engineering Times,* May 31, 1993, p. 30.
2. John Tuck, "The Ultimate MCM-L: Flip Chip on a PCB," *Manufacturing Market,* October 1992, p. 1.
3. Art Burkhart, "Recent Developments in Flip Chip Technology," *Surface Mount Technology,* July 1991, p. 41.
4. Barbara Gibson, "Flip Chips: The Ultraminiature Surface Mount Solution," *Surface Mount Technology,* May 1990, p. 23.
5. Koji Yamakawa, Kaoron Koiwa, and Yoshimi Hisatsune, "Maskless Bumping by Electroless Plating for High Pin Count, Thin and Low Cost Microcircuits," *The International Journal for Hybrid Microelectronics,* vol. 13, no. 3, 1990, p. 69.
6. Robert Lefort and Tat Contantinon, "Flip Chips Improve Hybrid Capability," *Hybrid Circuits Technology,* May 1990, p. 44.
7. Yutaka Isukada, "Low Temperature Flip Chip Attach Packaging on Epoxy Base Carrier," *Surface Mount International Proceedings,* 1992, p. 294.
8. George G. Harman, *Wire Bonding in Microelectronics,* ISHM Publications, Virginia, 1989.
9. Karl J. Puttliz, "An Overview of Flip Chip Replacement Technology on MLC Multichip Modules," *The Journal of Microcircuits and Electronic Packaging,* vol. 15, no. 3, 1992, p. 113.
10. R. K. Spielberger, et al., "Silicon-on-Silicon Packaging," *IEEE Transactions on Components, Hybrids, and Manufacturing Technology,* vol. CHMT-7, no. 2, June 1984, pp. 193–196.
11. E. Davidson, "The Coming of Age of MCM Packaging Technology," *First International MCM Conference Proceedings,* 1992, p. 103.
12. W. Pierce and J. Krusius, "The Fundamental Limits for Electronic Packaging & Systems," *IEEE Transactions on Components, Hybrids, and Manufacturing Technology,* vol. CHMT-10, 1981, p. 176.
13. J. Hagge, "Ultrareliable Packaging for Silicon-on-Silicon WSI," *ECC 38th Proceedings,* 1988, p. 282.
14. "Accepting Hurdles for MCM Use," *BPA,* May 1980.
15. I. Turlik and G. Adema, "Main Construction Approaches for MCM-D," Thin Film Multichip Modules, ISHM Press, 1992, pp 45–87.
16. T. Tessler and P. Garrow, "Overview of MCM Technologies: MCM-D," *ISHM 1992 Proceedings,* 1992.

17. P. Garrow, "Polymer Dielectrics for MCM Packaging," *Proceedings of IEEE,* vol. 80, 1992.
18. *NEC Report,* October 1993.
19. G. Geschwind, R. M. Clary, and G. Messner, "Comparison of MCM-D and PCB Costs," *Inside ISHM,* May/June 1991.
20. J. Caulfield, J. A. Benenanti, and J. Acocella, "Cost-Effective Interconnections for High I/O MCM-C-to-Card Assemblies," *Surface Mount Technology,* July 1993, p. 18.
21. Sndo Toshio, et al., "Design Optimization of Wiring Substrate in CMOS Based MCM," *IEEE/ECTC Proceedings,* 1992, p. 285.
22. Tessera, Inc.
23. Hewlett Packagard Equipment.
24. Joseph D. Leibowitz, "Improving Solder Joint Reliability in MCM-L's," *Electronic Packaging and Production,* October 1993, p. 56.
25. William Blood and Allison Dixon, "MCM-L: Cost Effective Multichip Module Substrates," *Inside ISHM,* March/April, 1993, p. 7.
26. Motorola Semiconductor Technical Data on QFP-MCML, Chandler, Arizona.
27. The Panda Project, Inc., Product Brochure.
28. P. Isaacson, "Dream This," *OEM Magazine,* May 1994, p. 100.
29. H. Lowe and W. Ford, "Overview of PCMCIA Packaging and Card Assembly Process" *Proceedings, National Electronic Packaging Conference, NEPCON West,* 1993, p. 737.
30. N. Chan, "PCMCIA Memory Card Design and Applications," Proceedings, NEPCON West, 1994, P. 1299.

2.12 Suggested Readings

Adema, Gretchen M., et al., "Flip Chip Technology: A Method for Providing Known Good Die with High Density Interconnections," *The International Journal of Microcircuits and Electronic Packaging,* vol. 17, no. 4, 1994, p. 352.

Bert, Eric, "Meeting the Challenges of PCMCIA Card Fabrication," *Electronic Packaging and Production,* April 1995, p. 38.

Kovac, Zlata et al., "Fine Pitch Wire Bonding for BGA Packages," *Advancing Microelectronics,* January/February 1995, p. 20.

Markstein, Howard W., "Applications for MCMs," *Electronic Packaging and Production,* April 1995, p. 48.

Mazzullo, Tony, "How IC Packages Affect PCB Design," *Surface Mount Technology,* February 1995, p. 115.

Vardaman, Jan, "Multichip Packages," *Advanced Packaging,* January/February 1995 p. 36.

3
Solder Materials

This chapter is intended to provide a general understanding of solder materials in terms of physical, metallurgical, and mechanical characteristics. The emphasis is put on the needs of the manufacturing sector.

3.1 Basis of Classification

Based on the requirements of various applications, solder materials can be viewed from four aspects:

- Alloy composition
- Land pattern that solder is to be applied to
- Chemistry/cleaning
- Physical form

One general definition of solders is the fusible alloys with liquidus temperature below 400°C (750°F). Common alloy systems include Sn-Pb, Sn-Ag, Sn-Sb, Sn-Bi, Sn-In, Sn-Pb-Ag, Sn-Pb-Sb, Sn-Pb-Bi, Sn-Pb-In, Sn-Ag-Sb, Pb-In, Pb-Ag, and Pb-Sb. Solder can be made in various physical forms including bars, ingots, wire, powder, preform in designated shape and dimension, ball, and paste. In addition to the elemental compositions and physical forms, the performance of the solder material is dominated by the specific land pattern that the solder will be applied on. For example, the surface mount PCB containing a land pattern of 20-mil (0.020-in, 0.50-mm) pitch demands different characteristics in paste from that having a 50-mil (0.050-in, 1.25-mm) pitch.

The detailed correlation will be discussed in Chap. 10. The general guidelines in separating the types or requirements of solders for the corresponding land patterns are

Standard 50- to 25-mil (0.050- to 0.025-in or 1.25- to 0.63-mm) pitch

25- to 15-mil (0.025- to 0.015-in or 0.63- to 0.38-mm) pitch

15- to 8-mil (0.015- to 0.008-in or 0.38- to 0.20-mm) pitch

Soldering is also distinguished by the type of chemistry or cleaning process. The chemistry of a flux/vehicle system and the after-soldering cleaning follow the same selection rules. Three main chemistries are mildly activated rosin- (RMA)-chemistry with solvent cleaning, water-soluble chemistry with water cleaning, and no-clean chemistry without cleaning.

3.2 Physical Properties

Although other properties can be contributing factors to the overall performance of solders, five physical properties are discussed in this section because of their special importance to today's and future electronic packaging and assembly:

- Phase transition temperature
- Electrical conductivity
- Thermal conductivity
- Coefficient of thermal expansion
- Surface tension

3.2.1 Phase transition temperature of common solder alloys

An *alloy* is a metallic material which is formed from mixing a combination of two or more elements in the liquid state and then allowing them to solidify into the solid state. The elements in an alloy system may have a different extent of mutual miscibility in either the liquid or solid state. The thermodynamic interaction between elements results in the complete, partial, or nil miscibility which in turn may form metallurgical phases. Generally, the resulting alloys fall into one of the following categories:

- Complete miscibility in solid and liquid states
- Complete miscibility in liquid state and partial miscibility in solid state
- Complete miscibility in liquid state and no miscibility in solid state
- Systems with formation of intermediate phases or intermetallic compounds

For practical applications, three distinct temperatures are often used:

- *Solidus temperature.* The temperature below which the solid phase is stable (above the solidus temperature, material consists of a liquid phase and solid phase).
- *Liquidus temperature.* The temperature above which the liquid phase is stable.
- *Plastic range.* The temperature range between liquidus and solidus for a given composition at which the alloy is neither solid nor liquid.
- *Eutectic temperature.* The temperature at which a single liquid phase is transformed into two solid phases.

Table 3-1 summarizes these important temperatures for common solder alloy compositions. To facilitate practical use, Table 3-1 is segmented according to fourteen alloy systems. Table 3-2 lists the solder alloys covered by the most current version of the American National Standards Institute (ANSI) J-STD-006 Interim Final Issue "General Requirements and Test Methods for Electronic Grade Soft Solder Alloys and Fluxed and Non-fluxed Solid Solders for Electronic Soldering Applications." Further discussion on standards and specification are covered in Chap. 17.

3.2.2 Electrical conductivity

3.2.2.1 Basic theory. By definition, electrical conductivity is the result of movement of electrically charged electrons or ions from one location to another under an electric field. The electron conductivity is primarily predominant in metals, whereas ionic conductivity is responsible for the conductivity of oxides and nonmetallic materials. The characteristics of electrical conductivity can be described in terms of electron energy band structure as shown in Fig. 3-1. For metals, there is a partially filled electron band or a finite concentration of electrons in the conduc-

Table 3-1. Phase Transition Temperatures of Solder Alloys

Alloy composition	Solidus °C	Solidus °F	Liquidus °C	Liquidus °F	Plastic range °C	Plastic range °F
Sn-Pb System						
100 Sn	232	450	232	450	0	0
80 Sn/20 Pb	183	361	199	390	0	0
70 Sn/30 Pb	183	361	193	380	10	19
63 Sn/37 PB (E)	183	361	183	361	0	0
60 Sn/40 Pb	183	361	190	375	7	14
50 Sn/50 Pb	183	361	216	420	33	59
40 Sn/60 Pb	183	361	238	460	55	99
30 Sn/70 Pb	185	365	255	491	70	126
25 Sn/75 Pb	183	361	266	511	83	150
15 Sn/85 Pb	183	361	288	550	105	189
10 Sn/90 Pb	268	514	302	575	34	61
5 Sn/95 Pb	308	586	312	594	4	8
100 Pb	328	620	328	620	0	0
Sn-Ag System						
100 Ag	961	1761	961	1762	0	0
96.5 Sn/3.5 Ag (E)	221	430	221	430	0	0
95 Sn/5Ag	221	430	240	464	19	34
Sn-Sb System						
100 Sb	630.5	1167	630.5	1167	0	0
99 Sn/1 Sb (E)	235	456	235	456	0	0
95 Sn/5 Sb	235	455	240	464	5	9
Sn-Bi System						
100 Bi	271.5	520	271.5	520	0	0
42 Sn/58 Bi (E)	138	281	138	281	0	0
Sn-In System						
100 In	156.6	313	156.6	313	0	0
70 Sn/30 In	117	241	175	346	58	105
52 Sn/48 In	118	244	131	268	13	24
60 Sn/40 In	113	235	122	252	9	17
48 Sn/52 In (E)	117	241	117	241	0	0
Pb-In System						
70 Pb/30 In	240	464	253	488	13	24
50 Pb/50 In	180	256	209	408	29	52
40 Pb/60 In	174	345	185	365	11	20
30 Pb/70 In	160	320	174	345	14	25

Table 3-1. Phase Transition Temperatures of Solder Alloys (*Continued*)

Alloy composition	Solidus °C	Solidus °F	Liquidus °C	Liquidus °F	Plastic range °C	Plastic range °F
Pb-Sb System						
95 Pb/5 Sb	252	485	295	563	43	78
90 Pb/10 Sb	252	485	260	500	8	15
88.9 Pb/11.1 Sb (E)	252	485	252	486	0	0
Sn-Au System						
100 Au	1063	1945	1063	1945	0	0
80 Au/20 Sn (E)	280	536	280	536	0	0
Pb-Ag System						
97.5 Pb/2.5 Ag (E)	303	578	303	578	0	0
Sn-Pb-Ag System						
62 Sn/36 Pb/2 Ag (E)	179	355	179	355	0	0
15 Sn/82.5 Pb/2.5 Ag	275	527	280	537		
10 Sn/88 Pb/2 Ag	268	514	290	554	22	40
5 Sn/90 Pb/5 Ag	292	558	292	558	0	0
5 Sn/92.5 Pb/2.5 Ag	287	549	296	564	9	15
5 Sn/93.5 Pb/1.5 Ag	296	564	301	574	5	10
2 Sn/95.5 Pb/2.5 Ag	299	570	304	579	5	9
1 Sn/97.5 Pb/1.5 Ag	309	588	309	588	0	0
Sn-Pb-Bi System						
60 Sn/14.5 Pb/25.5 Bi	96	205	180	356	84	151
43 Sn/43 Pb/14 Bi	144	291	163	325	19	34
34 Sn/20 Pb/46 Bi (E)	100	212	100	212	0	0
16 Sn/32 Pb/52 Bi (E)	96	205	96	205	0	0
Sn-Pb-Sb System						
85 Sn/10 Pb/5 Sb	188	370	230	447	42	76
5 Sn/85 Pb/10 Sb	245	473	255	491	10	18
5 Sn/92 Pb/3 Sb	239	462	285	545	46	83
Sn-Pb-In System						
70 Sn/18 Pb/12 In (E)	162	324	162	324	0	0
Pb-In-Ag System						
92.5 Pb/5 In/2.5 Ag (E)	300	572	300	572	0	0
90 Pb/5 In/5 Ag	290	554	310	590	20	36

NOTE: (E) = eutectic alloy.

Table 3-2. Phase Transition Temperature of ANSI/J-STD-006 Solder Alloys

Alloy composition	Solidus		Liquidus	
	°C	°F	°C	°F
Ag 06 Pb 94	304	579	380	716
Ag 03 Pb 97	304	579	304	579
Au 97 Si 03	685	685	363	685
Au 88 Ge 12	356	673	356	673
Au 82 In 18	451	844	485	905
Au 80 Sn 20	280	536	280	536
In 99	156	313	156	313
In 80 Pb 15 Ag 05	149	300	150	302
In 70 Pb 30	160	320	174	345
In 60 Pb 40	174	345	185	365
In 52 Sn 48	118	244	118	244
In 40 Pb 60	195	383	225	437
In 30 Pb 70	238	460	253	487
In 26 Sn 38 Pb 37	134	273	181	358
In 25 Pb 75	250	482	264	507
In 20 Sn 54 Pb 26	183	361	196	385
In 19 Pb 81	270	518	280	536
In 12 Sn 70 Pb 12	153	307	163	325
In 05 Pb 92 Ag 03	300	512	310	590
Sn 99	232	450	232	450
Sn 96 Ag 04	221	430	221	430
Sn 95 Sb 05	235	455	240	464
Sn 95 Ag 05	221	430	240	464
Sn 95 Ag 04 Cd 01	216	421	219	426
Sn 90 Pb 10	183	361	213	415
Sn 70 Pb 30	183	361	193	379
Sn 36 Pb 37	183	361	183	361
Sn 62 Pb 36 Ag 2	179	345	179	345
Sn 60/40	183	361	191	376
Sn 60 Pb 38 Cu 02183	183	361	216	421

tion band so that there is no barrier to exciting electrons to higher energy states. In semiconductors, the concentration of electrons in the conduction band depends on temperature and composition. The completely filled energy band is separated from a completely empty conduction band of higher energy states by a band gap. The band gap is not too large when comparing it with the thermal energy, so a few electrons are thermally excitable into the conduction band leaving electron holes in the

Table 3-2. Phase Transition Temperature of ANSI/J-STD-006 Solder Alloys (*Continued*)

Alloy composition	Solidus °C	Solidus °F	Liquidus °C	Liquidus °F
Sn 50 Pb 50	183	361	216	421
Sn 50 Pb 49 Cu 01	181	361	215	419
Sn 50 Pb 32 Cu 18	145	293	145	293
Sn 46 Pb 46 Bi 08	120	248	167	333
Sn 43 Pb 43 Bi 14	144	291	163	525
Sn 42 Bi 58	138	280	138	280
Sn 40 Pb 60	183	361	238	460
Sn 35 Pb 65	183	361	246	475
Sn 35 Pb 63 Sb 02	185	365	243	469
Sn 34 Pb 20 Bi 46	100	212	100	212
Sn 30 Pb 70	183	361	254	489
Sn 30 Pb 68 Sb 02	185	365	250	482
Sn 30 Cd 70	140	284	160	320
Sn 20 Pb 80	183	361	277	531
Sn 20 Pb 79 Sb 01	184	363	270	518
Sn 18 Pb 80 Ag 02	178	352	270	518
Sn 16 Pb 32 Bi 52	96	205	96	205
Sn 10 Pb 90	215	527	302	576
Sn 10 Pb 88 Ag 02	268	514	290	554
Sn 05 Pb 95	308	586	312	594
Sn 05 Pb 94 Ag 01	296	565	301	574
Sn 05 Pb 93 Ag 02	280	536	284	543
Sn 03 Pb 93	314	597	320	608
Sn 03 Pb 95 Ag 02	305	581	306	583
Sn 02 Pb 96 Sb 02	299	570	307	585
Sn 01 Pb 98 Ag 01	309	588	309	588

normally filled band. However, when the materials possessing the energy band are so large that thermal excitation is insufficient to move the electrons to the conduction band, the materials act as insulators. This lack of free electron movement results in the fact that ionically and covalently bonded materials, such as semiconductors and insulators, have much lower electrical conductivity, as shown in Fig. 3-2. When an electric field is applied to a material, electric current density J is related by

$$J = n\, Z_e\, V \tag{3.1}$$

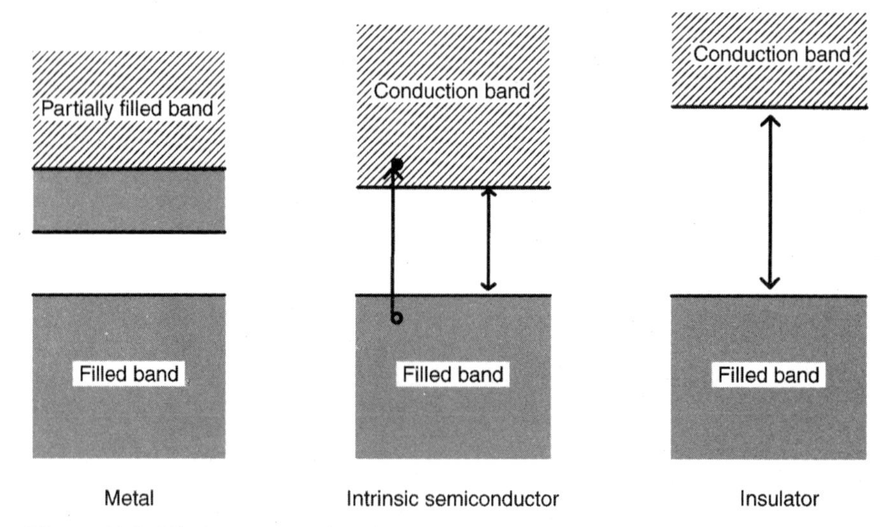

Figure 3-1. Electron energy band structure.

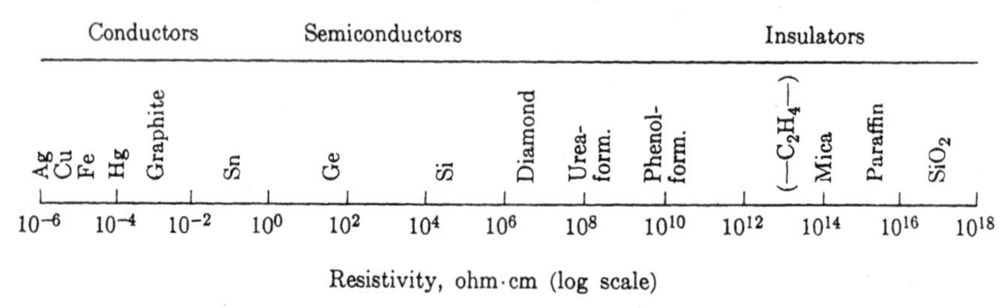

Figure 3-2. Magnitude of electrical conductivity of metal, semiconductor, and insulator.

where n = number of charge carriers in material
Z_e = charge carried by each
V = drift velocity

Electrical conductivity σ is defined by the relationship

$$\sigma = \frac{J}{E} \tag{3.2}$$

where E is the electric field strength. The drift velocity V is proportional to the acting electric field strength with the coefficient μ, which is defined as mobility of charge carriers:

$$V = \mu E \tag{3.3}$$

Substituting Eqs. (3.1) and (3.3) into Eq. (3.2) gives

$$\sigma = (nZ_e)\mu \tag{3.4}$$

Thus, electrical conductivity is the product of the concentration of charged carriers and charged carrier mobility. Because of the high mobility of electrons, several orders of magnitude greater than ionic mobility, metals with predominant electron movements exhibit high electrical conductivity.

The units for electrical conductivity are as follows:

$$\frac{1}{\text{ohm }(\Omega)\cdot\text{cm}} = \frac{\text{no. of carriers}}{\text{cm}^3}\,\frac{\text{coulombs (C)}}{\text{carrier}}\,\frac{\text{cm/s}}{\text{V/cm}}$$

$$= \frac{\text{no. of carriers}}{\text{cm}^3}\,\frac{\text{amperes (A)}\cdot\text{s}}{\text{Carrier}}\,\frac{\text{cm/s}}{\text{A}/\Omega/\text{cm}}$$

The conveniently used electrical property of materials is resistivity ρ, which is expressed in Ω[cdt]cm and is related to resistance as follows:

$$\text{Resistance} = \text{resistivity}\,\frac{\text{length}}{\text{area}}$$

$$= \Omega\cdot\text{cm}\,\frac{\text{cm}}{\text{cm}^2}$$

$$= \Omega$$

Electrical conductivity is the reciprocal of electrical resistivity.

$$\sigma = \frac{1}{\rho}$$

3.2.2.2 Electrical resistivity versus temperature. The total electrical resistivity is the sum of two components: the thermal component ρ_T, and the component due to lattice imperfection ρ_I.

$$\rho = \rho_T + \rho_I$$

The two components are the result of the disturbance of normal motions of electrons. There are two primary sources which hinder the motions of electrons: (1) lattice scattering, the collision of electrons with phonons (phonons are the quantified lattice thermal vibration energy), and (2) impurity scattering, the collision of electrons with solute atoms and lattice defects such as grain boundary, dislocations, point defects, and impurities. Lattice imperfections reduce the mean free path of electron motion. Since the electron mobility is proportional to the mean free path of electron motion, electrons collide with defects and lose momentum. Consequently, this results in a reduction in electron mobility and an increase in resistivity. Although these two components of resistivity respond differently to temperature and to the

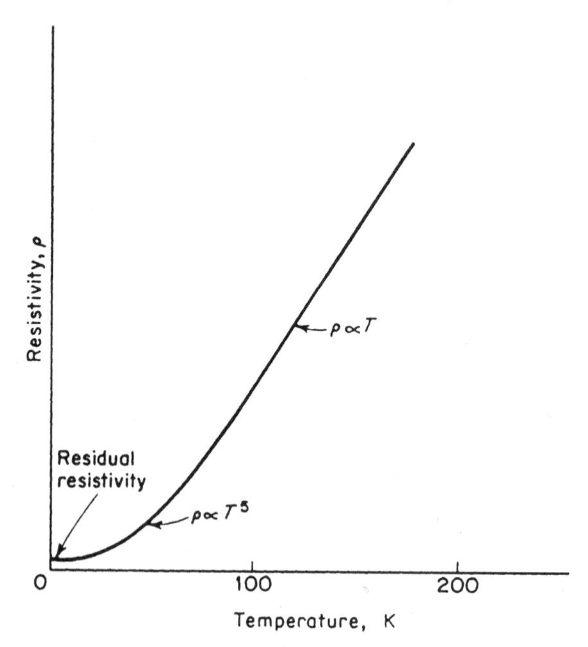

Figure 3-3. General relationship of electrical resistivity of metals versus temperature.

degree of lattice perfection, the general temperature dependency of the electrical resistivity as shown in Fig. 3-3 indicates that the resistivity of metals increases with increasing temperature. In terms of the two components, ρ_I is a function of the purity of material and is independent of temperature, and ρ_T, in contrast, reacts in a linear fashion to temperatures above 100°C; ρ_T is generally a function of T^5. The purity of the material has no bearing on the component of ρ_T. For metals where electrical conductivity is primarily contributed by electrons, the resistivity increases with increasing temperature for the reasons discussed above. In contrast, the electrical resistivity decreases (electrical conductivity increases) with increasing temperature for nonmetals where the ionic conduction predominates, such as for oxides and semiconductors as shown in Fig. 3-4.

3.2.2.3 Resistivity versus plastic deformation. Electrical resistivity generally increases with an increased amount of plastic deformation as exemplified in Fig. 3-5 for aluminum alloys. This is due to the internal structural changes in the metal, and such a distorted structure reduces the mean free path of electron movements. For metals, reduced mean free path correlates to the increased electrical resistivity. Therefore,

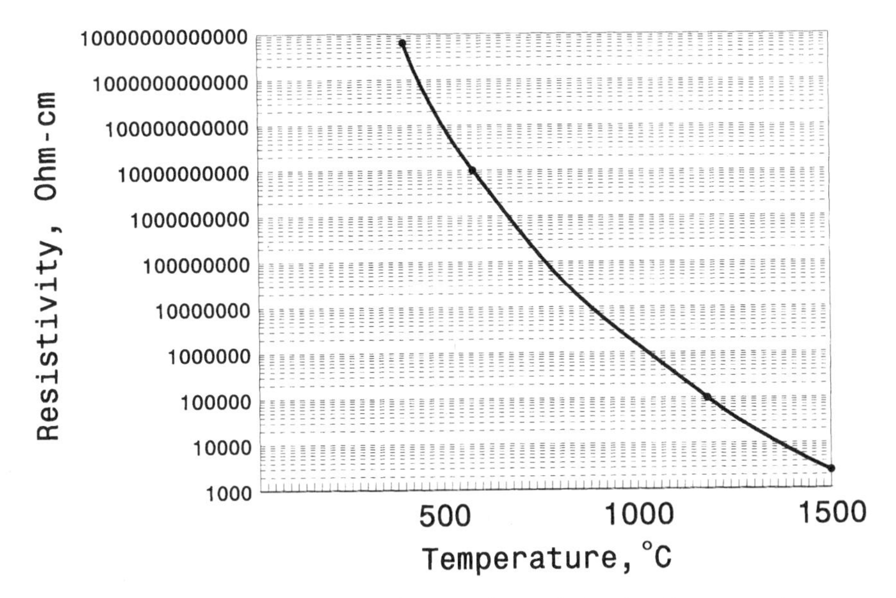

Figure 3-4. General relationship of electrical resistivity of oxides versus temperature.

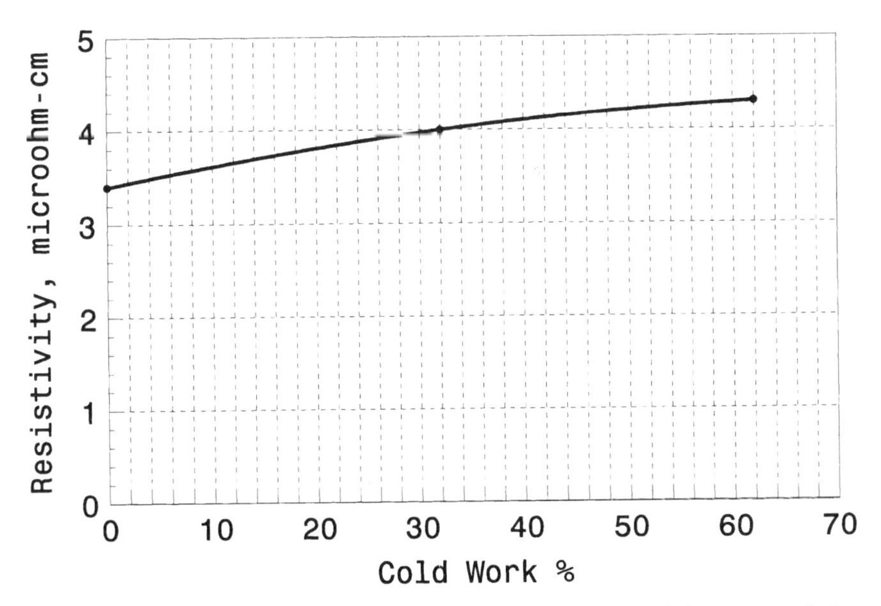

Figure 3-5. Example of electrical resistivity versus plastic deformation of aluminum Alloy 3003.

Table 3-3. Electrical Conductivity of Solder Elements and Alloys[1]

Material	Conductivity, $10^4\Omega^{-1}cm^{-1}$	
	0°C (32°F)	22°C (71.6°F)
Ag	66.7	62.1
Cu	64.5	58.8
Au	49.0	45.5
Al	40.0	36.5
Zn	18.1	16.9
Ni	16.0	14.3
Sn	10.0	9.1
Pb	5.2	4.8
Bi	1.0	0.8
63 Sn/37 Pb		6.9
60 Sn/40 Pb		6.9
50 Sn/50 Pb		6.4
40 Sn/60 Pb		5.9
30 Sn/70 Pb		5.5
20 Sn/80 Pb		5.2
10 Sn/90 Pb		4.9
5 Sn/95 Pb		4.8
62 Sn/36 Pb/2 Ag		6.8
1 Sn/97.5 Pb/1.5 Ag		3.5
42 Sn/58 Bi		2.6
96.5 Sn/3.5 Ag		9.4
95 Sn/5 Sb		7.0
48 Sn/52 In		6.8
70 Sn/18 Pb/12 In		7.2
30 Pb/70 In		5.2
40 Pb/60 In		4.1
92.5 Pb/2.5 Ag/5 In		3.2
90 Pb/5 Ag/5 In		3.3

as metals undergo the recovery process, electrical conductivity decreases. Table 3-3 lists the electrical conductivity of solder elements and alloys.[1]

3.2.2.4 Antimony effect. The presence of antimony (Sb) in Sn-Pb solders results in a decrease in electrical conductivity. Figure 3-6 qualitatively illustrates the relationship of Sn-Pb solders with no Sb, with 3% Sb based on Sn content, and with 6% Sb based on Sn content in their electrical conductivity over the entire range of Sn-Pb solder compositions.

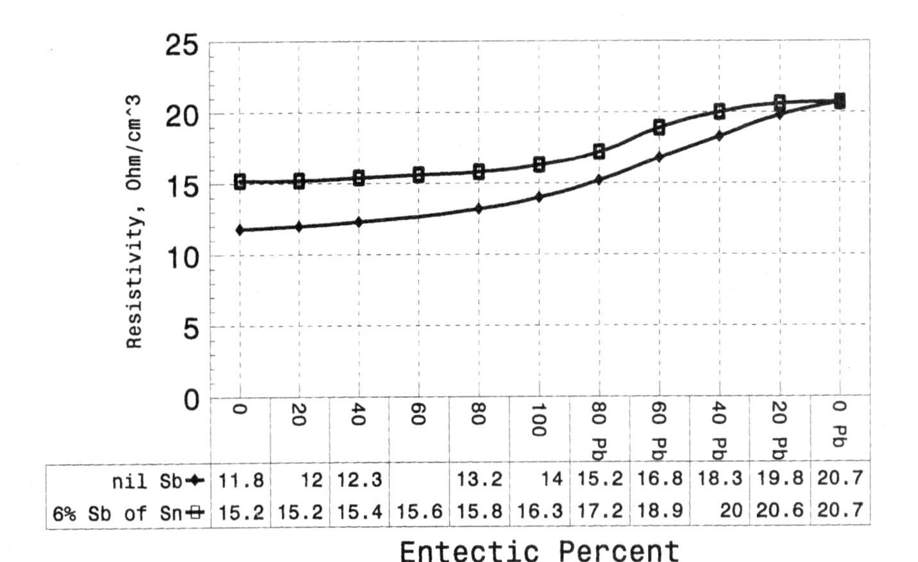

	0	20	40	60	80	100	80 Pb	60 Pb	40 Pb	20 Pb	0 Pb
nil Sb	11.8	12	12.3		13.2	14	15.2	16.8	18.3	19.8	20.7
6% Sb of Sn	15.2	15.2	15.4	15.6	15.8	16.3	17.2	18.9	20	20.6	20.7

Figure 3-6. Electrical resistivity of Sn-Pb solder versus antimony content.

3.2.3 Thermal conductivity

3.2.3.1 General definition. Thermal conductivity k is responsible for heat transfer through solids and is related to the properties of the material by

$$h = \frac{k}{C_p d}$$

where h = thermal diffusivity
C_p = heat capacity
d = density of material

Their units are related by

$$\frac{\text{cm}^2}{\text{s}} = \frac{[\text{calories (cal)} \cdot \text{cm}]/(^\circ\text{C} \cdot \text{s} \cdot \text{cm}^2)}{[\text{cal}/(\text{g} \cdot {}^\circ\text{C})]\ (\text{g}/\text{cm}^3)}$$

There are two main contributing components to thermal conductivity: electrons and phonons as expressed by

$$k_T = k_{\text{electron}} + k_{\text{phonon}}$$

Thermal conductivity of metals normally correlates well with electrical conductivity due to the fact that electrons which carry both thermal and electrical energy well are primarily responsible for thermal conductivity as well as electrical conductivity. For insulators, however, the phonon component predominates; for example, Al_2O_3 and BeO are good thermal conductors because the fast phonon transfers are facilitated by strong covalent bonds. This is the characteristic that makes both Al_2O_3 and BeO excellent substrates for electronic packaging.

3.2.3.2 Thermal conductivity versus temperature. Thermal conductivity of metals decreases with increasing temperature because of the role of electrons in metals. Figure 3-7 illustrates how copper conductivity responds to temperature.

The behavior of thermal conductivity in oxides above the Debye temperature is in congruence with that for metals. That is, thermal conductivity decreases with increasing temperature, which is opposite to the relationship of electrical conductivity with temperature. This is because the mean free path of phonons which are responsible for the thermal conductivity of oxides decreases with increasing temperature as shown in Fig. 3-8.

Figure 3-9 illustrates the thermal resistivity of common oxides (SiC, NiO, Al_2O_3, BeO), which are widely used in electronic packaging. Table 3-4 lists the available data for thermal conductivity of elements and alloys of interest to soldering.

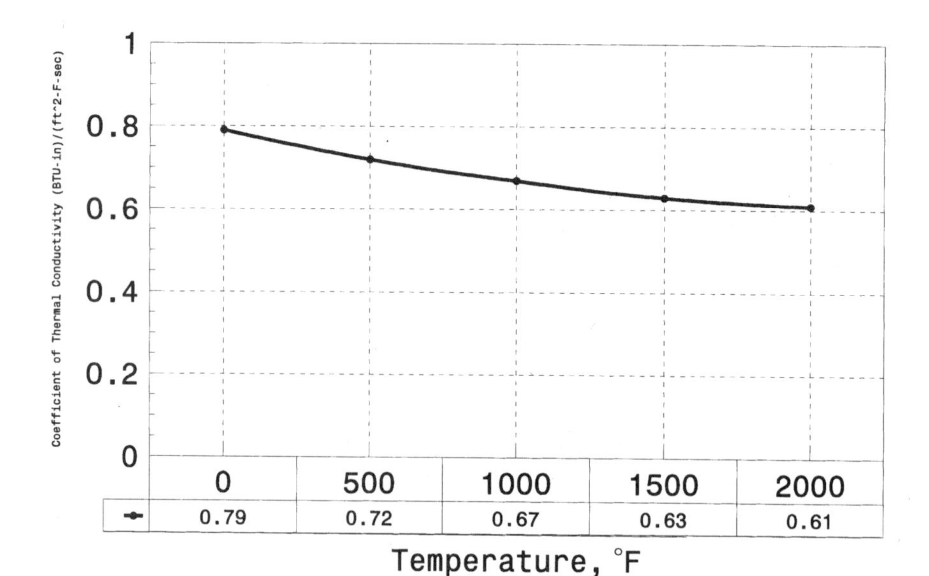

Figure 3-7. Thermal conductivity of Cu versus temperature.

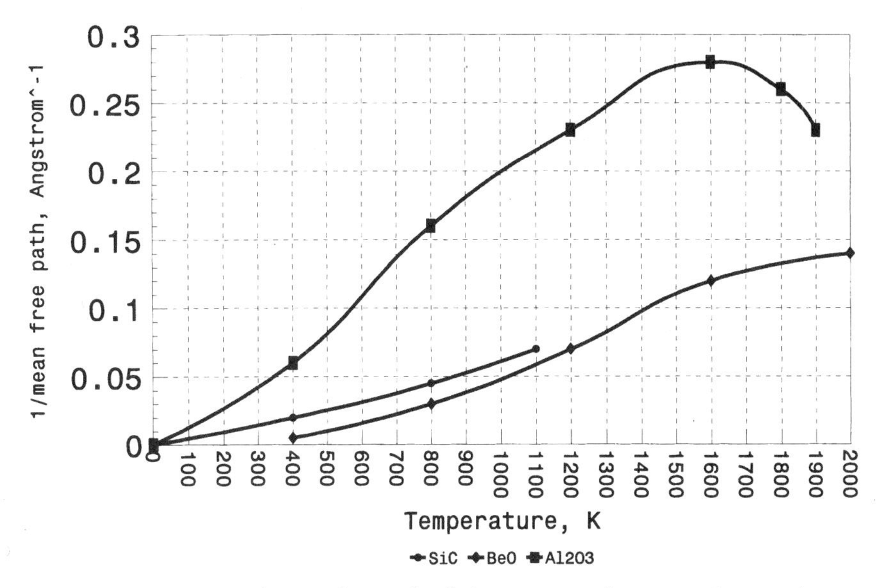

Figure 3-8. Reciprocal mean free path of phonon in oxides versus temperature.

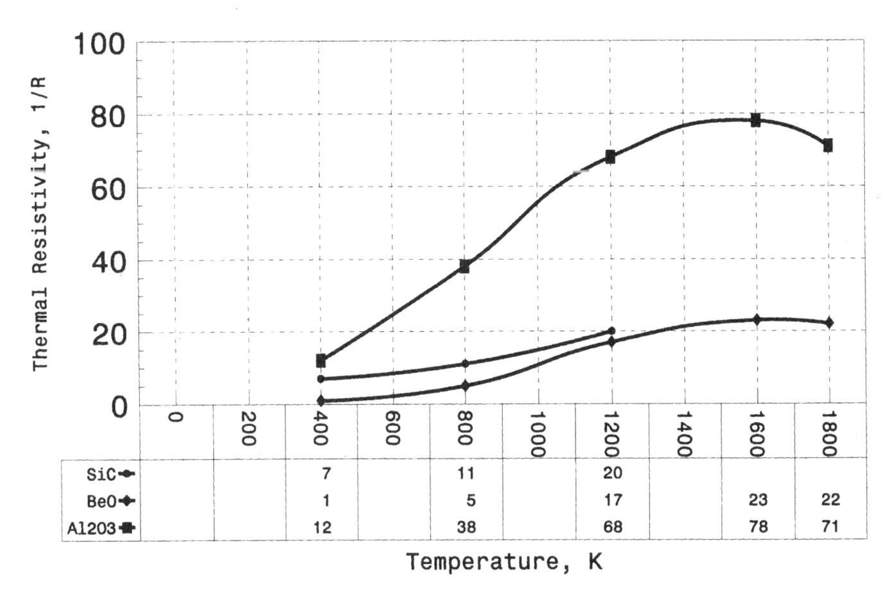

	0	200	400	600	800	1000	1200	1400	1600	1800
SiC			7		11		20			
BeO			1		5		17		23	22
Al203			12		38		68		78	71

Figure 3-9. Thermal resistivity of oxides versus temperature.

Table 3-4. Thermal Conductivity of Solder Elements and Alloys

Material	Thermal conductivity, W/(m · K) at 300 K
Diamond (CVD)	1300–2400
Ag	429.0
Cu	401.0
Au	317.0
BeO	250.0
Al	240.0
SiC	210.0
AlN	200.0
W	180.0
Zn	116.0
Ni	91.0
Fe	84–90
In	82.0
Si	82.0
Pd	72.0
Pt	72.0
Sn	66.0
Pb	35.0
Al_2O_3 (96%)	35.0
Sb	24.0
Bi	8.0
63 Sn/37 Pb	50.9
60 Sn/40 Pb	49.8
50 Sn/50 Pb	49.8
62 Sn/36 Pb/2 Ag	49.0
40 Sn/60 Pb	47.8
30 Sn/70 Pb	43.6
30 Sn/70 Pb	40.5
20 Sn/80 Pb	37.4
10 Sn/90 Pb	35.8
5 Sn/95 Pb	35.2
Alloy 42	15.6
Ag-filled die attach	1.3–5.0
Molding compound	0.66
BT epoxy	0.19
FR-4	0.11
Air	0.03

3.2.4 Coefficient of thermal expansion

Thermal expansion is commonly expressed with a coefficient in linear dimensional changes per unit dimension per degree of temperature change. The coefficient of linear thermal expansion is in in/(in · °F) or cm/(cm · °C).

3.2.4.1 CTE versus temperature. The specific volume of crystals increases with temperature. At sufficiently high temperatures, thermal expansion (increase in volume) of a material is the result of the increased amplitude of thermal oscillations of atoms (lattice vibration) about a mean position in a crystal structure lattice. The change in volume due to lattice vibrations is closely related to the increase in energy content. Like heat capacity, the coefficient of thermal expansion is sensitive to the change of temperature. As temperature increases, the amplitude vibration of the atoms about their mean positions increases; therefore, the coefficient of linear expansion increases. Figure 3-10 provides an example of the temperature dependency of the CTE for metals. The continuity in the relationship between the CTE and temperature is disrupted at the point where there is a change in the arrangement of the atoms and molecules of the material, such as at the melting point. Two major material properties dic-

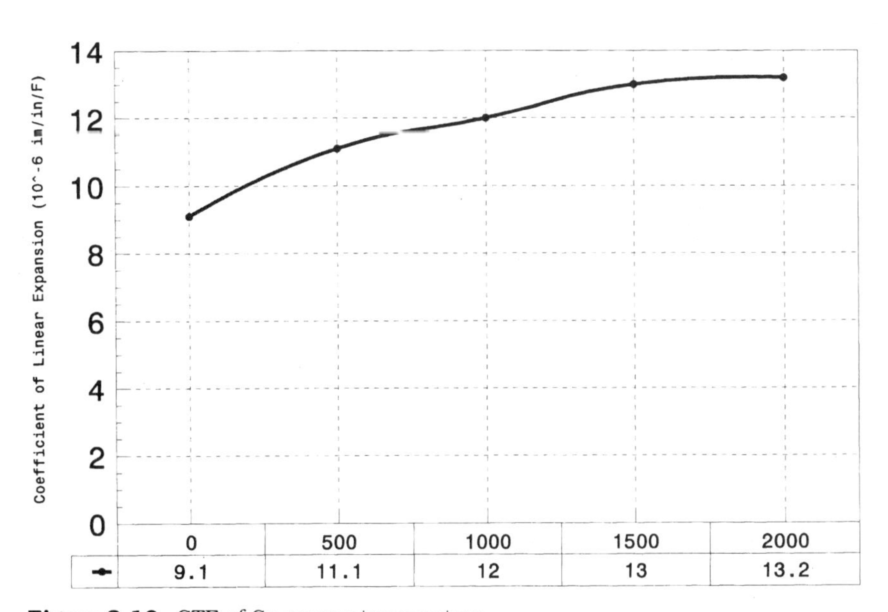

Figure 3-10. CTE of Cu versus temperature.

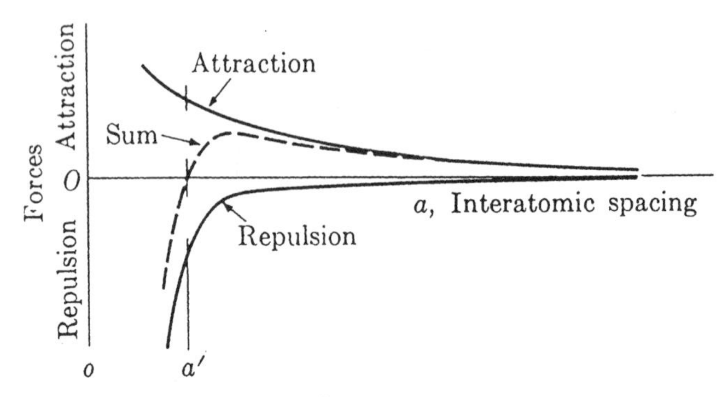

Figure 3-11. Interatomic distance versus energy.

tate the magnitude of the CTE: crystal structure and melting point. When materials have a similar lattice structure, the thermal expansions of the materials vary inversely with their melting points. This correlates to the fundamental relationship between interatomic energy which is the balance of interatomic attraction and repulsion versus interatomic spacing. The melting points and boiling points of materials are closely associated with the depth of the energy trough as shown in Fig. 3-11. A higher melting point generally corresponds to the deeper and more symmetrical energy trough, which in turn imparts less change in mean interatomic distance at a given change in temperature, and thus a smaller CTE.

Table 3-5 lists the melting point, crystal structure, and CTE of elements commonly used in electronic packaging and assembly.[2] Table 3-6 lists CTE of solders and substrates, and Figure 3-12 illustrates the space lattices of various crystal structures.

3.2.5 Surface tension

Surface tension is a key parameter related to wetting phenomenon, and thus solderability. The relative strength of attraction forces acting between molecules on the surface is weaker than that of molecular forces in the interior because of broken bonds at the surface. Thus the free surface of a material has a higher energy than the interior. Surface tension is a direct measure of the intermolecular forces acting at the surface.

For a liquid, the asymmetric force field results in a state of tension at the free surface, and the whole surface layer behaves as if it were a thin elastic film.

Surface tension is often expressed as surface energy. As shown in the following equation, free surface energy, which is the free energy per unit area of surface and is defined as the work necessary to increase

Table 3-5. CTE in Relation to Melting Point and Crystal Structure

Element	Melting point, °C	Crystal structure*	CTE, 10^{-6}/°C at 293 K (25°C
In	156.4	FC tetra	32.1
Sn	232	BC tetra	22.0
Bi	271.3	Rhombic	13.4
Pb	327.4	FCC	28.9
Zn	419.5	HCP	30.2
Sb	630.5	Rhombic	11.0
Al	660.2	FCC	25.0
Ag	960.5	FCC	18.9
Au	1063	FCC	14.2
Cu	1083	FCC	16.5
Si	1430	Diamond cubic	3.5
Ni	1455	FCC	13.4
Pd	1554	FCC	11.8
Pt	1773	FCC	6.8
W	3410	BCC	4.5

*BC = body centered; BCC = body-centered cubic; FC = face-centered; FCC = face-centered cubic; HCP = hexagonal close-packed.

Table 3-6. CTE of Solders and Substrates

Material	CTE, 10^{-6}/°C at 293 K
Diamond	0.8–1.7
Si	3.0
SiC	3.7–3.8
AlN	4.1–4.6
GaAs	5.9
Al_2O_3	6.4
BeO	8.0
Kovar	3.3–5.8
Alloy 42	5.3
FR-4	16
Cu	16–17
Polyimide	13
Epoxy/kevlar	3–8
Kevlar/cyanate Ester	1–3
Cyanate ester	15
BT epoxy	14(xy)
Molding compound	15
Ag-filled die attach	52
75 Sn/25 Pb	23.8
65 Sn/35 Pb	24.0
60 Sn/40 Pb	24.1
30 Sn/70 Pb	26.8
20 Sn/80 Pb	27.9
62 Sn/36 Pb/2 Ag	21.0
40 Sn/60 Bi	17.5
50 Sn/50 Bi	18.6

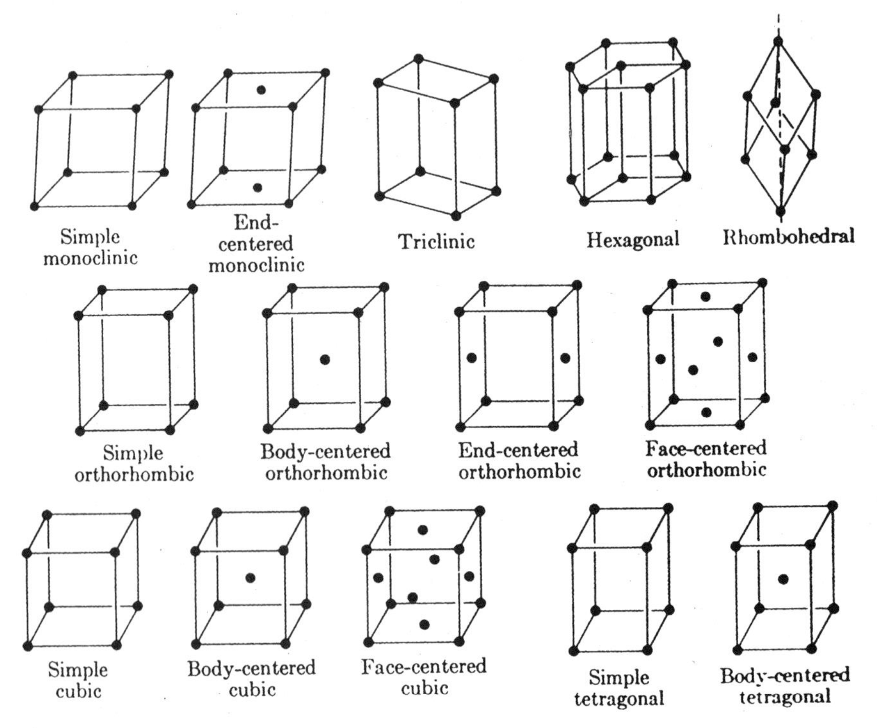

Figure 3-12. Space lattices of basic crystal structures.

the surface by unit area, is interchangeable with surface tension, the force acting normally across unit length.

$$\text{Surface tension} = \frac{\text{dyne (dyn)}}{\text{cm}} = \frac{\text{dyn} \times \text{cm}}{\text{cm} \times \text{cm}} = \frac{\text{erg}}{\text{cm}^2} = \text{surface energy}$$

In SI units

$$\frac{\text{Newtons (N)}}{\text{m}} = \frac{\text{N} \times \text{m}}{\text{m} \times \text{m}} = \frac{\text{joules (J)}}{\text{m}^2}$$

Surface energy and surface tension are numerically equal. The surface tension of most liquids is sensitive to the chemical compositions and the temperature. Generally, the surface tension decreases with increasing temperature as the following equation shows:

$$\gamma_T = \gamma_m + \frac{d\gamma}{dt}(T - T_m)$$

where γ_T is the surface tension at temperature T, γ_m is the surface tension at melting temperature T_m, and $d\gamma/dT$ is normally negative.

3.2.5.1 Surface tension versus melting temperature. An empirical correlation between the melting point and γ_m is derived as[3]

$$\gamma_m = \frac{CT_m}{V_m^{2/3}}$$

where $V_m^{2/3}$ is the molar surface area at the melting point.

3.2.5.2 Surface tension of common elements in solders. The empirical equations of surface tension for elements of interest are established.

Bismuth[4–6]

$$\gamma_T = 378 - 0.088(T - 270)\ \mathrm{mN/m}$$

$$\gamma_m = 378\ \mathrm{mN/m} \qquad \text{(at melting temperature, 270°C)}$$

Antimony[7–9]

$$\gamma_T = 382 - 0.063(T - 630)\ \mathrm{mN/m}$$

$$\gamma_m = 382\ \mathrm{mN/m} \qquad \text{(at melting temperature, 630°C)}$$

Indium[10–12]

$$\gamma_T = 561 - 0.096(T - 160)\ \mathrm{mN/m}$$

$$\gamma_m = 561\ \mathrm{mN/m} \qquad \text{(at melting temperature, 160°C)}$$

Lead[13–15]

$$\gamma_T = 465 - 0.96(T - 327)\ \mathrm{mN/m}$$

$$\gamma_m = 465\ \mathrm{mN/m} \qquad \text{(at melting temperature, 327°C)}$$

Silver[16–18]

$$\gamma_T = 944 - 0.150(T - 960)\ \mathrm{mN/m}$$

$$\gamma_m = 944\ \mathrm{mN/m} \qquad \text{(at melting temperature, 960°C)}$$

Tin[19–21]

$$\gamma_T = 574 - 0.140(T - 232)\ \mathrm{mN/m}$$

$$\gamma_m = 574\ \mathrm{mN/m} \qquad \text{(at melting temperature, 232°C)}$$

3.3 Mechanical Properties

3.3.1 Stress versus strain

3.3.1.1 General definition. The engineering stress-strain curve of an alloy generally follows the curve as shown in Fig. 3-13. As the stress (force per unit area) is applied, the alloy starts to deform elastically until it reaches the yield point where the metals begin plastic deformation at an almost constant stress. Eventually, the alloy starts strain-hardening with an increased stress necessary for additional deformation. The deformation as presented by strain is the length change as a percent of the original length. The elastic portion of the strain is reversible as shown in the straight line *o-a* of Fig. 3-13. The plastic strain is the permanent deformation to the alloy by stresses which exceed the elastic limit. The stress at the yield point *a* is the yield strength; beyond that the alloy starts to deform plastically. The yield strength is also an indication of the ability to resist plastic deformation as it is often represented by flow stress at which the metal will flow. The maximum point *c* of the curve is the maximum stress that an alloy can endure. For engineering purposes, the stress which produces a small amount of permanent deformation, a strain of 0.002, is defined as yield stress. It is well accepted that yield stress is favored by alloys with a fine grain size. A hypothetical curve is shown in Fig. 3-14 exhibiting the general relation of yield stress versus grain diameter—yield stress increases with decreasing grain diameter. The stress then decreases after reaching the maximum. There are two factors primarily responsible for the decreased stress:

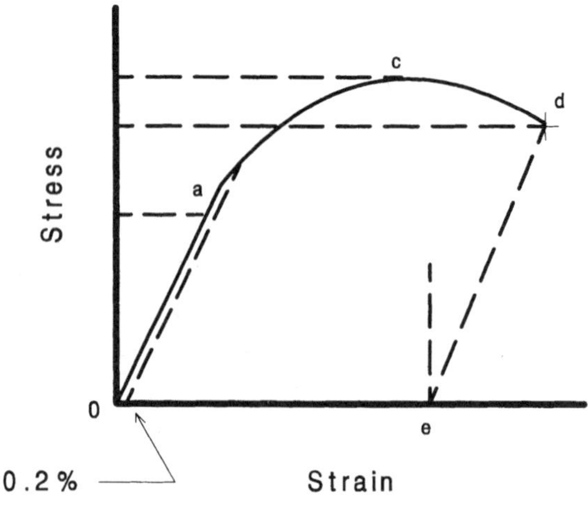

Figure 3-13. Engineering stress versus strain curve.

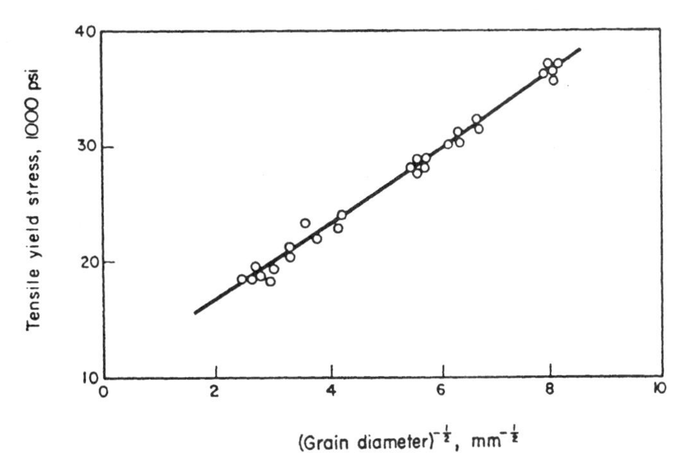

Figure 3-14. General relationship of yield stress versus grain diameter.

- Necking, reduction of the cross-sectional area of the alloy as plastic deformation proceeds
- Dynamic recovery process (see Sec. 3.4.6)

Another property reflected by the stress-strain relationship of a material is *ductility* which is defined as the amount of plastic deformation at the breaking point *d* or the reduction in area at the breaking point. The ductility is commonly expressed as elongation, or uniform elongation (defined as the uniform strain out to the point at which necking begins). Table 3-7 lists the uniform elongation of various solder alloys. Although stresses can be applied by tension, compression, or shear force, most alloys are weaker under shear than under tension or compression. Alloys tend to yield by slip; that is, one crystal plane displaces over the other as shown in Fig. 3-15. In the case of solders, shear strength is a more important property over tensile or compression because the majority of solder joints are subjected to shear stress during service.

Figure 3-16*a* and *b* shows the strengths of various solder alloys at 20°C and 100°C, respectively. At the temperature of 20°C, the 58 Bi/42 Sn alloy is ranked the highest in shear strength with 62 Sn/36 Pb/2 Ag, 96.5 Sn/3.5 Ag, 95 Sn/5 Sb, and 60 Sn/40 Pb ranked as second, and high-Pb alloy 1 Sn/97.5 Pb/ 1.5 Ag at the lowest. As the test temperature increases to 100°C, the strength of 58 Bi/42 Sn solder drops drastically due to its low melting temperature. Actual strength will depend to some extent upon the testing conditions, particularly, the

Table 3-7. Uniform Elongation of Solder Alloys

Alloy composition	Uniform elongation,* %
42 Sn/58 Bi	1.3
43 Sn/43 Pb/14 Bi	2.5
30 In/70 Sn	2.6
60 In/40 Sn	5.5
30 In/70 Pb	15.1
60 In/40 Pb	10.7
80 Sn/20 Pb	0.82
63 Sn/37 Pb	1.38
60 Sn/40 Pb	5.3
25 Sn/75 Pb	8.4
10 Sn/90 Pb	18.3
5 Sn/95 Pb	26.0
15 Sn/82.5 Pb/2.5 Ag	12.8
10 Sn/88 Pb/2 Ag	15.9
96.5 Sn/3.5 Ag	0.69
95 Sn/5 Ag	0.84
95 Sn/5 Sb	1.06
85 Sn/10 Pb/5 Sb	1.40
5 Sn/85 Pb/10 Sb	3.50
95 Pb/5 Sb	13.70

*Gage length = 2 in (50.8 mm).

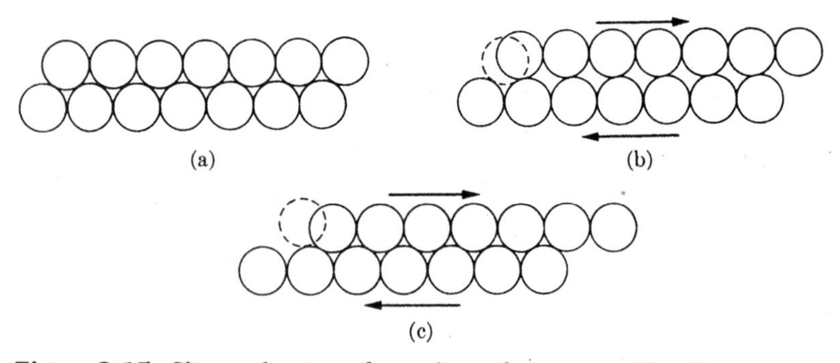

Figure 3-15. Slip mechanism of one plane of atoms over the other.

testing rate. The shear strength generally increases with increasing test rate. Figure 3-17 exhibits the shear strengths of Sn-Pb compositions over the entire range of tin and lead ratios. When adding lead to pure tin, the strength of the resulting alloy increases and reaches the maximum around the eutectic composition, 63 Sn/37 Pb.

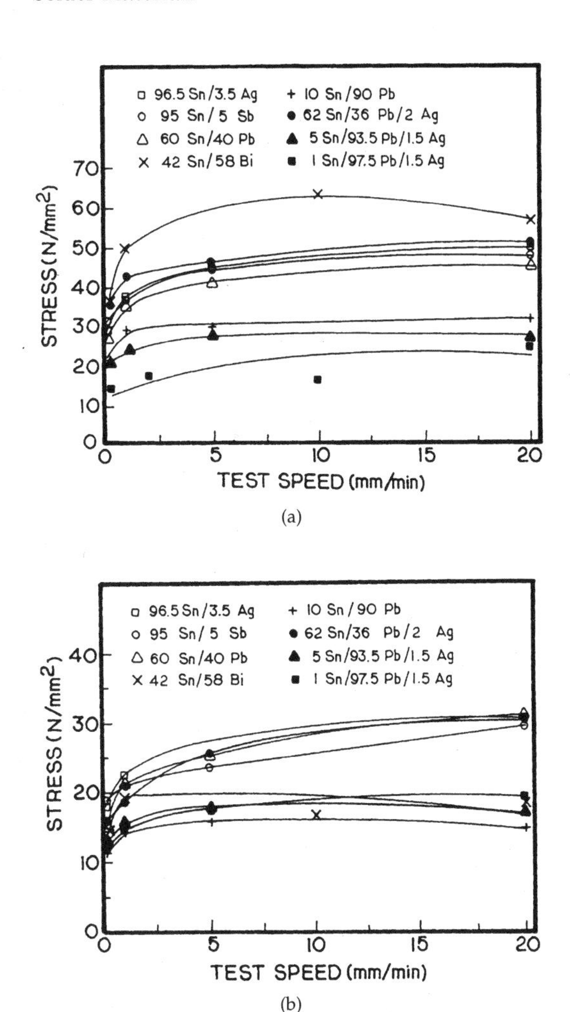

Figure 3-16. (*a*) Shear strength of solder alloys at 20°C. (*b*) Shear strength of solder alloys at 100°C.

3.3.1.2 Antimony effect. It is well established that the addition of Sb to Sn-Pb solder increases its strength as shown in Fig. 3-18. How much Sb may be added to solder is determined entirely by the amount of Sn present since Sb has negligible solubility in Pb. For practical purposes, Sb is considered insoluble in Pb. The rule of thumb for the Sb amount is not to exceed 6% Sn content which approaches the saturation point. When exceeding this point, Sb is rejected from solution in the form of relatively large cubical crystals (white) of high melting point, rendering the alloy viscous and gritty.

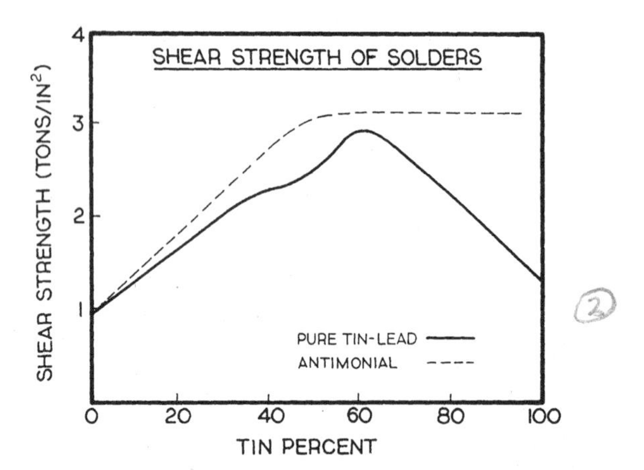

Figure 3-17. Shear strength of Sn-Pb compositions.

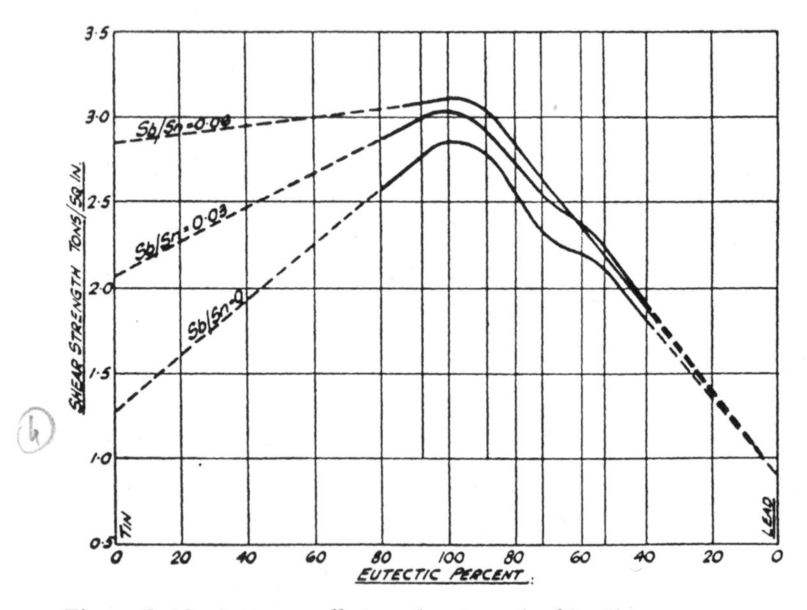

Figure 3-18. Antimony effect on the strength of Sn-Pb compositions.

3.3.2 Creep resistance

According to Webster's Dictionary, the definition of creep is to go very slowly; to enter or advance gradually so as to be almost unnoticed, to shift or gradually shift position. With respect to engineering behavior, creep has the same meaning and is defined as a time-dependent deformation of the solid under stress beyond its yield strength. It is a global plastic deformation where both temperature and stress are kept constant. Theoretically, the time-dependent deformation can occur at any temperature above absolute zero (0 K). In the low-temperature region ($T < 0.5\ T_m$, where T_m is the melting temperature), the creep strain is very limited and deformation normally does not lead to eventual fracture. The creep phenomenon occurs only at the active temperature ($T \gtrsim 0.5\ T_m$).

A typical creep curve, strain versus time, of alloys consists of three stages—primary, secondary, and tertiary. As shown in Fig. 3-19, there is an instantaneous strain being formed at the beginning. In the primary stage, the transient strain rate decreases rapidly from a very large initial value due to structural alterations in the alloy as it deforms as the result of strain hardening. This acts to retard the flow. The behavior is predominant for low-temperature and low-stress conditions. The secondary creep prevails at temperatures above $0.5\ T_m$ and correlates well with the self-diffusion process. The creep rate in this stage, however, reaches a steady state as a result of a balance between two competing processes: strain hardening and recovery (a softening effect). In addition, the stress rises due to the constant load imposed on the reduced cross-sectional area of the test specimen. When the softening effect and rising stress overcome the strain hardening, the tertiary stage begins, where the creep rate accelerates until the fracture occurs. A sudden occurrence of recrystallization can also induce the tertiary stage creep. The actual shape of the creep curve of an alloy depends on the stress applied and temperature exposed. The three stages often may not all be present. Generally, under high-temperature and high-

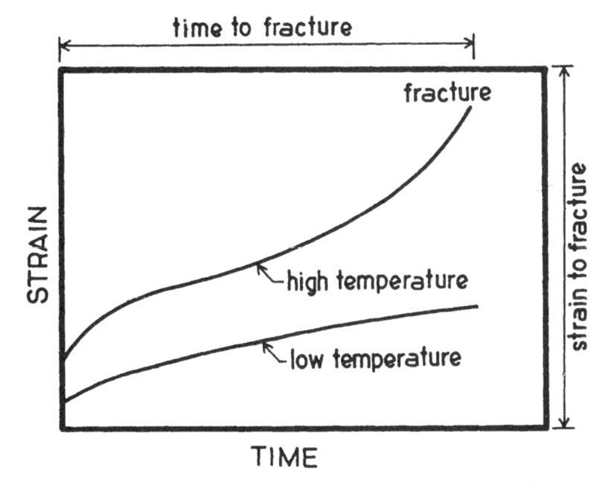

Figure 3-19. Typical creep curve.

stress conditions, the strain rate accelerates and the tertiary stage predominates. In the meantime, the primary stage is reduced and the secondary stage is not present. At low temperatures and low stresses, an extensive and well-defined secondary stage is normally observed. At intermediate temperatures and stresses, both primary and secondary stages are clearly defined.

In addition to stress and temperature, creep behavior depends on lattice defects, dislocation structure, dislocation movement, and its relationship with temperature and stress. The mechanism of creep is usually associated with the movement of dislocations. At low temperatures, deformation strain which is limited due to the dislocation movement may be hindered by grain boundary or impurity atoms. At high temperatures, the vacancies and atoms can move to and from dislocations and the grain boundaries serve as the source of vacancies and atoms. The dislocations can also climb out of the initial slip plane, and the grain boundaries may facilitate a dislocation climb. The role of grain boundaries is complex. Depending on the temperature effect on the movements of vacancies and atoms, grain boundaries can resist deformation as well as promote the deformation, since grain boundaries can serve as sources or sinks of vacancies. It is also generally observed that creep is often associated with an active change of internal structure such as void formation.

Creep can be measured in terms of creep rate which is defined as the strain produced per unit time during the period of steady elongation. In contrast to creep deformation, under constant load and temperature, an alternate test is the stress rupture where the material is deformed to complete fracture and the time to rupture is the information being monitored. The constant-structure and steady-state creep rate, the stress-rupture time and total stress-rupture time, and total stress-rupture ductility are all important design parameters. Among these, the steady rate of creep in the secondary stage is a convenient parameter to monitor in determining the useful life of the material.

Figures 3-20 through 3-25 are the creep curves for Sn-Pb alloys, Figs. 3-26 and 3-27 for Bi alloys, Figs. 3-28 and 3-29 for In-Sn alloys, Figs. 3-30 and 3-31 for In-Pb alloys, Figs. 3-32 through 3-35 for Sn-Pb-Ag alloys, Figs. 3-36 and 3-37 are for Sn-Ag alloys, Fig. 3-38 for Sn-Sb alloys, Figs. 3-39 and 3-40 for Sn-Pb-Sb alloys, and Fig. 3-41 for Pb-Sb alloys.[22]

For comparative purposes, Table 3-8 summarizes the relative creep resistance for the common solder alloys. It is ranked into five groups: low, low-moderate, moderate, moderate-high, and high.

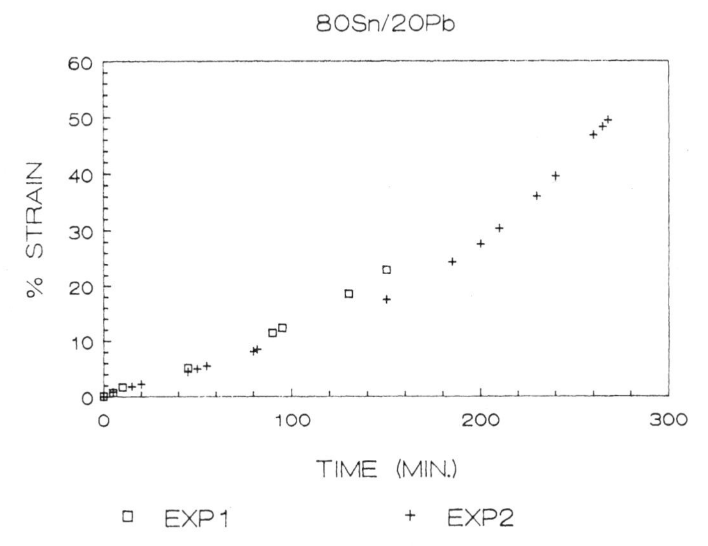

Figure 3-20. Creep curve of 80 Sn/20 Pb.

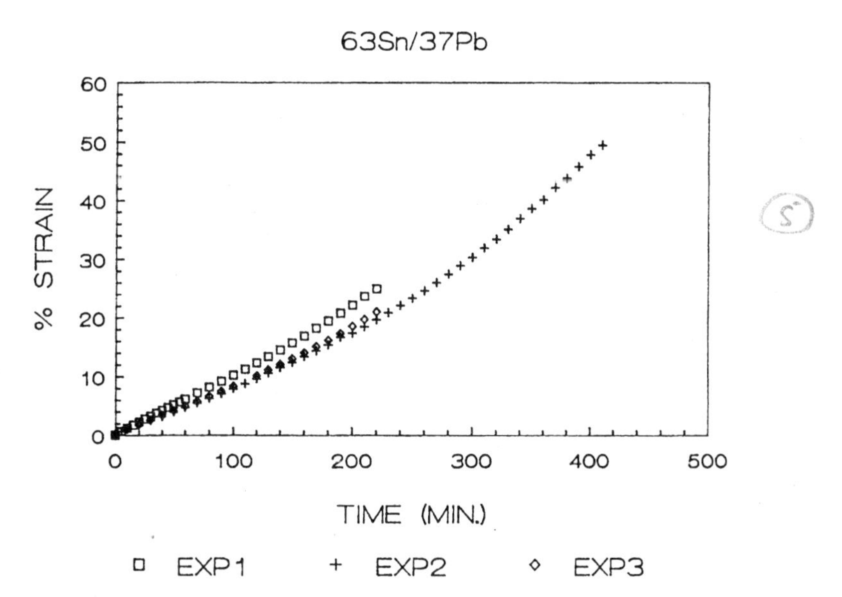

Figure 3-21. Creep curve of 63 Sn/37 Pb.

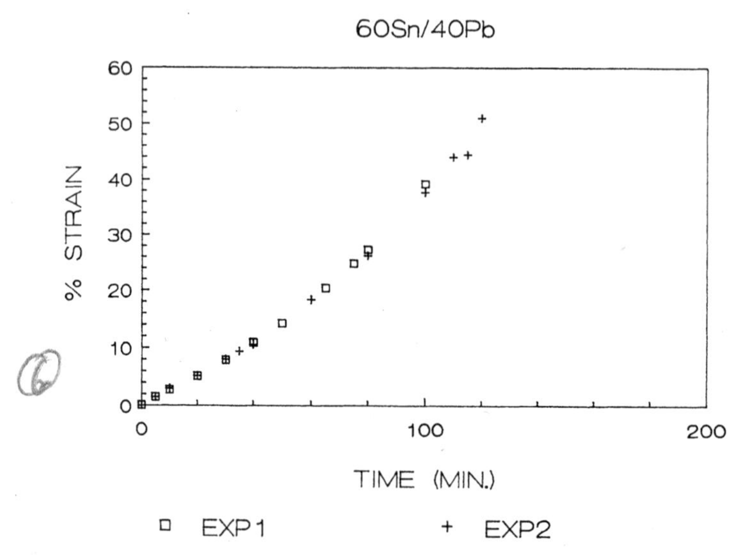

Figure 3-22. Creep curve of 60 Sn/40 Pb.

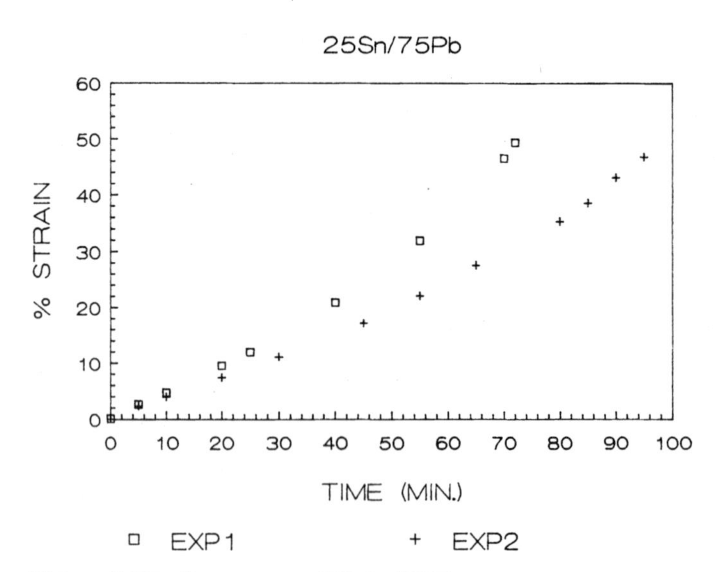

Figure 3-23. Creep curve of 25 Sn/75 Pb.

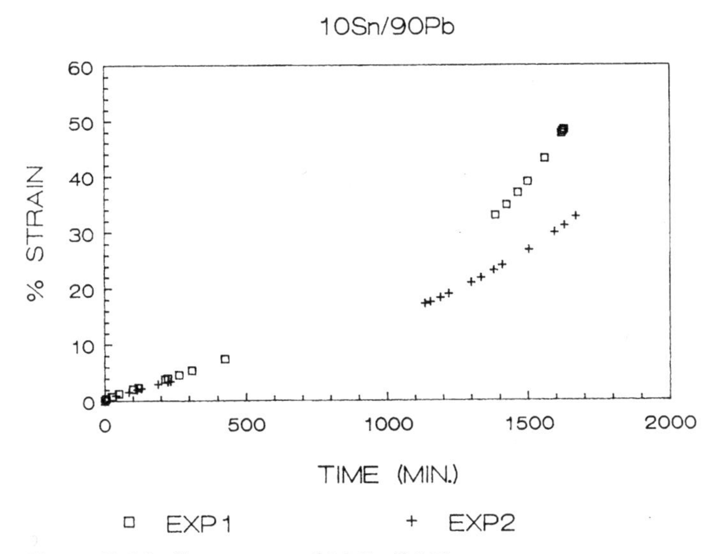

Figure 3-24. Creep curve of 10 Sn/90 Pb.

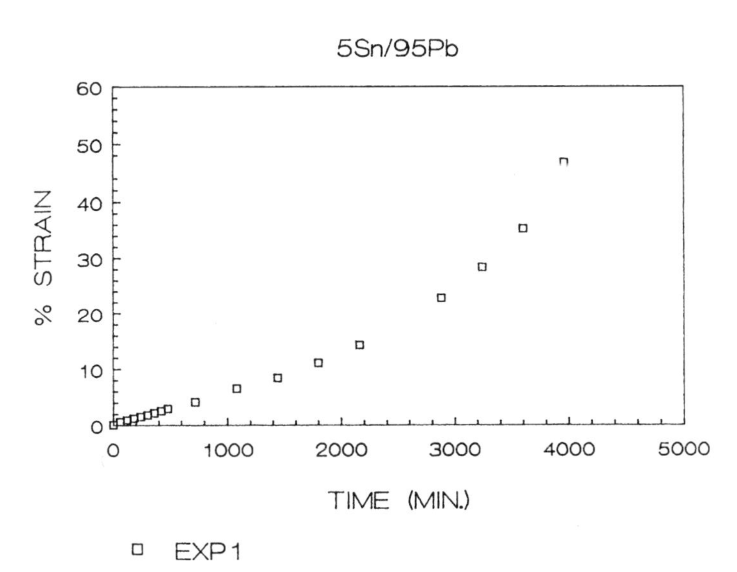

Figure 3-25. Creep curve of 5 Sn/95 Pb.

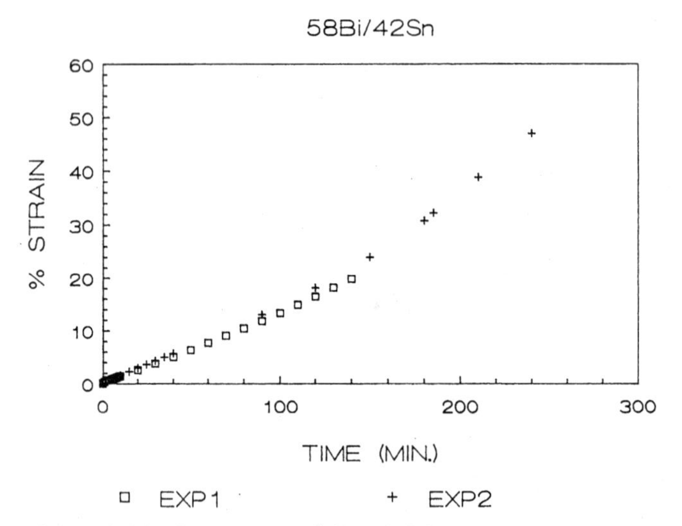

Figure 3-26. Creep curve of 58 Bi/42 Sn.

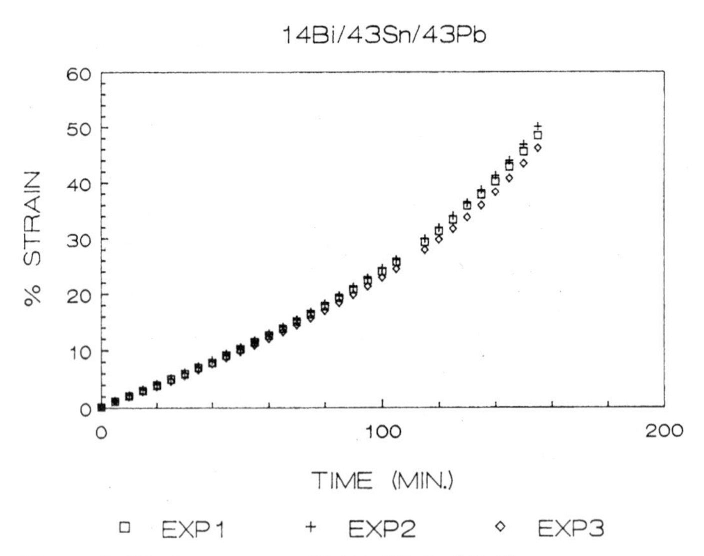

Figure 3-27. Creep curve of 14 Bi/43 Sn/43 Pb.

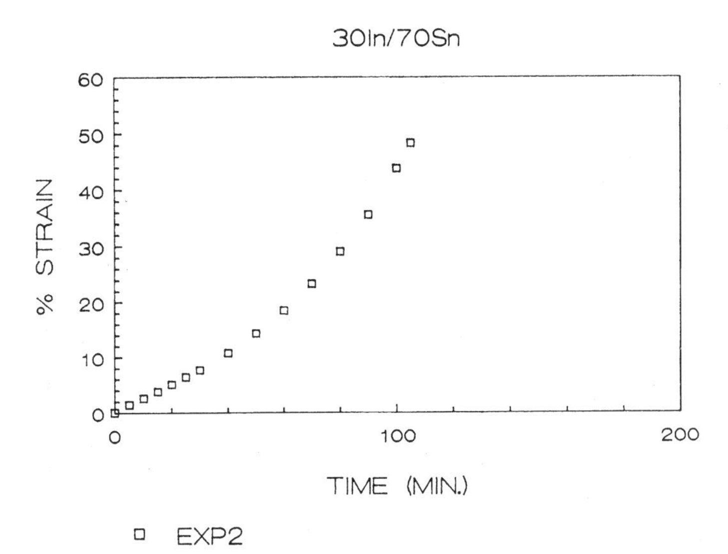

Figure 3-28. Creep curve of 30 In/70 Sn.

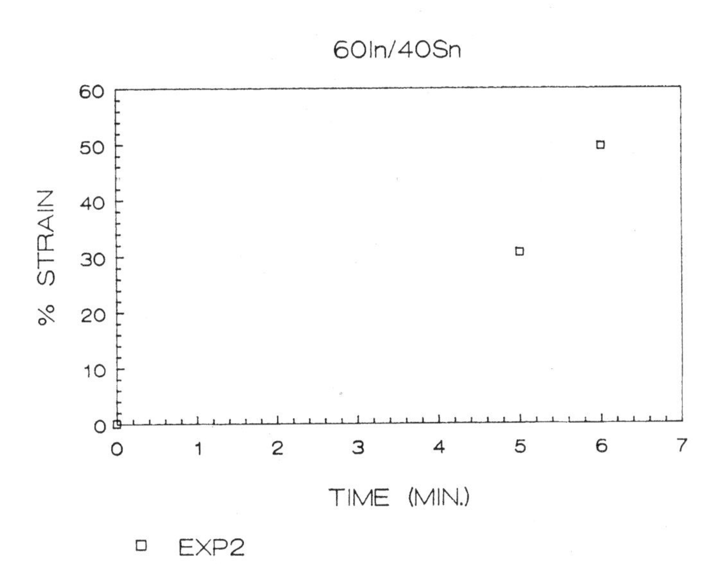

Figure 3-29. Creep curve of 60 In/40 Sn.

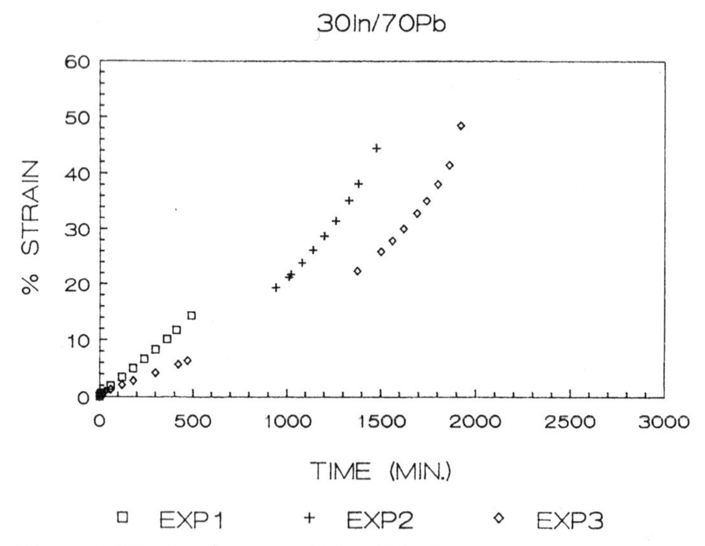

Figure 3-30. Creep curve of 30 In/70 Pb.

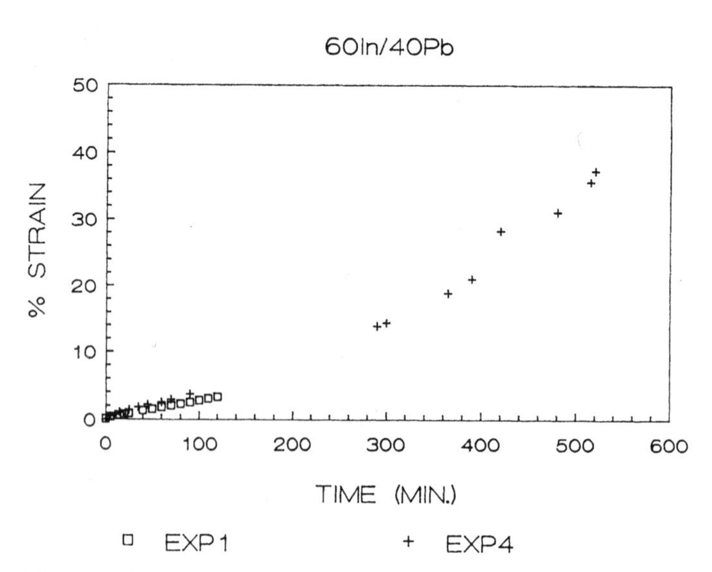

Figure 3-31. Creep curve of 60 In/40 Pb.

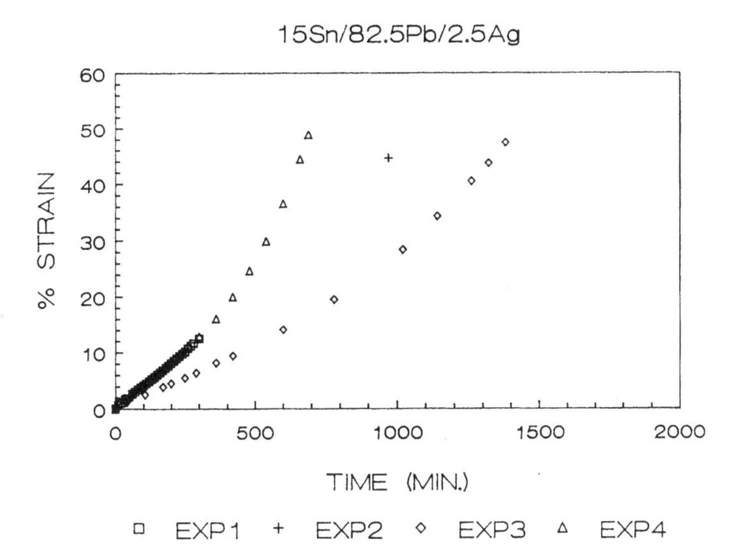

Figure 3-32. Creep curve of 15 Sn/82.5 Pb/2.5 Ag.

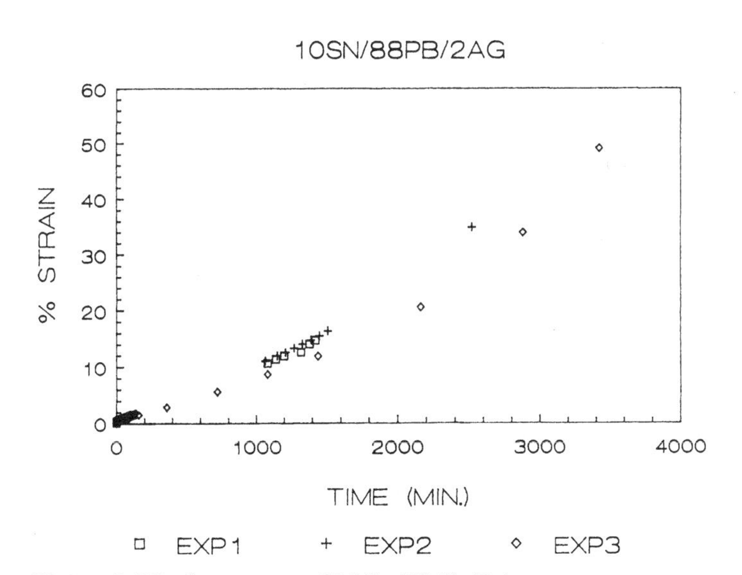

Figure 3-33. Creep curve of 10 Sn/88 Pb/2 Ag.

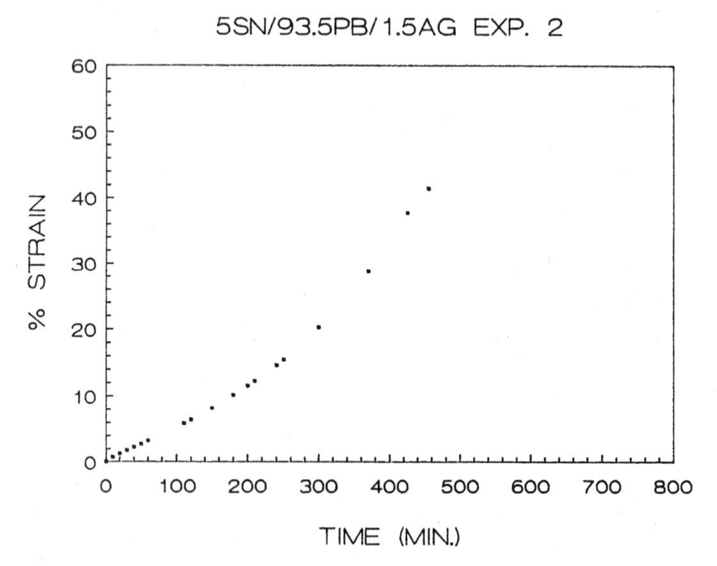

Figure 3-34. Creep curve of 5 Sn/93.5 Pb/1.5 Ag.

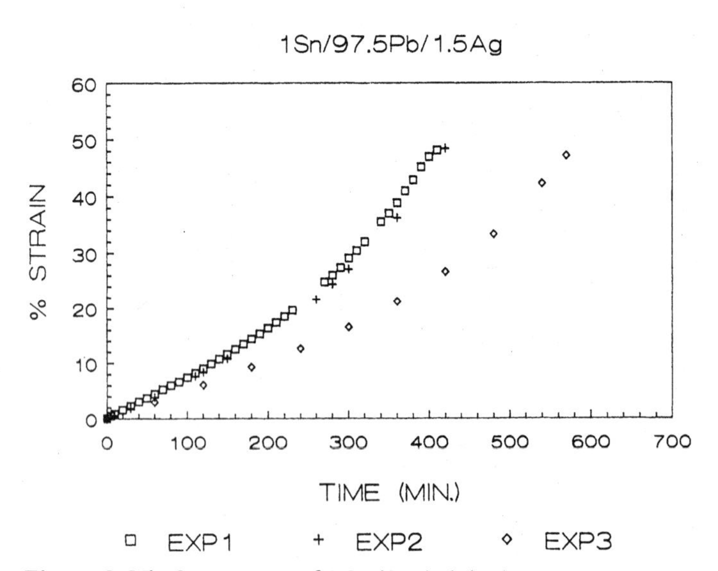

Figure 3-35. Creep curve of 1 Sn/97.5 Pb/1.5 Ag.

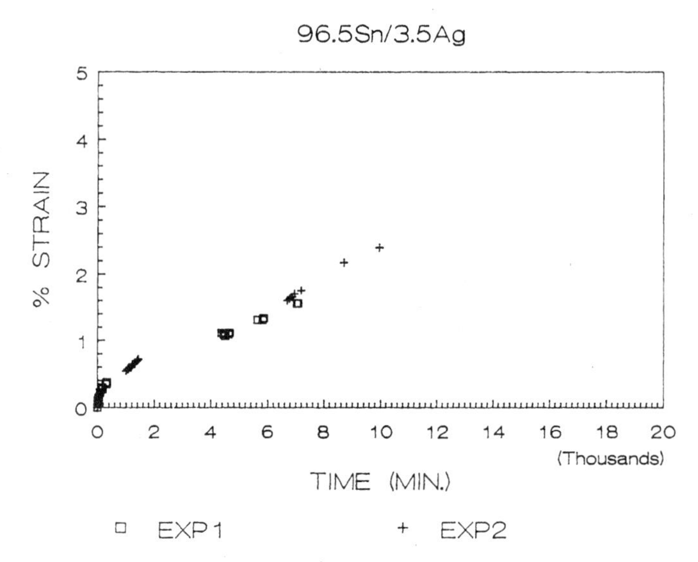

Figure 3-36. Creep curve of 96.5 Sn/3.5 Ag.

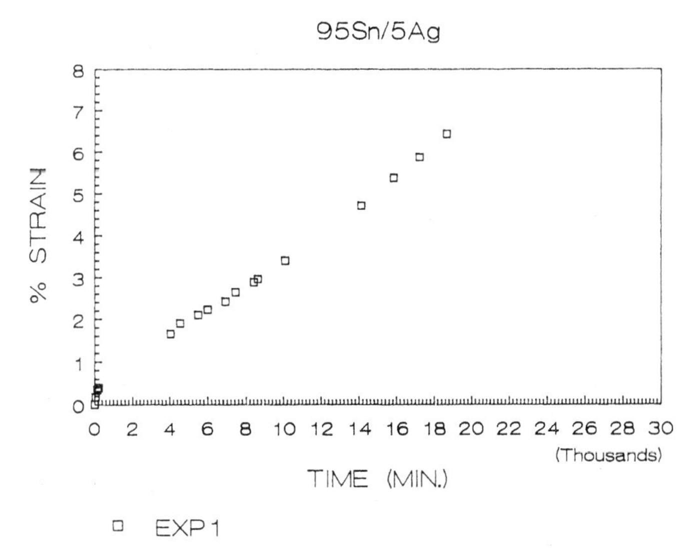

Figure 3-37. Creep curve of 95 Sn/5 Ag.

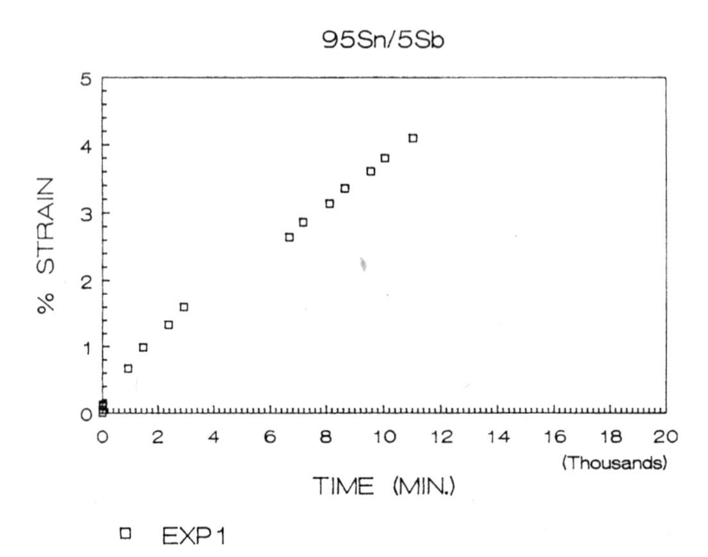

Figure 3-38. Creep curve of 95 Sn/5 Sb.

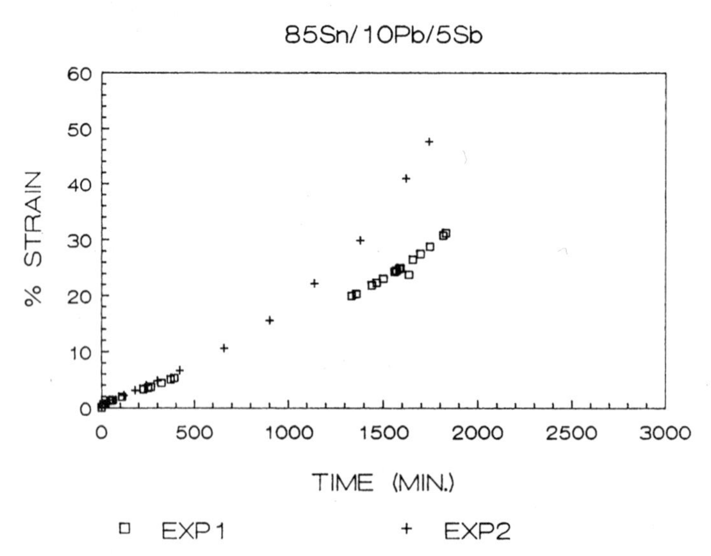

Figure 3-39. Creep curve of 85 Sn/10 Pb/5 Sb.

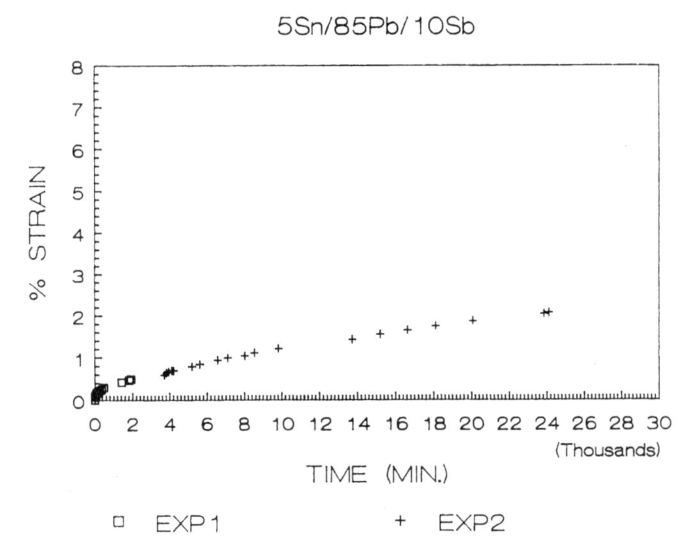

Figure 3-40. Creep curve of 5 Sn/85 Pb/10 Sb.

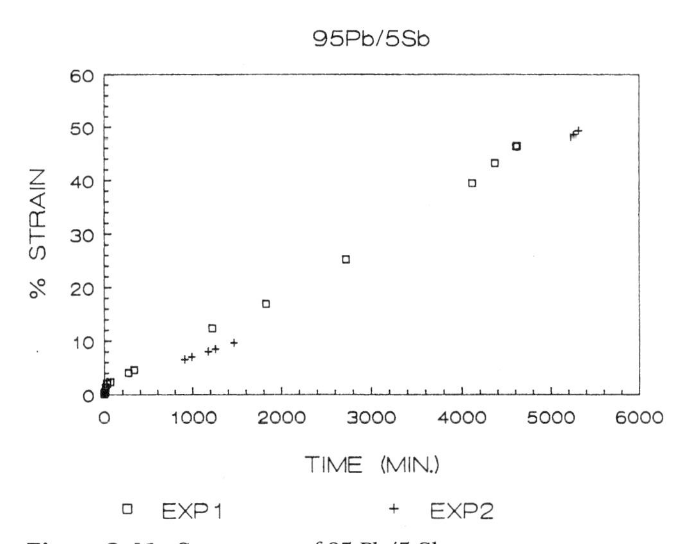

Figure 3-41. Creep curve of 95 Pb/5 Sb.

Table 3-8. Relative Creep Resistance of Solder Alloys

Alloy composition	Rank
42 Sn/58 Bi	Moderate
43 Sn/43 Pb/14 Bi	Low-moderate
30 In/70 Sn	Low
60 In/40 Sn	Low
30 In/70 Pb	Moderate
60 In/40 Pb	Moderate
80 Sn/20 Pb	Moderate
63 Sn/37 Pb	Moderate
60 Sn/40 Pb	Low
25 Sn/75 Pb	Low
10 Sn/90 Pb	Moderate
5 Sn/95 Pb	Moderate-high
62 Sn/36 Pb/2 Ag	High
15 Sn/82.5 Pb/2.5 Ag	Moderate
10 Sn/88 Pb/2 Ag	Moderate-high
5 Sn/93.5 Pb/1.5 Ag	Moderate
1 Sn/97.5 Pb/1.5 Ag	Moderate
96.5 Sn/3.5 Ag	High
95 Sn/5 Ag	High
95 Sn/5 Sb	High
85 Sn/10 Pb/5 Sb	Moderate
5 Sn/85 Pb/10 Sb	High
95 Pb/5 Sb	Moderate-high

Alloys of Sn-Ag, Sn-Sb, and 5 Sn/85 Pb/10 Sb impart high creep resistance. This is primarily attributed to solution hardening as substantiated by their high strength and low elongation as shown in Table 3-9.

In order to correlate creep to basic material characteristics, creep behavior is evaluated based on the melting point, ultimate tensile strength, microstructure, and metallurgical strengthening effect. These material characteristics are considered major factors affecting creep resistance and are briefly discussed below.

3.3.2.1 Effect of melting point (softening point). As the rule of thumb, the creep behavior of an alloy becomes significant when the homologous temperature is equal to or larger than 0.5. The homologous temperature is defined as the ratio of test temperature (service temperature) to melting temperature in degrees Kelvin (K). Thus while other parameters are equal, the alloy with the higher melting point is more creep-resistant. As we compare the creep rate among Sn-Pb composi-

Table 3-9. Ultimate Tensile Strength of Solder Alloys

Alloy composition	Ultimate tensile strength, ksi	0.2% Yield strength, ksi	0.01% Yield strength, ksi
42 Sn/58 Bi	9.71	6.03	3.73
43 Sn/43 Pb/14 Bi	5.60	3.60	2.77
30 In/70 Sn	4.67	2.54	1.50
60 In/40 Sn	1.10	0.67	0.53
30 In/70 Pb	4.83	3.58	3.08
60 Sn/40 Pb	4.29	2.89	2.06
80 Sn/20 Pb	6.27	4.30	2.85
60 Sn/40 Pb	4.06	2.06	2.19
25 Sn/75 Pb	3.35	2.06	1.94
10 Sn/90 Pb	3.53	2.02	1.98
5 Sn/95 Pb	3.37	1.93	1.83
15 Sn/82.5 Pb/2.5 Ag	3.85	2.40	1.94
10 Sn/88 Pb/2 Ag	3.94	2.25	2.02
5 Sn/93.5 Pb/1.5 Ag	6.75	3.85	2.40
1 Sn/96.5 Pb/1.5 Ag	5.58	4.34	3.36
96.5 Sn/3.5 Ag	8.36	7.08	5.39
95 Sn/5 Ag	8.09	5.86	3.95
95 Sn/5 Sb	8.15	5.53	3.47
85 Sn/10 Pb/5 Sb	6.45	3.63	2.62
5 Sn/85 Pb/10 Sb	5.57	3.67	2.26
95 Pb/5 Sb	3.72	2.45	1.98

tions, the decreased creep rate (or increased creep resistance) is largely attributed to the increased melting point (softening point) in the following order:

60 Sn/40 Pb, 80 Sn/20 Pb, 25 Sn/75 Pb > 10 Sn/90 Pb > 5 Sn/95 Pb

The same explains the higher creep resistance of 80 In/70 Pb than that of 60 In/40 Pb, and of 30 In/70 Sn than that of 60 In/40 Sn. Bismuth alloys, 42 Sn/58 Bi and 43 Sn/43 Pb/14 Bi, having high strength are prone to creep at or above ambient temperature (298 K). This is primarily due to their low melting point.

3.3.2.2 Effect of strength. Table 3-9 lists the ultimate tensile strength of various solder alloys. The high strength of 96.5 Sn/3.5 Ag, 95 Sn/5 Ag, and 95 Sn/5 Sb contributes to their high creep resistance.

3.3.2.3 Effect of microstructure. The rate of creep deformation highly depends on the microstructure of the alloy, as reflected by the grain (interphase) size and distribution. The grain size effect, however,

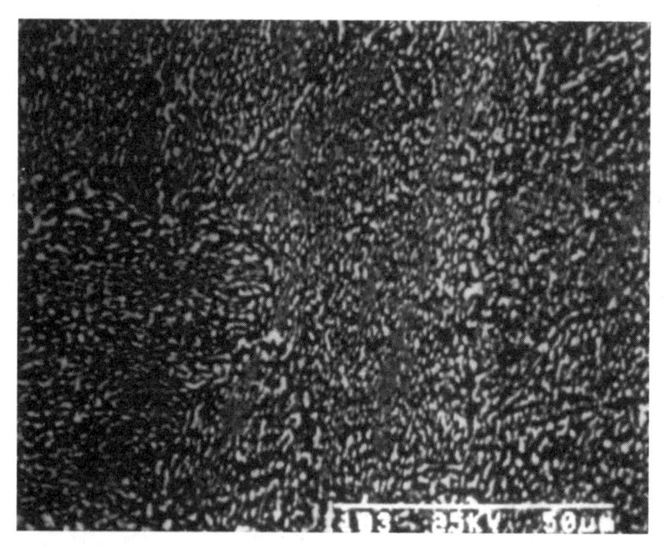

Figure 3-42. SEM microstructure of 63 Sn/37 Pb.

can be altered by the mechanism of plastic deformation. When the shear on grain boundaries is the controlling mechanism, the strain rate is inversely proportional to the square of the grain diameter.[23]

It is commonly recognized that in the low-temperature range, lower than $\frac{1}{2} T_m$, alloys normally fail by fractures passing through the interior of the grains or phases, so-called transcrystalline fractures. At high temperatures, intercrystalline fractures run along grain boundaries because the grain boundary shear under this condition is a predominant mechanism of plastic deformation. The microstructure of eutectic 63 Sn/37 Pb as shown in Fig. 3-42 favors creep resistance; thus a 63 Sn/37 Pb alloy has better creep resistance than noneutectic compositions.

Compositions of Sn-In have poor creep resistance as they are disadvantaged by their low melting point of the microstructural continuous phase. Compositions of In-Pb are ductile as exhibited by their elongation data in Table 3-7. Their single-phase microstructure and moderate melting point make them a moderate creep-resistant material. (See Chap. 6 for further discussions.)

3.3.2.4 Effect of metallurgical reaction. Solution hardening, precipitation hardening, and work hardening have proven to be the primary strengthening methodologies for metals. The strength of 96.5 Sn/3.5 Ag, 95 Sn/5 Ag, 95 Sn/5 Sb, and 5 Sn/85 Pb/10 Sb are benefited by solution hardening.

3.3.3 Fatigue resistance

The failure of alloys under alternating stresses is known as fatigue. The stress an alloy can tolerate under cyclic loading is much less than it is under static loading. Therefore, the yield strength, which is a measure of the static stress that an alloy will resist without permanent deformation, often does not correlate well with the fatigue resistance. The fatigue crack usually starts as a small crack, which under repeated applications of stress, grows in size. As the crack grows, the load-carrying cross section is reduced. Fatigue generally involves three stages: initiation of cracks, crack growth, and catastrophic failure. The fatigue fracture surfaces are divided into two areas with distinctly different appearances:

- Surface with a smooth appearance corresponding to the region where the crack grows slowly. The smooth texture is the result of surfaces rubbing against each other as the material is deformed back and forth through each stress cycle.
- Surface with rough and irregular texture or granular appearance corresponding to the early stage of fatigue when the material finally fractures and there is no rubbing action.

Figure 3-43*a* and *b* shows the fracture surfaces of 63 Sn/37 Pb. Figure 3.43*a* exhibits the early stage of fatigue fracture with a rough and irregular texture, and Fig. 3-43*b* shows the later stage of fatigue fracture with a smooth texture.[24] In contrast to global creep behavior, the lack of microscopic plastic deformation at small distances from the immediate fracture surface is generally a characteristic of fatigue. Fatigue failure is thus related to nonuniform plastic deformation. The irreversible strains developed under cyclic stresses are localized along slip planes, at grain boundaries, or around surface irregularities which can be compositional defects or geometrical defects such as microscopic notches. Since the free surface is usually more intensely stressed than any other location, the stress-concentrated surface will undergo plastic deformation when subjected to cyclic loads which would not deform the rest of the material. Thus, it is common for the fatigue failure to start at the free surface of alloys. To combat fatigue, a smooth and defect-free surface is one of the important design criteria. It is also observed that fatigue failure can occur at a very low temperature such as 4 K. At this extreme temperature, thermal energy cannot make any appreciable contribution to the mechanism of fatigue fracture. It is indicated that thermal activation is not always necessary to fatigue failure. In these cases, the diffusion process is obviously not the rate-controlling mechanism. Slip and twinning processes that are not ther-

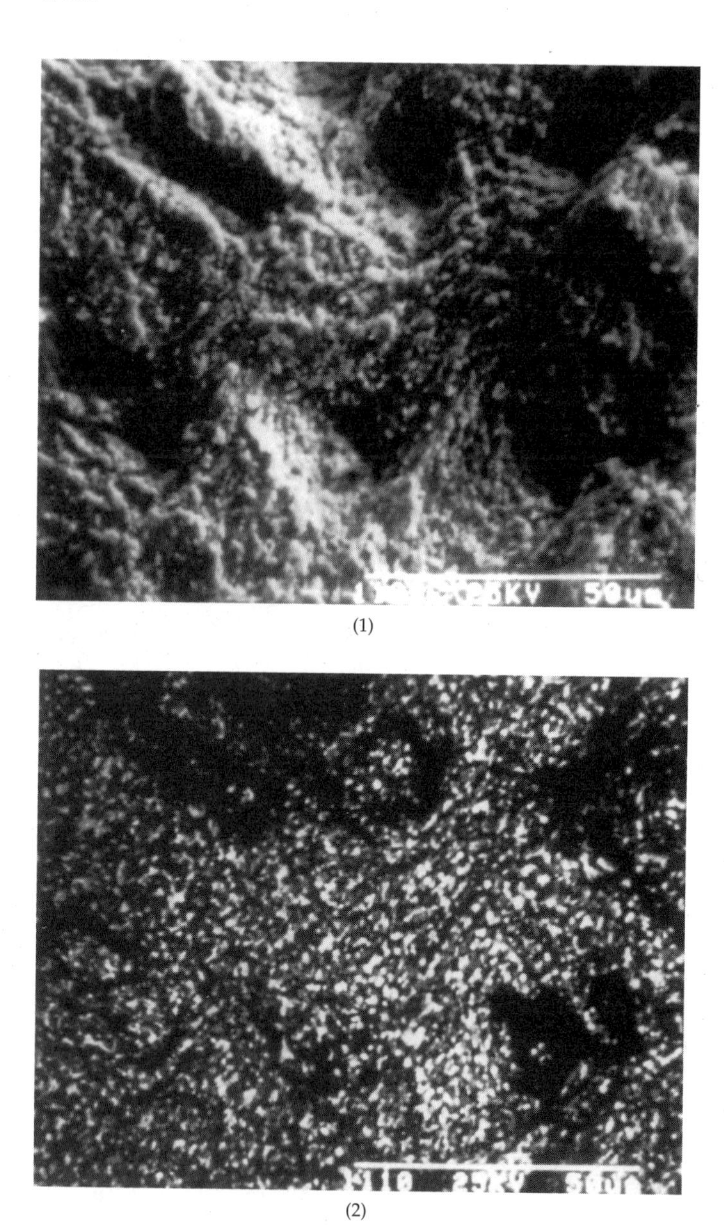

(1)

(2)

Figure 3-43. (*a*) Fractograph of 63 Sb/37 Pb at early stage of low cycle fatigue. (1) Secondary electron. (2) Back scattering.

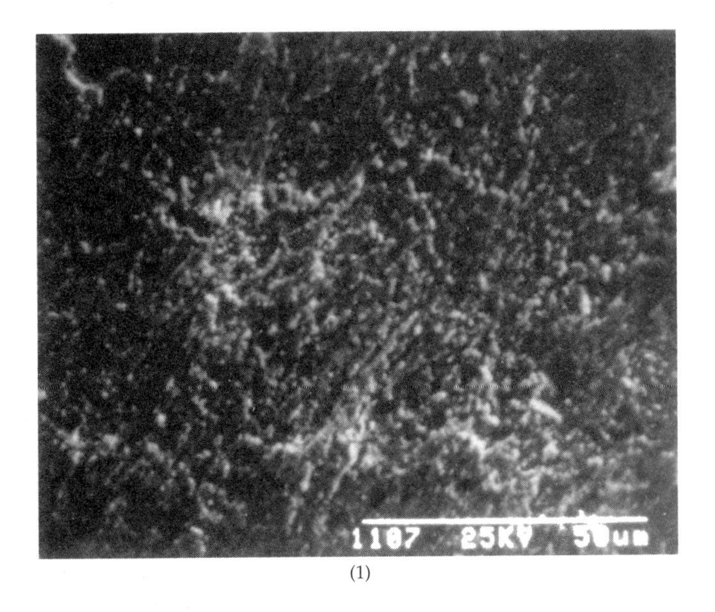

(1)

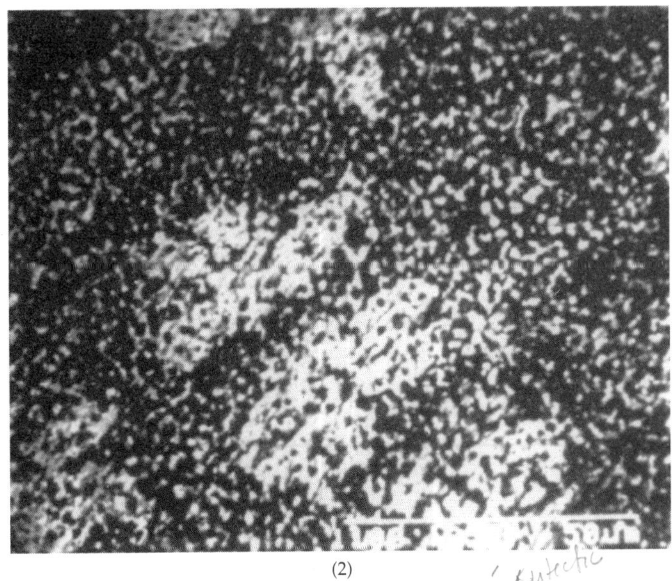

(2)

Figure 3-43. (*b*) Fractograph of 63 Sn/37 Pb at later stage of low cycle fatigue. (1) Secondary electron. (2) Back scattering.

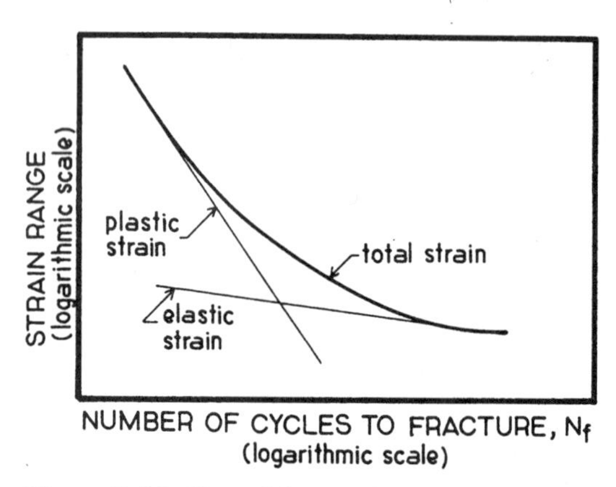

Figure 3-44. Typical fatigue (*S-N*) curve.

mally activated are likely to be responsible for the very low temperature fatigue.

The fatigue characteristics are generally represented by the decrease in strength versus the number of cycles (*S-N* curve) as shown in Fig. 3-44. The level of maximum stress before fracture is defined as the endurance limit which is normally related to 40 to 50 percent of ultimate tensile strength, and in many cases is very close to the yield stress. On the other hand, many alloys do not exhibit an endurance limit.

It is generally accepted that the decrease in strength under cyclic loading occurs according to the following events:

- Repeated cyclic stress results in slips and strain hardening locally.
- The localized strain-hardened area causes the reduction in local ductility and consequently the microscopic cracks.
- The submicroscopic cracks concentrate stresses until fracture occurs.

Solders in electronic packaging and assembly applications normally undergo low-cycle fatigue, which is defined as a fatigue life of less than 10,000 cycles, and are subjected to high stresses.

In addition to low-cycle fatigue which is conducted at a constant temperature, the thermomechanical fatigue is another test mode to characterize the fatigue behavior of solders. The test subjects the material to cyclic temperature extremes.

3.4 Metallurgy of Solders

3.4.1 Alloy behavior versus temperature

For electronic packaging and assembly applications, solder interconnections inevitably encounter temperature rising. This temperature change is a result of in-circuit function or the external ambient environment. From a material point of view, several events may occur as rising temperature is imposed:

- Increase in the mobility of atoms or vacancies
- Increase in the mobility of dislocations
- Increase in the concentration of vacancies
- Slip system change
- Creation of new slip system
- Occurrence of deformation at grain boundary
- Occurrence or acceleration of metallurgical reactions, such as recovery, recrystallization, grain coarsening, second-phase particle coarsening
- Enhancement of environment effect

3.4.2 Alloy property trend versus temperature

Figure 3-45 illustrates the trend in the change of major properties of alloys in relation to temperature change. As temperature rises, the internal strain can be gradually relieved, and the material softens as

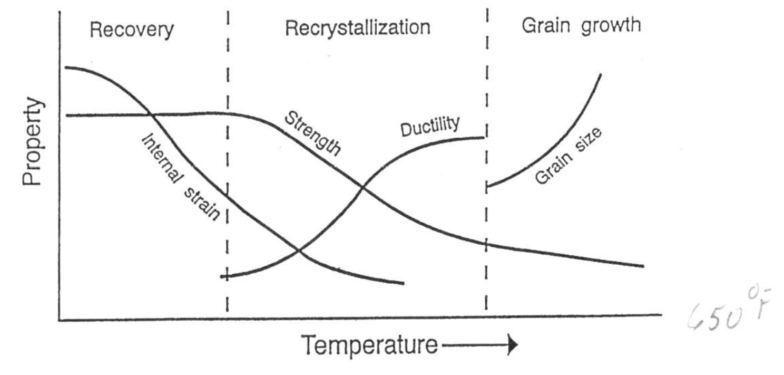

Figure 3-45. Qualitative change in alloy properties in relation to temperature.

reflected by the reduction in strength and increase in ductility. When a threshold temperature is reached, the grain starts to grow in size. Correlating with metallurgical processes, the internal strain relief is associated with the recovery process, and the decrease in strength and increase in ductility are related to recrystallization. After the recrystallization is completed and when the temperature is high enough, the strain-free grains (recrystallized) start to grow. The driving force for grain growth is the reduction in free energy by decreasing grain boundary area in forming larger grains.

3.4.3 Plastic deformation and plasticity

Plastic deformation refers to the irreversible deformation in the material, and plasticity is the ability of the material to be permanently deformed without fracture. From the failure point of view, materials can be classified as ductile, brittle, and porous. Usually, the failure of ductile material is preceded by slip, brittle material undergoes tensile failure, and porous material is often associated with compression failure. Most metals are considered ductile material. As a result of applied forces, the crystals which yield plastically will shear on a number of parallel planes. When this type of deformation occurs, the crystal is said to have undergone slip. The mechanism of slip normally involves a dislocation movement.

Metals undergo four types of plastic deformation:

- Slip
- Dislocation climb
- Shear on grain boundary
- Plastic flow inside the grains

During the process of deformation, the dislocations are created, and the thermal energy aids the movement of dislocations. Dislocations will move by overcoming the Peierls stress (the force holding a dislocation in its low-energy position in a lattice), or through dislocation jogs and dislocation climb. *Dislocation atmosphere,* which is defined as a situation where the excess solute atoms surround the dislocation as a result of the interaction between solute atoms and dislocations and which has the driving force to lower the free energy of crystal lattice, can also be thermally created at the cross slip.

Dorn[25] has derived the relationship between the creep rate and the stress as follows:

$$\dot{\gamma}_p = A \frac{D\mu b}{kT} \left(\frac{\tau}{\mu}\right)^n \left(\frac{b}{d}\right)^P \tag{3.5}$$

where $\dot{\gamma}_p$ = plastic strain rate (creep rate)
τ = shear stress
D = diffusion coefficient
μ = shear molecules
b = Burger's vector
d = grain size or interphase size
n = stress exponent
p = size exponent

The diffusion coefficient is given by

$$D = D_0 \exp\left(\frac{-\Delta H}{Kt}\right)$$

thus

$$\dot{\gamma}_p = A \frac{\mu b}{kT} \left(\frac{\tau}{\mu}\right)^n \left(\frac{b}{d}\right)^p D_0 \exp\left(\frac{-\Delta H}{kT}\right) \tag{3.6}$$

In the general Eq. (3.6), n, p, and ΔH are characteristic parameters of a particular controlling mechanism. Their relationship to the five main deformation mechanisms are established:

1. For high-stress regions corresponding to high strain rate and low temperature, the power law creep occurs, the dislocation climb is the rate-controlling factor, and the activation energy is lattice self-diffusion energy ΔH_L. The dislocation glide is essentially responsible for the strain developed. The strain rate depends on the stress applied. The stress exponent n falls in the range of 3 to 10, varying with crystal structure and solute/solvent distribution. The strain rate is independent of grain size. Equation (3.5) becomes

$$\dot{\gamma}_p = A \frac{\mu b}{kT} \left(\frac{\tau}{\mu}\right)^{3\text{–}10} D_o \exp \frac{-\Delta H_L}{kT} \tag{3.7}$$

2. For high-stress regions and elevated temperature, the power law breakdown prevails. In this case, the stress exponent exceeds 10. It is believed that dislocation cross slip participates in the process of dislocation climbs-controlled plastic flow, thus,

$$\dot{\gamma}_p = A \frac{\mu b}{kT} \left(\frac{\tau}{\mu}\right)^{>10} D_o \exp \frac{-\Delta H}{kT} \tag{3.8}$$

3. For low-stress regions corresponding to relatively low strain rate and high temperature, Eq. (3.5) is represented by

$$\dot{\gamma}_p = A\frac{\mu b}{kT}\left(\frac{\tau}{\mu}\right)^{1.6\text{–}2.4}\left(\frac{b}{d}\right)^z D_o \exp\frac{-\Delta H_{gb}}{kT} \tag{3.9}$$

where ΔH_{gb} is the grain boundary diffusion energy. The plastic strain rate is characterized by the stress exponent $n = 1.2$ to 2.4 and grain size dependency $p = 2$. The grain boundary sliding is the rate-controlling factor.

4. The plastic deformation is carried out by vacancy diffusion inside the grains of polycrystalline metal. Self-diffusion inside grains is conducted by atoms being carried away from the grain boundary where there is a net compressive stress over the grain boundary to the grain boundary where there is a net tensile stress. The plastic deformation in this case is a combined grain boundary sliding and grain matrix deformation. This is also called Nabarro-Herring creep. Its plastic strain rate follows the equation

$$\dot{\gamma}_p = A\frac{\mu b}{kT}\frac{\tau}{\mu}\left(\frac{b}{d}\right)^{2\text{–}3} D_o \exp\frac{-\Delta H_{NH}}{kT} \tag{3.10}$$

where ΔH is the diffusion energy of vacancies or atoms inside the grain. The plastic deformation rate is featured with stress exponent $n = 1$ and grain size exponent $p = 2$ to 3.

5. At very low stress regions, the linear viscous creep prevails. There is no grain size dependency and the lattice diffusion is the controlling factor. The plastic deformation follows Harper-Dorn equation.

$$\dot{\gamma}_p = A\frac{\mu b}{kT}\frac{\tau}{\mu} D_o \exp\frac{-\Delta H_L}{kT} \tag{3.11}$$

The lattice diffusion energy is commonly higher, about 2 times, than that of the grain boundary diffusion. It is also common that in the high-temperature range where recovery occurs, grain boundary shear is a predominant process which leads to intercrystalline fracture. The strain rate used in the testing can also be a factor which affects the fracture mode. The high strain rate, which is represented by the short test, introduces transgranular fractures and the slow strain rate (corresponding to low stress), represented by the long test, introduces intercrystalline fractures.

Figure 3-46 exhibits the relationship of plastic shear rate versus shear stress for 63 Sn/37 Pb solder.

3.4.4 Superplasticity

Superplasticity refers to alloys that have extreme extendability with elongation reaching 100 to 1000 percent of original length and are characterized by a grain size or interphase spacing of the order of 1

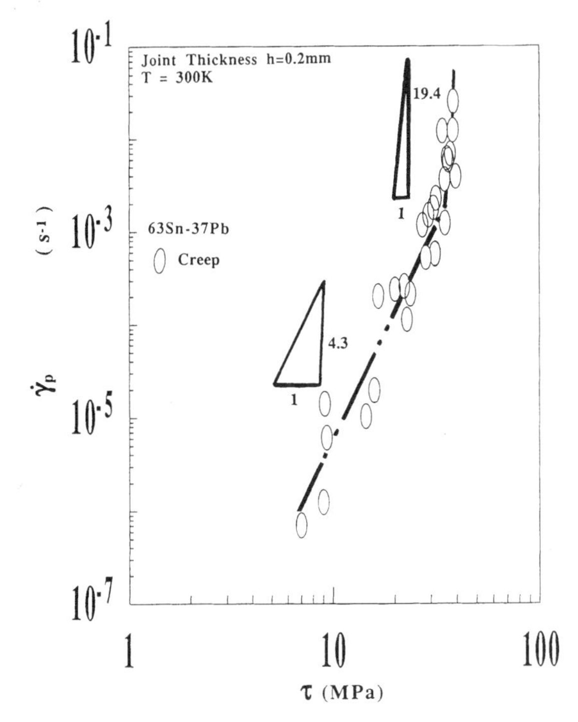

Figure 3-46. Plastic shear rate versus stress for 63 Sn/37 Pb solder.

μm and by high strain rate sensitivity. The limiting strain rate generally falls around 0.01 s^{1}[1], above which the superplastic behavior diminishes. The superplasticity is generally associated with fine-grain polycrystals where grain boundaries act as sources and sinks of vacancies. Superplastic alloy deformation takes place under low stress, which is inadequate for other plastic deformation processes, and high temperatures exceeding $0.4T_m$.[26]

At very low stresses and high temperature, the deformation is primarily carried out by vacancy creep which is a slow process, and creep rate varies with stress in a linear fashion.

$$\dot{\gamma}_p \propto \frac{\tau}{d^{2-3}}$$

3.4.5 Strain hardening

As a result of plastic deformation, the metal can be hardened. Along with the strain hardening are the increased tensile strength, increased yield strength, and reduced ductility.

As shown in the stress-strain curve of Fig. 3-13, the curve rises beyond the yield stress due to the strain-hardening phenomenon.

$$\tau = Kr^n$$

where n is the strain-hardening exponent. For a perfect elastic material, $n = 1$, and n approaches 0 for a totally plastic material. During plastic deformation the shear stress required to produce slip continuously increases with increasing shear strain. The increase in the stress required to continue further deformation due to previous plastic deformation is known as strain hardening or work hardening. As dislocation density increases with increasing strain, the dislocations intersecting with each other and interacting with other barriers impede dislocation motion through the crystal lattice. In other words, the hardening effect is caused by the mutual obstruction of dislocation gliding on intersecting systems, resulting in entanglement and immobilization of dislocations. The dislocations affect one another through their stress field. During work hardening, the shear strength is built up with the density of dislocations in accordance with

$$\tau = A\mu b\rho^{1/2}$$

where A = geometrical structure of dislocations
b = Burger's vector
ρ = dislocation density (total length of dislocation lines in unit volume)

3.4.6 Recovery and recrystallization

Both recovery and recrystallization are softening processes in metals. Being driven by thermodynamics, the material tends to release the strain energy stored in a deformed (cold-worked) material. This energy-releasing process which starts at a rapid rate and proceeds at a slower rate is called *recovery*. During the recovery stage, the physical properties that are sensitive to point defects tend to restore to their original values; however, there is no detectable change in the microstructure. The mechanism of recovery normally involves the annihilation of dislocations, the rearrangement of a given number of dislocations, or some form of atom and vacancy movement. Therefore, the recovery process is highly temperature dependent. When the material is under load, the recovery process can occur at relatively low temperatures which is referred to as *dynamic recovery*. The recovery process is opposite to strain hardening, and often is competitive to the strain hardening. Consequently, the flow stress required for producing a given deformation decreases as a result of recovery. The phenomenon is shown in Fig. 3-13.

Recrystallization usually occurs at a relatively high temperature and involves a larger amount of energy released from the strained material than the recovery process. During recrystallization, in addition to the energy release, a new set of essentially strain-free crystal structures is formed, which obviously involves both nucleation and growth processes.

The temperature required for the occurrence of recrystallization usually falls in the range of one-third to one-half of the absolute melting point of the material. The extremely pure metals have very rapid rates of recrystallization. The recrystallization temperature for solders is expected to be very low. At 100°C, recrystallization in 63 Sn/37 Pb has been often observed.

Figure 3-47 is the scanning electron microscope (SEM) micrographs of recrystallized 63 Sn/37 Pb solder. The factors affecting recrystallization are

- The amount of prior deformation
- Temperature and time
- Initial grain size

Figure 3-47. SEM micrograph of recrystallized 63 Sn/37 Pb solder.

- Composition
- Amount of recovery prior to the start of recrystallization

The larger amount of deformation offers more driving force to recrystallize; the foreign atoms to the alloy structure are believed to raise the recrystallization temperature by interfering with grain boundary movement, thus retarding recrystallization; at a given amount of deformation, the smaller initial grain size corresponds to a larger grain boundary area which increases the number of nucleation sites facilitating the nucleation rate, consequently resulting in finer recrystallized grains.

3.4.7 Solution hardening

In a binary alloy, the element in larger concentration is called the *solvent* and the element in smaller concentration is the *solute.* There are two types of solid solutions. When the added solute atoms have a similar size to that of the solvent atoms, the solute atoms have a great opportunity to directly substitute for the solvent atoms in lattice structure, forming a substitutional solution. However, when the size of the added solute atoms is much smaller than that of the solvent atoms, the interstitial solvent is formed in which the solute atoms enter the interstitial sites in the crystal structure instead of replacing solvent atoms. Figure 3-48 is the schematic of substitutional and interstitial solutions. In general, atoms that do not have a strong chemical affinity for each other tend to form solid solutions. The extent of solid solution to be formed depends on the crystal structure and atomic size. In order to have a complete solid solution over the entire range of compositions, the two types of atoms must have the same crystal structure. Another general rule of thumb is that the formation of a solid solution is

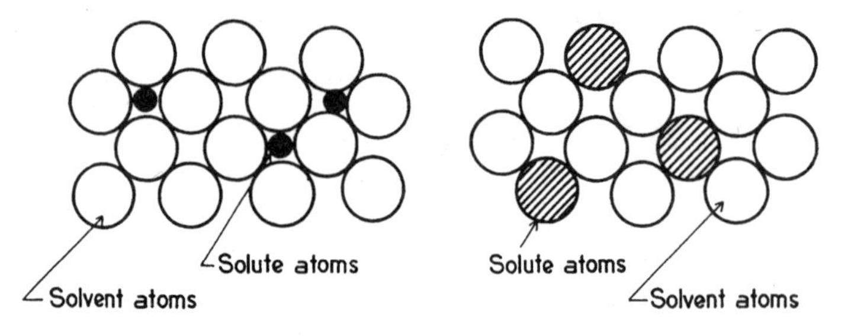

Figure 3-48. Schematic of substitutional and interstitial solutions.

Table 3-10. Atomic Size and Crystal Structure of Elements Commonly Used in Electronics

Element	Atomic radius, Å	Crystal structure
Pb	1.750	FCC
In	1.625	FC tetra
Bi	1.556	Rhombic
Sn	1.509	BC tetra
Sb	1.452	Rhombic
Ag	1.444	FCC
Au	1.441	FCC
Al	1.431	FCC
Pt	1.387	FCC
Pd	1.376	FCC
W	1.369	BCC
Zn	1.332	HCP
Cu	1.278	FCC
Ni	1.245	FCC
Si	1.176	Diamond cubic

favored when the size difference between the two types of atoms is less than 15 percent. When the size difference is larger than 15 percent, the extent of solid solution is usually limited to less than 1 percent. Table 3-10 lists the atomic radius and crystal structure of the elements commonly used in electronic packaging and assembly.

Here are some examples. In and Pb have an atomic size difference of less than 8 percent, and their crystal structures are isotopic. A continuous series of solid solutions with a miscibility gap is observed as shown in the In-Pb phase diagram of Fig. 3-49. Both Ag and Au have a similar atomic size and the same crystal structure; Au and Ag exhibit complete miscibility as shown in the phase diagram of Fig. 3-50. For the systems that have different crystal structures and a significant difference in atomic size, such as Sn-Ni, or Sn-Pd, the solid solubility of Ni in Sn or Pd in Sn at temperatures below 400°C, is practically nil;[27] these are evident in their phase diagrams of Fig. 3-51 and Fig. 3-52, respectively.

The effect of solid solution alloying resulting in an increase in the yield stress and the alteration of the stress-strain relationship is illustrated by the strengthening effect of Sb on Sn-Pb solders as shown in Fig. 3-18.[28] The Pb-Sb phase diagram of Fig. 3-53 indicates that the solid solubility of Sb in Pb at room temperature is negligible, and the Sb-Sn phase diagram of Fig. 3-54 exhibits that the solid solubility of Sb in Sn ranges from 7.9 weight percent (wt %) (7.7%) at 225°C to 4 wt %

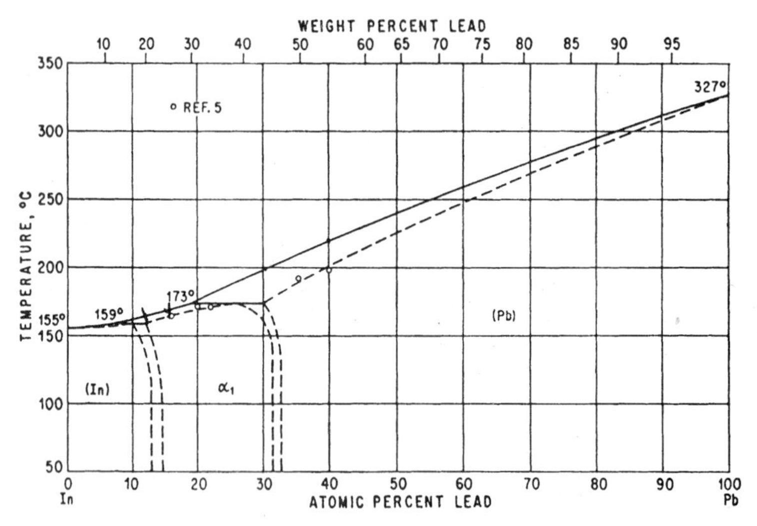

Figure 3-49. Phase diagram of In-Pb.

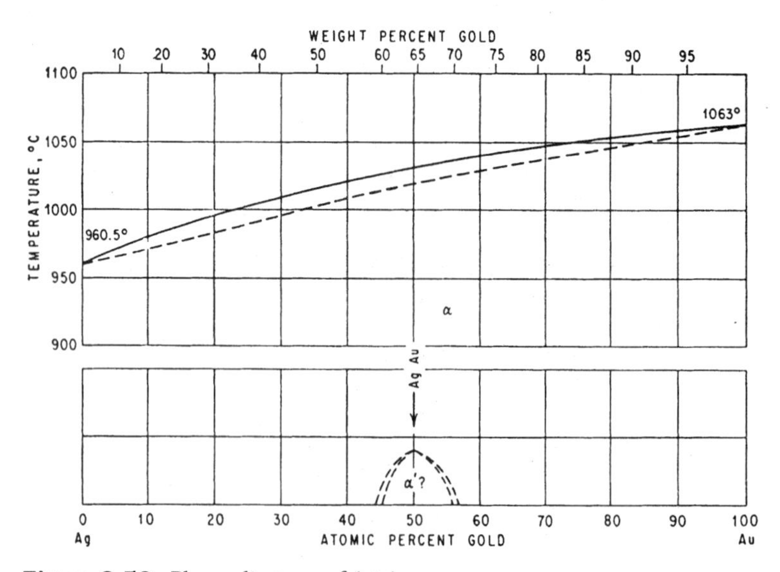

Figure 3-50. Phase diagram of Ag-Au.

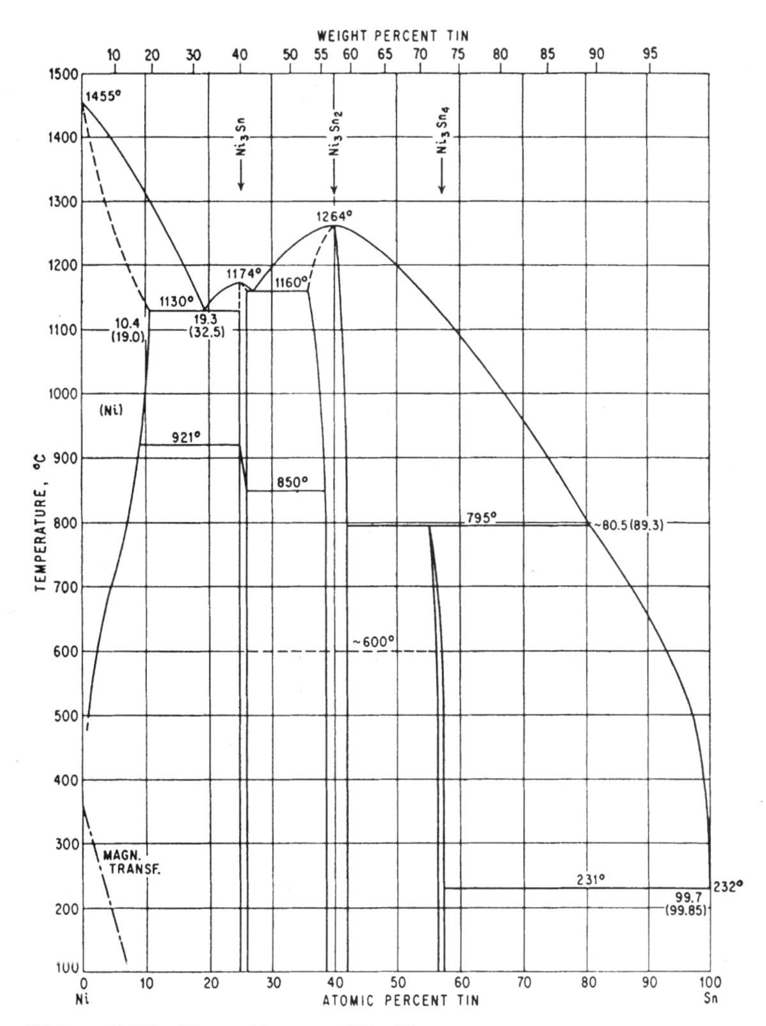

Figure 3-51. Phase diagram of Sn-Ni.

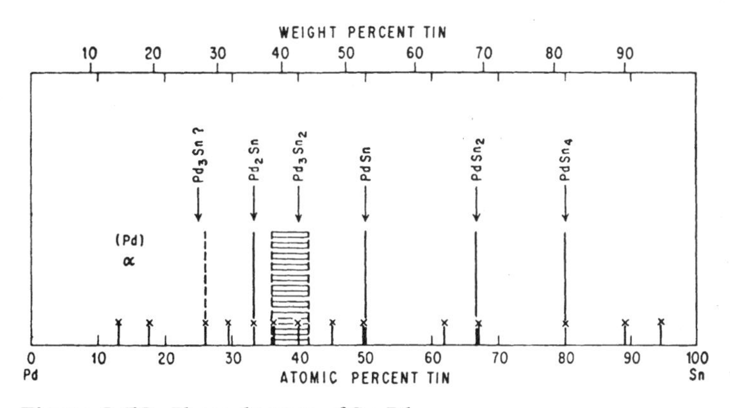

Figure 3-52. Phase diagram of Sn-Pd.

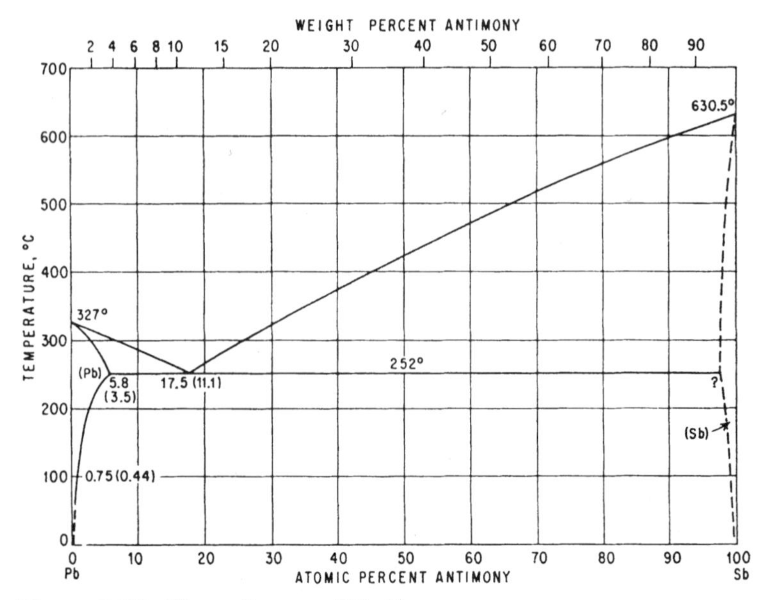

Figure 3-53. Phase diagram of Sb-Pb.

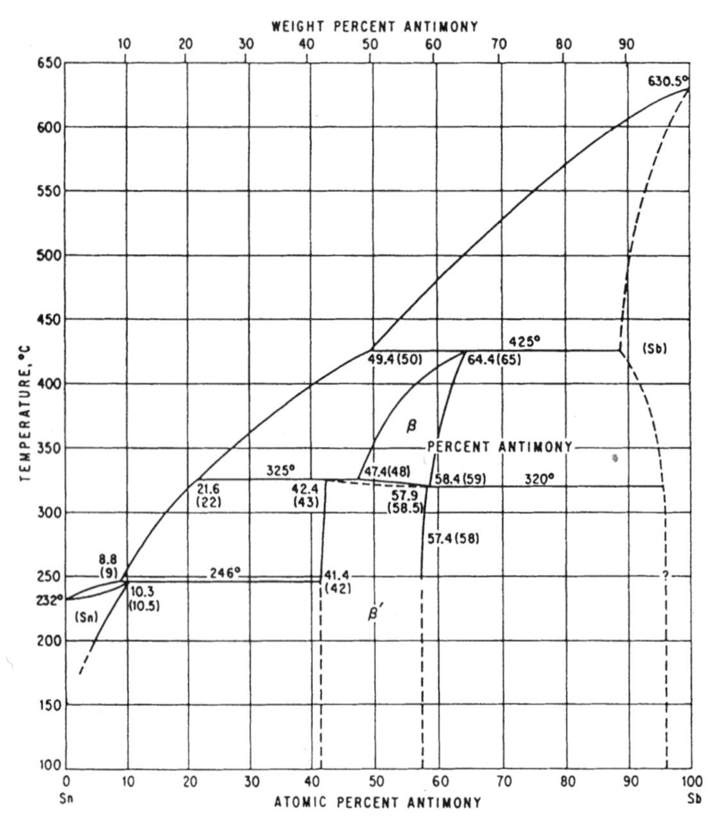

Figure 3-54. Phase diagram of Sb-Sn.

(3.9%) at 190°C to further reduced percentages at lower temperatures. The limit of solid solubility of Sb in Sn contributes to the strength of Sn-Pb solders.

3.4.8 Precipitation and dispersion hardening

Precipitation and dispersion hardening are another two strengthening tactics for metals through the formation of a second phase in the matrix. Precipitation hardening requires the formation of a solid solution in the alloy possessing the various extents of solubility of the second phase in the matrix over a range of temperatures. Dispersion hardening customarily deals with the totally insoluble second phase which is evenly distributed in the matrix.

As the temperature is lowered from the melting temperature, the solubility of solute in solvent decreases. When the temperature reaches below the solvus line (solid solubility curve of a phase diagram), the nucleation of second phase or precipitation occurs. The dwell time at the specific temperature contributes to the growth of the nucleus. The precipitation hardening determined by the nucleation and growth process is a function of temperature and composition (concentration of solute in solvent).

The initial stage of precipitation from a supersaturated solid solution coincides with age hardening where the precipitation rate is low as the melt is quenched to supersaturate the solid solution and then is reheated to an intermediate temperature for a reasonable length of time. The strengthening effect is a result of well-distributed fine precipitates which are capable of impeding the dislocation movement.

Since the supersaturation is a requirement of precipitation, the size and distribution of precipitates are sensitive to the competing nucleation and growth process. When precipitates are not well distributed and grow to a large size, overaging may occur, imparting a softening effect. Overaging often occurs at the grain boundary long before the precipitates in the matrix have had a chance to develop fully. In contrast, the dispersion hardening is not subject to overaging because there is very little solubility of the second phase in the matrix.

The degree of strengthening achieved by using second-phase particles depends on several factors, including the shape of the particles, average diameter of the particles, mean interparticle spacing, and the distribution of the particles. Optimum hardening effect is often achieved when there are a large number of well-dispersed fine particles.

3.5 Solder Alloy Selection Criteria

The following areas should be considered when selecting a solder alloy for a specific application:

- Temperature compatibility
- Mechanical properties
- Alloy-substrate compatibility
- Eutectic versus noneutectic compositions

3.5.1 Temperature compatibility

The phase transition temperatures, particularly the solidus and liquidus temperatures, are crucial to the performance of a solder joint. It is advisable to have the liquidus temperature be at least 2 times higher than the upper limit of the expected service temperature. Table 3-11 lists the worst-case use environment of electronics in various use categories.[29,30]

Table 3-11. Worst-Case Use Environment of Electronics[29,30]

Use category	Worst-case use environment					
	T_{min}, °C	T_{max}, °C	ΔT, °C	t_p, h	Cycles per year	Years of service
Consumer	0	+60	35	12	365	1–3
Computers	+15	+60	20	2	1460	~5
Telecommunication	−40	+85	35	12	365	7–20
Commercial aircraft	−55	+95	20	2	3000	~10
Industrial and automotive	−55	+65	20	12	185	~10
passenger compartment			40	12	100	
			60	12	60	
			80	12	20	
Military ground and ship	−55	+95	40	12	100	~5
			60	12	265	
Space leo*	−40	+85	35	1	8760	5–20
geo†				12	365	
Military a	−55	+95	40	2	500	~5
b			60	2	500	
c			80	2	500	
Avionics			20	1	1000	
Automotive, under hood	−55	+125	60	1	1000	~
			100	1	300	
			140	2	40	

*leo = low earth orbit (10–12 orbits/day).
†geo = geosynchronous orbit (1 orbit/day).

3.5.2 Mechanical and physical properties

Intrinsic shear strength, creep resistance, fatigue resistance, thermal conductivity, electrical conductivity, and coefficient of thermal expansion are important properties to the overall performance of solder joints. It should be noted that both electrical and thermal conductivity of solders decreases with increasing temperature as discussed in Secs. 3.2.2.2 and 3.2.3.2.

3.5.3 Alloy substrate compatibility

Thermodynamically, most tin-based solder alloys are metallurgically reactive to common substrates used for electronic packaging and assembly, with which solder is in contact. Table 3-12 lists some examples of compatibility limitations.

The selection of solder alloys depends on the kinetics of such metallurgical reactions at a given set of process conditions. Because of the metallurgical affinity, the process condition must be controlled to avoid the excessive extent of reaction between the solder and substrate. Further discussion in this area will be covered in Chaps. 5 and 13.

3.5.4 Eutectic versus noneutectic alloy

Upon cooling, a liquid composition of a binary or ternary system freezes at a lower temperature than all other compositions of the same system, and it freezes like a pure metal to simultaneously form a mixture of two solid phases. The point at which the three phases in the binary system coexist is called a *eutectic point* which corresponds to a single temperature (eutectic temperature) and a single composition

Table 3-12. Compatibility Limitation Between Solders and Substrates

Solder	Substrate
Sn-based	Cu
Sn-based	Au
Sn-based	Ag
In-based	Cu
Bi-Sn	Pb
Sb-containing	Brass, Zn

(eutectic composition). Its phase transformations are expressed by the thermally reversible reaction:

$$\text{Liquid solution} \underset{\text{heating}}{\overset{\text{cooling}}{\Longleftrightarrow}} \text{solid solution (1) + solid solution (2)}$$

A eutectic is therefore characterized by three features:

1. The eutectic temperature, which is the lowest temperature at which any liquid phase can exist.
2. The eutectic temperature above or below which the solid solubility decreases.
3. Above the eutectic temperature, any excess over the solid solution limit is liquid, and below the eutectic temperature, any excess over the solid solubility is solid.

As an example, 63 Sn/37 Pb forms a liquid solution at temperatures above 183°C, and results in a two-phase mixture. Each of the two phases is a terminal solid solution corresponding to 19.2 wt % Sn in Pb and 2.5 wt % Pb in Sn at the eutectic temperature. As the temperature goes above or below 183°C, the solid solubility of Sn in Pb or Pb in Sn decreases. In contrast to eutectic, other thermally reversible phase transformations occur in eutectoid, peritectic, and monotectic as repre-

$$\text{Eutectoid solid solution (1)} \underset{\text{heating}}{\overset{\text{cooling}}{\Longleftrightarrow}} \text{solid (2) + solid (3)}$$

$$\text{Peritectic liquid solution + solid solution (1)} \underset{\text{heating}}{\overset{\text{cooling}}{\Longleftrightarrow}} \text{solid solution (2)}$$

$$\text{Monotectic liquid solution (1)} \underset{\text{heating}}{\overset{\text{cooling}}{\Longleftrightarrow}} \text{liquid solution (2) + solid solution}$$

sented by

For a given binary system, the eutectic composition offers the liquid phase with the maximum fluidity.[31] Figure 3-55 demonstrates the relationship between fluidity and Sn-Pb compositions: The high fluidity can be an important attribute for certain applications such as penetrating into narrow gaps to form joints. On the other hand, low fluidity or a more viscous liquid phase, which is offered by a noneutectic compo-

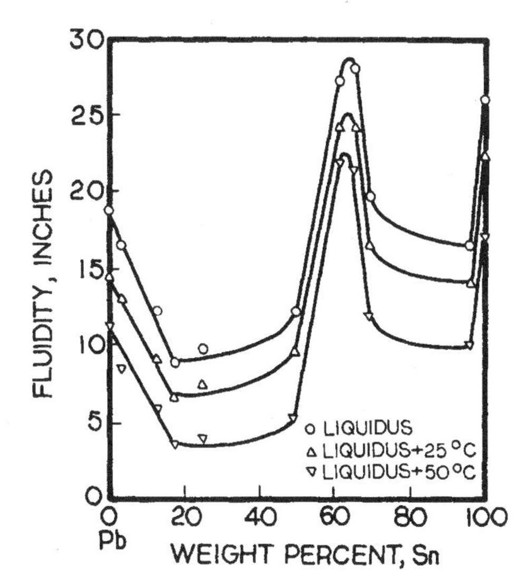

Figure 3-55. Molten solder fluidity versus Sn-Pb composition.

sition with a wide plastic (mushy) range, is desired for the purpose of avoiding running or dripping (working range) when applying the solder. The viscosity of the metal at the melting point η_m, relates to the molecular weight of metal and the molecular volume.[32,33]

$$\eta_m = A\frac{(MT_m)^{1/2}}{V^{2/3}}$$

$$= A\frac{T_m^{\ 1/2}\rho^{2/3}}{M^{1/6}}$$

where M = molecular weight of metal
V = molecular volume = M/ρ)
ρ = density

3.6 References

1. Matt and Jones, *Theory of the Properties of Metals and Alloys,* Oxford University Press, London, 1936, pp. 246–297.
2. Y. S. Toulonkian, et al., *Thermophysical Properties of Materials,* vol. 12, IFI/Plenum Data Company, New York, 1975.
3. C. L. Reynolds, P. R. Couchman, and F. E. Karsz, *Philos. Mag.,* vol. 34, no. 4, 1976, p. 659.
4. T. R. Hogness, *J. Am. Chem. Soc.,* vol. 43, 1921, p. 1621.

5. G. Lang, *J. Inst. Met.*, vol. 101, 1973, p. 300.
6. I. V. Kazakova et al., *Russ. J. Phys. Chem.*, vol. 58, no. 6, 1984, p. 932.
7. V. Somol and M. Beranek, *Sb Vysk. Sk. Chem.-Technol. Praze Anorg. Chem. Technol.*, 1984, B30, 199.
8. V. F. Ukhov et al., *Akad, Nauk SSR Ural. Nauchn. Tsentr. Tr. Inst.*, 1971, vol. 25, p. 30.
9. H. T. Grenaway, *J. Inst. Met.*, 1948, vol. 74, p. 133.
10. D. A. Melford and T. P. Hoar, *J. Inst. Met.*, 1956–57, 85, 197.
11. S. P. Yattsenko, V. I. Kononenko, and A. L. Sukhman, *High Temp.*, vol. 10, no. 1, 1972, p. 55.
12. N. Kong and W. Keck, *J. Less-Common Met.*, 1983, vol. 90, p. 299.
13. K. Mukai, *J. Jpn Inst. Met.*, vol. 37, no. 5, 1973, p. 482.
14. V. Somol and M. Beranek, *Hutn. Listy*, vol. 40, no. 4, 1985, p. 278.
15. L. Gourmiri and J. C. Joud, *Surf. Sci.*, vol. 83, no. 2, 1979, p. 471.
16. W. Krause, F. Sauerwald, and M. Micalke, *Z. Anorg. Chem.*, 1929, vol. 181, p. 353.
17. A. Kasama, T. Iida, and Z. Morita, *J. Jpn Inst. Met.*, vol. 40, no. 10, 1976, p. 1030.
18. S. K. Rhee, *J. Am. Ceram. Soc.*, 1970, vol. 53, p. 639.
19. N. A. Bykova and V. G. Schevchenko, *Tr. Inst. Khim. Ural. Nauchn. Tsentr. Akad, Nauk SSR*, 1974, vol. 29, p. 42.
20. Y. Matuyama, *Sci. Rep. Tokoki Imp. Univ.*, 1927, vol. 15, p. 555.
21. A. P. Passerone, F. Ricci, and R. Sandiorge, *J. Mater. Sci.*, vol. 25, 1990, p. 4255.
22. Jennie S. Hwang, "Solder Joint Reliability - Can Solder Creep?" *1989 International Symposium on Microelectronics Proceedings,* San Jose, Calif., p. 38.
23. C. Herring, *J. Appl. Phys.*, vol. 21, 1950, p. 437.
24. Jennie S. Hwang and Zerfeng Guo, "Strengthened Solder Materials for Electronic Packaging," *Surface Mount International Conference Proceedings,* 1993.
25. J. L. Lytlon, L. A. Shepard, and J. E. Dorn, *Trans. AIME,* vol. 212, p. 220, 1958.
26. W. A. Backofen, et al., *Metals Handbook,* "Ductivity," Chapter 10, American Society for Metals, 1968.
27. H. Nowatig, K. Schubert, and L. I. Dettinger, *Z. Metalkunde,* vol. 37, 1946, pp. 137–145.
28. Tin Research Publication No. 93, Columbus, Ohio.

29. Werner Engelmaier, "Thermal-Fatigue Life Predictions for Surface Mount Solder Joints," *Test and Measurement World,* vol 8, no. 10, October 1988, p. 14.
30. Werner Engelmaier, "The Importance of Continuous Monitoring for SMT Solder Joint Failure During Accelerated Fatigue Testing," IPC Spring Proceedings, April 1988.
31. C. V. Rogone, C. M. Adams, and H. F. Taylor, *Trans. Am. Foundryman's Soc.,* vol. 64, 1956, p. 650.
32. E. N. C. Andrade, "The Viscosity of Liquids," *Proc. Roy. Soc.,* vol. A-215, 1952, p. 356.
33. E. N. C. Andrade, "The Viscosity of Liquids," *Proc. Roy. Soc.,* vol. A-215, 1952, p. 36.

3.7 Suggested Readings

Barrett, C. S., *Structure of Metals,* McGraw-Hill, New York, 1952.
Dieter, G. E., *Mechanical Metallurgy,* McGraw-Hill, New York, 1961.
Hull, Derek, *Introduction to Dislocations,* Pergamon Press, Ltd, London, 1965.
Marin, J., *Mechanical Behavior of Engineering Materials,* Prentice-Hall, New Jersey, 1962.
Rady, G. S., *Materials Handbook,* McGraw-Hill, New York, 1951.

4
Soldering Chemistry

The chemistry involved in soldering provides two primary purposes; one is fluxing and the other is serving as a carrier (or vehicle) to the rest of the ingredients in a flux/vehicle composition. The function of cleaning is called *fluxing,* and the material used is the *flux.*

4.1 Purpose of Fluxing

The function of fluxes for soldering is to chemically clean the surface to be joined and to maintain the cleanliness of the surfaces so that a metallic continuity at the interface can be achieved to make sound solder joints. For solder pastes, fluxes, in addition to performing the aforementioned function, are to clean the surface of solder powder particles as well as obtaining a complete coalescence of the solder powder particles during reflow. Another important performance property is that the flux/vehicle system can assist in the transfer of heat to the solder joint during soldering. Therefore, broadly speaking, any chemicals capable of providing good reactivity with metal oxides and other nonmetallic compounds and capable of cleaning the metals which are commonly used in soldering could be potentially flux candidates.

From a chemical point of view, the inorganic chemicals, such as strong acids, strong bases, and certain salts are highly reactive and thus are not suitable for electronic applications. Organics containing active functional groups, such as carboxylic acid (-COOH) and amines (-NH_2, -NHR, -NR_2), can be good fluxing agents. Commonly used organics include aliphatic acids such as succinic acid, abietic acid, and adipic acid; aromatic acids such as substituted benzoic acids; aliphatic amines and their derivatives, such as triethanol amine, hydrochloride salts of amines, and hydrobromide salts of amines.

4.2 Strength of Fluxes[1,2]

The fluxing strength depends on the chemical activity of the flux agent and the properties of the flux/vehicle system as well as the external conditions. The factors include

- Functional group and molecular structure of flux agent
- Melting point and boiling point of flux chemicals
- Thermal stability in relation to soldering conditions
- Chemical reactivity in relation to soldering conditions
- Surrounding medium of flux agent including the rest of chemicals in a flux/vehicle formula
- Substrates to be fluxed
- Environmental stability (temperature, humidity)
- Soldering conditions (temperature versus time, atmosphere)

The effects of molecular structure and medium on the strength of acids and bases are classified as inductive, resonance, hydrogen bonding, solvation, hybridization, and steric effects. For the commonly adopted inductive effect, the electron-withdrawing groups adjacent to the carboxylic group of molecules enhance the acidity strength of the carboxylic group as a result of anion stabilization. Conversely, electron-releasing groups decrease the acidity. These effects are illustrated as follows:

$$G \leftarrow C(=O)OH \longrightarrow G \leftarrow C(O)(O)^{\ominus} + H^+$$

Electron-withdrawing group: increased strength.

$$G \rightarrow C(=O)OH \longrightarrow G \rightarrow C(O)(O)^{\ominus} + H^+$$

Electron-releasing group: decreased strength.

Table 4-1. Acidity Constants of Organic Carboxylic Acids

Chemical	K_a
CH_3COOH	1.75×10^5
$ClCH_2COOH$	136×10^{-5}
$Cl_2CHCOOH$	5530×10^{-5}
Cl_3CCOOH	$23{,}200 \times 10^{-5}$
$CH_3CH_2CH_2COOH$	1.52×10^{-5}
FCH_2COOH	260×10^{-5}
$BrCH_2COOH$	125×10^{-5}
⌬—COOH	6.3×10^{-5}
O_2N—⌬—COOH	36×10^{-5}
H_3C—⌬—COOH	4.1×10^{-5}

Table 4-1 lists the acidity constants of some carboxylic acids, exemplifying this structural effect. As can be seen, by the substitution of a hydrogen by electron-withdrawing groups such as -F, -Cl, -Br, the acidity is intensified. For aromatic acids, the substitution of hydrogen on the aromatic ring by an electron-withdrawing group such as NO_2 increases the acidity, and that by an electron-releasing group such as CH_3 reduces the acidity.

Similarly, basicity is influenced by the electron inductive effect which renders the free pair of electrons more or less available for sharing with an acid. Therefore, the electron-releasing group increases the basicity of amine by providing more availability of electrons and by stabilizing the ion, and the electron-withdrawing group decreases its basicity. The mechanism is illustrated as follows:

$$G \longrightarrow \underset{H}{\overset{H}{N}}: \xrightarrow{+H^+} G \longrightarrow \underset{H}{\overset{H}{N}}:H^+$$

Electron-releasing group: increased strength.

$$G \longleftarrow \underset{H}{\overset{H}{N}}: \xrightarrow{+H^+} G \longleftarrow \underset{H}{\overset{H}{N}}:H^+$$

Electron-withdrawing group: decreased strength.

Table 4-2. Basicity Constants of Amines

Chemical	K_b
Methylamine	4.5×10^{-4}
Dimethylamine	5.4×10^{-4}
Trimethylamine	0.6×10^{-4}
N-Propylamine	4.1×10^{-4}
Di-*N*-Propylamine	10×10^{-4}
Tri-*N*-Propylamine	4.5×10^{-4}
Aniline	4.2×10^{-10}
Methyl aniline	7.1×10^{-10}
Dimethylanilane	11.7×10^{-10}
o-Chloroaniline	0.05×10^{-10}
o-Bromoaniline	0.03×10^{-10}
o-Nitroaniline	0.00006×10^{-10}

The basicity of some amines as listed in Table 4-2 demonstrates the effect of electron induction. In addition to electronic effect, solvation and steric effects also play a role, as shown in the comparison of primary, secondary, and tertiary aliphatic amines. The basicity of the tertiary amine in most cases is hindered by the solvation and steric effect. Temperature is another factor affecting the strength of acids and bases; it can alter the expected strength, as predicted from the structural effects. It is evident that organic chemistry can be well utilized in this area. Because in-depth discussion is beyond the scope of this book, readers with a further interest in this area are referred to the Recommended Readings at the end of this chapter.

In addition to the strength of the flux agent, fluxing kinetics is another important aspect of fluxing performance. Regarding kinetics, temperature and time are two universal variables. Without getting into formulation specifics, the flux activation temperature and activation time during heating and soldering have to be designed to fit the process settings.

The activity of a specific flux agent is often affected by the surrounding medium, including the solvent system and other chemical ingredients being incorporated into the chemistry system. The effect is reflected in the fact that different fluxing capacities are often observed in the different formulas which contain the identical flux agent(s). Because of the effect of surrounding medium on the flux, a flux in a simple flux/solvent system that is excellent for a liquid flux used for wave soldering may not perform as well in a paste flux/vehicle system that makes up a solder paste used for surface mount. The performance of a flux agent highly depends on the operating temperature. For example, mild organic acid such as rosin is not chemically active at ambient room temperature but is chemically active enough at soldering temperatures. Each flux chemical has its characteristic activation temperature above which the chemicals become active in cleaning metal substrates through the chemical reaction expressed as

$$2R\text{-COOH} + MO \rightarrow (R\text{-COO}^-)_2\, M^+ + H_2O$$

where R-COOH represents a mild organic acid and MO a metal oxide. The flux obviously has to be chemically compatible with the substrates to be joined and be capable of removing oxides, debris, and other nonmetallic compounds that could hinder the wetting and the formation of metallic continuity. For example, the flux strength and design for copper or tin substrate may differ from that required for substrates such as nickel or silver-palladium compositions. The strength of a flux agent's fluxing activity which normally correlates well with its conductivity can be recognized from the result of well-established tests. Tests include specific resistivity value of a water extract, a corrosion test on copper mirror, and a halogen test. Generally, halide-containing formulas offer the strongest flux activity. This group of chemicals is also most corrosive to metals if they are not removed from the circuitry after use. The corrosive nature of halides to metals is well-established. Figure 4-1 illustrates the drastically increasing corrosion rate on copper with increasing halide content.

4.3 Types of Flux Chemistry

Rosin chemistry has been regarded as an effective and reliable fluxing chemistry. Rosin-based fluxes have been used successfully in electronic systems for many years. For readily solderable systems, rosin alone

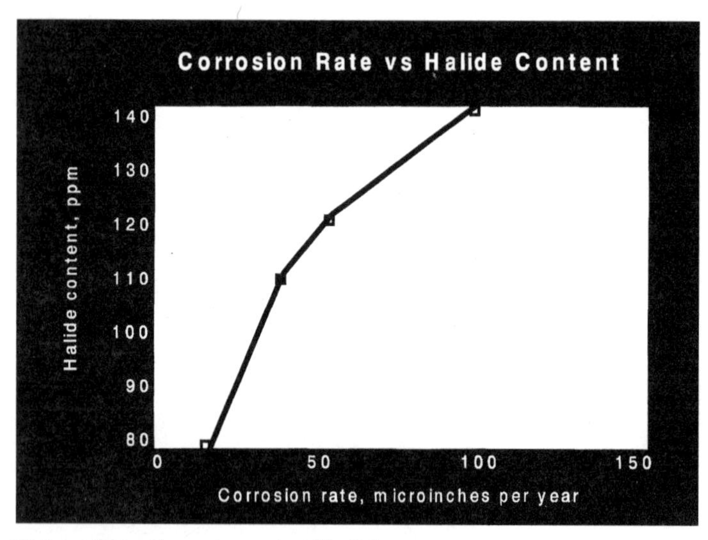

Figure 4-1. Corrosion rate of halides on copper.

provides adequate fluxing action. For less readily solderable systems, other more active chemicals are added (as activators) to enhance the fluxing activity. Traditionally the rosin-based fluxes are classified as type R, type RMA, and type RA. The R (rosin) flux is the weakest, containing only rosin without the presence of activators; the RMA (mildly activated) flux is a system containing both rosin and activators. The RA flux is a fully activated rosin or resin system, having a higher flux strength than the RMA type. The RMA type is most practical because it balances the activity level between that required for most solder surfaces and that of the minimal corrosive nature to circuitry. It should be noted that the RMA type is normally accepted as the system that does not consist of halide-containing chemicals. However, the compositions may vary with vendors. Some commercial grades classified under type RMA or its equivalent do contain halides. Therefore, the activity level of RMA grades may vary drastically between those containing and not containing halides.

For many years, rosin-based fluxes have been used successfully in conjunction with CFC solvents as the postsoldering cleaning agent. With the elimination of CFC solvents, two types of flux chemistries, water-cleaning and no-clean, are being adopted and gaining acceptance. Water-clean fluxes are defined as the fluxes which after soldering need to be removed using H_2O. This newly designed group of fluxes should be considered separately from the conventionally recognized organic acid (OA) fluxes, although most new compositions still contain OAs. This is because of the corrosive and hygroscopic nature of the conventional OA fluxes used in soldering in the past. The newly developed water-clean flux systems are neither corrosive, nor hygroscopic. Most formulas do not contain halides, in contrast to OA fluxes. They possess the stability and other application characteristics equivalent to the RMA type.

The water-clean type can be rosin-containing or rosin-free; the fluxing action and other performance characteristics are the result of a well-balanced mixture of mild organic acid and/or organic amines in a stable polymer/solvent matrix. The applications of water-clean chemistry will be discussed in Chap. 8.

The no-clean type is designed to not require cleaning after soldering. This demands that the flux residue after soldering meet the following criteria:

- The residue left on the assembly is not harmful to the electrical function during the service life of the assembly.
- The residue does not interfere with the subsequent process steps, e.g., bed-of-nails testing, conformal coating.
- The residue is aesthetically acceptable.
- There are no excessive solder balls left on the assemblies (no solder balls with sizes and characteristics which will interfere with electrical function during the service life of the assembly).

The chemistry of no-clean flux can be synthetic polymer/resin based or rosin-containing. The no clean chemistry is still a wide-open field in terms of chemical composition. Its applications will be discussed in Chap. 8.

4.4 Rosin Chemistry

Although newly developed water-clean and no-clean fluxes may or may not contain rosin, the chemistry of rosin deserves reviewing because of its effectiveness and proven reliability in electronics applications.

The beauty of the chemical group of rosins is the combined properties of relatively inert nature at ambient temperature and yet fluxing ability at elevated temperatures which make it an ideal flux for electronics soldering. In addition, its tacky nature in various solvent systems makes it an excellent tackifier in holding electronic components in place during the surface mount process.

Rosin is a natural resin obtained from pine trees. It is classified as gum rosin, wood rosin, and tall oil rosin. Gum rosin is the residue obtained from the oleoresin collected from living trees after separation from turpentine oil; wood rosin is the residue of the distillate extracted from tree stumps; and tall oil rosin is the product separated from fatty acid by fractional distillation of tall oil in pulping processes.

Rosin consists of several rosin acids, rosin acid esters, rosin anhydrides, and fatty acids. The major components of an unmodified rosin are abietic acid, isopimaric acid, neoabietic acid, pimaric acid, dihydroabietic acid and dehydroabietic acid with the following chemical structures, and general formula $C_{19}H_{29}COOH$ or $C_{19}H_{27}COOH$ or $C_{19}H_{31}COOH$.

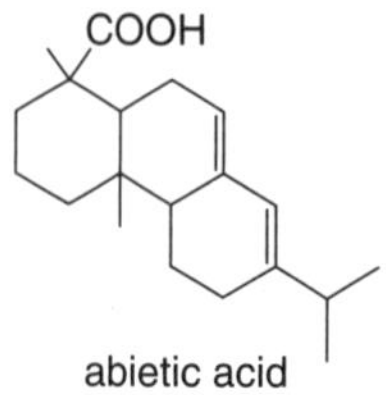

COOH COOH

CH_3

$CH_2{=}CH_2$

abietic acid 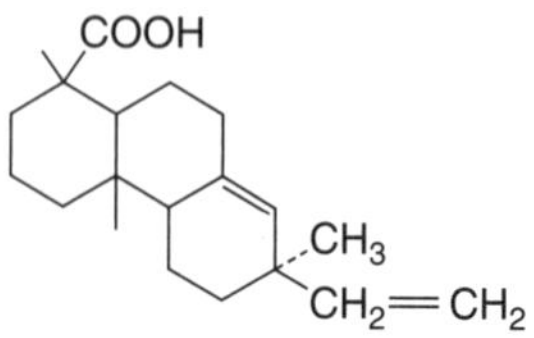 isopimaric acid

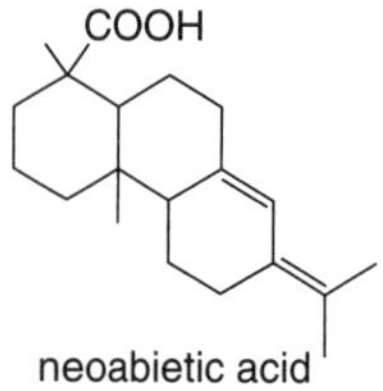

COOH COOH

$CH{=}CH_2$

CH_3

neoabietic acid 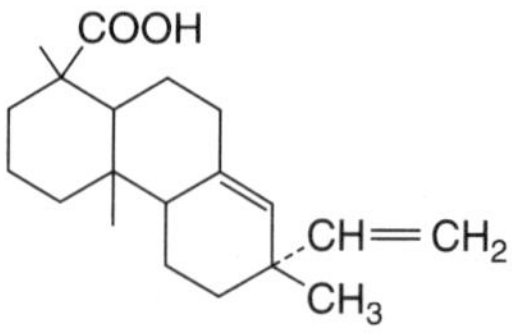 pimaric acid

COOH COOH

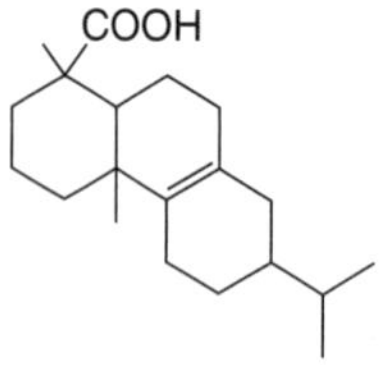

dihydroabietic acid 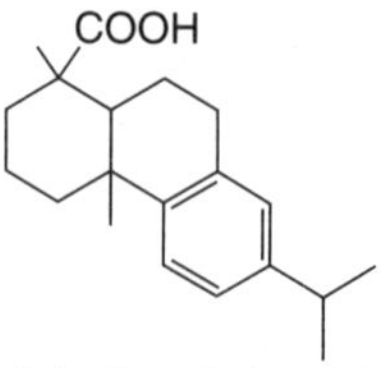 dehydroabietic acid

All these components are characterized by one carboxyl group, a condensed three-ring structure, and a double bond or conjugated double bond, except the fully hydrogenated acid. The reactive sites are the carboxyl group and conjugated double bond or the double bond. Therefore, the rosin can be readily modified through organic disproportionation, hydrogenation, polymerization, saponification, esterification, and the Diels-Alder reaction, as represented by the following reactions of abietic acid, which is most prevalent among the different compounds in natural rosins.[1]

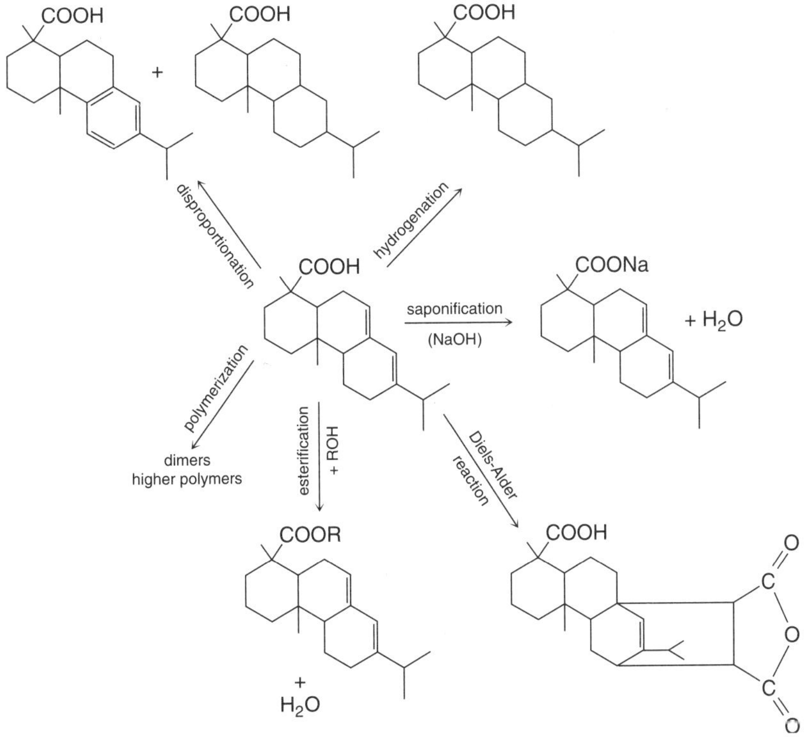

In view of these readily occurring reactions, rosins and rosin derivatives are available in many structures such as disproportionated rosin, rosin esters, hydrogenated rosin esters, polymerized rosins, and polymerized rosin esters, in addition to physically modified rosins. Crystallization and oxidation are two phenomena that often occur in the unmodified rosins. Crystallization can increase the melting point of the rosin, and oxidation normally degrades the properties of rosin.

Tall oil rosin is more susceptible to crystallization. Avoiding excessively heating the rosin at low temperatures (below 140°C) minimizes the phenomenon. The resistance to crystallization and oxidation can be built into the rosin during manufacturing, to assure its quality.

In addition, at elevated temperatures (above 300°C) the rosin may undergo the decarboxylation. Each rosin possesses different responses to temperature. Figure 4-2 illustrates the changes in two different rosins, as represented by weight loss with time at two different tem-

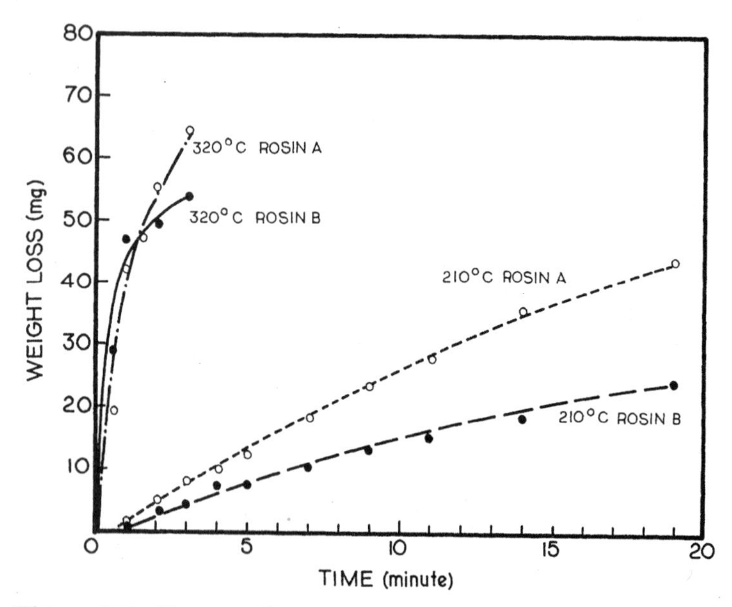

Figure 4-2. Changes of rosins in weight loss with time under conduction heating.

peratures under conduction heating. At a temperature of 210°C, both rosins change with the time of heat exposure in air, and rosin *A* is more vulnerable to the temperature change as reflected in the weight loss than is rosin *B*, showing that rosin *B* has a better heat resistance. At a relatively high temperature, the two rosins degrade at a much faster rate and in a similar manner. Figure 4-3 is the response of rosins *A* and *B* to infrared heating. The heating source also affects the thermal behavior of chemicals. Atmosphere soldering with a high flow rate of nitrogen in the heating chamber may alter the thermal decomposition of chemicals from that under ambient air.

Under normal soldering conditions, W. L. Archer et al. report that isomerization among the rosin acids occurs, and no polymerization is found in the rosin acid as indicated by the lack of significant change in molecular weight distribution before and after heating as shown in Fig. 4-4.[3] The physical and chemical behaviors, however, of the rosin acids are expected to vary with the environment, such as the presence of strong acid, strong base, and the solvent system.

A typical rosin effective in soldering possesses the following characteristics:

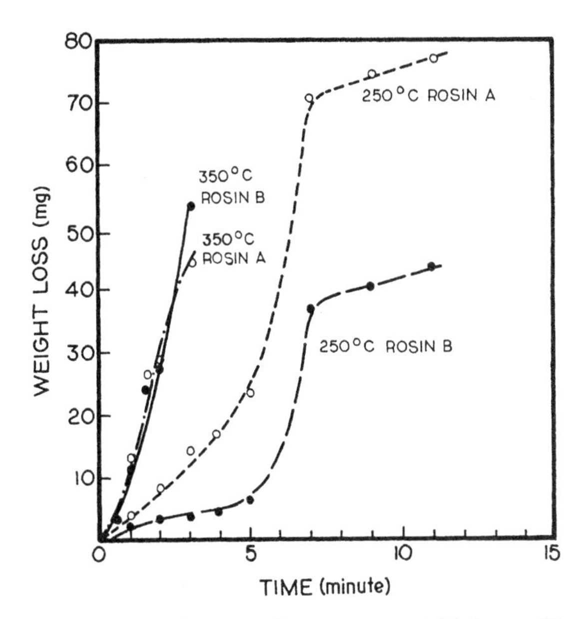

Figure 4-3. Changes of rosins in weight loss with time under infrared heating.

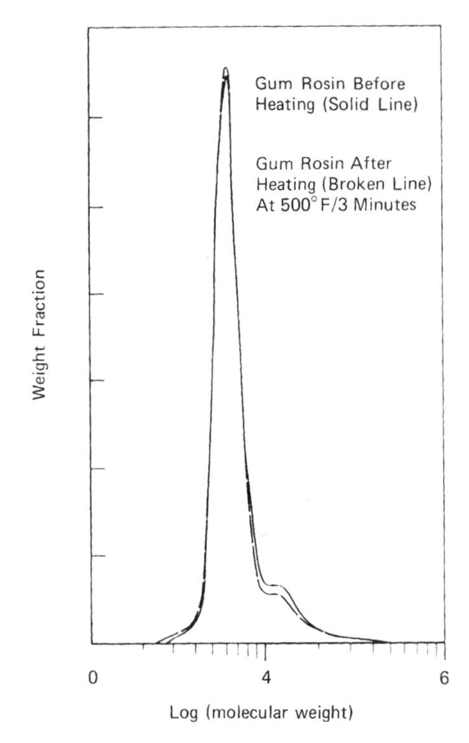

Figure 4-4. Molecular distribution of a gum rosin before and after heating.

Insoluble matter in toluene (maximum percent by weight)	0.05
Softening point, °C minimum	70
Acid number	160
Saponification number, minimum	166

The carboxylic group of rosin acid is the primary source of acidity and fluxing activity. Although a higher acid number is expected to have higher fluxing strength, the fluxing activity in solder paste is usually complicated by other ingredients in the flux/vehicle system.

With respect to solubility, most rosins and their derivatives are soluble or partially soluble in aromatic hydrocarbons, esters, ketones, and some alcohols. The extent of solubility in a specific solvent system affects the physical properties of the resulting solution, as well as chemical reactivity in some cases.

4.5 Alternative Methods of Fluxing

Flux can be incorporated into the solder paste which is composed of fluxes, vehicle carrier, and solder powder, or it can be applied as a liquid flux by a spraying or foaming technique directly onto the component or solder pad surfaces. Fluxing can also be performed in the gas phase by supplying a proper chemical atmosphere (reducing atmosphere) to the soldering assembly (for further discussion see Chap. 9). In addition, the metal surface can be cleaned by plasma with a mixture of fluorine-containing gaseous ions as discussed in Sec. 9.5.

4.5.1 Reduced oxide soldering activation[4]

Reduced oxide soldering activation (ROSA) is a process in which the oxides of Sn, Sn-Pb, and Cu surfaces are removed without using a flux. Rather, the oxides are reduced to metallic surface in an aqueous solution containing highly reducing vanadous ions that can be continuously regenerated via an electrochemical process in a closed-loop system. Thus, no significant waste chemicals are produced, nor is there any flux residue to be removed. The chemical reduction and oxidation reactions involved in this process can be represented as

Reducing agent $$4V^{2+} \rightarrow 4V^{3+} + 4e^- \tag{4.1}$$

Oxides to be reduced $$SnO_2 + 4H^+ + 4e^- \rightarrow Sn + 2H_2O \tag{4.2}$$

where V = generic vanadous ion. The regeneration of the reducing agent involves

Cathode $$4V^{3+} + 4e^- \rightarrow 4V^{2+} \tag{4.3}$$

Anode $$2H_2O \rightarrow 4H^+ + O_2 + 4e^- \tag{4.4}$$

Overall, the operation of the ROSA process is equivalent to the charging and discharging of a battery. The net reaction by combining chemical and electrochemical reactions (4.1) through (4.4) is

$$SnO_2 \rightarrow Sn + O_2$$

Test results indicate that the aged copper surface after ROSA treatment with a treatment time of 5 s delivers a much improved wettability as measured by wetting balance technique.

It is reported that the ROSA treatment usually can be completed within 10 s and is energy efficient. It is also reported that the residues left after the treatment do not adversely affect the solderability and are relatively stable. The development work has demonstrated that the ROSA process in lieu of conventional fluxing may lend its applications to the fluxless soldering under inert blanket atmosphere or to the enhancement of solderability of components and printed circuit boards.

US Army research or other research institutes plan to continue studying the practical issues of the process. Issues include, for example, the allowable exposure time under ambient air after the ROSA treatment in relation to solderability and its effect on various materials of an assembly.

4.6 Classes of Flux

Fluxes can be classified by various manners, these include

- Chemical composition
- Activity level
- Specific measurable performance parameter
- Selected standard

The activity level can be measured by the aggregate data of water extract resistivity, copper mirror test, halide test, and surface insulation test. Because of the complexity and density of modern circuitry coupled with the inherent nature of each test, any single test is not adequate to draw a definitive conclusion. Further discussion on each of these tests will be covered in Chap. 17.

Regardless of the class of flux being adopted or the standard being adhered to, the permitted level of flux activity should be specified in accordance with the requirements of the assembly in terms of manufacturability, service conditions and reliability. In the recently published ANSI standard J-STD-004 "Requirements for Soldering Fluxes," fluxes are identified in 24 types as shown in Table 4-3.[5]

To correlate these new flux designations to the conventional well-established rosin-based fluxes, ANSI/J-STD-004 offers the following guidelines:[4]

Table 4-3. Flux Identification[5]

Flux materials of composition	Symbol	Flux activity levels % Halide	Flux type
Rosin	RO	Low (0%)	L0
		Low (<0.5%)	L1
		Moderate (0%)	M0
		Moderate (0.5–2.0%)	M1
		High (0%)	H0
		High (>2.0%)	H1
Resin	RE	Low (0%)	L0
		Low (<0.5)	L1
		Moderate (0%)	M0
		Moderate (0.5–2.0%)	M1
		High (0%)	H0
		High (>2.0%)	H1
Organic	OR	Low (0%)	L0
		Low (<0.5%)	L1
		Moderate (0%)	M0
		Moderate (0.5–2.0%)	M1
		High (0%)	H0
		High (>2.0%)	H1
Inorganic	In	Low (0%)	L0
		Low (<0.5%)	L1
		Moderate (0%)	M0
		Moderate (0.5–2.0%)	M1
		High (0%)	H0
		High (>2.0%)	H1

New flux designations	Conventional designations
L0 type fluxes	"equivalent to" All R, some RMA, some low solids no-clean
L1 type fluxes	"equivalent to" Most RMA, some RA
M0 type fluxes	"equivalent to" Some RA, some low solids no-clean
M1 type fluxes	"equivalent to" Most RA
33 type fluxes	"equivalent to" Some water-soluble
H0	No equivalent
H1 type fluxes	"equivalent to" Most water-soluble and synthetic activated

Table 4-4 outlines the test requirements versus flux activity classification offered by ANSI/J-STD-004.

4.7 Selection Consideration Between Water-Clean and No-Clean

With the increased need for water-clean and no-clean processes, it is important to understand the substance and the spirit of each process. The following provides a capsule view on comparing these two processes.

4.7.1 Key characteristics of the water-clean process

Water-soluble fluxes have been used in wave soldering and structural (nonelectronics) soldering for years, yet their use for making surface mount and fine-pitch assemblies is relatively new. Three common material/process systems involving the water-clean process are water-saponifiable, water-soluble, and hydrocarbon/emulsion/water. The efficiency of water saponification is sensitive to the amount of the saponifier charged into the cleaner and to the compatibility of the saponifier with the flux or paste being used. The hydrocarbon/emulsion/water process consists of two cleaning steps: solvent followed by water. Both systems allow the continued use of the well-established RMA chemistry. This is certainly considered to be a convenience as well as a comfort. However, the water-soluble system is the simplest because it does not require foreign chemicals to be added to the cleaner.

With significant advances in the development of water-soluble fluxes and pastes, today's water-soluble systems are on a par with conventional RMA grade in all key performance areas including printability,

Table 4-4. Test Requirements versus Flux Activity Classification

		Qualitative halide		Quantitative halide		Conditions for passing
Flux type	Copper mirror	Silver chromate (Cl, Br)	Spot test (F)	(Cl, Br, F)	Corrosion test	500 mΩ SIR requirements
L0 L1	No evidence of mirror breakthrough	Pass Pass	Pass Pass	0.0% <0.5%	No evidence of corrosion	Both cleaned and uncleaned
M0 M1	Breakthrough in less than 50% of test area	Pass Fail	Pass Fail	0.0% 0.5–2.0%	Minor corrosion acceptable	Both cleaned and uncleaned
H0 H1	Breakthrough in more than 50% of test area	Pass Fail	Pass Fail	0.0% >2.0%	Major corrosion acceptable	Cleaned

dispensability, solderability, cleanability, and physical and chemical stability. In terms of cleaning, the water-clean process has no principal differences from the solvent-clean process. Its cleaning efficiency depends on the design and features of the cleaner as well as the execution and control of the cleaning process. The compatibility between soldering temperature profile and the chemical and physical properties of the specific flux or paste remains to be a critical point to achieve optimum soldering results and the highest level of cleanliness.

The closed-loop recycle system, which manages the gross amount of wastewater and waste-disposal, has been extensively developed, although continued improvement in its efficiency and cost reduction is expected. The resulting cleanliness after the water-clean process is able to demonstrate acceptable reliability of electronic assemblies. As expected, the cleaning efficiency as measured by the surface insulation resistance (SIR) test increases with increasing water temperature and/or water pressure, when all other parameters are equal.

4.7.2 Key characteristics of the no-clean process

For no-clean solder flux or paste, the established tests and criteria for RMA grades with regard to chemical and physical properties are still applicable. In conjunction with visual examination, the ionic contaminant test and the SIR test on standard test boards and SIR test and the accelerated life test on the assembly in question will provide a good indication about the reliability of the soldered assembly. But these tests should be conducted for assemblies that have undergone the expected reflow/soldering process. Overall, the no-clean system is essentially application-specific. The best results come from not only the ultimate performance of the individual process, materials, and equipment involved, but also the best compatibility among them.

Although no-clean is poised as an elegant approach with its obvious superiority by deleting one process step, five questions need to be asked in order to determine which process, to-clean or not-to-clean, is most appropriate.

- Whether the residue is considered harmful to the electrical function during the projected service life of the assembly
- Whether the residue interferes with the subsequent process steps, e.g., bed-of-nails testing, conformal coating
- Whether the residue causes aesthetic objection
- Whether the bare boards or components need cleaning

- Whether the process produces solder balls exceeding the acceptable level

If the answer to any of these questions is yes, then the no-clean process may not be the preferred choice.

With respect to the level of residue, it is generally expected that the smaller amount of residue is often associated with higher demands in process control and the decreasing level of oxygen which can be present during soldering.

The use of protective and reactive atmosphere with oxygen content ranging from 10 to 2000 parts per million (ppm) in relation to ambient atmosphere with 23% oxygen generally facilitates solderability in addition to potentially providing other benefits. (Controlled atmosphere soldering will be discussed in Chap. 9.) However, it is an added cost. The benefit in relation to the cost is obviously to be assessed: Understanding each targeted performance and its relation to the benefit of controlled atmosphere would prevent any misinterpretation of test results. Misinterpretation of the results is not unheard of in this area.

4.7.3 Comparison of water-clean and no-clean processes

The direct comparison between the water-clean and no-clean processes will be covered in Sec. 8.8.

4.8 Chemistry in Relation to Other Performance Parameters

The chemistry of liquid flux is relatively straightforward. However, for solder paste, three other physical and chemical properties are important to the viability of a solder paste. These three properties are tackiness, open time, and thermal behavior.

4.8.1 Tackiness

For most surface mount processes, tackiness is one of the requirements for paste performance as it is needed to hold components and devices in position and to accommodate the lag between paste deposition on the board and reflow. The solder paste must possess adequate tackiness in its fresh state; it also needs to maintain the tackability after the paste is exposed to ambient atmosphere for a period of time required for a specific application. The chemistry and chemical makeup is directly related to the tackiness capability. The interim can be expressed in tack force and tack time. The IPC-TM-650 test method 2.4.44 offers a test procedure. Essentially, the tack force is measured by a stainless-steel test probe with a 5.1 ± 0.13 mm diameter bottom surface which is smooth, flat, and aligned parallel to the plane of the subject test specified by bringing the probe in contact with the specimen at a rate of 2.5 ± 0.5 mm/min and applying a force of 300 ± 30 g to the specimen. After 5 s following the application of this force, the peak force required to break the contact by withdrawing the probe at the same rate is the tack force. Tack time can be obtained by recording the time reaching 80 percent of peak force, or the time over which the peak value is maintained or for the tack force to decline to 80 percent of its peak value. In any of these measurements, a good correlation between the measurements under the standard test and the actual performance in tack time and ability requires further work. The acceptable criteria in tack time vary with specific manufacturing operation. In practice, the required tack time/tack force is the time lag permitted or the time lag that the paste can provide under a given manufacturing production flow and ambient conditions so that components and devices are held securely.

4.8.2 Open time

In contrast to tackability, *open time* refers to the paste's ability to maintain its soldering performance without degradation during the surface mount manufacturing process, including the paste exposure time on printer and the lag between printing and reflowing. Any degradation may reflect on the decreasing performance in printability or in solder balling or wetting.

4.8.3 Thermal property

Most ingredients in flux/vehicle systems are organic in nature. Whether the ingredients are in the liquid or solid phase, they are sensitive to thermal effect at the soldering temperature. The phase transition (solid → liquid → vapor), chemical decomposition (covalent bond breaking), and chemical reaction (covalent bond breaking and formation) are common phenomena occurring in chemicals as temperatures rise. The molecular structure, functional group, chemical reactivity, and molecular weight are factors to be considered.

The thermal effect on individual chemicals can be measured by several techniques. The instrumentation includes differential scanning calorimeter (DSC), thermomechanical analysis (TMA), thermogravimetric analysis (TGA), and differential thermal analysis (DTA). As a function of temperature, DSC measures the heat associated with the transition of materials and thus is very useful in determining the boiling point, melting point, softening point, glass transition, phase transition, and heat of reaction. It is a convenient tool to monitor thermal stability of materials. Providing information about thermal stability, DTA measures temperature and semi-quantitative calorimetric properties. It is capable of measuring relatively high temperature properties. Measuring dimensional change as a function of temperature or time, TMA generates information about the mechanical behavior of the material under external load, in response to temperature. Monitoring weight change as a function of temperature or time, TGA provides data on material degradation and decomposition. It is especially useful in providing compositional information. Although very limited information on solder chemistry in this regard is available, thermal analysis instrumentation is useful to solder users and solder researchers as a quality control and material characterization tool.

It should be noted that the thermal effect on flux/vehicle composition becomes pronounced when high-temperature solder alloys (requiring higher than 300°C peak temperature to reflow) are used. At this temperature, most organics undergo some physical and chemical changes. These temperature-induced changes in organics in turn affect the residue cleanability and the properties of the resulting solder joint.

The compatibility of thermal behavior of flux or flux/vehicle chemistry with the soldering temperature profile (temperature versus time, and peak temperature) is important to the effectiveness of fluxing and cleaning. Excessive heat causing early degradation (consumption) of fluxes results in solder balling, wetting problems, or cleaning difficulty. Insufficient heat may leave a higher-than-expected residue level for the no-clean system; it may also cause inadequate flux activation resulting in poor solderability. Further discussion on the effect of heating (temperature profile) will be covered in Chap. 12.

4.9 References

1. Jennie S. Hwang, *Solder Paste in Electronics Packaging—Technology & Applications for Surface Mount, Hybrid Circuits and Component Assembly,* Chap. 3, Van Nostrand Reinhold, New York, 1989.
2. Jerry March, *Advanced Organic Chemistry, Reactions, Mechanisms, and Structure,* McGraw-Hill, New York, 1968.

3. Wesley L. Archer, T. D. Cabelkas, and J. J. Nalazek, The Dow Chemical Company, private communication.
4. M. Tench, D. Hillman, and G. Lucey, "Environmentally Friendly Closed-Loop Soldering," *1993 International CFC and Halon Alternatives Conference*, Washington, D.C.
5. ANSI/J-STD-004, 2nd Interim Final.

4.10 Recommended Readings

Breslow, Ronald, *Organic Reaction Mechanisms—An Introduction*, W. A. Benjamin, Inc., New York, 1969.

Morrison, Robert T., and Roberet N. Boyd, *Organic Chemistry*, Allyn and Bacon, Boston, 1973.

5 Solderability

The solderability on a substrate, in a broad sense, is related to the ability to achieve a clean, metallic surface on substrates during the dynamic heating process so that good wetting of molten solder on the surface of the substrates can be formed. For applications using solder paste, solderability is affected by the ability to clean the solder powder during the dynamic reflow process to achieve a complete coalescence of solder particles so as to minimize the formation of solder balls.

5.1 Scientific Principle

5.1.1 Wetting theory

Wetting is an important surface phenomenon, which is driven by the relative thermodynamic surface energy and interfacial energy before and after a wetting process. The free surface of a solid, as is well known, has a higher energy than the interior of the solid due to broken bonds at the surface. In soldering, we deal with metallic surfaces which possess high surface energy (assuming a clean and pure state) as compared with oxides or organic materials. The energy magnitudes of three basic materials are compared below:

Metals	>200 cal/cm^3
Organics (nonmetals)	50–200 cal/cm^3
Fluorocarbons or low hydrocarbons	<50 cal/cm^3

As shown in the following equation, free energy is expressed as ergs per square centimeter, where an erg is equivalent to a dyne-centimeter in physical magnitude. That is, free energy has the units of work, which are force times the distance through which it acts—in this case, dyne times centimeters. Hence, the surface energy and surface tension are used interchangeably in terms of physical magnitude as discussed in Sec. 3.2.5.

For a system at constant temperature T and pressure P,

$$\left(\frac{\delta G}{\delta A}\right)_{P,T} = \gamma$$

where G = free energy
A = area
γ = surface tension

The thermodynamic condition for wetting or spreading to occur is

$$\Delta G < 0$$

Figure 5-1 depicts the spreading of a liquid with negligible vapor pressure on a solid surface. Thus,

$$-\left(\frac{\delta G}{\delta A}\right)_{P,T} = \gamma_{sv} - (\gamma_{ls} + \gamma_{vl} \cos \theta)$$

where γ_{ls}, γ_{vl}, γ_{sv} are liquid-solid, liquid-vapor, and solid-vapor interfacial tension, respectively.

Letting $-(\delta G/\delta A)_{P,T}$ be the spreading coefficient S, we have

$$S = -\left(\frac{\delta G}{\delta A}\right)_{P,T} = \gamma_{sv} - (\gamma_{ls} + \gamma_{vl} \cos \theta)$$

Therefore, for spread to occur,

$$\gamma_{sv} - (\gamma_{ls} + \gamma_{vl} \cos \theta) > 0$$

or

$$\gamma_{sv} > \gamma_{ls} + \gamma_{vl} \cos \theta$$

It should be noted that the above spreading condition holds true when

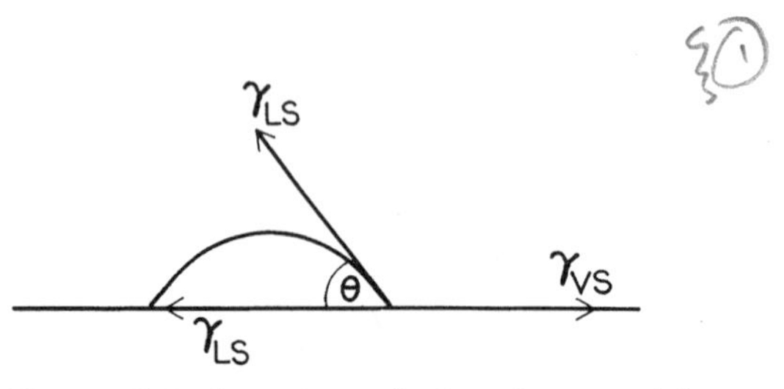

Figure 5-1. Spreading of a liquid on a solid surface.

the system does not involve significant chemical or metallurgical reactions at the interface.

As a general guideline, for a system where liquid is to wet the solid substrate, the spreading occurs only if the surface energy of the substrate to be wetted is higher than that of the liquid to be spread. In other words, the relative magnitude of the surface energy of the liquid and the surface energy of the solid substrate are key factors in determining the spreading and wetting.

The spreading of liquid tin-lead solder on clean copper surfaces with zinc-ammonium-chloride flux has been studied.[1] The area of spread for tin-lead solders with varying compositions is shown in Fig. 5-2. The spread increases with increasing tin content and reaches a maximum at 50% lead/50% tin and then decreases with a further increase in tin content. The study also showed that the spread may be affected by foreign additives. The addition of silver up to 3% of the tin content did not influence the amount of spread, but the addition of antimony up to 6% of the tin content caused a marked reduction in the area of spread.

This phenomenon can be explained by the competition between surface tension and the relative affinity of liquid metal for the copper substrate. Beyond a certain concentration of lead, the inferior affinity of lead for copper substrate predominates and thus reduces spreading. As the temperature increases, the solder with the maximum spread shifts to that with a higher tin content, as shown in Fig. 5-3. With the general phenomena in mind that surface tension decreases as temperature increases and that wetting ability increases as temperature increases, the shift of the solder with maximum spread to one containing more tin further displays that the tin-copper affinity surpasses the surface tension factor as temperature rises.

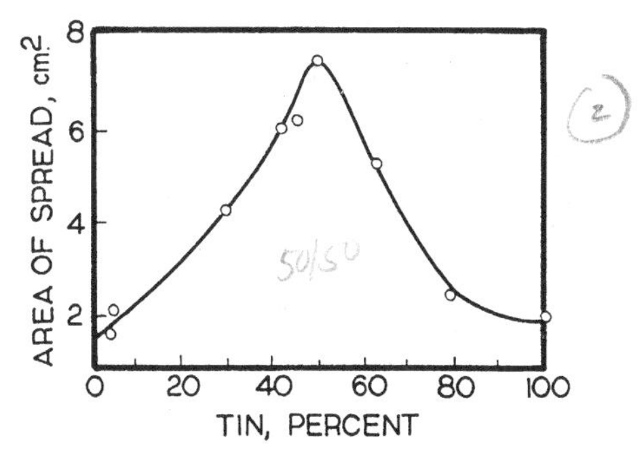

Figure 5-2. Relative spreading of Sn-Pb compositions.

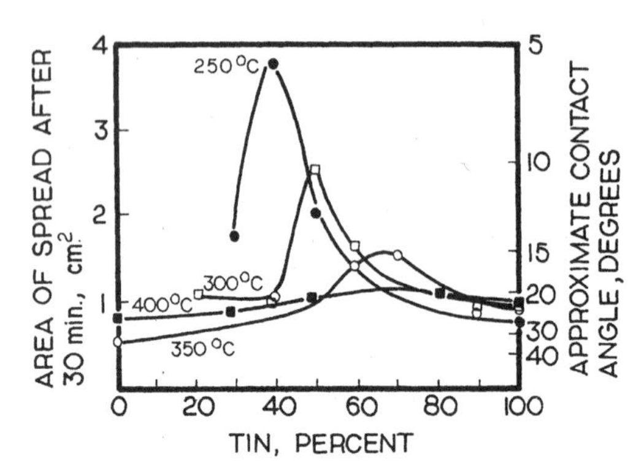

Figure 5-3. Maximum of area spread of Sn-Pb compositions versus temperature.

Another important phenomenon is that a polycrystalline metal may be wetted by a liquid metal not only at the interface surface but also along the grain boundaries. The latter may occur when the interfacial energy of solid-liquid is smaller than the grain boundary energy. This process may often cause embrittlement of the solid substrate.[2]

The conditions for wetting the solid surface are

$$\gamma_{sv} > \gamma_{ls} + \gamma_{vl} \cos \theta$$

and

$$\gamma_{ls} < \tfrac{1}{2} \gamma_{gb}$$

where γ_{gb} is the grain boundary energy.

Thus, the condition for wetting the surface without wetting the grain boundaries is

$$\tfrac{1}{2} \gamma_{gb} < \gamma_{ls} < \gamma_{ls} - \gamma_{vl} \cos \theta$$

5.1.2 Metallurgical reaction

The bonding process between the bonding material and the substrates to be bonded, whether the bonding material is metal- or polymer-based, relies on one of the following mechanisms:

1. Chemical reaction
2. Metallurgical reaction
3. Mechanical interlocking (molecular scale)
4. Diffusion
5. Physical adsorption through surface layer

Table 5-1. Intermetallic Compounds

System	Thermodynamically stable intermetallic compounds
Sn-Ag	Ag_3Sn
Sn-Au	$AuSn$, $AuSn_2$, $AuSn_4$
Sn-Pd	Pd_3Sn, Pd_2Sn, Pd_3Sn_2, $PdSn$, $PdSn_2$, $PdSn_4$
Sn-Pt	Pt_3Sn, $PtSn$, Pt_2Sn_3, $PtSn_2$, $PtSn_4$
Sn-Cu	Cu_6Sn_5, Cu_3Sn, Cu_4Sn, $Cu_{31}Sn_8$
In-Cu	Cu_9In_4, Cu_3In
Sb-Zn	$ZnSb$, Zn_4Sb_3, Zn_3Sb_2
Sb-Sn	Sn_3Sb_4, Sn_4Sb_5 $SnSb$, Sn_3Sb_2, Sn_2Sb

For bonding between solder and the metal substrate to occur, the metallurgical reaction is considered the primary bonding mechanism. That is, the active element of solder, primarily tin, reacts with the substrate to form a thin bonding layer; for example, Sn-Cu and Sn-Ni metallurgical intermetallic compounds are formed during the soldering process. Some thermodynamically stable intermetallic compounds between Sn-Pb solder and metals are listed in Table 5-1.

However, it should be cautioned that a proper amount of metallurgical reaction is necessary to form sound solder joints, yet any excessive amount can be detrimental (see Sec. 13.1 for further discussion).

5.2 Factors to Be Considered

In addition to the intrinsic wetting ability of solder on the specific substrate surface, other process, material, and ambient conditions are equally important to solderability. Areas of concern are

1. Condition of substrates to be soldered
2. Flux compatibility with substrate
3. Flux/vehicle chemistry versus reflow process
4. Reflow temperature profile
5. Quality and characteristics of solder paste versus the soldering process

5.2.1 Substrates

Common metallic substrates on the board and component sides to be joined include

Board side	Component side
Tin/lead/copper	Tin-lead/copper
Copper + antioxidant	Tin–lead/Alloy-42
Nickel	Copper
Palladium	Silver
Palladium-silver mixture	Palladium-silver mixture
Platinum-silver mixture	Nickel
Palladium-gold mixture	Gold
Gold	Palladium/nickel/copper
Gold/nickel/copper	Palladium/copper
	Silver/nickel/copper

The intrinsic wettability of molten solder on the common substrates is ranked in the following descending order:

$$\text{Tin} > \text{tin-lead} > \text{copper} > \text{palladium-silver} > \text{nickel}$$

5.2.2 Flux strength and compatibility

As metals are in contact with ambient air, the corresponding oxides, sulfides, or other nonmetallic compounds are inevitably formed. In order to achieve good solderability, external methods must be used to remove the nonmetallic compounds on the surface. Traditionally, the chemical means have been prevalent, that is, to use the chemical reactions such as rosin-based chemistry or the water-soluble or no-clean chemistry to clean the substrates. The factors affecting fluxing ability and the types of fluxes are discussed in Chap. 4.

5.2.3 Manufacturing process

5.2.3.1 Reflow profile. The reflow profile setting and methodologies will be discussed in Chap. 12. Both preheating and the peak temperature including the setting temperature and time duration are the parameters to be monitored. The wetting ability is directly related to the peak temperature; the higher the temperature, the better the wettability. While other conditions are equal, the longer dwell time can further enhance the wetting to some extent. In the preheating stage, the range of temperature and the time at the temperature directly affect the activity of flux, and thus the solderability.

5.2.3.2 Other factors. The time lapse between the printing step and reflow, or the time that solder paste is exposed to the environment,

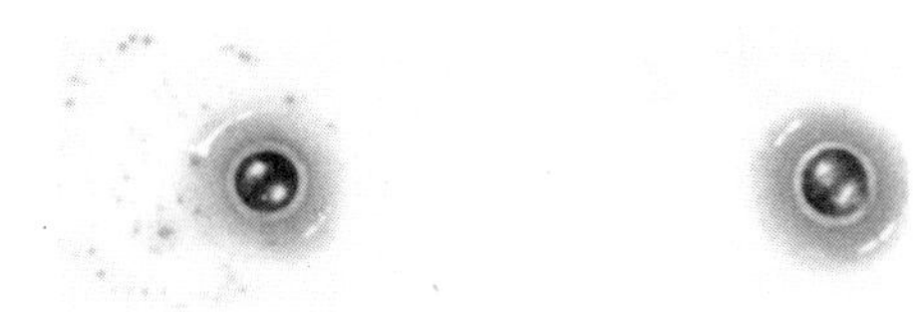

Figure 5-4. Solderability degradation of a paste after being exposed to incompatible conditions.

can also affect solderability. This is usually called the *open time* or *exposure life* of paste. It is determined by the built-in chemical makeup in the paste and by the ambient conditions (humidity and temperature). Figure 5-4 shows solderability degradation of a paste having been exposed under ambient temperature (75 ± 5°F, 24 ± 2°C) and humidity (65 ± 5%) for the time duration exceeding its capability (left) in comparison with the fresh paste (right). It is advisable to assure that the paste is compatible with the ambient environment that the paste is intended to be exposed to and that the performance capability of the paste matches what is required for the application.

5.2.4 Solder powder effect

Since the specific characteristics of solder powder require a certain level of flux for a given system, solder powder can be a contributor to the overall solderability. Details will be covered in Secs. 10.4.3 and 10.4.4.

5.3 Components

5.3.1 Factors

The solderability of component leads can be affected by the following factors:

Base metal of component leads

- Copper
- Copper alloys
- Alloy 42 (41–42.5 nickel, balance iron)
- Kovar (29% nickel, 17% cobalt, 53% iron, 1% others)

Coating composition

- Various ratios of tin and lead
- Silver-nickel-copper

- Palladium/nickel/copper
- Gold
- Palladium-silver

Coating technique

- Electroplating
- Electroplating-fusing
- Hot-dipping

Thickness of coating and storage condition

- A proper thickness which is compatible with the time and condition of the storage time, temperature, humidity

Process after lead coating

- Molding, burn-in, lead-trimming, and forming

The coating on leads or pads generally serves two purposes:

1. Solderability
2. Protection

A tin-lead coating thickness in the range of 0.00025 to 0.00045 in (6.3 to 11.3 μm) is commonly adopted and found to be suitable for most assembly processes and conditions. Too thin a coating was often found to be associated with poor solderability. For most coating processes, the cost determines the upper limit of coating thickness. Nonetheless, the ideal coating thickness depends on practical factors and the specific system. The composition of tin-lead coating can vary in a wide range of tin-lead ratios, from 5% tin/95% lead to 95% tin/5% lead. A high tin content provides better wettability. However, the composition with high tin content also promotes metallurgical reaction, such as intermetallic compound formation between the coating and the underneath base metal. For a given system, the rate and extent of the reaction depends on other external conditions, namely, time and temperature. When the reaction proceeds to the extent that the entire coating thickness of either the whole area or partial area is consumed, the solderability would be jeopardized.

The nature of the base metal for the leads can affect the solderability in a direct or indirect manner. Among the common lead materials, when the coating is porous, leads made of Alloy 42 and Kovar may experience more deterioration with age than copper due to potential moisture permeation through the porous surface. With the nature of Alloy 42 and Kovar, the leads made of either metal may have less solderability even though the coating surface is as intact as those made of copper.

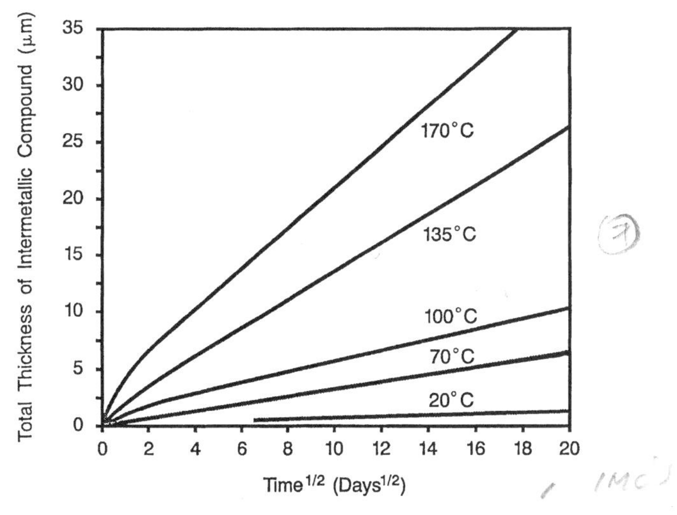

Figure 5-5. Growth of intermetallics of 60 Sn/40 Pb on Cu at various temperatures.

The two basic types of coating techniques are electroplated coating and molten solder dip. Each of these two processes has its inherent merits and drawbacks. The plating process normally provides more uniform thickness; however, the coating is often porous. The molten dip process produces a thicker and denser coating, yet lacks coat uniformity.

Figure 5-5 shows the growth of intermetallics at 60 Sn/40 Pb and Cu coupling at a temperature range of 20 to 170°C (68 to 237°F).[3] Figure 5-6 shows the growth of Sn-Cu intermetallics at room temperature for a range of coating compositions from 100% Sn to 30 Sn/70 Pb.[4] It is indicative that the growth rate increases with increasing temperature for a given composition. Also the growth rate increases with increasing amount of Sn content at a given temperature.

For a diffusion-controlled growth, a new phase grows out of another phase by a simple transfer of atoms of a single component. The new phase possesses a composition different from that of the old phase. Based on Fick's first law for one-dimensional growth, the interface position X, equivalent to the thickness of the new phase, varies with the square root of the time, and the growth velocity V varies inversely as the square root of time.

$$X = A_1\sqrt{Dt}$$

$$V = A_2\sqrt{D/t}$$

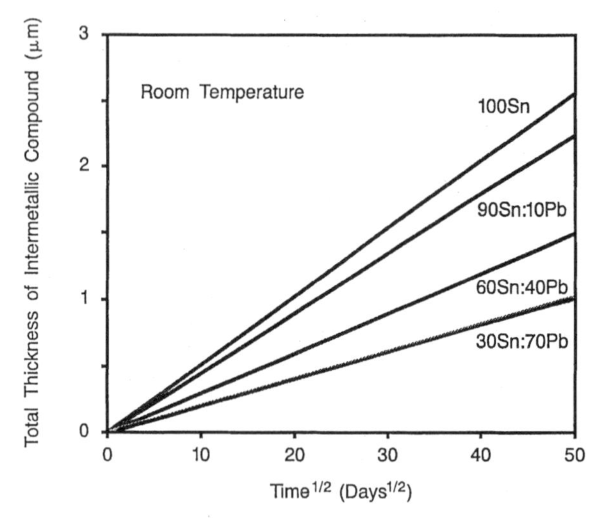

Figure 5-6. Growth of intermetallics of Sn/Pb compositions on Cu.

where D = diffusion coefficient
t = time
A_1 and A_2 = concentration constants

The data shown in Figs. 5-5 and 5-6 indicate that the formation of Cu-Sn intermetallics at the Sn-Pb and Cu interface follows a diffusion-controlled process. The properties that affect the solderability of coated leads can be grouped into five areas: intermetallics, surface oxides, surface porosity, other surface contaminants, and surface uniformity. Each of the three common coating techniques (electroplating, electroplating-fusing, and hot-dipping) has inherent characteristics. Electroplated leads with a copper base are normally associated with the lower initial level of the intermetallic layer between the copper base and the tin-lead plating. These intermetallics tend to grow with time. The growth may eventually consume all tin content in the coating and expose copper-tin intermetallics. This is the process that should be prevented by controlling storage conditions, coating thickness, and composition of the coating. In addition to intermetallics, surface conditions such as presence of oxides, porosity, and contaminants are other factors to be considered for solderability of leads. The extent of surface porosity and presence of oxides on the electroplated solder coating is expected to be greater than when using the hot-dipping technique. A fusing step following electroplating normally lowers the surface porosity, and subsequent oxidation, in the meantime, burns off

Table 5-2. Lead Coating versus Solderability

Factors	Electroplated	Electroplated and fused	Hot-dipped
Intermetallics	Continue to grow (Cu_3Sn)	Continue to grow (Cu_3Sn, Cu_6Sn_5)	Initial formation (Cu_6Sn_5)
Surface contaminants	Presence of organics likely	Organics burned off	No organics
Surface porosity	High	Low	Low
Surface oxides	High	Low	Low

other contaminants. The plating process provides more uniform thickness than the dipping process. The relative characteristics of these three coating techniques in relation with the intermetallics, surface oxide, surface porosity, other surface contaminants, and surface uniformity are summarized in Table 5-2.

5.3.2 Solderability of palladium-coated leads

An alternative to tin-lead solder coating, palladium-plated leads or lead frames have been developed.[5] The process plates the lead frame instead of the molded IC packages as with the coventional process using tin-lead plating. It involves

1. Cleaning
2. Activation step 1
3. Proprietary nickel layer formation
4. Activation step 2
5. Palladium plating

The first activation step reduces surface oxides and prepares a surface for the deposition of a nickel layer. The second activation step provides a surface for palladium plating for good adhesion. The palladium coating is very thin in the range of 3 to 5 microinches (0.076 to 0.127 μm); it may go down to 1 microinch (0.025 μm) for further cost savings. In comparison with the tin-lead dipping process, the following findings were reported related to the palladium-plated lead frames or leads as compared with tin-lead coating:

No effect on the thermal impedance

No effect on the reliability of the gold wire bonds

Superior adhesion to mold compounds

Less corrosive chemicals used

No effect on the conductivity of Pd-contaminated solder by up to 2% Pd content

Improved lead coplanarity

Equivalent solder joint strength

Palladium reacts with Sn but is relatively inert toward Pb. Figure 5-7 summarizes the dissolution rate of Pd in molten 60 Sn/40 Pb solder at various temperatures.[6,7] Based on the known dissolution rate in conjunction with the use of a very thin layer, a complete dissolution of Pd coating in solder is expected. Thus the underlying fresh, clean Ni surface can be easily wetted by solder. However, it should be noted that some reflow profiles may not provide the conditions necessary to completely remove Pd coating from the leads during the reflow process. The reflow profile, particularly the dwell time at the peak temperature,

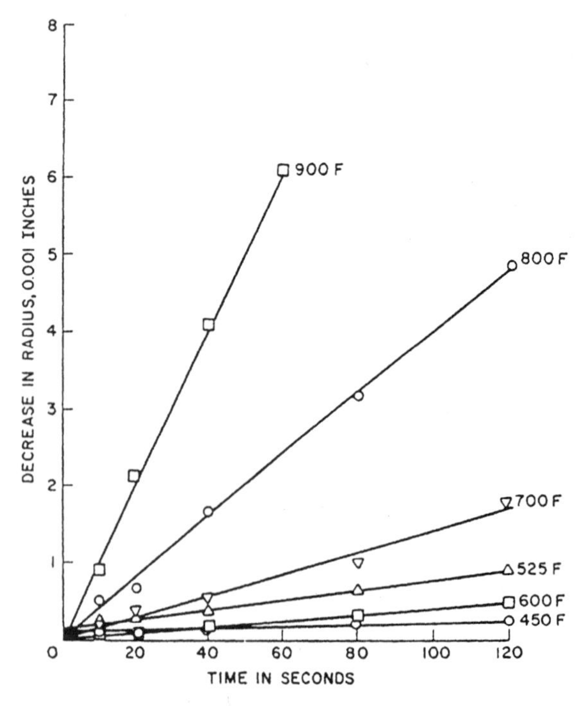

Figure 5-7. Dissolution rate of palladium in molten 60 Sn/40 Pb.

needs to be long enough to assure the complete dissolution of Pd. Incomplete removal of Pd may interfere with the wetting process.

It should also be noted that the appearance of a solder joint or solder fillet configuration on Pd-coated substrate may be different from that on Sn-Pb coated substrate. Solder may not easily cover the toe or top of the foot area.

5.3.3 Case study: Solderability and wetting versus nature of coating on component leads

Description

On a regular surface mount production run of high-end computer circuit boards, poor wetting or nonwetting was recorded as the reason of reject, resulting in a high defect rate and repair rate.

The type of circuit board was polyimide-based, composed of mixed components with the finest pitch of 20 mil (0.55 mm). The solder paste used was a water-soluble, fine-pitch product. The paste was printed through a stencil thickness of 6 mil (0.15 mm) and then reflowed by a convection oven. An aqueous cleaner with deionized water at 135°F, flowing at 3.5 gal/min in a cascade pattern, was used for after-reflow cleaning. The production reported that all steps of the process were maintained the same as in previous production runs (nothing had changed).

Examination

The production engineer-in-charge examined the rejected boards and found that the poor wetting or nonwetting occurred on one type of QFP, which was later identified to one component manufacturer's specific shipments.

Analysis and remedy

First, the engineer contacted the solder paste manufacturer to report the defective performance and to jointly identify the cause of the problem. Since the poor wetting occurred only on one type of QFP and all other conditions were reportedly the same, the effort was directed toward the component. In the meantime, the reflow profile was modified in an attempt to boost solderability. Changing the reflow profile did not solve the problem.

It was found that the component leads were plated with palladium, not tin-lead. Once this was identified, the paste manufacturer was asked to provide a solder paste which would be compatible with palladium substrate to allow for adequate wetting. The paste was designed, and the solderability problem was resolved.

Moral

The wetting ability of solders highly depends on the condition and the nature of the substrate. In today's industry, the solderability of components varies with the quality of as-produced components and with the subsequent shelf life and storage conditions, thus resulting in inconsistent solderability. Normally, a good fluxing chemistry is prepared for handling such a variation. However, as the metallurgy of the leads is altered, it may sometimes render the solderability unacceptable even with a well-designed fluxing system.

5.4 Board

5.4.1 Factors

Several factors affect the solderability of solder pads of PCBs:

- Pad surface composition

  Copper
  Tin-lead–coated copper
  Antioxidant-coated copper
  Gold/nickel/copper
  Palladium-nickel-copper

- Surface conditions

  Oxides, sulfides content
  Organic contaminants
  Intermetallics
  Other contaminants

- Thickness of coating

  Determination of a proper thickness

- Storage condition

  Time
  Temperature
  Humidity

The required coating thickness may vary. The proper thickness is the one which is compatible with the time and condition of storage to avoid the excessive formation of intermetallics and the exposure of intermetallics to the ambient environment. Generally, the lower the temperature and the humidity, the less degradation of solderability with time.

5.4.2 Major function of PCB surface finish

1. Solderability protection
2. Contact/switch
3. Wire bonding
4. Solder joint interface

5.4.3 Types of PCB surface finish

Hot air solder leveled SnPb (HASL) has been successfully used as the PCB surface finish for surface mount and mixed PCBs. As the industry continues to evolve, the following driving forces are developed for HASL alternatives in order to meet the requirements of

1. Increased demands for flat and uniform solder pads
2. Increased demands for consistent thickness of surface finish
3. Obtaining the same metal system and process for contact/switch
4. Less thermal stress process for temperature-vulnerable PCBs, such as PCMCIA.
5. Eliminating Pb.

Alternatives to HASL include

1. Electroplated Au/Ni
2. Electroless Au/electroless Ni
3. Immersion Au/electroless Ni
4. Immersion Sn
5. Immersion SnPb
6. Electroplated SnPb (reflowed or non-reflowed)
7. Electroplated Pd/Cu
8. Electroless Pd/Ni
9. Electroplated Pd/Ni

10. Electroplated SnNi alloy
11. Organic coating

5.4.3.1 Basic process. Three basic techniques to deposit metallic surface finish are

1. Electroplating
2. Electroless plating
3. Immersion

Inherently, electroplating utilizing electric current is able to economically deposit thick coating up to 0.000400″, depending on metal and process parameters. Electroless plating, requiring the presence of a proper reducing agent in the plating bath, converts metal salts into metal and deposits them on the substrate. The immersion plating process, in the absence of electric current and the reducing agent in the bath, deposits a new metal surface by replacing the base metal; in this process, plating stops when the surface of base metal is completely covered, thus only a limited coating thickness can be obtained through immersion process. For both electroless and immersion process, the intricate chemistry and the control of kinetics are vital to the plating results. Further, the designed process parameters and chemistry including pH and chemical ingredients must be compatible with the soldermask and PCB materials.

5.4.3.2 Metallic system. Available metallic surface finish on copper traces include Sn, SnPb alloy, SnNi alloy, Au/Ni, Au/Pd, Pd/Ni, and Pd. The systems containing noble or seminoble metals, such as Au/Ni, Au/Pd, Pd/Ni, Pd/Cu are capable of delivering the coating surface with uniform thickness. Those systems imparting a pure and clean surface also provide wire bondable substrate. In addition, wire bonding generally requires thicker coating, namely, more than 0.000020″. A unique feature of an Au/Ni system is its stability toward elevated temperature exposure during the assembly process as well as during its subsequent service life. When in contact with molten solder of SnPb, SnAg, or SnBi, surfaces coated with Sn and SnPb are normally associated with better spreading and lower wetting angle than others. Of the metallic systems, those containing an Ni interlayer are expected to possess the more stable solder joint interface; and in these systems, solder is expected to wet on Ni during reflow since noble metals are readily dissolved in solder. The concentration and distribution of noble metals in solder need to be noted in order to prevent any adverse effects in solder joining integrity. For a phosphorus-containing plating bath, a balanced concentration of phosphors in electroless Ni plating is needed. When the P content is too high, wettability suffers; and when it is too low, thermal-stress resistance and adhesion strength are sacrificed.

Another characteristic which is important to solderability is the porosity on the surface. Thinner coating is more prone to porosity-related problems although the surface density and texture can be controlled by the chemistry and kinetics.

5.4.3.3 Organic coating. Benzotriazole has been well recognized as an effective Cu and antitarnish and antioxidation agent for decades. Its effectiveness, attributable to the formation of benzotriazole complex, is largely limited to ambient temperature.

As temperature rises, the protective function disintegrates. Azole derivatives, such as Imidazole (m.p. 90°C, b.p. 257°C) and Benzimidazole (m.p. 170 C, b.p. 360°C) have been used to increase the stability under elevated temperature. SMT assembly of mixed boards involve three stages of temperature excursion—reflow, adhesive curing, and wave soldering. The reflow step, however, is considered to be potentially most harmful to the intactness of organic coating because it is the step with the highest temperature and longest exposure time.

Although the performance of organic coating varies with the formula and process, the general behavior of organic coating falls in the following regimen:

- Need compatible flux (generally more active flux).
- May need more active flux in wave soldering for mixed boards.
- Thicker coating is more resistant to oxidation and temperature, but may also demand more active flux.
- Organic coating needs to be performed as the last step of PCB fabrication.
- At temperatures higher than 70°C, coating may degrade. However, the degradation may or may not reflect on solderability.
- May be sensitive to PCB prebaking process (e.g., 125°, 1 hr to 24 hrs).
- For no-clean chemistry, may require N_2 atmosphere or higher solids content in no-clean paste.
- Steam aging test is not applicable.
- Not suitable for chip on board where wire bonding is rquired.

Nonetheless, when fluxing activity and process are compatible, organic coating can be a viable surface finish for PCBs. An additional bonus effect is that the bare copper appearance of the organic coated surface enhances the ease of visual inspection of peripheral solder fillets.

5.4.3.4 Relative cost. Cost among the alternative systems varies with the specific process and application. Yet the general trends follow in descending order:

- Electroless Au/Ni, electroplated Au/Ni
- Immersion Au/electroless Ni
- Electroless Pd/Ni
- Electroplated Pd/Ni
- Electroless Pd/Cu
- Immersion Pd/Cu
- Lead-free HASL
- SnPb HASL, SnPb plate (reflow)
- Immersion SnPb, organic coating, immersion Pd/Cu

5.4.4 Comparison of PCB surface finish systems

When selecting an alternative surface finish for PCB assembly, the key parameters in terms of solderability, ambient stability, high temperature stability, suitable for the use as contact/switch surface, solder joint integrity, and wire bondability for those assemblies that involve wire bonding, as well as cost, are to be considered.

Table 5.3 summarizes the relative performance of PCB surface finish systems. Regardless of other deficiencies, however, HASL provides the most solderable surface. When comparing metallic systems with HASL, the HASL process subjects PCBs to high temperature (above 200°C), producing inevitable thermal stress in PCB. HASL is also not suitable for wire bonding. To make a choice in replacing HASL, many variables are to be assessed. Understanding the fundamentals behind each variable in conjunction with setting proper priority of importance among the variables for a specific application is the way to reach the best balanced solution.

Table 5-3. Relative Performance of PCB Surface Finishes

HASL	Au/Ni	Pd/Ni	Pd/Cu	Organic
Pros:				
Most solderable	Uniform thickness Wire bondable Most stable to T	Uniform thickness Wire bondable	Uniform thickness Wire bondable	Uniform thickness Low cost Easy inspection
Cons:				
Nonuniform thickness	Higher cost	Higher cost	Higher cost (thicker coating)	Unsuitable for COB
Potential IMC problem				Flux and reflow process sensitive
Unsuitable for COB				Hi T degradation
PCB exposed to hi T				Cu reaches upper limit in solder bath Required as a last board fabrication step

5.4.5 Case study: Additional flux versus solderability degradation

When encountering a solderability and wetting problem on a production site, one convenient remedy is to use additional flux. Frequently, it indeed solves the problem. However, adding more flux may sometimes aggravate the problem. The following provides an example.

Description

The conventional PCB with FR-4, copper traces, Sn-Pb coating, and hot-air leveling was found to have an unexpected solderability problem. The board underwent conventional surface mount assembly using solder paste and oven reflow. The resulting solder fillets were inspected and found to be satisfactory, except that the solder did not flow on the pad far enough. A liquid flux organic acid (OA) was then applied to the specific solder pads in an attempt to promote flow and wetting. After the assembled board was applied with additional flux, it then went through a second reflow under a routine temperature profile. The solder did not flow better as was wished. Rather, the portion of the solder pads onto which the solder was intended to flow revealed bare copper.

Tests and examination

1. The visual examination under a microscope showed that the board, which was not applied with additional flux and not treated in any fashion, exhibited an acceptable surface of Sn-Pb coating on the solder pads.
2. The board which was applied with additional OA flux showed the exposed Cu pads as well as signs of corrosion after the reflow process.
3. When subjecting the as-received board to the same reflow process, no exposed Cu was detected and the Sn-Pb coating remained intact.
4. The receding pattern of Sn-Pb coating pads did not appear to be a typical dewetting.
5. It was found that applying a known-good solder paste on the pads with exposed Cu, which was created by the additional flux and reflow, could not obtain wetting through the reflow process.
6. The quality of the known-good paste was confirmed by obtaining good wetting on the as-received board.

7. All reflow processes were conducted under the same profile and conditions.

Analysis and conclusion

The additional flux applied on the solder pads with the objective of promoting flow and wetting was detrimental to the solderability. The solderability degradation was attributed to the chemically induced erosion on thin Sn-Pb coating which was activated by the high temperature of the reflow process. The eroded solder pad resulted in exposed Cu and poor solderability. The problem pad likely consisted of Cu-Sn oxide compounds and Cu-Sn chemical complexes.

Adding flux may not necessarily enhance wetting or resolve a solderability problem. The flux should be pretested to assure its compatibility and its performance.

5.5 Tests and Criteria

Solderability and wettability have been successfully evaluated by visual examination with or without visual aids (magnifier, optical microscope, etc.). The extent of solderability is categorized by three phenomena:

Wetting. The formation of a relatively uniform, smooth, continuous, and adherent film of solder to a metal substrate. Wetting angle generally is smaller than 75°.

Dewetting. A condition that results when molten solder coats a surface and recedes to leave irregularly shaped mounds of solder that are separated by areas that are covered with a thin film of solder; metal substrate is generally not exposed.

Nonwetting. The partial adherence of molten solder to a surface that it has contacted; metal substrate remains exposed.

The ANSI/J-STD-002 specification is used for evaluating the solderability of components,[8] and ANSI/J-STD-003 for the solder pads of PCBs.[9]

5.5.1 Components

Several tests are suggested:

1. Dip and look for leaded and leadless components
2. Wetting balance for leaded and leadless components

3. Globule for leaded components
4. Microwetting balance

5.5.1.1 Dip and look.[8] The criteria for the dip and look test specify that all leads or terminations shall exhibit a continuous solder coating free from defects for a minimum of 95 percent of the surface area of any individual leads or terminations.

5.5.1.2 Wetting balance.[8] Wetting balance can generate a quantitative measurement of wetting activity by monitoring wetting force versus time. The speed of wetting is evaluated by measuring the time taken to reach a specified point, and the extent of wetting is reflected by the magnitude of the wetting force. Figure 5-8 represents a typical wetting curve. As the specimen is dipped into the molten solder bath at a specified speed, the major parameters involved are the interfacial force between the molten solder and the perimeters of the specimen, the force due to archimedian buoyancy, and the geometry and volume of the specimen. The forces involved as shown in Fig. 5-9 are expressed as

$$F_w = CF_g \cos\theta - F_b$$
$$= CF_g \cos\theta - \rho g v$$

where F_g = interfacial force
F_b = buoyancy force
θ = contact angle

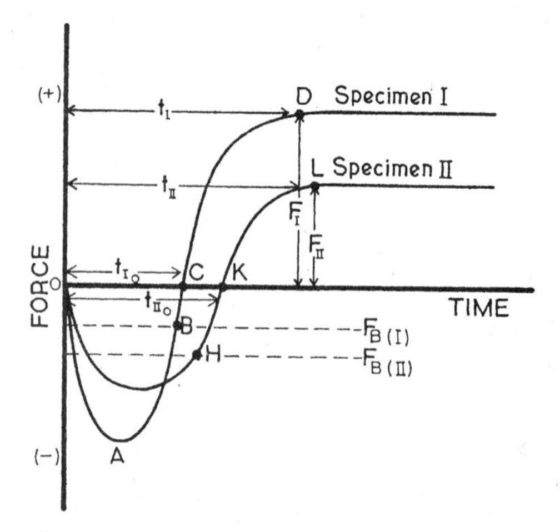

Figure 5-8. Typical wetting curve.

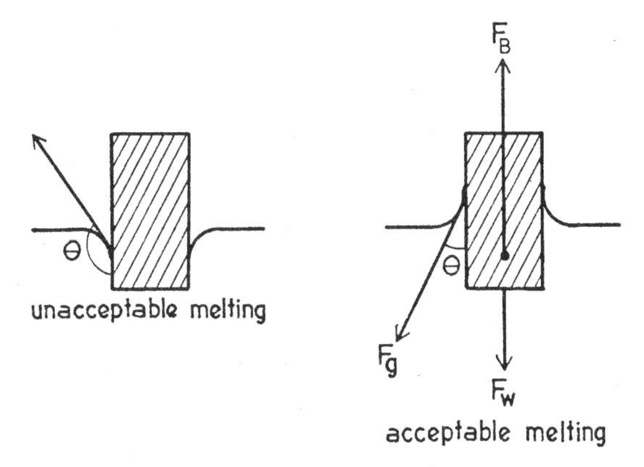

Figure 5-9. Wetting force in relation to buoyancy force and interfacial force.

ρ = density of molten solder
g = gravitational acceleration
v = immersed volume of specimen
F_w = resultant wetting force

As shown in Fig. 5-8, when the specimen is immersed in the molten solder bath at the preset depth, but prior to the commencement of wetting as indicated from 0 to *A*, an upward force is exerted on the specimen. Wetting starts at *A* and proceeds through *C*, then reaches equilibrium at *D*. At the point of *B*, the only force is that of upward buoyancy, and at *C* the upward buoyancy and downward wetting forces are balanced, resulting in a net zero force. The correction for buoyancy can be obtained by

$$\text{Buoyancy force} = dgV$$

where d = density of solder at 240°C (473°F) = 8.15 g/cm^3
g = acceleration of gravity, 9810 mm/s^2
V = immersed volume of specimen (width×thickness×immersion depth), cm^3

The induction period as shown in the curve between 0 to *A* depends on the size, conductivity, and surface condition of the specimen. During this period, the specimen is heated up to reach a wetting condition. Useful information can be derived from the curve by comparing the initial wetting time t, the equilibrium wetting time t_I, the wetting force F_I, and other specified parameters as called for. The criteria for the magnitude of wetting time and wetting force should be determined for a specific application and/or components. There is no universal number to be followed. Under a given set of conditions (temperature, specimen dimension, and speed and depth of immersion), the criteria for solderability can be readily established by the wetting balance technique.

The suggested criteria for acceptance and rejection given in ANSI/J-STD-002 are summarized here.

For leadless components. Coefficient of wetting > 150 N/s, where the coefficient of wetting is the ratio of the equilibrium wetting force to the time needed to reach equilibrium.

For leaded components. The time for across buoyancy corrected zero, as shown in Fig. 5-8, should be less than 1 s from the start of the test. The wetting force should reach 200 N/mm before 2.5 s after the start of the test. The wetting force should remain above 200 N/mm at 4.5 s from the start of the test.

It should be noted that the suggested criteria are based on the test procedure and test specimen specified. As test parameters and specimen change, the results are expected to change.

5.5.1.3 Microwetting balance.[10] This technique also monitors wetting force in relation to time; however, it uses a small solder globule, rather than a solder bath. The small globule is able to reduce the height of solder wetting, therefore, accommodating the small termination or leads of components. The termination of a 1206 capacitor in contact with a 200-mg solder globule and a single J-lead of PLCC package using a 25-mg solder globule are shown in Figs. 5-10 and 5-11, respectively.[11]

It was found that the microwetting balance technique using solder globules provided increased resolution and that the decreased size of globule offered a further increase in the testing sensitivity as shown in Fig. 5-12.[10]

5.5.2 Printed circuit boards

Although the majority of PCBs are tin-lead–coated copper, PCBs made of base copper treated with antioxidant and with other metal system coatings have become available.

In contrast to tin-lead–coated solder pads, the bare copper PCBs, which are treated with antioxidant system coatings, may behave differently in terms of wetting and spreading. It is often observed that molten solder may not flow as far as on boards with a tin-lead coating. Chemically, the effective antioxidant system is benzotriazole or its

Figure 5-10. Globule microwetting test fixture for PLCC J-lead.

Figure 5-11. Globule microwetting test for 1206-capacitor termination.

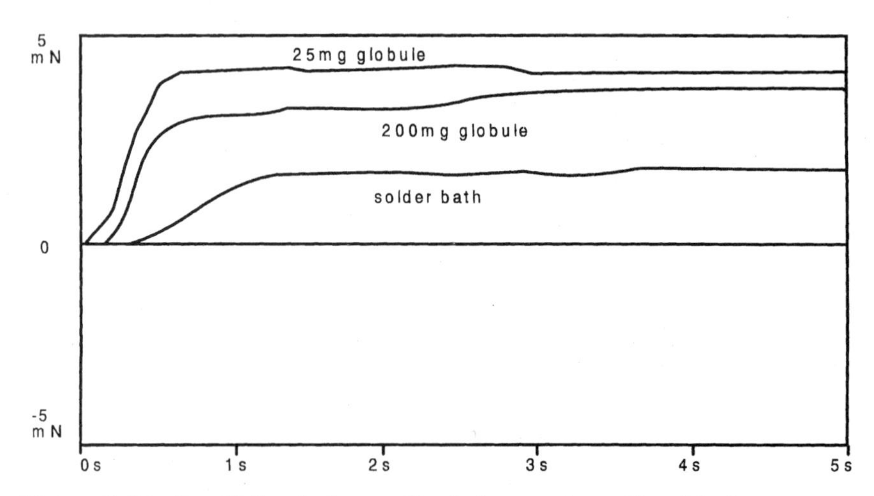

Figure 5-12. Sensitivity of micro wetting balance using solder globules.

derivatives. Its protecting ability is attributed to the effective formation of a copper-benzotriazole complex.[12,13]

The test methods included in ANSI/J-STD-003 are

1. Edge dip
2. Rotary dip
3. Solder float
4. Wave solder
5. Wetting balance

For surface mount solder pads, the acceptance criterion is that a minimum of 95 percent of the surfaces being tested shall exhibit good wetting using the edge dip, rotary dip, solder float, and wave solder tests.

Using the wetting balance test, some criteria are suggested:

1. The wetting time for the wetting curve to cross the corrected zero axis after the start of the test should be less than the maximum wetting time allowed for a sample with a given thickness. The suggested criteria are shown in Fig. 5-13.

2. A maximum wetting force taken after correction for buoyancy should be greater than the minimum acceptance force for a sample with a given thickness. Figure 5-14 provides the correlation between wetting force and thickness.

3. Dewetting should be less than 5 percent of the wettable surface area which had been immersed in solder during the test.

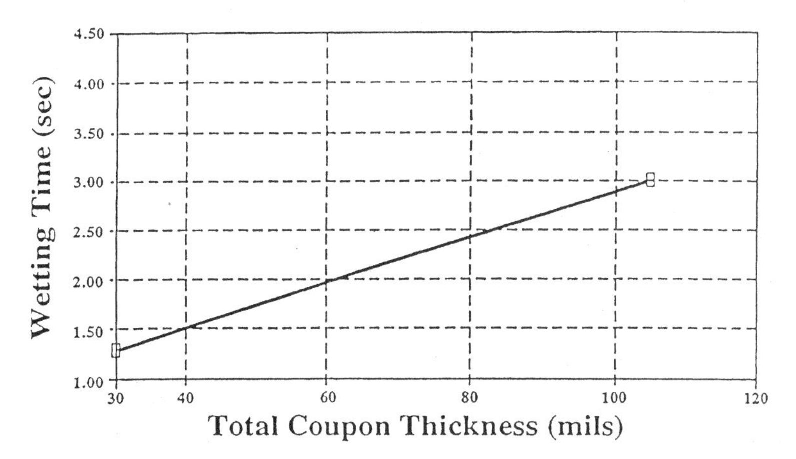

Figure 5-13. Suggested wetting time acceptance criteria versus test coupon thickness.

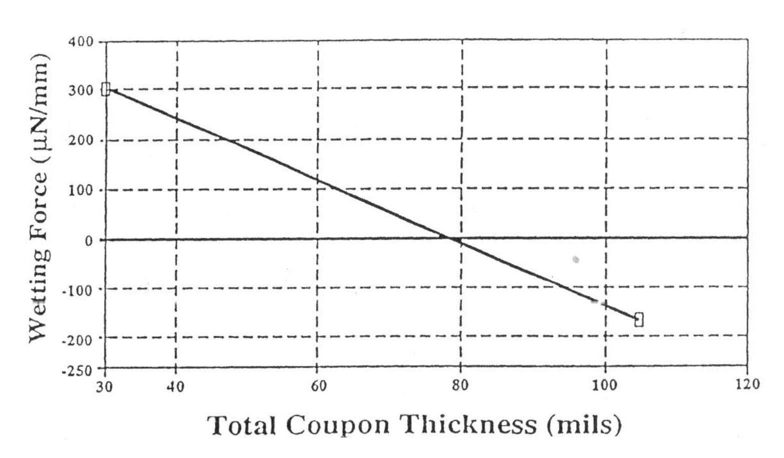

Figure 5-14. Suggested wetting force acceptance criteria versus test coupon thickness.

5.5.3 Other test methods

5.5.3.1 Sequential electrochemical reduction analysis (SERA). The SERA test is a recently developed methodology.[14,15] This chronopotentiometric method primarily engages in the reduction of surface oxides in sequence according to their electromotive reduction potentials. It is capable of determining the type of oxides present on the surface as well as the quantity. The system is comprised of three electrodes—an inert counter electrode at the anode, the specimen at the cathode, and the satu-

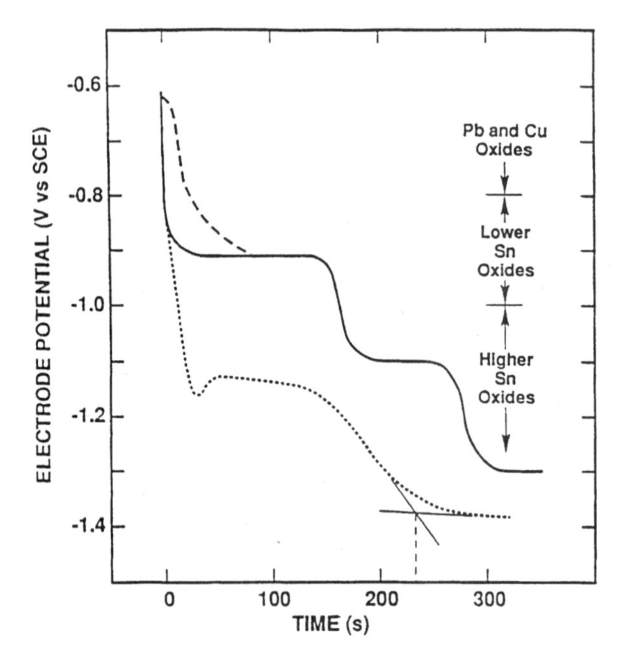

Figure 5-15. Typical SERA curve.

rated calomel reference electrode. They are immersed in an electrolyte of sodium borate/boric acid buffer solution with a pH of 8.4. A constant current is applied between the specimen and the inert electrode, and the cathode potential is then monitored as a function of time relative to the reference electrode.

For Sn-Pb–coated Cu substrate, a typical SERA curve, electrode potential versus time as shown in Fig. 5-15,[14] is evolved as the cathode voltage initially decreases to a plateau that corresponds to the reduction of the most readily reduced oxide, that is, the oxide having the smallest negative reduction potential. After the first oxide is completely reduced, the voltage is decreased to the value required for reducing the next oxide. This process of voltage drop and plateau forming continues until all oxides are reduced and a steady voltage that corresponds to hydrogen evolution from water hydrolysis is reached.

In practice, the ideal curve as illustrated in Fig. 5-15 may not be obtained because the oxides may not be immediately exposed to the electrolyte or they may not be in a discrete form. The resolution and accuracy of the curve also depend on other process parameters, such as current density and hydrogen evolution effect which often renders an overestimate in the quantity of oxides.

Table 5-4. Hydrogen Evolution Potential for Various Metals[14]

Metal	Electromotive potential
Palladium	−0.57
Platinum (99.95%)	−0.71
Nickel	−0.87
Gold	−0.90
Kovar	−0.95
Copper-beryllium	−0.95
Stainless steel (304)	−1.00
Copper	−1.05
Silver (99.9%)	−1.06
Titanium	−1.18
Brass	−1.20
63 Sn/37 Pb	−1.25
Tin (99.999%)	−1.27
Aluminum	−1.33
Lead (99.95%)	−1.35

Table 5-4 provides the steady-state hydrogen evolution potentials measured at $-20\ \mu A/cm$ in a berate buffer solution (pH 8.4) for various metals in reference with the saturated Calomel electrode.[14] The results obtained have shown that all oxides associated with a Sn-Pb-Cu system including intermetallic oxides can be detected by the SERA technique.[14] The technique is expected to provide nondestructive and quantitative data. It is relatively operator-independent and specimen-geometry–independent and is suitable for all component types and boards.

The SERA test is expected to be capable of determining the solderability level on the production floor or of being used as a research tool to understand solderability in relation to the amount, type, and morphology of oxide compounds that are commonly encountered in soldering electronic packages and assemblies.

5.5.3.2 Test for fine-pitch components. A test was designed for surface mount IC packages; the test involves the following steps:[16]

1. Print the proper amount of paste onto a ceramic substrate according to the land pattern of IC leads to be tested using a stencil.
2. Place the leaded IC package on the printed pattern.
3. Reflow the assembly.
4. Inspect and evaluate.

The reflow process can be selected so that it closely simulates the actual production process. The wetting phenomenon and solderability of component leads can then be visually examined as by the dip-and-look test. It can deliver the results for conditions that are closer to the actual production reflow. For palladium-coated leads, it is reported that the test also remedied the deficiency of the traditional dip-and-look test as a result of insufficient heat.

In addition, the test is able to reveal solder balling performance under the reflow conditions by examining the surface of ceramic board.

5.6 References

1. G. L. J. Bailey and H. C. Watkins, "The Flow of Liquid Metals on Solid Metal Surfaces and Its Relation to Soldering, Brazing and Hot-Dip Coating," *J. Inst. Metals*, vol. 80, 1951–1952, p. 57.
2. C. S. Smith, *Trans. Am. Inst. Meth. Engrs.*, vol. 175, 1984, p. 15.
3. J. P. Kay, *Trans. Inst. Metal*, vol. 54, 1976, p. 68.
4. D. Unsworth, *Trans. Inst. Finishing*, vol. 51, 1973, p. 85.
5. Donald C. Abbott, et al., "Palladium as a Lead Finish for Surface Mount Integrated Circuit Packages," *IEEE Trans. Components, Hybrids, Manu. Tech.*, vol. 14, no. 3, September 1991, p. 567.
6. G. M. Bouton and W. G. Bader, *Proc. Electr. Components Conf., IEEE*, 1968, p. 135.
7. R. P. Elliott, *Constitution of Binary Alloys*, First Supplement, McGraw-Hill, 1956.
8. ANSI/J-STD-002, 1992.
9. ANSI/J-STD-003, 1992.
10. Sy Loke, "Solderability Testing of Surface Mounted Components Using the Microwetting Balance," *Proc. 3rd Int. Microelectronics and Systems 1993 Conf.*, Kuala Lumpur, Malaysia.
11. *Billiton Witmetaal and G.E.C. Research Limited Company*, The Meniscograph Solderability Tester, Netherlands.
12. D. Chadwick and T. Hashemi, "Adsorbed Corrosion Inhibitors Studied by Electron Spectroscopy: Benzotriazole on Copper and Copper Alloys," *Corrosion Sci.*, vol. 18, 1978, p. 39.
13. Mau Kei Ho, "Copper Surface Finish Promotes Solderability," *Electr. Packaging, Prod.*, October, p. 28, 1987.
14. D. Morgan Tench, Dennis P. Anderson, and Paula Kim, "Solderability Assessment via Sequential Electrochemical Reduction Analysis," *J. Applied Electrochem.*, vol 24, 1994, p. 1.

15. D. Hillman, et al., "Sequential Electrochemical Reduction Analysis (SERA)—Solderability Assessment/Enhancement Technology," *Soldering and Surface Mount Tech.*, no. 13, p. 43, November 1993.
16. Douglas W. Romm and Willie R. Reynolds, "New Test Reveals Fine Pitch Bridges," *Surface Mount Tech.*, January 1994, p. 34.

6
Microstructure of Solders

6.1 General Definition and Characterization

Solders are generally polycrystallines that consist of an aggregate of many small crystals or grains. Most solder compositions contain multiple phases; for example, eutectic tin-lead solder has two phases: lead-rich and tin-rich. *Microstructure* can be defined as the internal structure that is composed of individual homogeneous and physically distinct phases formed and distributed according to given thermodynamic and kinetic conditions. In terms of dimensional range, microstructure correlates to the size of the phases and grains measured in micrometers. Therefore it takes a magnification of 100 to 5000 times to examine these structural features. A structure with features finer than grains and phases is referred to as a *submicrostructure,* which displays the atomic arrangement within grains and is associated with a dimensional scale of angstroms (Å). Coarser than microstructure, the next higher scale, is *macrostructure,* which is in the dimensional range of millimeters. The studies in these dimensional levels of materials correspond to scientific and engineering disciplines, as shown in Fig. 6-1.

Different techniques and instruments are required to obtain the necessary resolution of different levels of dimensions. In line with the spatial resolution limits which are associated with each technique as sum-

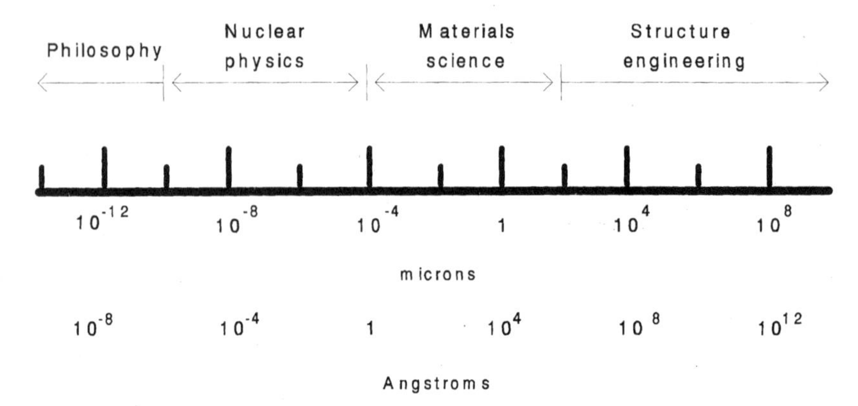

Figure 6-1. Dimensional level of materials versus scientific and engineering discipline.

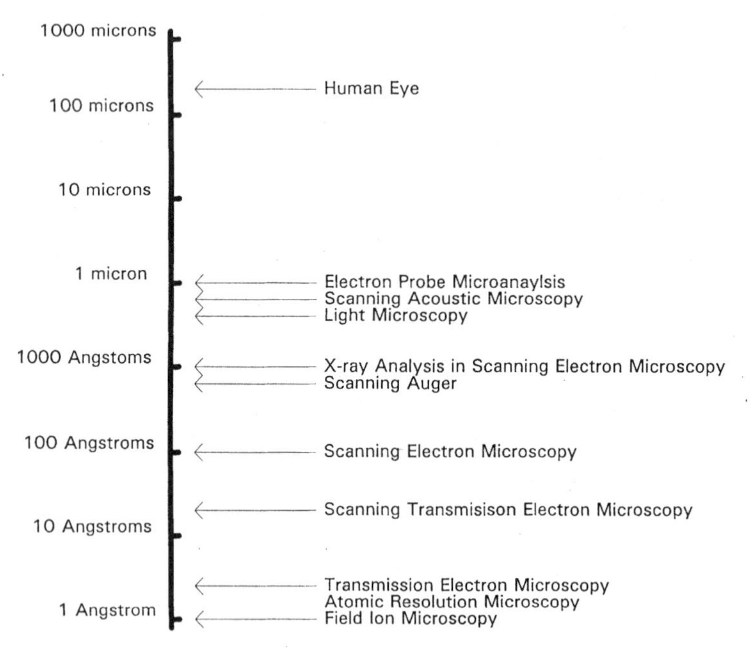

Figure 6-2. Spatial limit versus analytical technique.

marized in Fig. 6-2, the characterization of microstructure can utilize either optical or electron microscopy or both. Specifically, the light microscope and scanning electron microscope (SEM) are appropriate equipment and techniques to be used.

For light microscopy, the solder specimen has to be carefully prepared through metallography involving successive grinding and polishing with ascending levels of fineness of abrasive particles that are bonded on papers or used as slurry on a cloth-covered wheel. The size of abrasive particles applied can range from 23 μm (400 grit) to several micrometers and to submicrometers. With the polished surface, the specimen then requires an etching process that involves immersing the specimen in a specially formulated etching solution. An etching solution dissolves the surface layers of atoms, however, in different extent, depending on the nature of the phases, the orientation of the phases, and crystal boundaries. The specimen may also interact differently with different etching solutions. The resulting uneven surface reflects light into the objective lens of a light microscope in various ways, revealing the contrasts and structural features. It is obvious that the formula of the etching solution and the preparation of the specimen are vital to the image and resolution of the structure. For harder-to-prepare specimens, artifacts as a result of preparation techniques could occur.

In contrast, SEM often requires little sample preparation for viewing the microstructural features when the sectioning (cutting) is carefully performed. In SEM, a source of electrons is focused, in a vacuum, into a fine beam. When the electron beam impinges on a specimen surface, the electrons penetrate the surface of the specimen and interact with the atoms, emitting electrons and photons from or through the surface. The electron signals are produced from specific emission volumes within the specimen. The signals include secondary electrons, backscattered electrons, x rays, Auger electrons, and photons of various energies. The SEM images are produced from three types: secondary electron image, backscattered electron image, and x-ray spectroscopy. The secondary electrons and backscattered electrons are produced by different mechanisms and are distinguished by their energies. Emitted electrons having energies less than 50 electronvolts (eV) are conventionally considered secondary electrons. Most of the emitted secondary electrons are produced within the first few nanometers of the surface. Backscattered electrons are those with energies greater than 50 eV. They are scattered elastically by the nucleus of an atom, in contrast to inelastic scattering by atomic electrons as occurs with secondary electrons. Another electron interaction in the SEM occurs when an electron collides with a specimen atom and causes the ejection of a core electron. The excited atom will decay to its ground state by emitting a characteristic x-ray photon or an Auger electron. The x-ray emission signal is detected by the energy-dispersive x-ray analyzer. The analysis of the x ray emitted from the specimen provides

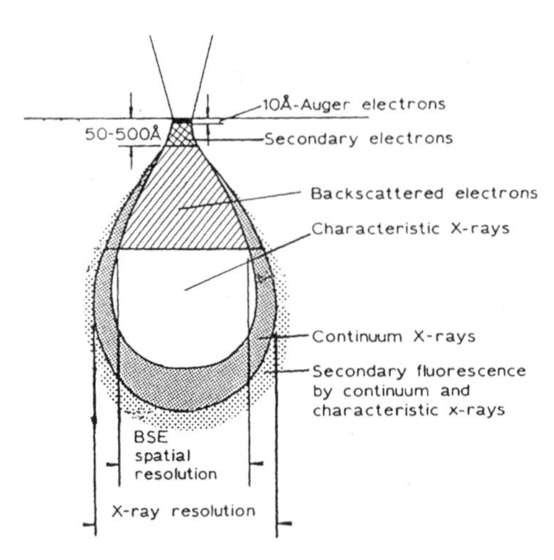

Figure 6-3. Range and spatial resolution of various electron signals.

qualitative and semiquantitative compositional information. The signals obtained from specific emission volumes within the sample are a function of the incident electron beam energy ranging from 1 to 50 keV and of the atomic number of the elemental composition. Figure 6-3 illustrates the range and spatial results of various signals.[1] The spatial resolution of secondary electrons is approximately the same as the electron probe diameter, whereas the spatial resolution of the backscattered electrons and x-ray signals is usually much larger than the electron probe diameter. As the surface slope changes, the number of secondary electrons produced changes as well. The image obtained as a result of signal variations in secondary electron emission from the near surface of a specimen reflects the surface topography. The amount of backscattering electrons, however, depends on the atomic number. The amount of scattering electrons increases with increasing atomic number. Thus, the region of higher atomic number corresponds to higher backscattered electrons. Figure 6-4 displays the relationship of the backscattered electrons with the atomic number.[2] This results in an image with dark, light, and intermediate shades. The region with the highest atomic number appears the lightest, and that with the lowest atomic number is the darkest; the intermediate values are various shades of gray. For example, the SEM image of the microstructure of a 63 Sn/37 Pb solder joint is normally composed of a light (lead-rich) phase and a dark (tin-rich) phase; and the SEM image of Sn-Sb solders exhibits much less distinct phases than that of Sn-Pb solders. In com-

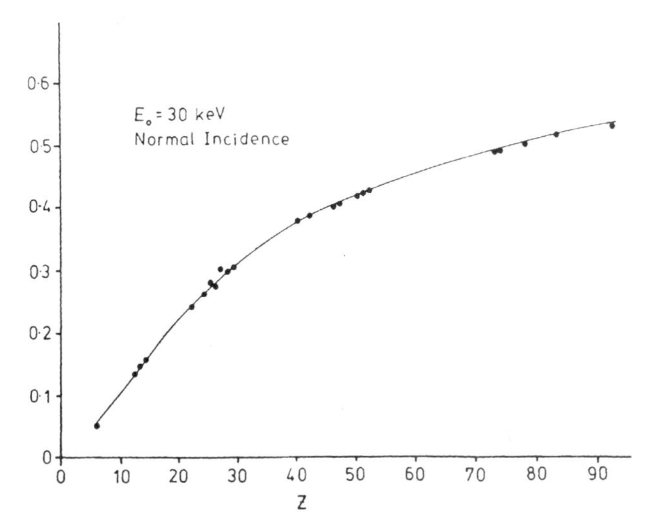

Figure 6-4. Backscattered electron (η) versus atomic number (Z).

parison, the image of secondary electrons is not a sensitive function of atomic number. However, both electron signals provide valuable information about the microstructure of materials. Generally, secondary electron images facilitate the visibility of surface texture, and backscattered electron images enhance the contrast of the phase structure. Combining the information and observation from both secondary and backscattered electron images, the morphology in micrometer scale will become better understood.

Furthermore, microstructures obtained from either SEM or metallography/optical microscopy are complementary to each other. Since solders are relatively soft alloys that consist of multiple phases with significantly different hardness, it may not be always straightforward in sample preparation. For consistency and convenience, microstructures are presented in the text by using SEM technique only.

6.2 Correlation to Phase Diagram

Thermodynamically, the microstructure is expected to correlate well with the phase diagram. However, when the solidification process deviates from equilibrium conditions, anomalies will occur. Furthermore, the heating, cooling, and associated metallurgical reactions and interfacial boundaries of solder joints may also affect the development of the microstructure, resulting in various size, shape, and distribution of thermodynamically equilibrium phases as well as nonequilibrium phases as discussed in Secs. 6.3 and 6.4. Nonetheless, phase diagrams provide fundamental data for the specific alloy system of interest.

There are basically four groups of binary phase diagrams, based on the mutual elemental solubility in solid state and liquid state and on the formation of intermediate metallurgical phases.

1. *System with complete solid and liquid miscibility between two elements.* The phase diagram consists of a liquid region (L), a solid region (S), and a coexisting liquid-solid region ($L + S$) as shown in Fig. 6-5. At a given composition A_x, the alloy is a liquid at temperature T_L, a solid at T_S and a two-phase $L + S$ at T_1. The concentration of the two phases at temperature T_1 and composition A_x is estimated by the Lever Law: the ratio ad/ab is the proportion of the liquid phase. Alloy examples are In-Pb (Fig. 3-49) and Ag-Au (Fig. 3-50) systems.

2. *System with complete liquid miscibility and no solid miscibility between two elements.* As shown in Fig. 6-6, the phase diagram is characterized by two triangular two-phase regions ($S_A + L$ and $S_B + L$), a horizontal solidus line (CED), and a eutectic point E at which three phases (S_A, S_B, L) coexist. Since there is no mutual solubility in the solid state, the properties of these alloy systems cannot be altered by heat treatment.

3. *System with complete liquid miscibility and partial solid miscibility between two elements.* Figure 6-7 contains three two-phase regions ($S_A + L$, $S_B + L$, $S_A + S_B$), solid solution of A in B (S_B), solid solution of B in

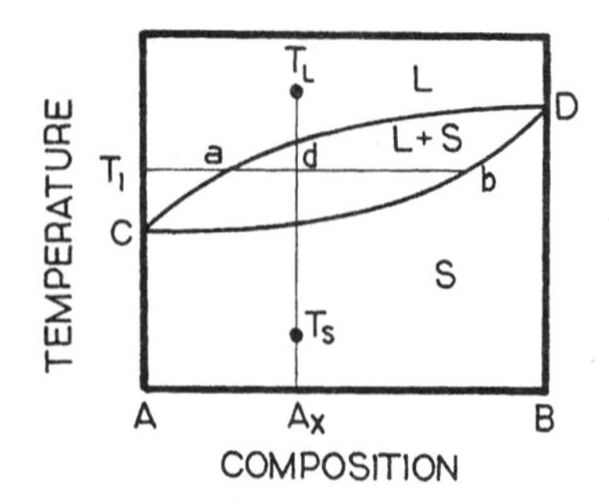

Figure 6-5. General phase diagram of complete solid and liquid miscibility.

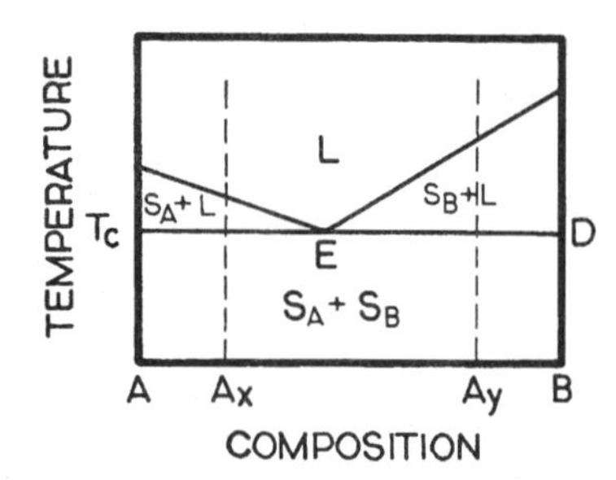

Figure 6-6. General phase diagram of complete liquid miscibility and no solid miscibility.

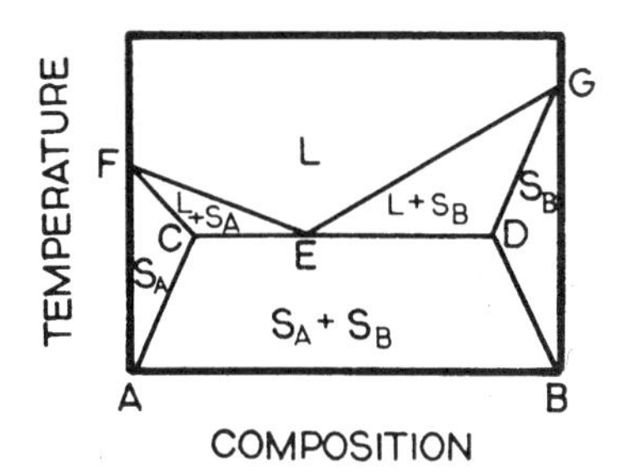

Figure 6-7. General phase diagram of complete liquid miscibility and partial solid miscibility.

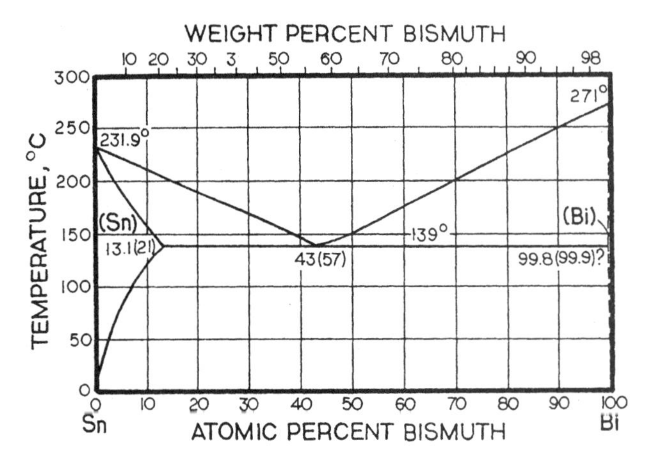

Figure 6-8. Phase diagram of Sn-Bi.

A (S_A), liquid phase, and a eutectic point, constituting the phase diagram. Examples are the Sn-Bi, Sn-Pb, and Sb-Pb systems as shown in Figs. 6-8, 6-9, and 6-10, respectively.

4. *Systems containing intermediate phases.* These more complex phase diagrams include intermetallic compounds, allotropy (different crystalline states), peritectoid, and eutectoid. Systems that fall under this group are Sn-Sb (Fig. 3-55), Sn-Ni (Fig. 3-51), Sn-Pt (Fig. 6-11), and Sn-In (Fig. 6-12).

Take the Sn-Pb system as an example, Sn and Pb have mutual solubility in both liquid and solid states. At the eutectic composition, 63 Sn/37 Pb, and at the eutectic temperature 183°C, the mutual solubility

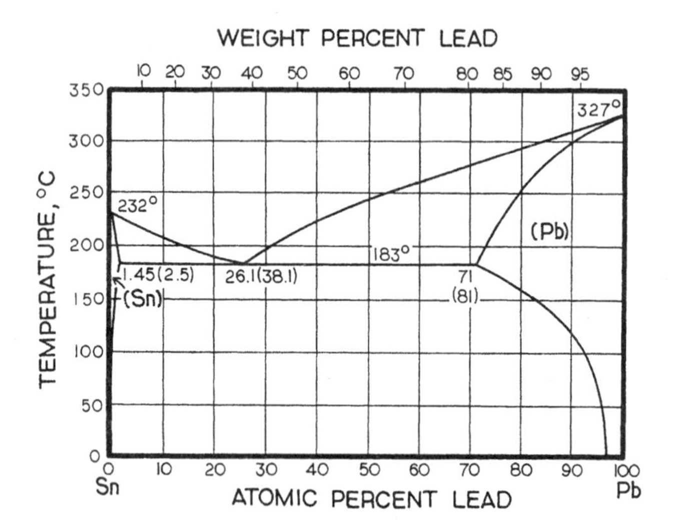

Figure 6-9. Phase diagram of Sn-Pb.

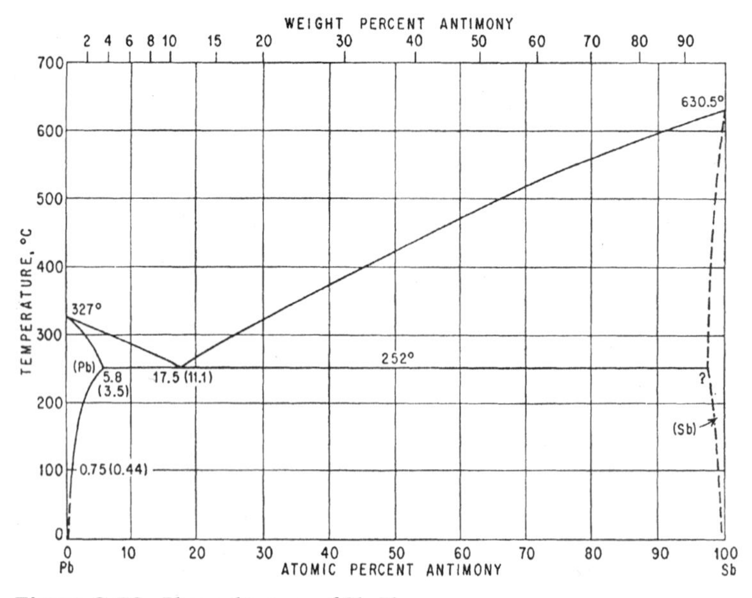

Figure 6-10. Phase diagram of Sb-Pb.

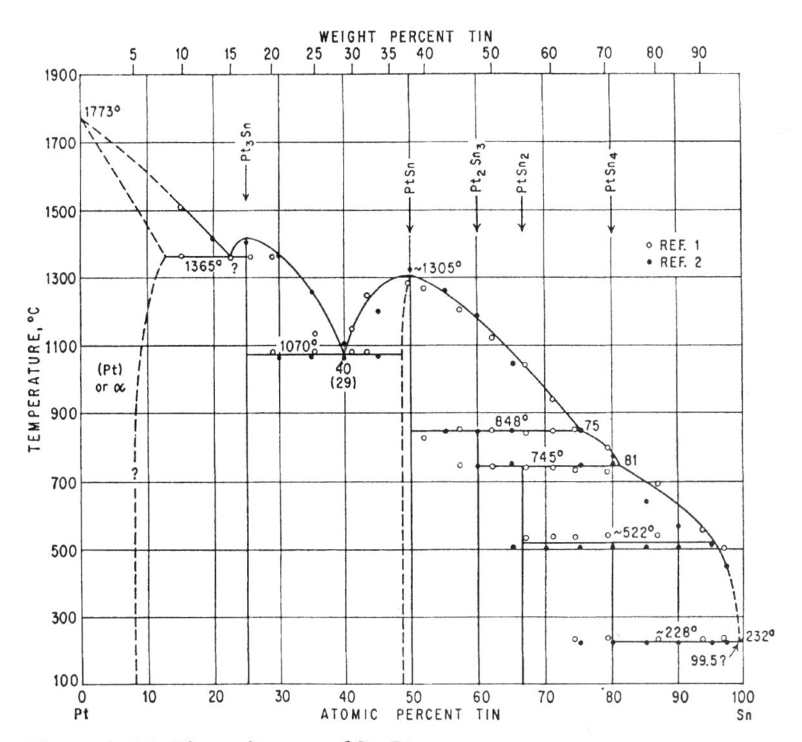

Figure 6-11. Phase diagram of Sn-Pt.

reaches the maximum: Pb in Sn is 2.5 wt% and Sn in Pb is 19 wt%. As the temperature increases or decreases, the solid solubility decreases. At room temperature, 63 Sn/37 Pb is composed of two phases: tin-rich phase and lead-rich phase as shown in the SEM micrograph of Fig. 6-13. It should be noted that the microstructure of 63 Sn/37 Pb solder joints formed under practical conditions is often found to deviate appreciably from that of the equilibrium state. The eutectic equilibrium structure consists of eutectic colonies, in each of which alternating lamellae of Pb-rich α-phase [face-centered cubic (FCC)] crystal structure and Sn-rich β-phase [body-centered tetragonal (BCT)] platelets exist. The deviation is not only due to the nonequilibrium condition, but also to the presence of interfacial boundaries and the metallurgical reaction at the boundary that are expected to separate the behavior of the solder joint from that of the bulk solder.

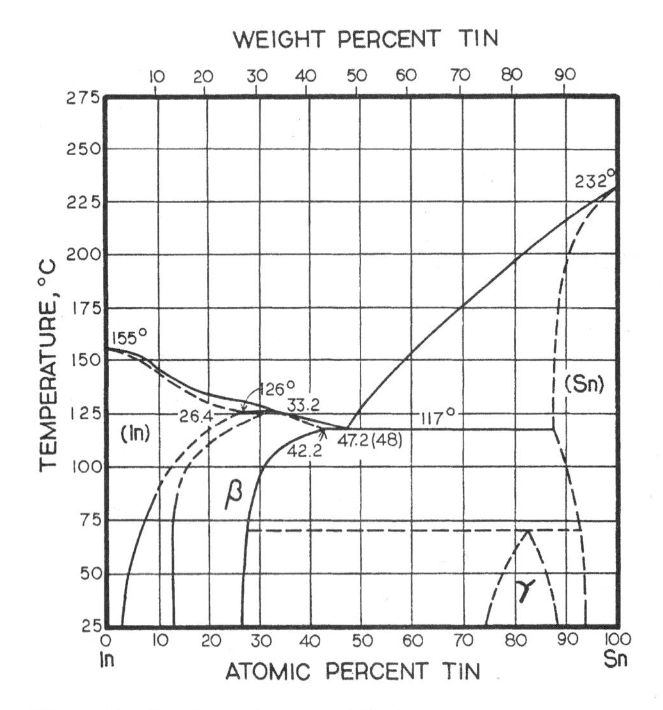

Figure 6-12. Phase diagram of Sn-In.

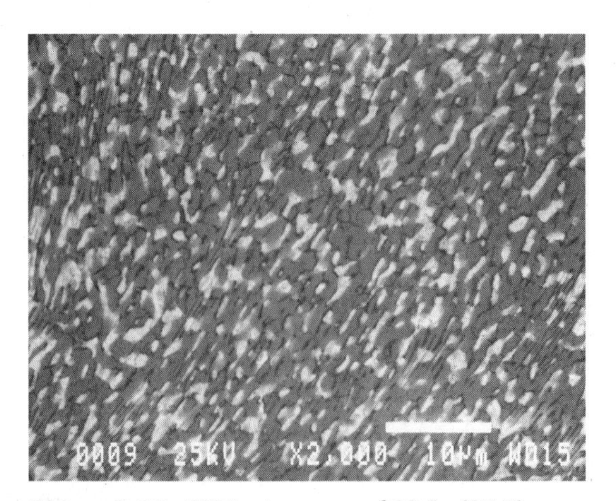

Figure 6-13. SEM microgram of 63 Sn/37 Pb.

6.3 Heating Effect

For an assembly which is prone to the formation of intermetallic compounds at the interface or in the intrinsic solder composition, a prolonged heating during reflow (soldering), particularly at peak temperature at which solder is in a molten state, may produce the excessive intermetallic compounds at the interface or in the interior of the solder joint. As the temperature is high enough and when the solder is in the liquid state, intermetallic compounds formed at the interface may continue to grow and migrate toward the interior bulk solder. In extreme cases, intermetallics may emerge onto the free surface of the solder (opposite to the interface). Figure 6-14 shows the microstructure of the cross section of the resulting solder joint of 63 Sn/37 Pb on Cu substrate after prolonged exposure at peak temperature, showing the additional needlelike dark phase as compared with Fig. 6-13. The needlelike dark phase was identified as Cu-Sn composition by semiquantitative x-ray dispersion analysis as shown in Fig. 6-15. The examination on the free surface of the same solder joint indicated the presence of Sn-Cu intermetallics as shown in Fig. 6-16.

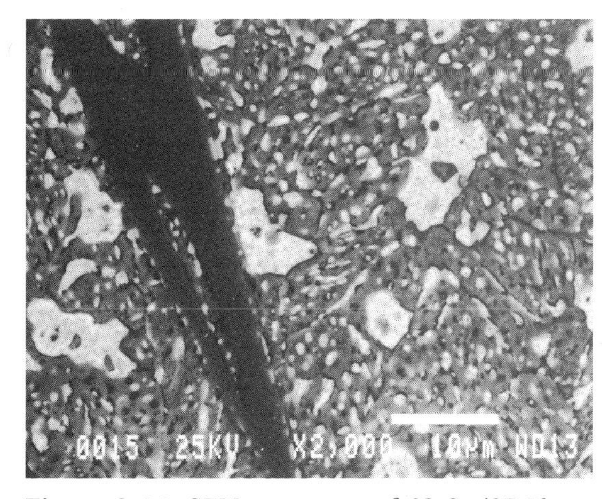

Figure 6-14. SEM microgram of 63 Sn/37 Pb on Cu solder joint after excessive heating at peak temperature.

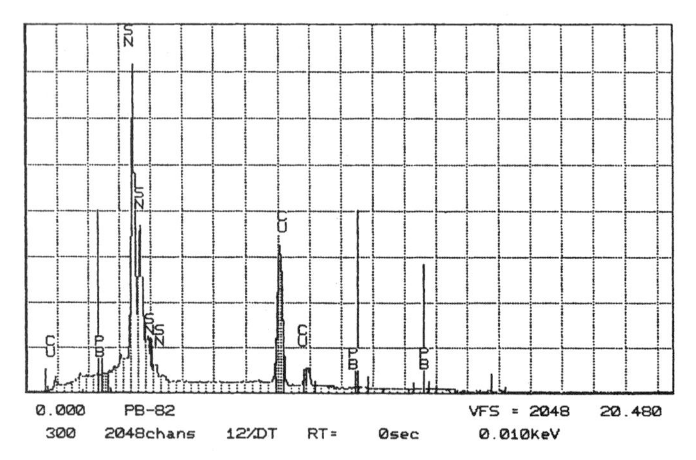

Figure 6-15. X-ray analysis of needlelike dark phase of Fig. 6-14.

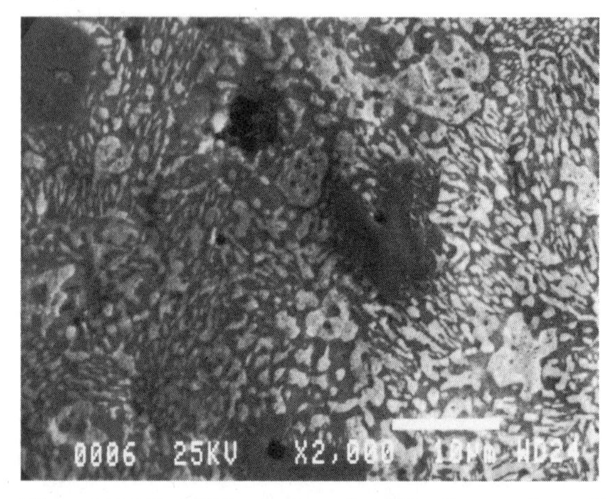

Figure 6-16. SEM micrograph of the free surface of the same solder joint as in Fig. 6-14.

6.4 Cooling Effect[3]

The cooling effect on microstructures is illustrated by conducting tests under a series of process conditions in conjunction with the examination of SEM micrograph images from backscattered electron signals. Solder joints of 63 Sn/37 Pb on Cu substrate were reflowed under identical conditions and then solidified in five different manners corresponding to four cooling rates as measured above 100°C: 0.1°C/s, 1°C/s, 50°C/s, and 230°C/s, respectively. In addition, the fifth group of solder joints was solidified in an uneven cooling rate conducted in two steps resulting in an average cooling rate of 12°C/s. Scanning electron microscope micrographs were obtained in two magnifications, 500× and 1000×, in order to examine a sufficient area as well as the detailed microstructure features.

At the slowest cooling rate being tested, 0.1°C/s, the microstructure as exhibited in Fig. 6-17 was rather inhomogeneous, consisting of small and distinctly wavy eutectic colonies (distribution of rodlike and globular Pb-rich crystals), discontinuously faulted eutectic colonies, and large primary Pb-rich proeutectic dendrites (light phase) situated along the interface. Globally speaking, the size of colonies was on the

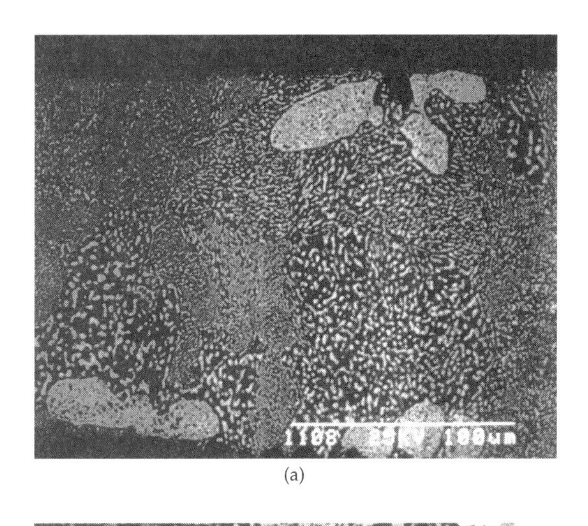

(a)

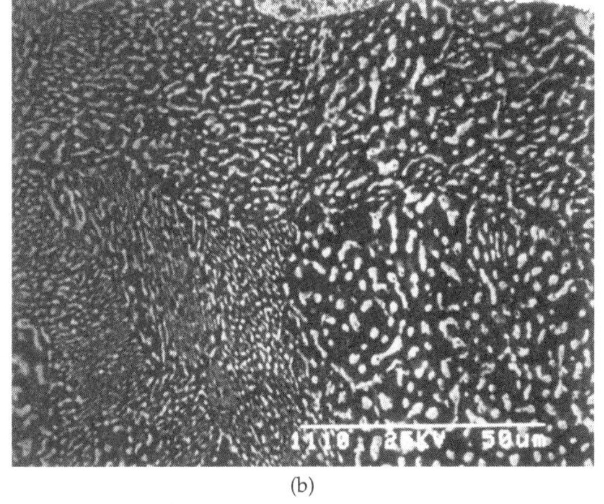

(b)

Figure 6-17. SEM microstructure of 63 Sn/37 Pb on Cu solder joint cooled at 0.1°C/s.

order of solder layer thickness. As the cooling rate increased from 0.1 to 1.0°C/s, the eutectic colonies degenerated into a more faulty pattern, being more wavy, discontinuous, or granular; in the meantime, the large globular Pb-rich phase gradually developed toward the center of the solder joint, as shown in Fig. 6-18*a* and *b*.

As the cooling rate increased further, the size of the degenerated colonies became even smaller, the contour of the colonies became less apparent, and the Pb-rich particulates in smaller size and greater numbers were scattered in the solder joints. The microstructures at fast cooling rates, 50°C/s and 230°C/s, are shown in Fig. 6-19*a* and *b*, and

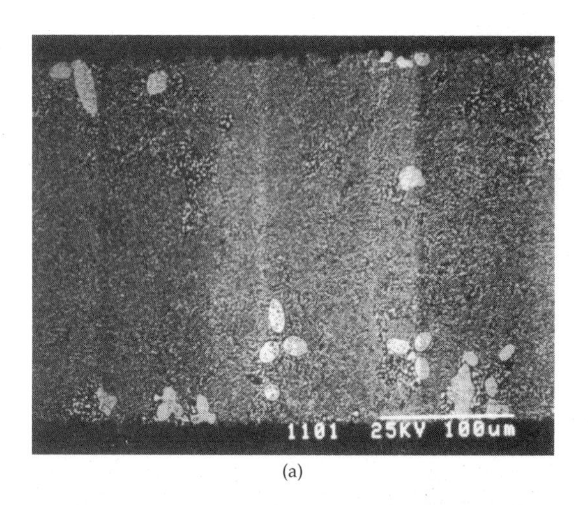

(a)

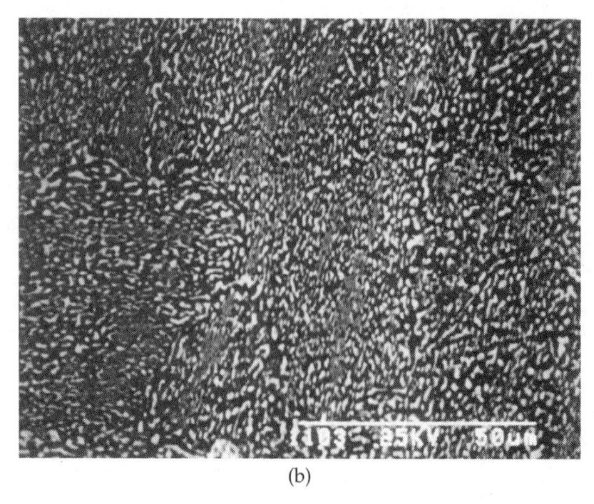

(b)

Figure 6-18. SEM microstructure of 63 Sn/37 Pb on Cu solder joint cooled at 1.0°C/s: (*a*) 300× (*b*) 1000×.

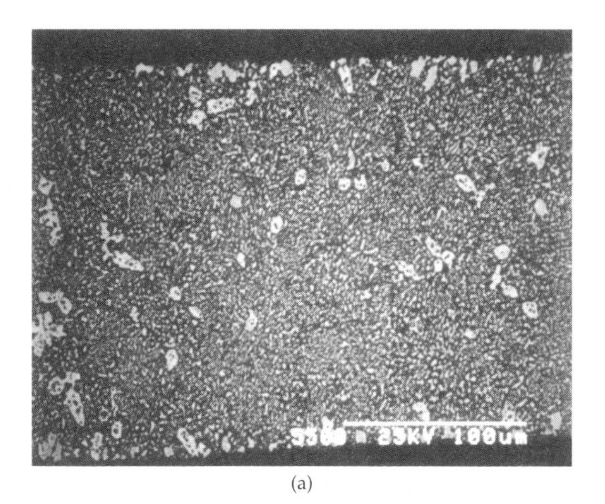

(a)

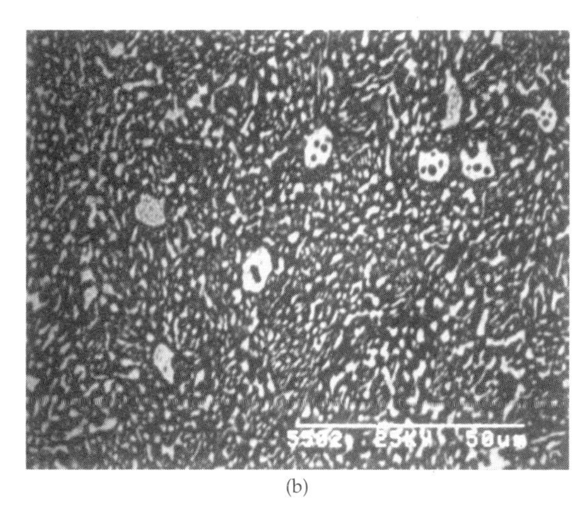

(b)

Figure 6-19. SEM microstructure of 63 Sn/37 Pb on Cu solder joint cooled at 50°C/s: (*a*) 300× (*b*) 1000×.

Fig. 6-20*a* and *b*, respectively. Along the interface, all solder joints, regardless of the cooling rate at which the solder joints were formed, were found to have protruding intermetallics, as shown in Fig. 6-21.

As a function of increasing cooling rate, the microstructure exhibited the following tendencies:

- The faulty eutectic colonies became dominate, taking over the distinct colonies; the Pb-rich crystals inside the colonies became more

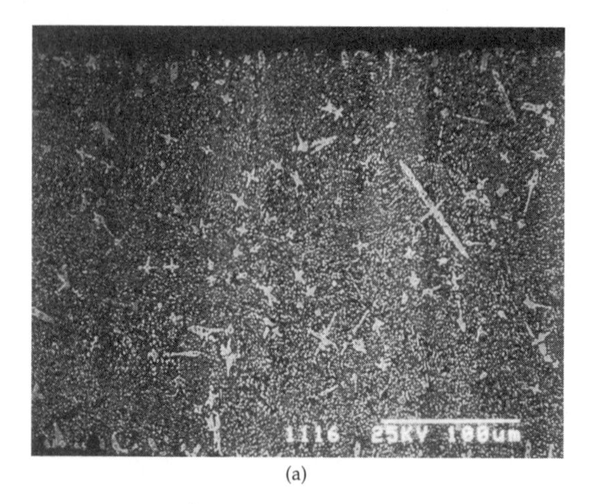

(a)

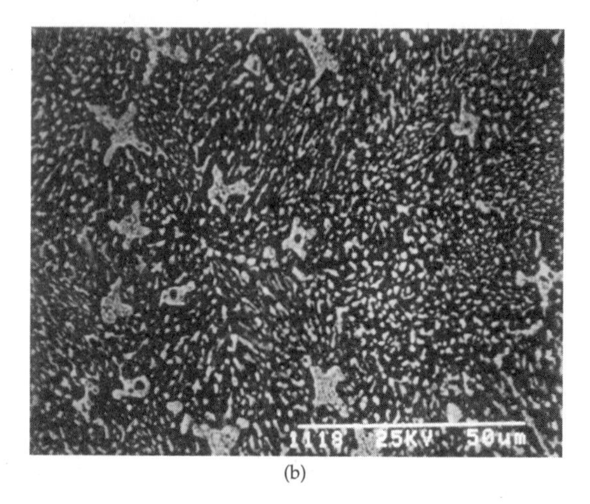

(b)

Figure 6-20. SEM microstructure of 63 Sn/37 Pb on Cu solder joint cooled at 230°C/s: (*a*) 300× (*b*) 1000×.

globular and finely dispersed, forming degenerated eutectic structures.

- The degree of degeneration of lamellar eutectic colonies increased.
- The size of the eutectic colonies decreased.
- The grain size of Pb-rich proeutectic dendrites decreased.
- The Pb-rich proeutectic dendrites became smaller and tended to break away from the interface and move toward the interior of the solder joint.

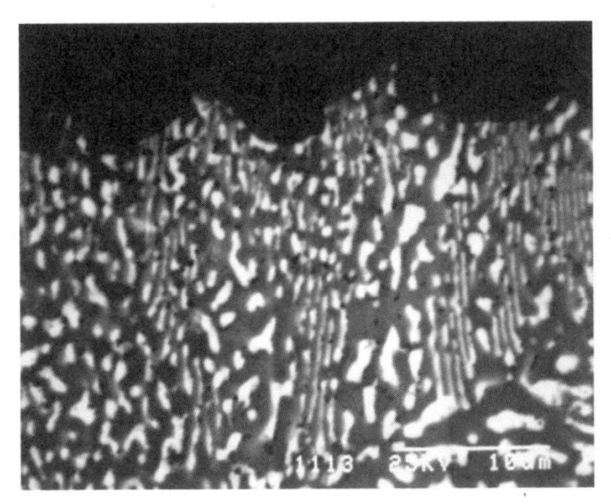

Figure 6-21. SEM micrograph showing interface of solder joint on Cu.

The solder joint made by an uneven two-step cooling process resulted in a highly inhomogeneous microstructure as shown in Fig. 6-22*a* and *b* for near the center of the solder joint and in Fig. 6-23*a* and *b* for near the free surface of the solder joint. The gradient in microstructure across the solder joint was noted by comparing Fig. 6-22*a* with Fig. 6-23*a*. The microstructure near the center of the solder joint was closer to that at a slower cooling rate and that near the surface resembling that at a faster cooling rate. The relation of microstructure with mechanical properties of solder joints will be discussed in Chap. 14.

It is generally accepted that the faster cooling rate creates a finer-grain microstructure in bulk solders. This general rule is often complicated by the interfacial boundary and metallurgical reaction at the interface of solder joints. The nature of the substrate and its metallurgical affinity to the specific solder composition can affect the development of the microstructure of the solder joint.

6.5 Overaging Effect

Metallurgically, overaging denotes the phenomenon of alloy weakening (softening) after being held for too long a period at a given temperature. For solder applications, overaging may be affected by the following variables:

- Time
- Temperature

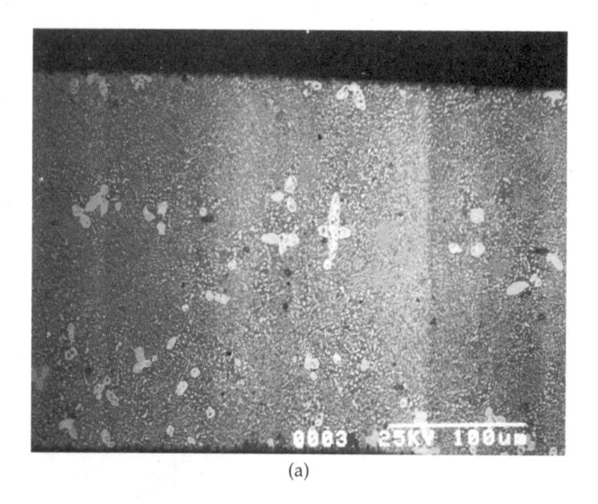

(a)

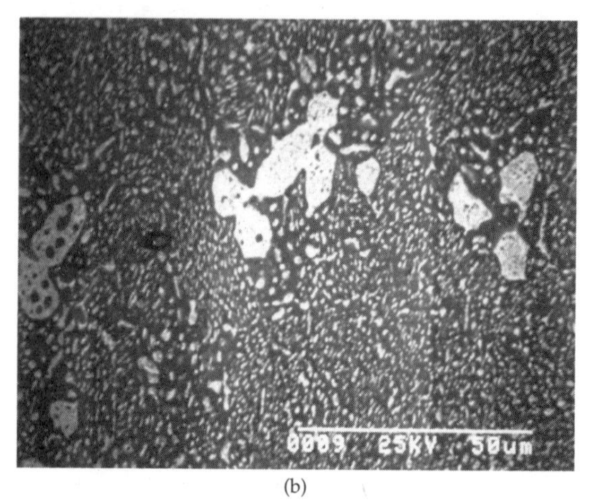

(b)

Figure 6-22. SEM microstructure of 63 Sn/37 Pb on Cu solder joint after uneven cooling—near the center. (*a*) 300× (*b*) 1000×.

- Solder composition
- Metallurgical reaction between solder and substrate metal

Microstructure serves as an indicator of internal change. Overaging is often reflected in grain coarsening in the interior of the solder joint and new phase formation in the interior of the solder joint and/or at the interface.

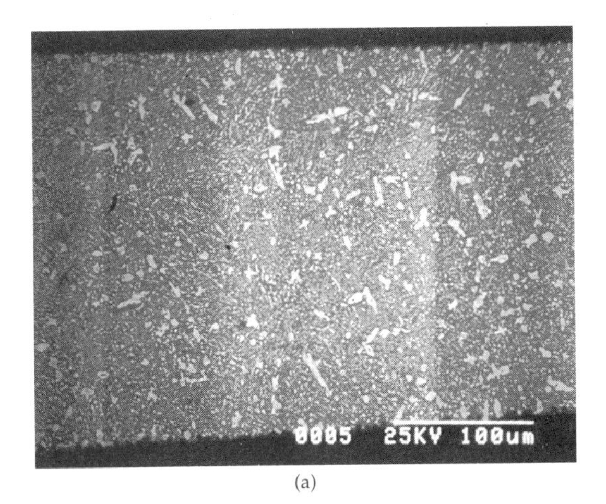

(a)

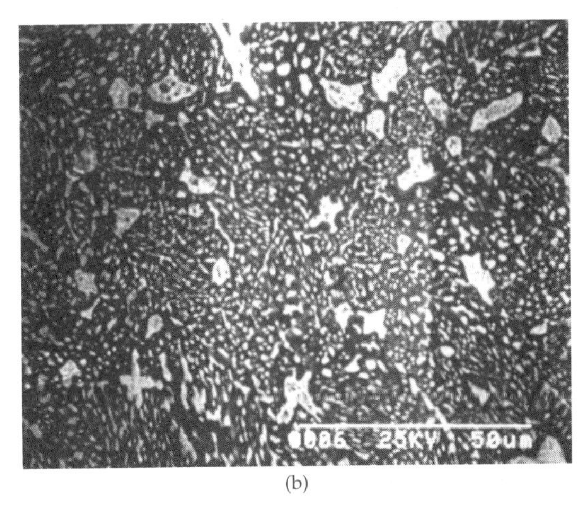

(b)

Figure 6-23. SEM microstructure of 63 Sn/37 Pb on Cu solder joint after uneven cooling—near the free surface. (*a*) 300× (*b*) 1000×.

6.6 SEM Micrographs

Microstructures of solder joints which are formed on Cu substrate for common solder systems are exemplified in the indicated figures.

Sn-Pb system

70 Sn/30 Pb	Fig. 6-24
60 Sn/40 Pb	Fig. 6-25

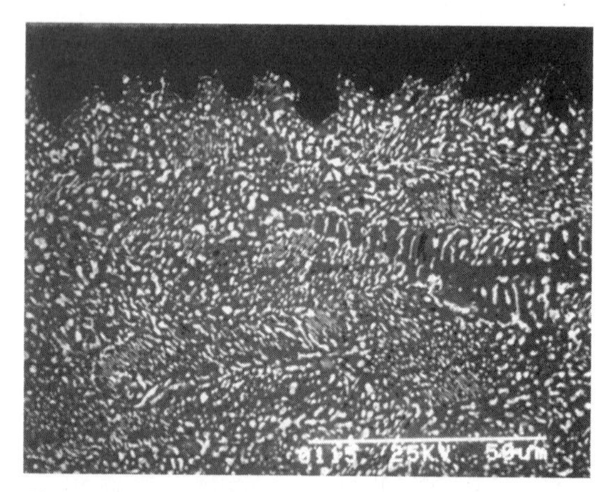

Figure 6-24. SEM microgram of 70 Sn/30 Pb.

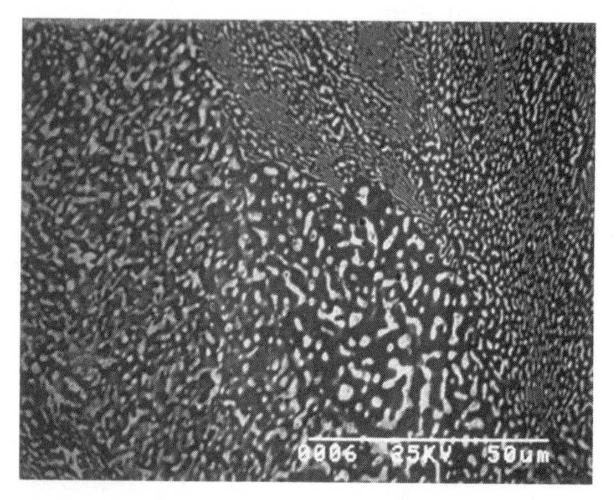

Figure 6-25. SEM microgram of 60 Sn/40 Pb.

50 Sn/50 Pb	Fig. 6-26
40 Sn/60 Pb	Fig. 6-27
10 Sn/90 Pb	Fig. 6-28
5 Sn/95 Pb	Fig. 6-29

The near eutectic composition 60 Sn/40 Pb exhibits eutectic colonies as shown in Fig. 6-25, resembling 63 Sn/37 Pb (Fig. 6-17). As lead content increases, the proeutectic lead-rich phase (light globules) merges and

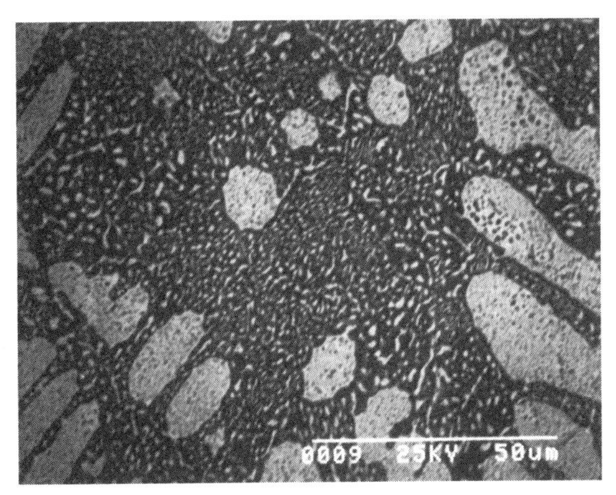

Figure 6-26. SEM microgram of 50 Sn/50 Pb.

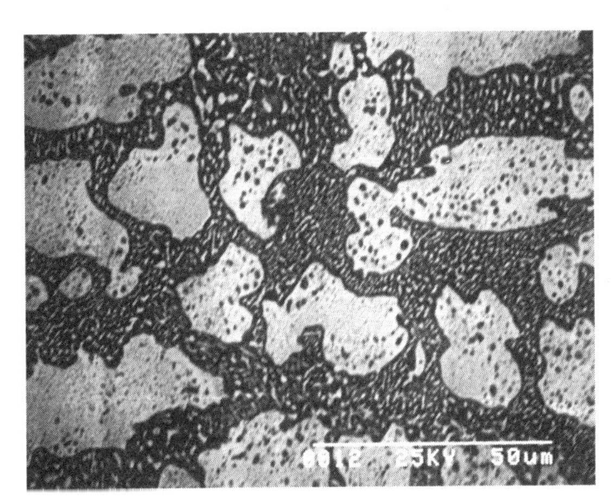

Figure 6-27. SEM microgram of 40 Sn/60 Pb.

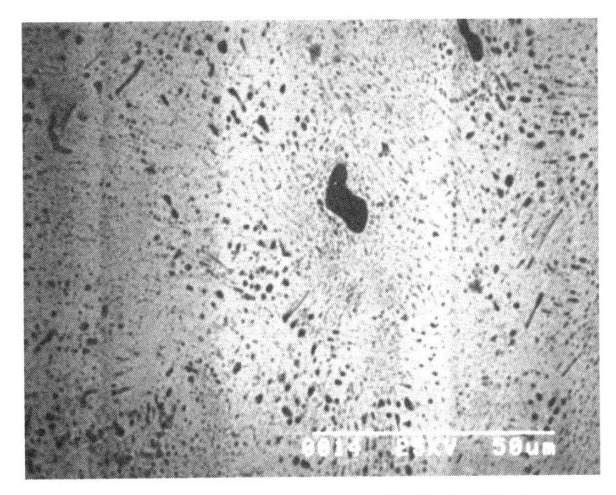

Figure 6-28. SEM microgram of 10 Sn/90 Pb.

Figure 6-29. SEM microgram of 5 Sn/95 Pb.

grows at the expense of eutectic matrix as exhibited in Figs. 6-26 and 6-27. The microstructures of high-lead compositions are generally composed of a primary lead-rich phase with tin precipitates throughout the matrix. For 10 Sn/90 Pb and 5 Sn/95 Pb, the microstructure becomes noncharacteristic with lead phase predominating under SEM backscattered electron imaging.

Sn-Pb-Ag system

62 Sn/36 Pb/2 Ag   Fig. 6-30

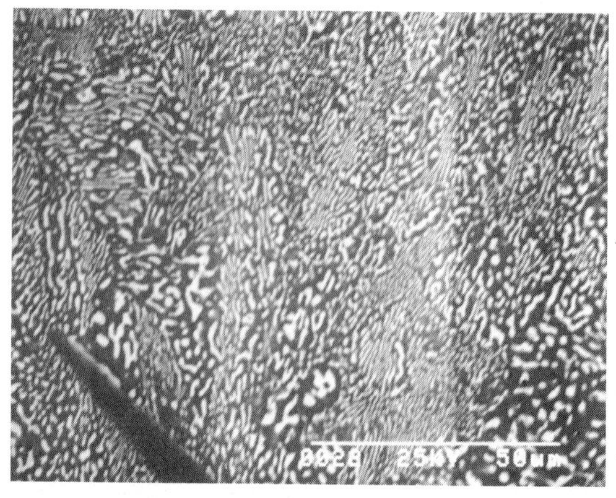

Figure 6-30. SEM microgram of 62 Sn/36 Pb/2 Ag.

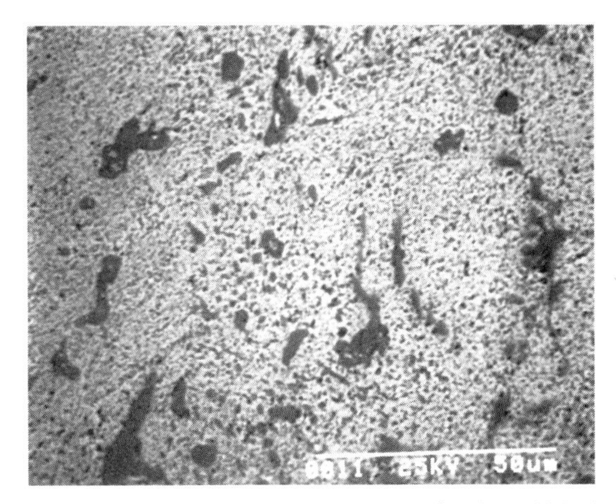

Figure 6-31. SEM microgram of 15 Sn/82.5 Pb/2.5 Ag.

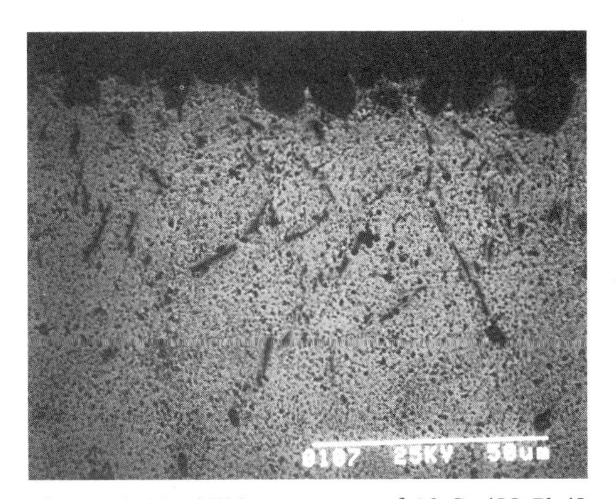

Figure 6-32. SEM microgram of 10 Sn/88 Pb/2 Ag.

15 Sn/82.5 Pb/2.5 Ag	Fig. 6-31
10 Sn/88 Pb/2 Ag	Fig. 6-32
5 Sn/92.5 Pb/2.5 Ag	Fig. 6-33
1 Sn/97.5 Pb/1.5 Ag	Fig. 6-34

The 62 Sn/36 Pb/2 Ag composition displays the eutectic structure. For the 15 Sn/82.5 Pb/2.5 Ag composition, the dark phase is identified as primarily Sn-Ag intermetallics and the Sn-rich phase is as indicated in x-ray analysis of Fig. 6-35.

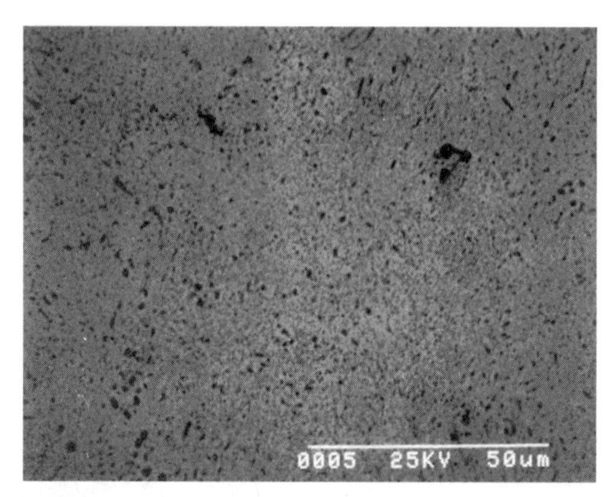

Figure 6-33. SEM microgram of 5 Sn/92.5 Pb/2.5 Ag.

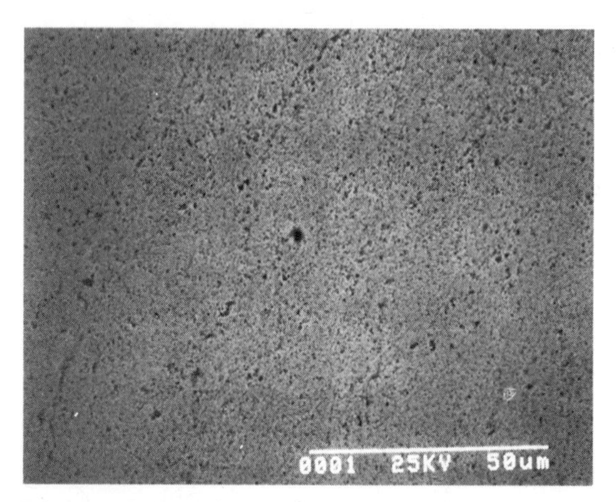

Figure 6-34. SEM microgram of 1 Sn/97.5 Pb/1.5 Ag.

Sn-Bi system

42 Sn/58 Bi Fig. 6-36

A characteristic eutectic structure for the eutectic composition 42 Sn/58 Bi is shown in Fig. 6-36 with the light phase being Bi-rich and the dark phase being Sn-rich.

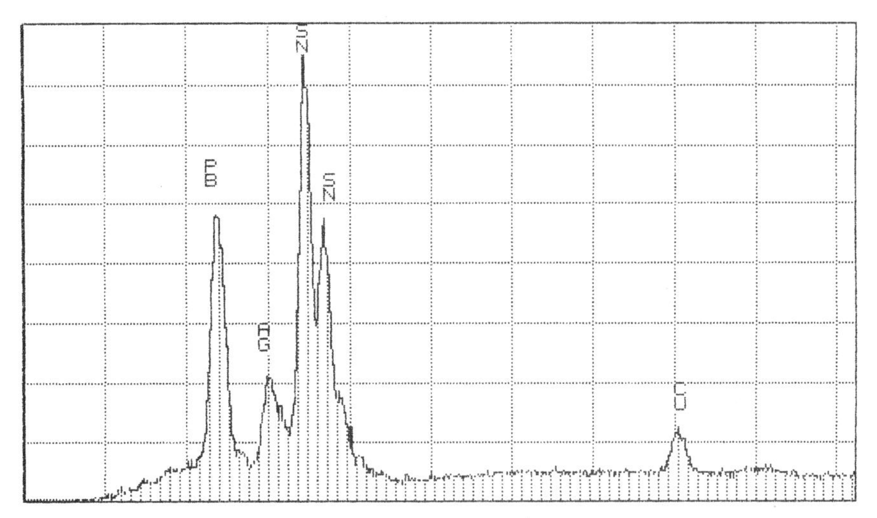

Figure 6-35. X-ray analysis of dark phase in 15 Sn/82.5 Pb/2.5 Ag microstructure.

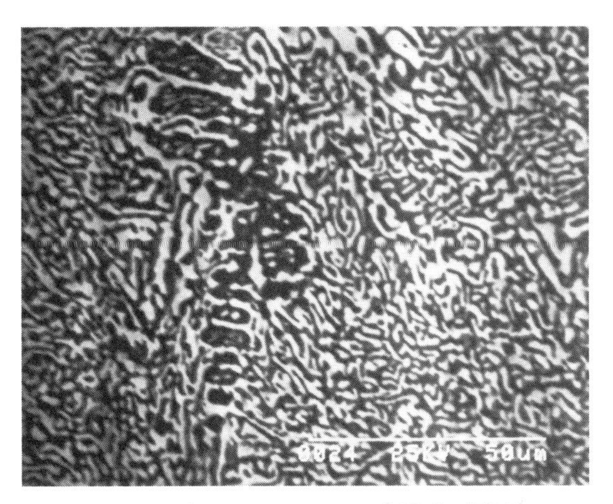

Figure 6-36. SEM microgram of 42 Sn/58 Bi.

Sn-Pb-Bi system

43 Sn/43 Pb/14 Bi	Fig. 6-37
15.5 Sn/32 Pb/52.5 Bi	Fig. 6-38

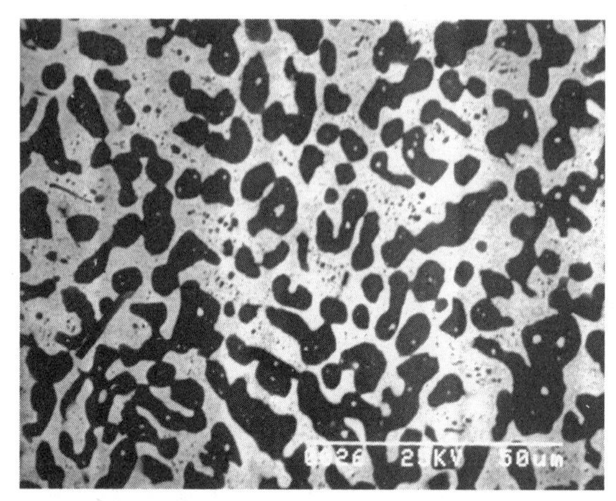

Figure 6-37. SEM microgram of 43 Sn/43 Pb/14 Bi.

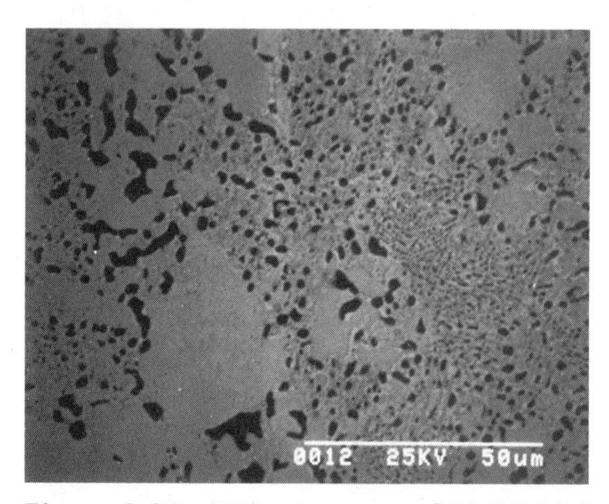

Figure 6-38. SEM microgram of 15.5 Sn/32 Pb/52.5 Bi.

Figure 6-36 of 43 Sn/43 Pb/14 Bi exhibits an Sn-rich phase in the dark area and a mix of the Pb-rich phase and the Bi-rich phase in the light area. The matrix in the gray shade of Fig. 6-38 for the 15.5 Sn/32 Pb/52.5 Bi composition is primarily the Bi-rich phase and the dark precipitates are the Bi-Sn–rich phase.

Sn-Ag and Sn-Sb system

Microstructures of Sn-Ag and Sn-Sb containing a minor percentage of Ag or Sb under SEM backscattered electron imaging are noncharacteristic.

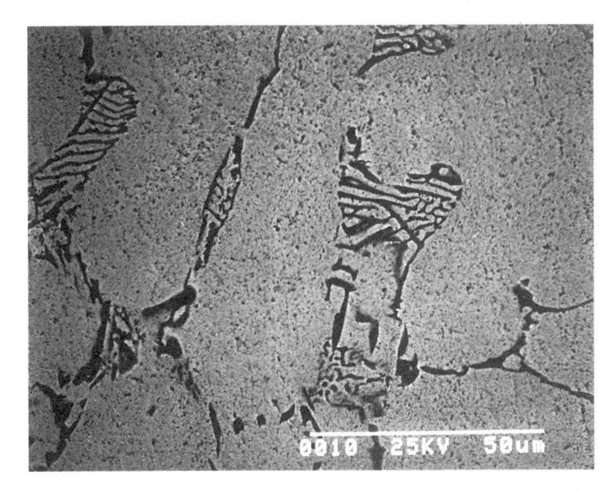

Figure 6-39. SEM microgram of 3.5 Sn/89.5 Pb/7 Sb.

Sn-Pb-Sb system

35 Sn/89.5 Pb/7 Sb	Fig. 6-39

Sn-In system

30 Pb/70 In	Fig. 6-40

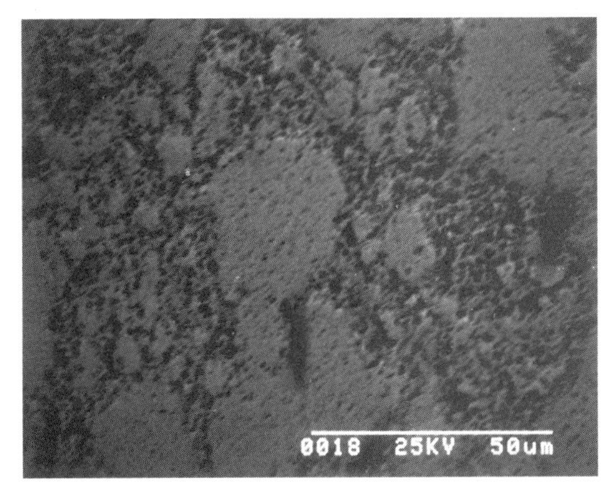

Figure 6-40. SEM microgram of 52 Sn/48 In.

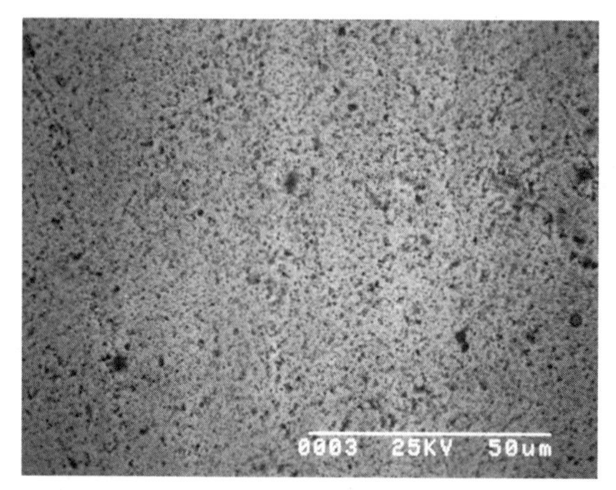

Figure 6-41. SEM microgram of 30 Pb/70 In.

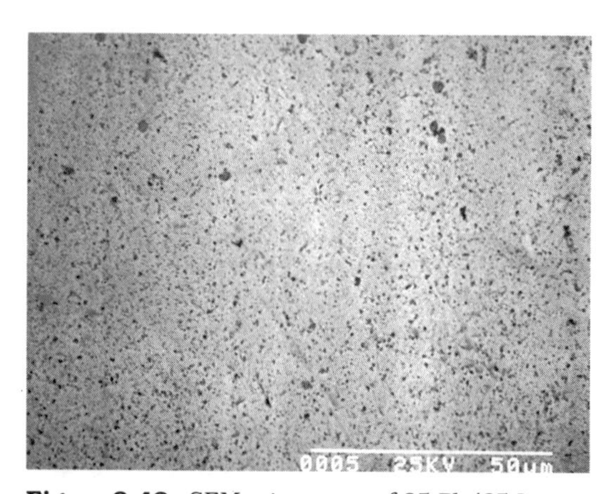

Figure 6-42. SEM microgram of 35 Pb/65 In.

Pb-In system

30 Pb/70 In	Fig. 6-41
35 Pb/65 In	Fig. 6-42
40 Pb/60 In	Fig. 6-43

Compositions of Pb-In due to their mutual solubility primarily form a Pb-In solid solution.

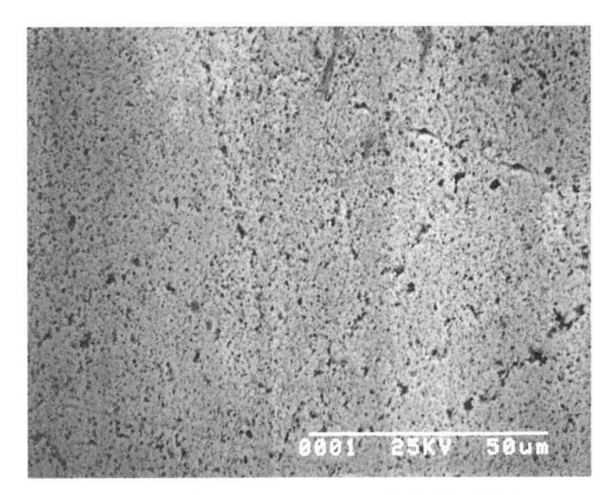

Figure 6-43. SEM microgram of 40 Pb/60 In.

Sn-Pb-In system

37.5 Sn/37.5 Pb/25 In   Fig. 6-44

X-ray analysis of Fig. 6-45*a* and *b* indicates that the light domain contains Pb and Sn and the dark matrix contains Sn. Presumably, Sn and In are indistinguishable in this analysis; the Pb solid solution with Sn and In constitutes the light domain, and the Sn-rich phase constitutes the dark matrix domain.

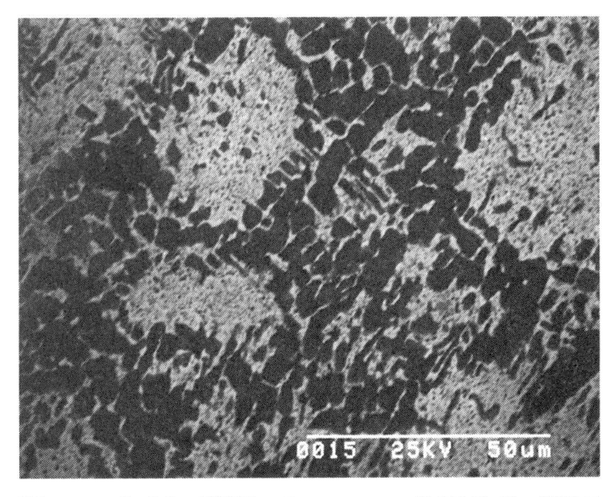

Figure 6-44. SEM microgram of 37.5 Sn/37.5 Pb/25 In.

PEAK LISTING

	ENERGY	AREA	EL. AND LINE
1	1.482	353	AL KA OR BR LA?
2	2.381	6242	PB MA
3	3.280	1560	K KA
4	3.491	622	SN LA
5	9.152	280	UNIDENTIFIED

NCSU Analytical Instrumentation Facility TUE 05-DEC-:2 04:16
Cursor: 0.000KeV = 0

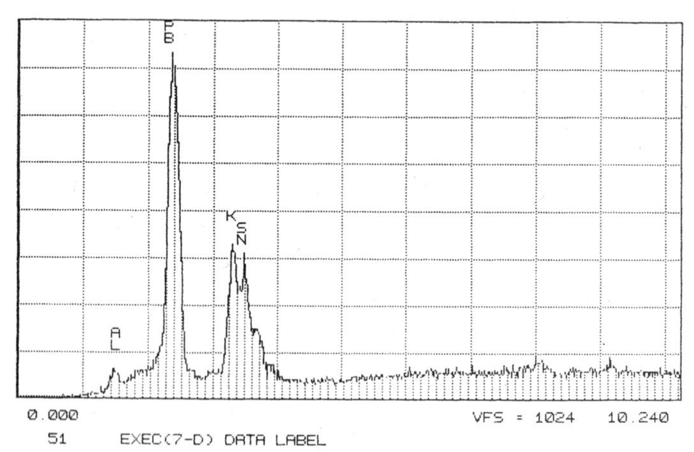

(a)

PEAK LISTING

	ENERGY	AREA	EL. AND LINE
1	3.443	12998	SN LA
2	3.702	4200	SN LB

NCSU Analytical Instrumentation Facility TUE 05-DEC-:2 04:13
Cursor: 0.000KeV = 0

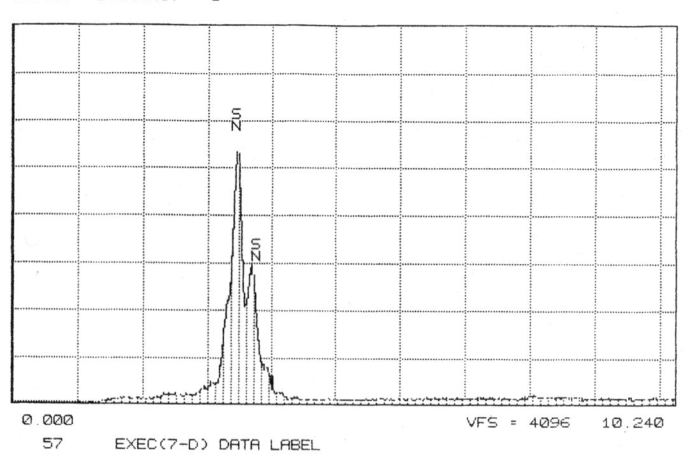

(b)

Figure 6-45. X-ray analysis of 37.5 Sn/37.5 Pb/25 In solder: (a) light domain; (b) dark domain.

6.7 Fractography

Microstructure study of fracture surfaces of 63 Sn/37 Pb solder joints corresponding to the early and later (faster) stage of isothermal low-cycle crack growth provides information about the failure mode and mechanism. Through secondary electron signals as shown in Fig. 3-43 (a-1), the early low-cycle fatigue crack growth is irregular and rough with many secondary cracks. Examination of the backscattered electron image [Fig. 3-43 (a-2)] reveals that these rough facets reconcile with the dimension of eutectic colonies and that the secondary cracks are associated with the colony boundaries. This suggests that fatigue crack units or plastic zones at the tip of fatigue cracks are limited within the scale of the eutectic colony. At the later stage of fatigue crack growth, the fracture surfaces are flat with shear-deformed morphology as shown in Fig. 3-43 (b-1), and fatigue cracks propagate transgranularly as exhibited in Fig. 3-43 (b-2). This indicates that the plastic zone at the tip of the fatigue crack is intense and on a larger scale than the colony structure, when the fatigue crack grows longer at the later stage. This results in crack propagation without the preference of microstructure. Another feature to be noted is that large primary Pb-rich dendrites appear on the fracture surface [Fig. 3-43 (b-2)], indicating that the fatigue crack paths are indeed through the coarsened bands near the Cu-solder interfaces.

6.8 Importance of Microstructure

Although it is generally accepted that the faster cooling rate creates a finer-grain structure in bulk solders, this general rule is often complicated by the interfacial boundary and metallurgical reaction at the interface of the solder joints. The nature of the substrate and its metallurgical affinity to solder composition can affect the development of the microstructure of the solder joint. It would not be a surprise to see the microstructure of a 63 Sn/37 Pb solder joint interfacing with Ni substrate differ from that with Cu substrate

If we assess the mechanical properties of a solder joint by the commonly established techniques, then shear strength, creep, isothermal low-cycle fatigue, and thermomechanical properties are the top four parameters. For the eutectic solder composition, the shear strength of the solder joint is improved by a very slow cooling rate which results in the formation of a near-equilibrium lamellar eutectic structure. However, on the other hand, the strength is also enhanced by using a very fast cooling rate as a result of grain size refining. For plastic deformation under creep mode, creep resistance depends on the operating mechanism. When the lattice or vacancy diffusion process is the predominating step, creep resistance is often lower with finer microstructure. This is due to the increased vacancy concentration as a result of the faster cooling rate. Under an isothermal fatigue environment, the relation of microstructure and fatigue resistance is more complex. Nonetheless, the homogeneity in the microstructure is most important to low-cycle fatigue resistance. With thermal cycling, the increased fatigue resistance is often associated with the decreased size of grains (phases).

For solder joints, the two most information-revealing parameters are elemental composition and microstructure. For a given solder composition, the microstructure in the form of a quality microgram is "a picture that is worth a thousand words." It provides sights and insights into the state of solder joint integrity.

6.9 References

1. Joseph Goldstein, "Metallography—A Practical Tool for Correlating the Structure and Properties of Materials," *ASTM Special Technical Publication 557*, ASTM, 1974, p. 86.
2. Heinrich, *X-Ray Optics and Microanalysis*, Plenum Press, N.Y., 1966, p. 159.
3. Jennie S. Hwang and Zhenfeng Guo, "Reflow Cooling Rate vs. Solder Joint Integrity," *Proceedings, NEPCON-West*, 1994, p. 1095.

6.10 Suggested Readings

Goldstein, Joseph I., and Harvey Yakowitz, eds., *Practical Scanning Electron Microscopy*, Plenum Press, New York, 1975.

International Tin Research Institute, Publications on Solder Microstructure, 1992.

7 Solder Paste Technology

7.1 General Description

Solder paste, by one definition, is a homogeneous and kinetically stable mixture consisting of three functional components—solder alloy powder, flux system, and vehicle system—which is capable of forming metallurgical bonds at a given set of soldering conditions and can be readily adapted to automated production in making reliable solder joints. Each of the three components serves a distinct function:

Solder alloy. Forms permanent metallurgical bonds between metallic surfaces

Vehicle. Provides suspending power for alloy particles and designated paste rheology

Flux. Cleans surfaces to be joined and solder alloy particles to form metallic continuity, and thus good metallurgical bonding

In addition, the flux and vehicle must be compatible with and be able to complement each other to form a shelf-stable matrix capable of delivering effective fluxing.

Both the vehicle and flux are fugitive in nature at the completion of the soldering process, either partially escaping during the heating (reflow) stage through volatilization, decomposition, and reaction;

being left as inactive residue; or being removed during the subsequent cleaning step. Nevertheless, they are crucial to the formation of a reliable, permanent bond. On a permanent basis, the alloy powder part is the only functional component in the final metallurgical bond.

Solder paste, because of its deformable viscoelastic form, can be applied in a selected shape and size and can be readily adapted to automation. Its tacky characteristic provides the capability of holding parts in position without additional adhesives before the permanent metallurgical bond is formed. The metallic nature of solder paste offers relatively high electrical and thermal conductivity. The combined features in adoption-to-automation, tackiness, and high conductivity make solder paste the most viable material for surface mount assembly manufacturing. It provides electrical, thermal, and mechanical interconnections for electronic packages and assemblies. The following sections summarize the performance and quality control parameters, as well as key chemical and physical properties. For complete coverage on the subject of solder paste, readers are referred to Ref. 1.

7.2 Multidisciplinary Technology[1]

Paste technology is an interplay of several scientific disciplines as depicted in Fig. 7-1. From the concept of paste technology, many com-

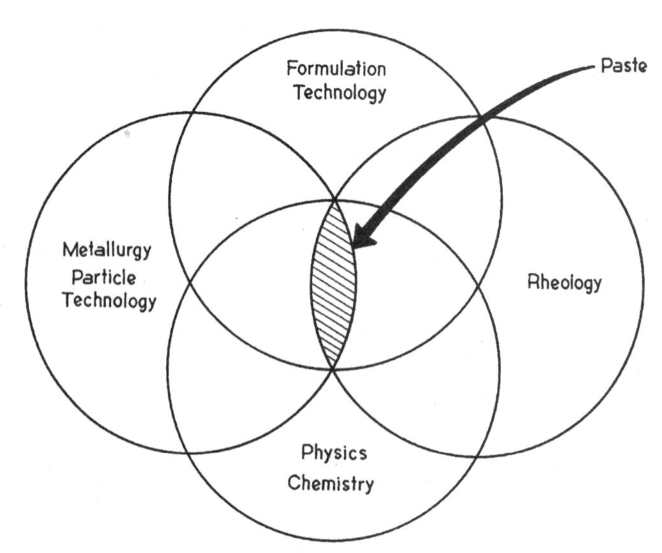

Figure 7-1. Paste technology.

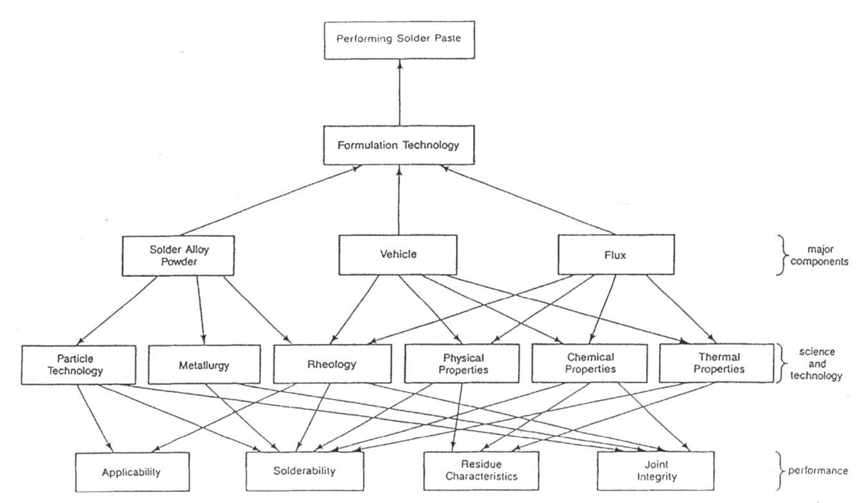

Figure 7-2. Solder paste technology correlating performance and scientific disciplines.

mercial product lines are derived. These include thick-film materials, polymer thick-film products, conductive adhesives, electromagnetic interference (EMI) shielding materials, brazing pastes, and other products that are composed of metallic or oxide particles uniformly distributed and imbedded in the organic/polymeric matrix. Each of these product lines has its unique performance requirements and processing parameters in its making process and its end-use function. However, one thing in common is paste technology.

Figure 7-2 is intended to summarize solder paste technology in one chart. It is hoped that this chart reflects the essence of solder paste technology by displaying the correlation of starting components and final performance of solder paste, and the role of fundamental sciences and technologies.

In addition to metallurgy and particle technology, the physical, chemical, thermal, and rheological properties of vehicle/flux systems, are equally important to the performance and characteristics of the resulting paste. Furthermore, what ingredients are to be composed (i.e., formulation) and how the formula is put together (i.e., processing) are crucial to the properties and parameters of vehicle/flux systems. With respect to performance, solder paste is categorized into four areas: applicability, solderability, residue characteristics, and joint integrity. *Applicability* refers to the ability of a paste to be adapted to a specific paste-applying technique such as dot-dispensing, screen printing, or stencil printing. *Solderability* is intended to be used in a broad sense—the ability of a paste to wet the surfaces to be joined with complete coalescence of solder powder particles and to achieve a reliable metallurgical bond. Solderability is discussed in Chap. 5. Residue characteristics cover the physical and chemical properties of the resulting chemical mixture after soldering (e.g., corrosivity, activity, tackiness, hardness, compatibility with cleaning process). Solder joint integrity is the ultimate performance of the solder joint after the soldering process in terms of mechanical properties, resistance to adverse environment, and its compatibility with service conditions. Each of these areas is discussed in subsequent chapters.

A working paste under real-world conditions is quite complex in nature. Its complexity and variability are further augmented by the dependency of performance parameters upon the variables of paste handling and the soldering process. With this in mind, it is apparent that an understanding of the fundamental technologies involved is a necessity, as indicated by the interconnecting lines of Fig. 7-2. Through the understanding of metallurgy, the solder alloy is selected with the consideration of solderability and joint integrity. Through particle technology, the size distribution, the shape, and the morphology of the alloy powder are considered to formulate the desired paste applicability and solderability.

Through a good understanding of chemistry, the chemical properties of solder paste (e.g., reactivity with solder powder and surfaces to be joined and reactivity in relation to temperature) which affect solderability, residue characteristics, and joint integrity can be better understood. In addition, the functional groups and the structure of chemicals in relation to a specific performance characteristic can be correlated and anticipated in principle. The physical properties of solder paste, including the surface and interfacial phenomena of individual ingredients and of the system as a whole, have a significant effect on the paste performance in solderability and residue properties.

The rheology, not only as a result of a designed composition of a flux/vehicle system and solder powder but also as a result of paste processing, controls the paste applicability, solderability, and even joint integrity. Thermal properties, such as stability versus temperature and reactivity versus temperature, contribute to the residue characteristics and solderability.

It is worth noting that the paste is considered mostly as being kinetically stable, rather than thermodynamically stable. This is in contrast to a true solution or to other multicomponent systems, such as microemulsions. Therefore, formulation and processing are crucial to the consistency and properties of the paste, and in most cases, even to its rheology and shelf stability.

A viable solder paste should be constituted by utilizing fundamental technologies in selecting starting raw materials, in anticipating the interactions among these raw materials, and in understanding interrelations between starting materials and end-use performance. Solder paste should be handled and used with an understanding of its characteristics, techniques, and technologies involved to assure its optimum performance.

7.3 Performance Parameters

Performance parameters can be grouped into three states: paste state, soldering state, and postsoldering state. Important parameters to be considered are listed for each state.

Paste state

- Paste stability and shelf life
- Dispensability through fine needle
- Screen printability
- Stencil printability
- Tack time
- Exposure life
- Quality and consistency

Soldering state

- Compatibility with substrates
- Solder balling phenomenon
- Wicking phenomenon
- Dewetting phenomenon
- Wettability
- Flow property before molten
- Residue corrosivity
- Residue cleanability

Postsoldering state

- Solder joint appearance
- Solder joint voids
- Solder strength
- Solder joint microstructure
- Solder joint integrity versus mechanical fatigue/creep
- Solder joint integrity versus thermal fatigue/creep

7.4 Chemical and Physical Properties

The chemistry and chemicals related to soldering are discussed in Chap. 4. In addition to chemical properties, physical properties of the flux and vehicle system also affect the solder paste performance. The basic physical properties to be considered include

- Melting point
- Boiling point
- Softening point
- Glass transition temperature
- Vapor pressure
- Surface tension
- Viscosity
- Miscibility and dispersibility

All these properties are more or less controlled by intermolecular forces, although they are not of great importance in the formation of chemical bonds. For example, the tendency of a molecule to volatilize from a liquid is a function of a total transitional energy which in turn depends on temperature. The boiling point is dependent on the relation of such transitional energy and the cohesive energy as a result of intermolecular interactions. For a polymer, its molecular weight plays an important role in the occurrence of decomposition or volatilization as temperature rises; it is also important to the tackiness, viscosity, and rheology of the resulting system. The viscosity of a polymer in solution is a function of molecular weight and its distribution in addition to other characteristics of polymer and solvent being used. The viscosity is related to molecular weight by

$$\eta = KM^a$$

where K and a are empirical constants and M is the molecular weight.[2] For chemicals in solid form, the melting point, in addition to cohesive energy, is influenced by the orderliness of molecules which in turn is expressed as entropy. Although the relation of melting point and boiling point to chemical structure is complicated, in general, the higher boiling point is associated with a higher melting point, and symmetrical molecules melt at higher temperatures due to the low entropy of fusion, when other factors are the same. The melting point of polymers generally increases with increasing molecular weight. The softening point also increases with molecular weight and increased crystallinity. For amorphous polymers, the softening temperature is near the glass transition temperature, whereas for highly crystalline polymers it is close to the melting point. The glass transition temperature T_g and the melting point T_m are normally related as follows:

For unsymmetrical polymers $$\frac{T_g}{T_m} \cong \frac{2}{3}$$

For symmetrical polymers[3] $$\frac{T_g}{T_m} \cong \frac{1}{2}$$

In solder paste, the vapor pressure of every ingredient affects more than one aspect of paste performance, including the selection of reflow method and the void development in the solder joint. The boiling point of the liquid phase as a whole or of an individual liquid ingredient can alter the flux activity and the residue characteristics. It may also affect the compatibility with a specific reflow method. The melting point or softening point of the solid phase as a whole or of individual solid ingredients has a direct impact on the flux activity.

7.5 Rheological Flow Property

The flow property, which is practically expressed as cold slump or hot slump, is a major contributor to a common manufacturing defect—pad bridging. In addition to paste deposition performance during the dynamic printing or dispensing operation which will be discussed in Chap. 10, the slump or not-to-slump behavior is an important paste property. Ideally, the paste is desired to retain the original shape after deposition before the formation of the final solder joint. This requires that the paste have a zero cold slump under ambient temperature between the printing/dispensing step and reflow as well as a zero slump when the temperature rises during reflow. Figure 7-3 illustrates the relative flow property of two pastes. Paste 1 (left side of the coupons) retained its shape when the reflow process started at t_0 and continued to heat up through t_1, t_2, and the t_3 and to molten stage until the solder solidified at t_s. Paste 2, however, spread in its area as the temperature rose.

It should be noted that spreading in the paste state and in the molten state is dominated by different driving forces. In the paste state, the interparticle forces are dominating, and in the molten solder

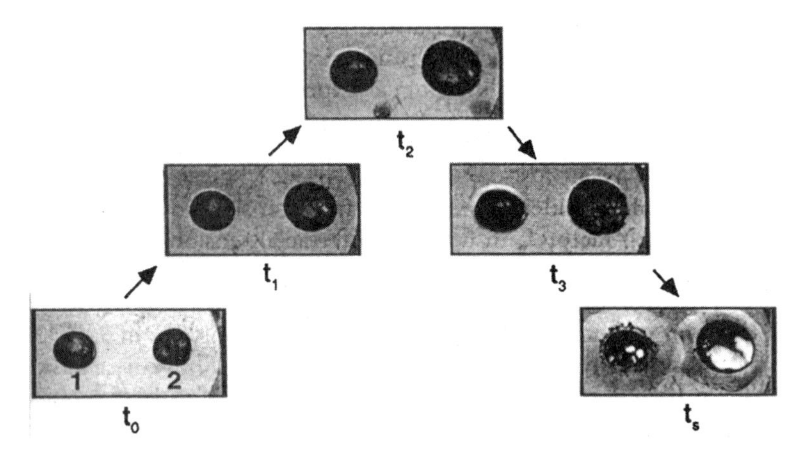

Figure 7-3. Paste slump comparison during reflow.

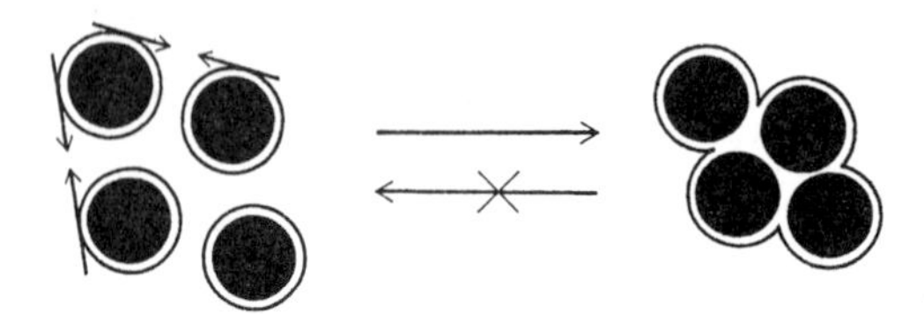

Figure 7-4. Paste slump versus cohesive force between particles.

state, the relative interfacial tension (force per unit area) between the flux/vehicle liquid, substrate, and molten solder, governs the spreading. Therefore, the objective is to design a flux/vehicle system capable of intimately wetting the surface of solder particles while providing adequate surface tension in the liquid phase so that the solder particles are imbedded in the matrix of the flux/vehicle system with a high cohesive force, as illustrated in Fig. 7-4. When the cohesive force F_c in such a system is not overcome by the combination of gravity force F_g and thermal disturbance F_t, the paste is able to retain its shape without slumping, as expressed by

$$F_c > F_g + F_t$$

In the molten solder state, the interfacial tension dominates, and the spreading factor S, in terms of interfacial tensions, may be expressed as follows:

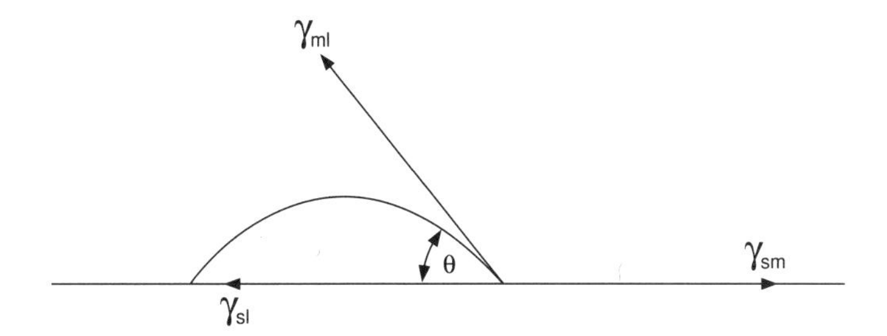

$$S = \gamma_{sl} - (\gamma_{sm} + \gamma_{ml} \cos \theta)$$

where γ_{sl} = interfacial tension of flux/vehicle and substrate
γ_{sm} = interfacial tension of molten solder and substrate
γ_{ml} = interfacial tension of molten solder and flux/vehicle

To illustrate the effect, the difference in the spreading factor between a system A (S_A) and a system B (S_B) using Antonow's approximation is

$$S_A - S_B = [\gamma_{sl} - (\gamma_{sm} + \gamma_{ml} \cos \theta)]_A - [\gamma_{sl} - (\gamma_{sm} + \gamma_{ml} \cos \theta)]_B$$

$$\cong (\gamma_{lB} - \gamma_{lA})(1 - \cos \theta)$$

where γ_{lB} = surface tension of liquid (l) for system B
γ_{lA} = surface tension of liquid (l) for system A.

Thus, when

$$\gamma_{lB} > \gamma_{lA}$$

$$\therefore S_A > S_B$$

and when

$$\gamma_{LB} < \gamma_{LA}$$

$$\therefore S_A < S_B$$

The spreading factor in a system with a relatively higher surface tension is smaller than in one with a lower surface tension.

Since most solder pads, as on PCBs or on hybrid circuits, are individual pads surrounded by an unwettable surface (solder mask or bare alumina surface), it is perceived that the molten solder should have retracted back to its own pad after reflow due to its high surface tension. This perception does not always hold as in the cases where the solder volume exceeds the tolerance of the pad area or where a component exerts sufficient weight on the solder. Again, the balance of the forces is the criterion. Furthermore, when we examine the spread phenomenon which causes pad bridging or binding between two solder joints, the primary factor is a result of hot slump in the paste state before the solder melts. *Hot slump* is defined as the shape change when the paste sinks in height and expands in area in response to a temperature rise before the solder melts. As the paste deposit between adjacent pads is in physical contact due to hot slump, it is likely to cause solder bridging between pads. The likelihood of this depends on the volume of solder applied in relation to the pad area and the spacing between pads, as well as on the weight of the component exerted on the paste deposit per unit area. Thus, as the circuitry becomes more intricate and shrinks in size, it is more prone to bridging problems.

By monitoring the thermodynamic parameters of the flux/vehicle system, the flow property can be controlled.[4] It is indicated that the flow property is directly related to the surface tension of the liquid phase. Figure 7-5 indicates a correlation between the surface tension and the flow restrictivity of paste while other parameters are held constant.

The flow restrictivity as expressed in percentage is conveniently chosen as the ratio of the initial area of the paste deposit before heat is applied to the final area at the completion of heating in the paste state and in the molten state, respectively:

$$\text{Flow restrictivity} = \frac{\text{initial area}}{\text{final area}} \times 100\%$$

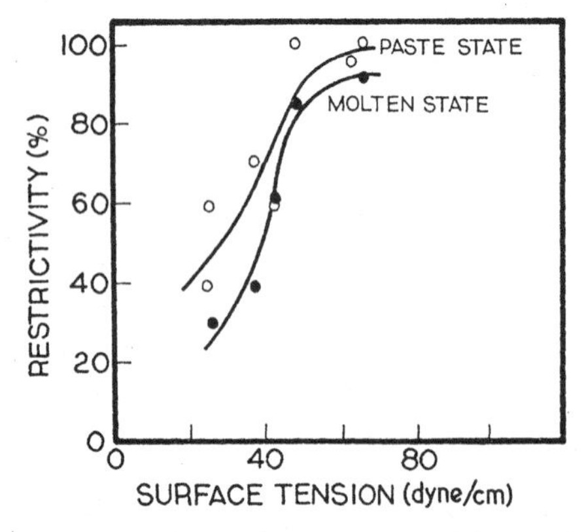

Figure 7-5. Surface tension of flux/vehicle system versus paste flow restrictivity.

Table 7-1. Effect of Surface Tension of Flux/Vehicle System on Flow Restrictivity

Surface tension, dyne/cm	Restrictivity, %					
	Conduction		Infrared reflow		Vapor phase	
	Paste	Molten	Paste	Molten	Paste	Molten
66.5	96	93	95	90	100	100
26.3	59	30	54	41	57	47

To compare the systems at extreme conditions, Table 7-1 lists the data of the effect of surface tension on the flow restrictivity. The table also demonstrates the results of different reflow methods. The two systems distinguished by the thermodynamic parameter display a significant difference in flow restrictivity in all three reflow methods.

Pad bridging can be essentially eliminated by using paste with a built-in flow restrictivity. The occurrence of pad bridging in relation to flow restrictivity is shown in Fig. 7-6, indicating that high flow restrictivity reduces pad bridging.

For pastes possessing low flow restrictivity, the extent of bridging is highly sensitive to the pressure exerted onto the paste deposit and onto the print thickness. Figures 7-7 and 7-8 indicate the effects of pressure and print thickness, respectively. Therefore, in order to minimize pad bridging, the selection of the flow property in paste should be compatible with the specific devices and components to be used. Increasing print thickness further aggravates the bridging probability of pastes that do not possess the flow restrictivity. With the growing adoption of the fine-pitch pattern in surface mount technology, it is critical to use a paste possessing a high flow restrictivity.

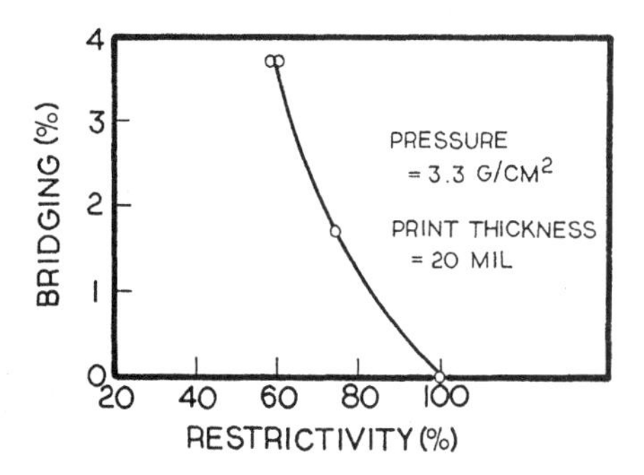

Figure 7-6. Pad bridging versus paste flow restrictivity.

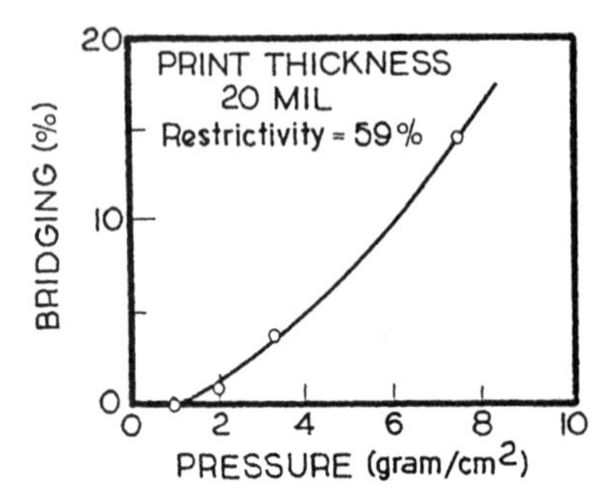

Figure 7-7. Pad bridging versus pressure applied on paste deposit.

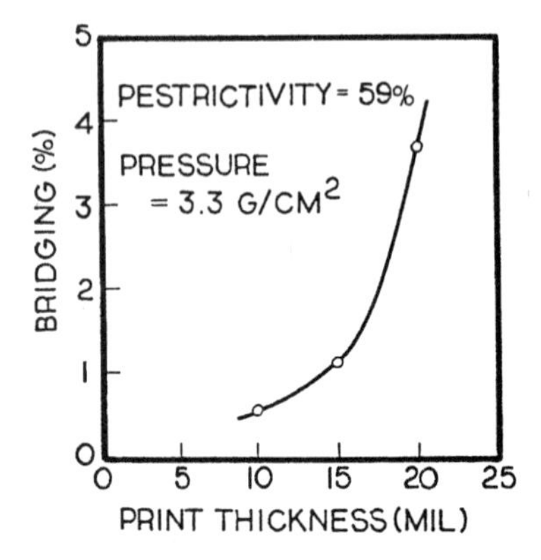

Figure 7-8. Pad bridging versus thickness of paste deposit.

Solder paste slump also contributes to both types of solder balling: large solder balls that are only associated with small capacitors and resistors and the rest of the solder balling phenomena as will be discussed in Sec. 13.3.

7.6 Formulation

Four states are involved in the formulation of a solder paste.

- *Design stage.* Utilizing fundamental technologies in chemistry, physics, and materials science for the targeted performance
- *Formulation state.* Making up a formula through a systematic and meticulous process

- *Fine-tune stage.* Working for the best balance and/or the specific performance
- *Scale-up stage.* Setting up a process to produce the formula consistently and reproducibly

The first step in formulating a paste product is to define the performance objective to be achieved. With a clear objective in mind, a paste can be designed to meet the performance parameters by utilizing the fundamental technologies, by understanding how the selected raw materials contribute to the characteristics of a paste, and by anticipating any synergistic or antagonistic interactions among raw materials. A product involves many performance parameters, and some of them may be tradeoffs. For example, high metal content is beneficial to the solder joint in volume, void, and residue, yet may make the paste more prone to crusting and difficult to apply. A highly viscous paste may improve flow control against temperature, yet causes the paste to be difficult to apply. Using highly active fluxing chemicals may improve solderability in some cases, yet its use may leave more corrosive residue. In such a case, improving the solderability by selecting proper ingredients without the use of highly active fluxing chemicals is a skill. Increasing flux content does not always improve solderability in terms of wetting and/or solder balling elimination.

The ability to achieve an ultimate balance among the performance parameters and to prioritize them for a specific group of applications is the key. Once a prototype formula capable of providing all functions is constituted, the next step is to fine tune the formula to coincide with specifications or any specific values designated.

With the accomplishment of the development of a specific composition formula, it is equally important to develop a reproducible process for making a paste with consistent performance. It is not an exaggeration, but an indication of the importance of the role of the process, to state that the identical composition of a formula can produce different products when the process is allowed to vary. The required skill is to be able to utilize the technologies to make the best balance of all desired performance parameters and to meet the performance requirements in the real world, as well as to develop a controlled process in making end-use products.

As an example, a typical rosin mildly activated (RMA) solder paste formula may contain 12 ingredients which provide various designated functions as shown in Table 7-2.

For designing water-soluble pastes, the primary distinction from traditional RMA grade is to make the after-reflow residue water-cleanable under prevailing manufacturing environments. The key to formu-

Table 7-2. Typical System of RMA Paste

Ingredient	Function
Rosin(s)	Rosin system for designated softening point, acid number, thermal stability, fluxing, tackiness
Nonhalogen activator(s)	Activator system for achieving fluxing action over the temperature range of interest and for desired rheology
Solvent(s)	Solvent system to provide solubility, rheology, temperature, and chemical compatibility
Binder(s)	Providing desired viscosity, rheology, and tackiness
Fluxing modifier	Stabilizing and modifying flux
Rheology modifier	Contributing to targeted rheology

lating a successful water-soluble paste is to have an intact chemical matrix that is water-cleanable, but the composition does not necessarily comprise ingredients that are all water-soluble.

The unique demand to no-clean paste is to minimize the amount of after-reflow residue. Two logical approaches are (1) to minimize the total solid contents in a flux/vehicle system and (2) to maximize the metal load in the paste. The solid content of the flux/vehicle system is, to a certain extent, related to the fluxing efficiency which commonly reflects in the amount of solder balls occurring during reflow. As noted before, fluxing efficiency is not solely related to the amount of fluxes. The rest of the chemical system which serves as a protecting barrier from ambient air during reflow also plays an important role. The relation of solid content in the paste to atmosphere is covered in Chap. 9. The acceptability of the metal load in the flux/vehicle system depends on

- Physical and chemical nature of the flux/vehicle system
- Metal particle size and distribution
- Surface condition of the metal particles
- Process of paste-making

Formulating a good solder paste not only requires knowledge and experience but also demands a high level of patience and good instinct.

7.7 References

1. Jennie S. Hwang, *Solder Paste in Electronics Packaging—Technology and Applications for Surface Mount, Hybrid Circuits and Component Assembly,* Von Nostrand Reinhold, New York, 1989, Chaps. 2 and 3.
2. Paul J. F. Flory, "Molecular Weights and Intrinsic Viscosities of Polyisobutylenes," *J. Am Chem. Soc.*, vol. 65, 1943, pp. 372–382.
3. Boyer, *J. Appl. Phys.*, vol. 25, 1954, p. 825.
4. Jennie S. Hwang and Ning-Cheng Lee, "A New Development in Solder Paste with Unique Rheology for Surface Mounting," *Proc. Int. Symp. Microelectr. (ISHM)*, 1985.

8 Aqueous-Clean and No-Clean Manufacturing

Issues and concerns for environmental and health safety have been predominant since the beginning of the decade. Ozone depletion caused by CFCs tops the list of concerns. The subject has received worldwide attention and prompted concerted efforts to safely replace the ozone-depleting chemicals. As part of the global effort to eliminate the use of these chemicals, the board assembly and soldering industries have examined various materials and process alternatives which will replace CFCs as the cleaning agent for component and board manufacturing. Aqueous-clean and no-clean chemistries are identified and gaining acceptance in the manufacturing sector of the industry. Based on different perspectives and performance parameters, both aqueous-clean and no-clean methods are being implemented from separate perspectives.

8.1 Aqueous-Clean Chemistry

The material/process systems involving water for cleaning include

- Water-saponification system
- Organic solvent-emulsion-water system
- Water-soluble system

The following outlines the key features which distinguish one system from the others.

8.1.1 Water-saponification system

The system consists of RMA-type flux or paste in conjunction with a water saponifier for cleaning. The cleaning medium of water is added with saponifier which converts the water-insoluble rosins or rosin derivatives into water-soluble salts as presented by the following reactions:

$$\text{R-COOH} + \text{NaOH} \rightarrow \text{RCOO}^- \text{Na}^+ + \text{H}_2\text{O}$$

$$\text{R-COOH} + \text{RNH}_2 \rightarrow \text{RCOO}^- \text{NH}_3\text{R}^+$$

The saponifier acts as a strong base with its basicity reacting with weak rosin acids, forming water-soluble salts. Most commercial saponifiers are in the category of low-molecular-weight amines.

The effectiveness of water saponification is sensitive to the amount of saponifier charged into the cleaner. It also depends on the compatibility of the saponifier with the total formula of the flux or paste being used and on the control and consistency of the level of saponifier. The recommended dosage of saponifier in water generally falls within the range of 5 to 10 wt%.

8.1.2 Organic solvent-emulsifier-water system

In this system, the flux or paste chemistry is of the RMA type, and the cleaning process involves both organic solvent and water. The solvent primarily is hydrocarbon-based with low volatility. Typical examples are

Bioact EC 7

- 90% Limonene
- 10.0% nonionic surfactant

Axarel 38

- 60–85% mixed aliphatic hydrocarbons
- 10–30% straight-chain aliphatic esters
- 1–5% branched-chain aliphatic esters
- 8–13% nonionic surfactant

- 1–5% aliphatic alcohol

The solvent system contains a surfactant which acts as an emulsifier to render hydrocarbon solvent miscible with water, leaving a clean surface after rinsing with water.

8.1.3 Water-soluble system

The system is composed of a flux or paste chemistry which is water-soluble and cleanable with plain water without requiring any additives.

8.1.4 Comparison of systems

Among the three systems of the above, their strengths and weaknesses are assessed in four areas: material requirement, soldering process requirement, cleaning requirement, and cleanliness and reliability.

In material requirement, both the water-saponification and organic solvent-emulsifier-water systems can continue using RMA-type chemistry, lending convenience and comfort due to the well-established performance and reliability of RMA chemistry. Water-soluble flux has been used in wave soldering and structural soldering for years, but its use for making surface mount assemblies is relatively new. However, with the concerted effort from material manufacturers and the collaboration between material manufacturers and users, today's water-soluble chemistry possesses similar physical and chemical properties and is capable of delivering an equivalent performance to RMA chemistry. In the cleaning process, the organic solvent-emulsifier-water system requires two-step cleaning: solvent followed by water. Users are required to handle the procedures and regulations of both solvent and water. In terms of cleanliness and reliability, when handled properly, all three systems are able to offer acceptable performance. Table 8-1 summarizes the relative features of the three systems.

Combining the performance, process, reliability, and cost, the comparison leads to an overall rating with the water-soluble system being the first choice.

8.1.5 General characteristics of solder paste

Chapter 7 discusses the general aspects of solder paste. Special notes for solder paste based on water-soluble chemistry are

Table 8-1. Comparison of Aqueous-Clean Options

	Water-soluble	Water saponification	Organic solvent-emulsifier-water
Material characteristics	Comparable to RMA, more latitude for solderability	Modified RMA	RMA
Reflow process	Essentially the same	Essentially the same	Essentially the same
Cleaning agent	Water	Water and saponifier	Solvent and water
Reliability	Essentially the same	Essentially the same	Essentially the same
Performance/cost preference	First choice	Third choice	Second choice

- Chemistry for eutectic Sn-Pb solder provides similar performance as RMA, in some cases exceeding the RMA's capability in accommodating solderability variation.
- Chemistry for high-temperature (>250°C) soldering requires a different chemical makeup.
- Same principle and practice as established for RMA in printing apply.
- Attention should be paid to reflow temperature profile in preheating, peak temperature, and dwell time at peak temperature to avoid overheating which may cause cleaning difficulty.
- Same principle and practice as established for RMA solvent cleaning apply.
- High level of cleanliness, visually or by ionic contaminant measurement, can be achieved. For example, less than 0.18 $\mu g/in^2$ ionic contaminants was reported by means of ion chromatography.

8.1.6 Polyglycol

Some existing specifications specify that flux or paste shall contain no polyglycol materials, and defines polyglycol as materials with polyether linkages, such as polyethylene glycol, or as a material generally derived by the reaction of organic acids, amines, alcohols, phenols, or water with ethylene or propylene oxides or their derivatives. They further define that this family of materials includes, but is not limited, to polyethylene glycol, polypropylene glycol, and a wide range of polyglycol surfactants.

Under the designation of polyglycol, which is also commonly known as polyethylene glycol, polyoxyethylene, or polyether glycol, the general chemical formula is

$$HOCH_2\,(CH_2OCH_2)_n\,CH_2OH$$

Their molecular weight normally falls in the range of 200 to 6000. As their derivatives are included, the family expands to the chemicals which are normally the products of direct esterification with fatty acids, to polyethenoxy derivatives of higher fatty acids, and to polyethenoxy derivatives of higher fatty alkylolamides and higher fatty acids. This is not only a large family of chemicals but also a class of chemicals which has a great commercial importance. The practical merits of this group of chemicals are established in many fields of applications for various functions such as base solvents, plasticizers, softeners, lubricants, ointment bases for pharmaceuticals and cosmetics, and binders for industrial products.

A paste flux or solder paste, which can deliver a suitable rheology and stable enough rheological structure to meet the electronics packaging/assembly needs, is normally composed of a number of key ingredients for separate and specific functions. A surfactant system or a surfactant often cannot be spared. The family of polyglycols and their derivatives constitutes a majority of the nonionic surfactant class.

Generally, the properties of polyglycols and their derivatives vary with the number of glycol units, the hydrocarbon chain length, the end-functional groups, and the molecular weight. The nature of interfacial interaction can be physical adsorption, ionic adsorption, and chemisorption. The extent of the interfacial interaction, between polyglycols/derivatives and any substrates, depends on several factors. These are the properties of a polyglycol including the molecular structure and in situ orientation, the polarity of the substrate, the surface morphology of the substrate, the interfacial thermodynamics, and the operating conditions. Thus, the chemistry and physics of such surface interaction by a large family of chemicals is far too complex and too vast a field to be concluded by a few experiments.

With this degree of complexity, even a specific polyglycol which exhibits undesirable characteristics under a given set of conditions does not make a whole large family a culprit. Furthermore, the interacting behavior of a pure polyglycol is quite different from that where a polyglycol is incorporated in a chemical system. A proper analogy perhaps is that an ideal flux chemical in one chemical environment (paste formula *A*) may act as a poor flux in another chemical environment (paste formula *B*).

Sufficient scientific evidences to substantiate the harmfulness of the use of polyglycols and their derivatives to electronic assemblies are lacking. Users are urged to evaluate the specific chemical makeup under specified conditions to determine the product's suitability.

8.2 Aqueous-Clean Process

8.2.1 Manufacturing process

The majority of today's and future manufacturing processes in board assembly using the water-clean method may fall into one of the following production flows:

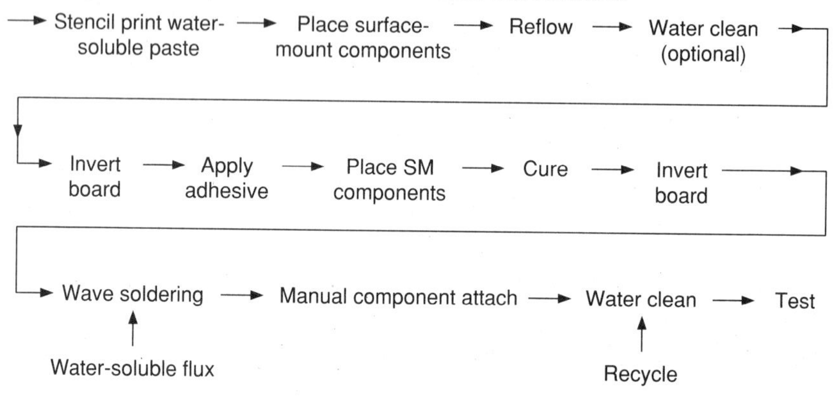

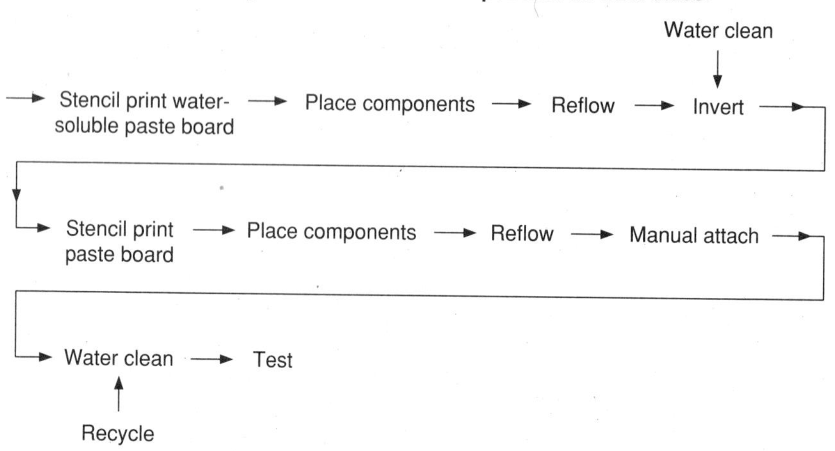

8.2.2 Equipment

Water cleaners range from small-colume batch cleaners to continuous in-line systems. Most cleaners are featured with wash, rinse, and dry functions. Some are equipped with mechanical aids to facilitate the cleaning of hard-to-remove residues and contaminants between narrow spaces. These include ultrasonic agitation with specified frequency and conditions (see Sec. 8.2.3), and centrifugal cleaning systems with rotation speeds up to 1000 revolutions per minute (r/min).[1]

8.2.3 Residue and cleaning

Residues of water-soluble chemistry which are left on or around solder joints after soldering need to be cleaned. The residues from liquid flux can be simple with a small amount of organic acid, high boiling solvents, surfactant, and reaction products. Residues from paste may contain a mixture of ingredients, including polar organics, nonpolar organics, ionic salts, and metal salts.

Key steps in a typical cleaning process are pre-rinse, wash, rinse, final rinse, drying. The parameters that affect the cleaning efficiency include

- Temperature of water
- Spray pressure
- Spray angle
- Wash time
- Flow rate
- Agitation aid

When a compromise among the parameters is needed, higher temperature and higher spray pressure often play more important roles than the flow rate and spray angle. In addition, mechanical agitation aids in dislodging the foreign matters from the board and the clearance between the component and board. Techniques, such as ultrasonic field and centrifugal energy, provide effective mechanical force. Centrifugal force, which is directly proportional to the square of angular velocity, and its parallel direction to the board assist the cleaning operation. Ultrasonic cleaning utilizes *cavitation,* which is defined as the implosion of microscopic vapor cavities within the solution and is induced by the changing pressure differentials in the ultrasonic field. The pressure differentials occur by the exchange between negative and positive pressure in a liquid region. When the liquid with negative pressure is created, its boiling point drops and many small vapor bubbles are formed. As the pressure changes to positive, the small bubbles implode with great violence. The mechanical sound wave is generated by a high-frequency electric energy wave through a transducer. The cavitation phenomenon provides mechanical agitation and a scrubbing effect.

The effectiveness of ultrasonic cleaning depends on the cavitation intensity which in turn is controlled by the magnitude of power or pulse width and by dissolved air in solution. The effect of dissolved air has been illustrated and tested, indicating that the dissolved air can act as an acoustical screen and energy absorber.[2] A deaeration step is needed to remove the air in order to obtain the true vaporous cavitation. It is suggested that a high audible noise level with a pronounced hissing sound and minimal visible bubbles in solution coupled with violent surface activity are the signs of ultrasonic efficiency. It is also suggested that the temperature of the solution is a factor in ultrasonic efficiency. The desirable temperature is approximately in the range of 80 to 98 percent of the boiling point of the solution.

Because of the concern that ultrasonic cleaning may damage wire bonding or other chip components, compatible process parameters are to be identified. Some guidelines are proposed:

Power of ultrasonic cleaner	<30 W/L
Ultrasonic range	30 to 66 kHz
Cleaning time	3 min/cycle for 5 cycles (not to exceed a total of 15 min)

8.2.4 Cost consideration

The initial investment in the equipment is represented in Table 8-2 in comparison with CFC solvent cleaner and organic-solvent cleaner.

Tables 8-3 through 8-5 provide three examples of the operating cost of water-clean operations.

Table 8-2. Initial Investment of Cleaner

Type	5-gallon	25-gallon	In-line*
CFC-113	$16,000	$21,000	$180,000–200,000
Solvent-water	$50,000	$62,000	$250,000
Water	$38,000	$46,000	$120,000 (+$35,000 closed loop)

*Continuous to the whole production line (not running as a separate unit).

Table 8-3. Water-Clean Annual Operating Cost—Case 1[3]

Type	Water	Heat	Total
Recycled	$ 75	$ 655	$ 729
No-cycled	$1728	$3787	$5515

NOTE: Operating hours = (30 h/wk) (48 wk/yr), flow rate = 4 gal/min, water/sewage fee = $0.005/gal, cost of power = $0.05 kWh, operating temperature = 140°F, tap water temperature = 50°F, evaporation loss = 4%.

Table 8-4. Water-Clean Annual Operating Cost—Case 2[4]

Type	Water	Heat	Consumables	Total
Closed loop	$1700	$4500	$19,000	$25,200

NOTE: Operating hours = 1440 h/yr, flow rate = 4 gal/min, water/sewage rate = $0.005/gal, cost of power = $0.050 kWh, operating temperature = 150°F, tap water temperature = 60°F.

Table 8-5. Water-Clean Annual Operating Cost—Case 3[5]

Type	Water	Heat	Consumables	Total
Closed loop	$120	$18,000	$13,600	$31,720

NOTE: Water/sewage fee = $0.0017/gal.

8.2.5 Wastewater handling

The concerned areas from wastewater when using a water-cleaning process are

- pH
- Temperature
- Heavy metal
- Biological oxygen demand (BOD)
- Chemical oxygen demand (COD)
- Suspended solids
- Toxic chemicals on hazardous substances
- Lead content

Table 8-6. Wastewater Pretreatment Regulations

	Average limitations (ppm), mg/L						
Category	Northeast	Mid-Atlantic	Southeast	Midwest	South-central	Southwest	Northwest
Temperature	160°F	147°F	145°F	127° F	132°F	143°F	147°F
Grease, oil, fat	100	84	100	125	133	283	100
pH	5.5–9.5	5.3–10.0	5.5–10.0	5.7–9 .5	5.8–9.8	5.5–11.2	5.5–10.5
BOD	275	317	383	250	275	300	300
COD	700	450	580	—	—	—	900
Suspended solids	300	333	250	258	288	300	350
Copper	3.1	3.1	1.4	6.0	2.5	6.9	2.5
Lead	1.9	0.9	0.4	4.8	2.2	2.5	1.4
Nickel	2.3	3.3	1.2	5.3	2.8	8.2	2.7
Chromium	3.3	2.9	1.4	7.7	3.7	4.3	3.7
Silver	2.5	1.8	0.2	0.8	2.1	3.6	0.5
Cadmium	0.3	0.6	0.1	0.4	0.4	5.6	1.4
Zinc	2.9	4.2	1.5	6.4	4.0	11.8	3.4
Fluorides	—	—	14.0	—	—	10.0	15.0
Total toxic compounds	5	1.42	2.1	3.1	2.1	1.0	1.4

Depending on the makeup of the wastewater in relation to the above categories, discharge of wastewater directly into public sewage may be prohibited. The limits of each specific category vary with local regulations. Table 8-6 summarizes the limits of the categories in geographic regions in the United States. Beyond the limits, pretreatment before discharging is required. Techniques to treat wastewater include

- Filtration to capture gross particulates, typically using a 20-μm filter
- Carbon adsorption to remove organic contaminants
- Ion exchange to remove suspended heavy metal ions through chelating resin, exchange H for cations with cation bed, and OH for anions with anion bed

For high-solids water, membrane technology (reverse osmosis), by which porous membranes remove ions, colloids, and organics, is effective. However, mixed resin beds (chelating resin, cation bed, and anion bed) via ion exchange deliver the highest purity level. The combined use of membrane technology and ion exchange resin beds have the ultimate results in efficiency and purity.

The closed-loop recycling in water-clean systems employs the above techniques and technologies to achieve a nearly zero-discharge water-cleaning process. As an example, Fig. 8-1 is a schematic of zero-discharge water recycling process.

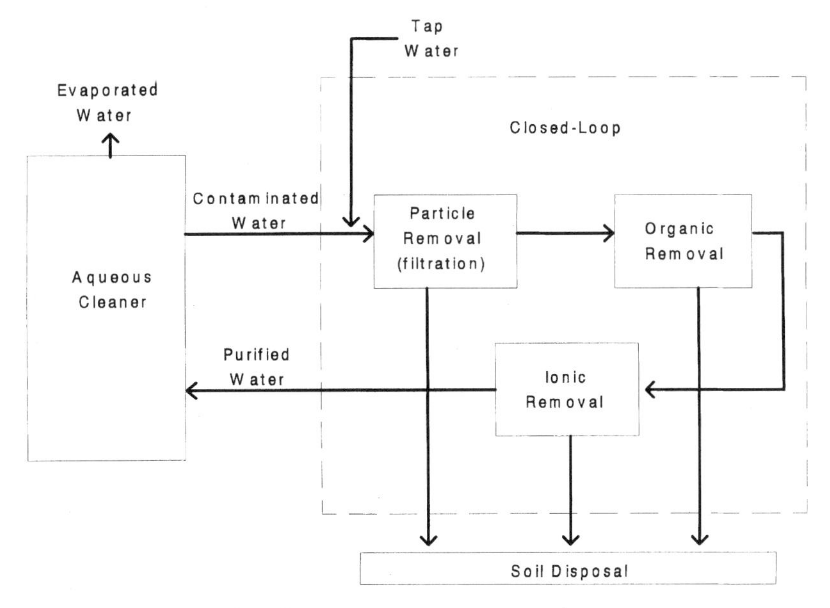

Figure 8-1. Schematic of zero-discharge water recycling process.

8.3 Cleanliness and Reliability of Water-Clean Systems

Guidelines for cleaning process requirements in accordance with the current ANSI/J-STD-001 standard are as follows:

- If cleaning is required during and after processing, parts, subassemblies, and final assemblies shall be cleaned within a time frame that permits appropriate removal of contaminants.
- When cleaning is required, flux residue shall be removed as soon as possible, preferably within 1 h after soldering.
- Cleaning of assemblies shall be done as necessary to remove particulate foreign matter and flux residues.
- After cleaning, assemblies shall be free of dirt, lint, solder splash, dross, etc. Solder balls must neither be loose, nor violate minimum electrical design clearance.
- Surface cleaned *shall be free* of visual evidence of flux residue and other ionic or nonionic contaminants (class 3 assemblies).
- Assemblies shall contain less than 1.56 $\mu g/cm^2$ NaCl equivalent ionic or ionizable flux residue.
- Surface insulation resistance test results shall have a minimum resistance of 100 MΩ after soldering and/or cleaning.
- The user and manufacturer shall agree upon the cleaning required and the tests for cleanliness.
- Ultrasonic cleaning is permissible:
    1. On bare boards or assemblies provided only terminals or connectors without internal electronics are present.
    2. On electronic assemblies with electrical components, provided the contractor has documentation available for review showing that the use of ultrasonics does not damage the mechanical or electrical performance of the product or components being cleaned.

Figures 8-2 through 8-4 demonstrate the SIR test results for 1206-chip components, 1812-chip components, and 100-lead QFP (25-mil pitch), respectively.[6] In this study, the cleaning process was conducted at a water temperature of 50 to 65°C, spray pressure of 3 to 4 bars, and total cleaning time of 6 min. The initial water conductivity was less than two microsiemens (μS)/cm. The SIR tests were performed in an 85°C/85% relative humidity (RH) chamber with a bias of 10 V between the parallel traces. The SIR results were recorded periodically under 10

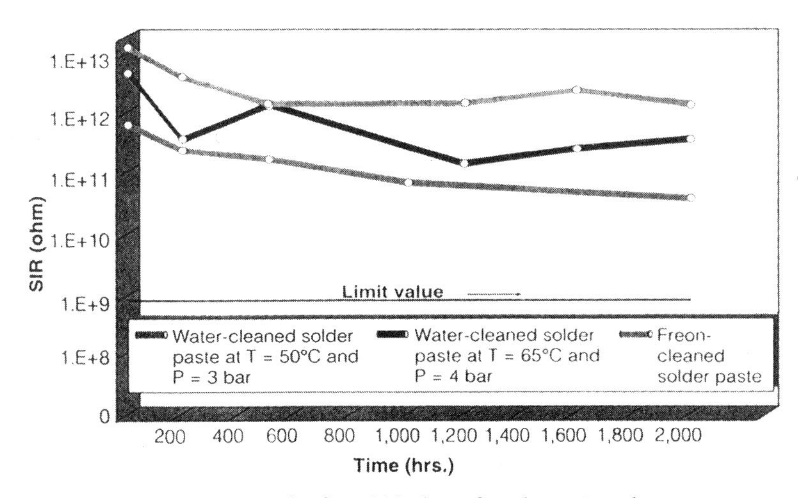

Figure 8-2. SIR test results for 1206 chip after the water-clean process.

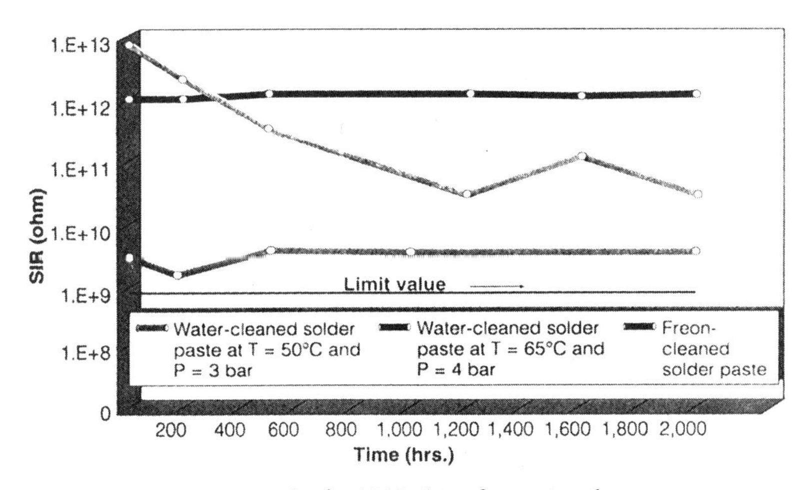

Figure 8-3. SIR test results for 1812 chip after water-clean process.

V after 2 h conditioning at ambient environment. Figure 8-2 indicates that a higher water temperature and pressure somewhat improved the cleaning efficiency for 1206-chip components. However, increasing the water temperature from 50 to 65°C and increasing the pressure from 3 to 4 bars significantly improved the cleanliness for the 1812-chip components as shown in Fig. 8-3. This is attributed to the fact that the

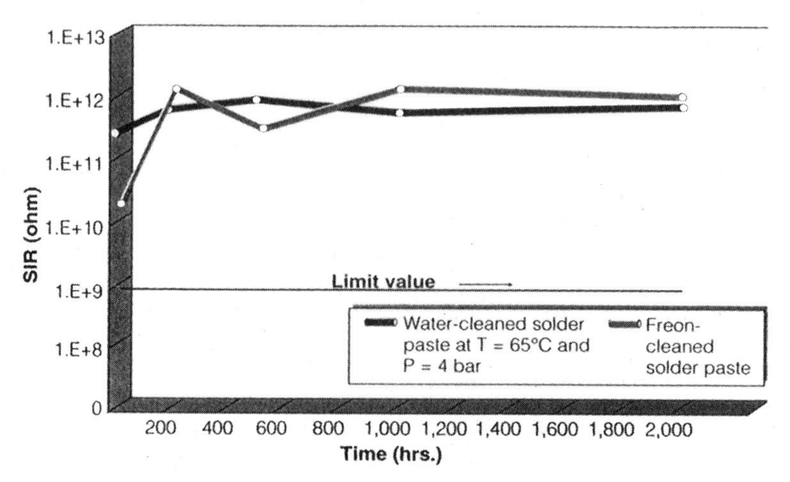

Figure 8-4. SIR test results for 100-lead QFP chip after water-clean process.

height of 2 mm for 1812-chip components mounted at a 1-mm distance apart demands more vigorous conditions than the 1206-chip components that have a 0.4- to 0.7-mm height. For a large component, 100-lead QFP, the cleaning condition of 65°C and 4 bars delivered an excellent level of cleanliness as reflected by the SIR test results of Fig. 8-4 (SIR comb pattern with parallel traces at a distance of 25 mils was imbedded under each QFP). All test results met the cleanliness criteria, indicating the viability of water-clean systems. It should be noted that the clearance of 100-lead QFP normally falls in the range of 4 to 12 mils (0.10 to 0.30 mm), whereas the 1206- and 1812-chip components usually fall below 3 mils. The requirements of cleaning process conditions and the resulting cleanliness not only depend on the capability of the equipment and its process parameter settings but also on the dimension and configuration of components and the density and layout of the populated board. A more vigorous cleaning condition is required to achieve the desired cleanliness when the following situation exists:

- Components with larger area
- Components with increased height
- Lower clearance between the underneath of the component and the board surface
- Assembled board containing high-density component population

Table 8-7 represents one manufacturer's study on the ionic contamina-

Table 8-7. Ionic Contamination Level

Cleaning agent	Contamination level, μg/in²		
	Before cleaning	Deionized water cleaning	Deionized water plus saponifier cleaning
Sodium	2.5–5.1	1.4–2.1	1.4–2.6
Ammonium	1.7–2.1	1.3–2.0	0.5–3.2
Organic Acid	32–92	None detected	0–12

tion level of an aqueous-clean system comparing before and after water-cleaning and water-saponifier cleaning for assemblies using water-soluble solder paste.[7]

8.4 No-Clean Chemistry

8.4.1 Criteria

The primary objective when cleaning assemblies has been regarded to be the removal of any foreign ingredients left on the circuit board after the heat excursion of the soldering process (residues). These residues are originated from the chemical makeup of solder paste, which is designed to perform a fluxing action during soldering so that good wetting and sound solder joints can be formed. This solder paste makeup is also designed to deliver the desired rheological behavior for consistent solder-paste placement on the designated solder pads. These residues are considered harmful if they jeopardize the functional reliability of solder joints or circuitry during their service life. Residues can also interfere with subsequent process steps, such as testing and coating, or may be aesthetically undesirable. Therefore, deciding whether to clean or not to clean is based on the targeted reliability, the process involved, and the aesthetics desired. This, in turn, depends on the design and service function of the assembly.

8.4.2 Performance requirements

To eliminate cleaning, the solder paste and soldering process must

- Flux effectively to make sound solder joints without forming solder balls
- Leave a minimal amount of residue after soldering

- Have left-on residue that is translucent, aesthetically acceptable, and nontacky; and that will not interfere with bed-of-nails testing or impede subsequent coating (if applicable); and that stays inert without detectable mobility under the influence of temperature, humidity, and voltage bias
- Have a soldering system that meets economic and practical feasibility

8.4.3 Test requirements

Common solder paste tests in chemical and physical characteristics continue to apply to no-clean systems. The industry's established test parameters and methods can be used to assess the quality and properties of the assemblies. These include ionic contaminant tests and visual examination. However, the tests for no-clean systems have one difference. These tests should be conducted after reflow or soldering. The solder paste chemical makeup measurement in terms of ionic mobility must also be taken after exposure to a specified reflow condition, not before exposure. This procedure is designed to target the characteristics of the residue left on the board, not the as-is solder paste chemistry.

8.5 No-Clean Process

8.5.1 Manufacturing technique

The flowchart illustrating the manufacturing steps in Sec. 8.2.1 also applies to the no-clean process except that the cleaning step is deleted. The reflow process can be approached via four options that involve different levels of requirements in terms of solder paste, soldering equipment, and reflow oven atmosphere using the RMA-grade and ambient-air oven as the reference.

Option	Solder paste	Reflow oven	Reflow atmosphere
Option 1	Tailored RMA or designed no-clean	Existing	Ambient air
Option 2	Designed no-clean	Retrofitted	Protective
Option 3	Designed no-clean	New	Protective
Option 4	Designed no-clean	New	Reactive

Option 1 is to use a solder paste that can be adapted to existing reflow equipment with ambient air reflow, producing good solder joints and benign residue. The second method is to use a combination of no-clean solder paste and a retrofitted reflow oven that is capable of containing inert gas and excluding O_2 at a reasonable level, e.g., 500 to 2000 ppm. Using an inert gas, such as N_2, broadens the working process parameters. The third option involves using a combination of a low-residue no-clean paste and reflow equipment that is capable of containing inert gas and maintaining the O_2 level below 500 ppm during the dynamic reflow process. The last option is to expect the solder paste to have a fugitive composition that essentially leaves no residue after reflow. The paste is to be used in conjunction with a reactive atmosphere. When using an atmosphere as a part of the process, additional operating parameters are to be monitored and controlled. The areas in operating parameters, the role of humidity and temperature, the optimum oxygen level, and the cost of using an N_2 atmosphere are covered in Chap. 9.

Each option has pros and cons as summarized here.

Option	Merits	Limitations
1 (ambient air)	Convenient, no gas handling, no new capital expenditure	Moderate residue
2 (N_2, $O_2 \geq 500$ ppm)	More latitude for solder paste, design, and solderability variation	Moderate-low residue
3 (N_2, $O_2 < 500$ ppm)	Reduced residue level	Capital expenditure Production process control
4 (reactive, $O_2 < 10$ ppm)	Minimal or no residue	Capital expenditure Production process control Cost

Its selection relies on the consideration of and priority setting in practicality, economics, and reliability.

8.5.2 Residue versus reflow process

The amount of residue left under a no-clean operation depends on the

- Design and makeup of solder paste
- Reflow atmosphere
- Reflow process parameters, particularly temperature and gas flow rate

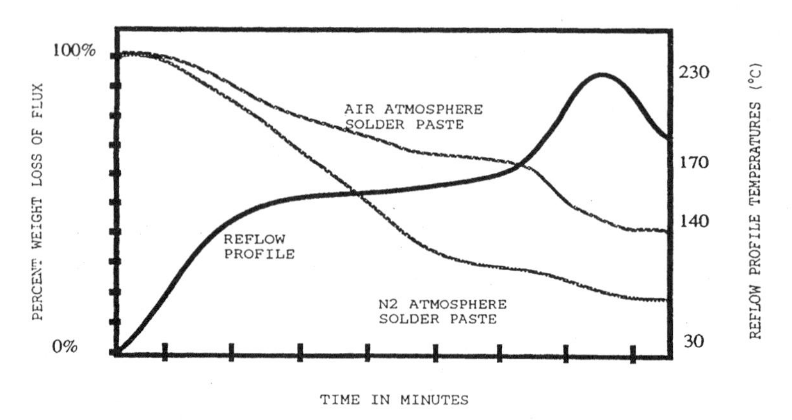

Figure 8-5. Residue formation of a no-clean paste versus reflow temperature profile.

Figure 8-5 illustrates the formation of residue in relation to the reflow temperature profile.[8] As the paste is heated up from ambient temperature, the volatiles in the paste start to escape and continue to lose weight with the maximum loss occurring at the peak temperature. In most cases, N_2, due to its added flow rate, facilitates the weight loss, resulting in less residue.

8.6 Cleanliness and Reliability of the No-Clean System

The use of a properly designed no-clean solder paste coupled with the compatible reflow process can produce assemblies which meet the functional and reliability criteria for most applications. The amount of benign residue depends on the reflow parameters including atmosphere, the level of process control, and the quality and design of solder paste.

It is found that the residue may contribute to the noise level and signal degradation, resulting in a lower signal-to-noise ratio; the effect becomes more noticeable at frequencies greater than 1 MHz. Figure 8-6 reveals the waveform at 5 MHz for (*a*) low residue (17%) no-clean paste and (*b*) high residue (65%) no-clean paste.[9]

Another uncertainty related to the no-clean system is the cleanliness of components and bare boards. With the elimination of the clean step, boards and components must meet the cleanliness requirement. However, this is not always the real world encountered. One measurement on the contamination level of incoming bare boards is listed in Table 8-8.[7]

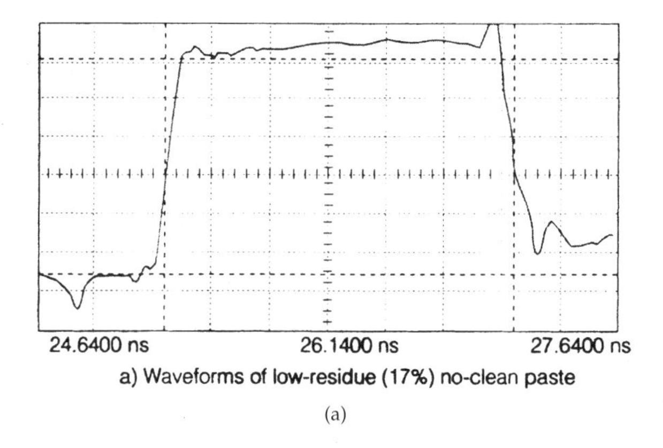

(a)

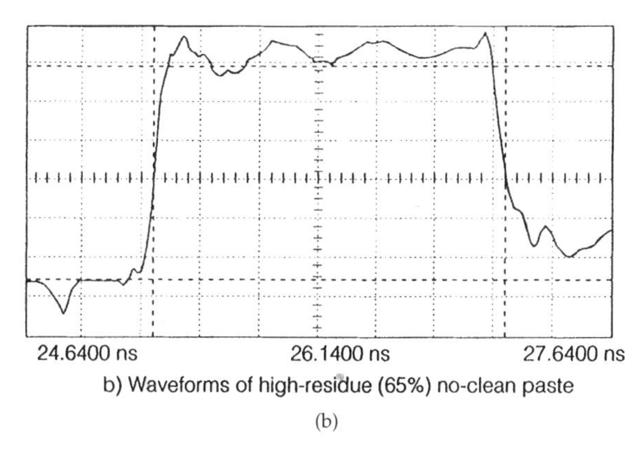

(b)

Figure 8-6. Electrical wave form at 5 MHz for (*a*) low residue, (*b*) high residue no-clean paste.

Table 8-8. Bare Board Contamination Level, $\mu g/in^2$

Chloride	Sodium	Ammonium
4.9	6.6	2.6

8.7 No-Clean Case Study

Description

A PCB manufacturer encountered the field failure of its products which were used in office equipment. The PCB was composed of mixed through-hole and surface mount components with finest spacing in 25-mil (0.63-mm) pitch. The board was produced by using a no-clean solder paste from a major solder paste supplier and by being reflowed under a convection reflow oven. The paste reportedly passed the criteria of ionic contamination and SIR tests. The residue of solder paste after reflow was inspected and recorded as being of small quantity and colorless.

Examination

Dark brownish stuff was found between some of the solder pads of the failed circuit board. It was observed that the residue not only had grown to a larger amount but also changed from colorless to dark brown. It was concluded that the residue linked the circuit pads, causing an electrical short.

Analysis

The obvious growth between the pads led to the conclusion that the electromigration had occurred, resulting in a metallic bridge between the pads. And the bridging material was the combined solder residue and the metallic dendrites which likely contained tin-lead and/or copper.

The contamination which induced the electromigration came from one or a combination of the following sources:

- Solder paste
- Components
- Board

Assuming that the solder paste was benign in nature as the manufacturer had assured and the tests indicated, the contamination must have come from components or the board. Its origin could be either from an as-received condition or from the subsequent handling during assembly.

Most steps in board fabrication involve highly ionic chemicals (sulfates, chlorides, bromides, sulfonates, hydrofluoric acid, ammonium ion). These steps include epoxy-glass fabrication, copper foil forming, through-hole etching, electroless copper deposition, and electrolytic copper plating on the hole, tin-lead electroplating, and chemical etching.

If any of the components or the board was not clean, the contaminants may be rendered active during the service life of the circuit board.

Moral

No-clean flux or solder paste can only take care of one source of contamination and does not assure the total elimination of potentially problematic uncleanness carried over from other sources. This is considered to be a major drawback to the implementation of the no-clean process.

8.8 Comparison between Water-Clean and No-Clean Systems

With the proper cleaning process and reflow parameters, a water-soluble process can produce clean assemblies in both function and appearance. In addition, the nature of its chemistry imparts wider fluxing latitude, better accommodating the inherent variations in solderability of components and boards. It requires initial equipment capital, added operating costs in energy and water consumption, as well as the cost of consumables for a closed-loop recycle system.

No-clean (air) systems eliminate one process step, clearly an economic advantage. It should be noted that the cleaning process has been perceived as a step to remove residues from solder flux or paste, yet it actually has provided the cleaning function for components and boards for many operations without being noticed. It is not unusual for boards to contain higher amounts of ionic contaminants before fluxing and soldering than after. The level of as-received contamination may exceed the acceptable level, since most steps in board fabrication and component plating involve highly ionic chemicals.

For the no-clean system that requires soldering under a protective atmosphere such as N_2, the cost of N_2 may offset or exceed the savings from the no-clean process, depending on the unit cost of N_2 which varies with the location. Other factors that may also complicate the assessment of a no-clean system are solder ball effect and the acceptability of residue appearance.

Figure 8-7 compares the reliability performance of solder joints made with water-soluble and no-clean solder paste and then subject to thermal cycling (1 cycle/45 min at 40 to 125°C).[10] The results indicate that solder joints perform similarly without distinction between water-soluble and no-clean systems for the first 2000 cycles. Their fatigue resistance performance starts to diverge after 2000 cycles with water-

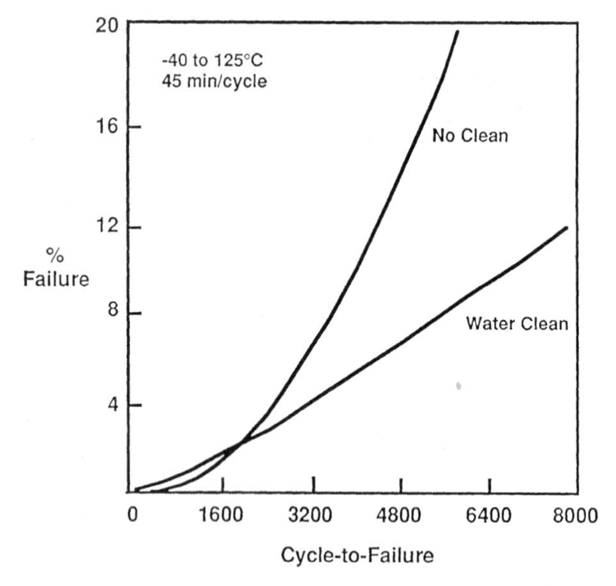

Figure 8-7. Thermal fatigue performance of solder joints made with water-soluble and no-clean paste.

Table 8-9. Comparison of Water-Soluble versus No-Clean (Air) Methods

	Water-soluble	No-clean (air)
Merits	Clean assembly in function and appearance Latitude for solderability variation	One less process step Less operating and capital expenses
Drawbacks	Extra step of process cleaning Operating cost: water, energy, and consumables Initial capital expenditure	Unable to remove contaminants from board and components Often demand higher level of process control Uncertainty in solder ball effect Appearance issue Possible limits for high-frequency application and/or uses that demand extraordinary extension of fatigue life

soluble solder joints experiencing less percentage failure than no-clean solder joints.

Nonetheless, both water-clean and no-clean routes are viable application systems. A basic understanding of the principle behind each practice and compliance with application requirements are essential to the successful implementation of either manufacturing system. Table 8-9 summarizes the comparison between general features of water-clean and no-clean.

8.9 References

1. Accel Product Bulletin, Accel Company, Los Altos, CA.
2. James B. Halbert, "Solvent Cleaning of SMDs with Boiling and Quiescent Ultrasonics," *Surface Mount Compendium*, IEPS, p. 49.
3. John Russo, "A Total Solution Approach for Water, Waste Water, and Waste Disposal for Aqueous/Semi-Aqueous PCB Cleaning," *Proceedings NEPCON-West*, 1992, p. 1049.
4. Edward Sanchez, "A Viable Alternative for the Defense Electronics Industry: Aqueous Cleaning," *Circuits Assembly*, June 1993, p. 66.
5. Kevin Borek, "Clean Break from CFCs," *Circuits Assembly*, February 1992, p. 43.
6. G. Castelli and G. Specchialli, "The Reality of Aqueous Cleaning," *Circuits Assembly*, June 1992, p. 44.
7. Anderson and J. A. Di-Giacomo, "Manufacturing Experience with Water-Soluble Solder Paste," *Proceedings SMI*, 1991, p. 483.
8. Richard Brooks, "Qualification and Implementation of Low-Residue Solder Pastes in a Six Sigma Environment," *Proceedings NEPCON-West*, 1993, p. 293.
9. Allan Beikmohamadi, "Post-Reflow, No-Clean Solder Paste Residue and Electrical Performance," *Circuits Assembly*, March 1994, p. 52.
10. J. Lau et al., "Reliability of 0.4 mm Pitch, 256-Pin Plastic Quad Flat Pack No-Clean and Water-Clean Solder Joints," *Proceedings IEEE/EIA Electronic Components and Technology Conference*, June 1993, p. 39.

8.10 Suggested Readings

Candeas, Barrie G., and Daniel Z. Gould, "Implementing a New PCA Cleaning Solvent," *Circuits Assembly*, June 1994, p. 38.

Chang, Edward K., Mark J. Kirschner, and Sean M. Adams, "A Study of Several No-Clean Fluxes Under Different Atmospheres," *Circuits Assembly*, March 1994, p. 32.

de Klein, Frank J., "Advantages of Protective Atmosphere Control for No-Clean Solder Paste," *Proceedings NEPCON-West*, 1995, p. 972.

Economou, Manthos "No-Clean Rework," *Circuits Assembly*, April 1995, p. 38.

Frazier, Janice, and Robert L. Jackson, "The Chemical Design and

Optimization of Non-Rosin Water-Soluble Flux Solder Paste," *Proceedings SMI,* 1991, p. 503.

Gutierreo, Sam, Richard Komm, and Cheryl Tulkoff, "Making a Transition from Solvents to Water to No-Clean: A Roadmap for Success," *Proceedings SMI,* 1993, p. 621.

Kearns, Philips, et al., "Suitability to a No-Clean Process—A Successful Transformation," *Proceedings NEPCON-West,* 1995, p. 1868.

Quynhgiao N. Le, and Yu Shen Kuo, "Implementation of Water-Soluble Flux Wave Soldering for High Reliability Avionics Electronics," *Proceedings SMI,* 1991, p. 450.

Maltby, Peter F., "Aqueous Cleaning Design Considerations," *Proceedings 3rd International Microelectronics and Systems 1993 Conference,* Kuala Lumpar, Malaysia, p. 237.

Maxwell, John, "No-Clean Solder Paste Experiments Using Reflow Soldering," *Proceedings MI,* 1992, p. 685.

Olson, Roger, and Joseph Yira, "Relative Compatibility of Parylene Conformal Coatings with No-Clean Flux Residues," *Proceedings ISHM,* 1993, p. 56.

Raleigh, Carl, and Jan Glesler, "Judging No-Clean Pastes," *Surface Mount Technology,* May 1994, p. 56.

Raleigh, Carl J., and Steven M. Scheifers, "Cleanliness Requirements for Flip-Chip-on-Board Assembly," *Proceedings SMI,* 1993, p. 638.

Raleigh, Carl, and Cindy Melton, "No-Cleaning Soldering Study," *Proceedings NEPCON-West,* 1991.

Roney, Rick, "The Best Way to Clean Water-Soluble Fluxes," *Surface Mount Technology,* March 1994, p. 84.

Schleckser, Jim, "CFC Replacement Costs," *Circuits Assembly,* June 1993, p. 54.

Shiloh, P. John, and Volker Liedke, "Flux-Free Soldering," *Proceedings NEP-CON-West,* March 1994, p. 251.

Spira, Jeff, "A Closed-Loop Wastewater System," *Surface Mount Technology,* March 1994, p. 102.

Sohn, John E., et al., "How Clean is Clean: Effect of No-Clean Flux Residues and Environmental Testing Conditions on Surface Insulation Resistance," *Journal of SMT,* January 1995, p. 20.

Tautscher, Carl, "The True Cost of No-Clean," *Surface Mount Technology,* February 1995, p. 105.

Weidman, Michael, and Brad Nunn, " A New Cleaning Evaluation Benchmark," *Circuits Assembly,* June 1994, p. 30.

Wilk, Lisa F., and P. E. Capaccio, "Treatment /Recycle of Wastewaters from Alternative Cleaning Processes," *Electronic Packaging and Production,* June 1994, p. 46.

Woodgate, Ralph, "No-Clean the Simple Way, Part I," *Circuits Assembly,* April 1995, p. 44.

Young, Charles, "SMT Aqueous Cleaning with Real Parts, SIR Patterns and Data," *Proceedings NEPCON-West,* 1992, p. 1027.

9 Controlled Atmosphere Soldering

Heating under a controlled atmosphere has been used in various electronic and nonelectronic applications such as thick-film firing, brazing, sintering, and surface metallization. Soldering at high temperatures (high-lead solder) using a hydrogen-nitrogen atmospheric blend has been adopted for manufacturing semiconductor devices and passive components. Yet soldering under a controlled atmosphere for printed circuit assemblies is a relatively recent development. One of the driving forces to consider when soldering under a controlled atmosphere in lieu of ambient air is the use or the facilitation of the use of a no-clean process in response to the need to eliminate the use of CFCs in the cleaning process.

Controlled atmosphere soldering can generally be classified as either reactive or protective. The reactive atmosphere is expected to aid the function of the flux system or to replace the flux that is incorporated in the solder paste reflow or is used in wave soldering. The reactive atmosphere may be oxidizing or reducing to the specific material under a given condition. The protective atmosphere is inert toward the soldering material and assembly; however, it serves as a blanket which prevents the substrate from oxidation during soldering. It should be noted that, for convenience, the mechanism behind the fluxing is often considered to be the reduction of metal oxides. In actual interaction, other phenomena, such as chemical erosion or dissolution of oxides, and foreign contaminants, may occur.

9.1 Potential Benefits and Phenomena

In order to obtain good solderability and quality solder joints, a metallic continuity at the interface(s) between the solder and substrate must be formed during soldering. When using solder paste, a complete coalescence of solder powder particles has to occur in synchronization with the formation of metallic continuity at interfaces. For wave soldering, the atmosphere is expected to protect the molten solder wave and to protect or interact with the surface of substrates. The interactions affect the physical and chemical phenomena in terms of volatilization, thermal decomposition, and surface/interfacial tension. The use of an inert or reducing atmosphere may potentially improve the following phenomena:

For reflow

- Solderability
- Extent of solder balling
- Solder paste residue
- Solder joint appearance
- Board discoloration
- Preservation of solderability for second-step reflow
- Process window
- Overall yield and quality

For wave soldering

- Solderability
- Hole-filling capability
- Dross content
- Flux residue

The performance results, however, highly rely on the specific atmospheric composition, operating process, and ambient temperature and humidity. At a given atmosphere composition and process, the performance, particularly solderability, further depends on the condition of the assembly components.

Operating parameters that directly affect the soldering results include

- Gas flow rate
- Resulting in-oven water vapor pressure or dew point
- Belt speed of oven operation
- Operating temperature
- Resulting in-oven oxygen level
- Internal gas flow pattern

9.2 Atmospheric Composition and Characteristics

9.2.1 Type of atmosphere

Broadly, the gases or chemicals that can provide oxidizing or reducing potential in relation to the specific metal/metal oxide system and that can generate sufficient vapor pressure at an operating temperature are expected to contribute to the overall function of the atmosphere.

The following lists some commonly used atmospheres. Their corresponding nominal compositions are summarized in Table 9-1.

Table 9-1. Compositions of Atmospheric Gases

	Composition, %						
Atmosphere	Carbon dioxide (CO_2)	Oxygen (O_2)	Carbon monoxide (CO)	Hydrogen (H_2)	Methane (CH_4)	Nitrogen (N_2)	Trace elements
Air	—	21.0	—	—	—	78.1	0.9
Nitrogen	—	—	—	—	—	99.8–100	0–0.2
Hydrogen	—	—	—	99.8–100	—	—	0–0.2
Dissociated methanol	—	—	33.3	66.7			
Dissociated ammonia	—	—	—	75.0	—	25.0	
Exothermic gas (air/gas = 6/1)	5.0	—	10.0	14.0	1.0	70.0	
Endothermic gas (air/gas = 2.4/1)	—	—	20.0	38.0	0.5	41.5	

- Dry air
- Nitrogen
- Hydrogen
- Nitrogen-hydrogen blends at different ratios
- Dissociated methanol-nitrogen blends at different ratios
- Dissociated ammonia
- Exothermic gas
- Nitrogen-dopants at different concentrations
- Organic chemicals

At high temperatures the thermal cracking of methanol essentially yields hydrogen and carbon monoxide as represented by the chemical equation

$$CH_3OH \Leftrightarrow 2H_2 + CO$$

At low temperatures (below 800°C; 1472°F), side reactions may occur leading to the formation of H_2O, CH_4, CO_2, and C. Figure 9-1 shows the resulting components of methanol decomposition over a range of temperatures.[1] The composition of exothermic and endothermic gases also varies with the air/gas ratio.

Each component of atmospheric gases may function as an oxidant or reducing agent depending on the temperature and its oxidation/reduction potential relative to that of the materials involved. Among the components of common atmospheric gases, oxygen, water vapor, and

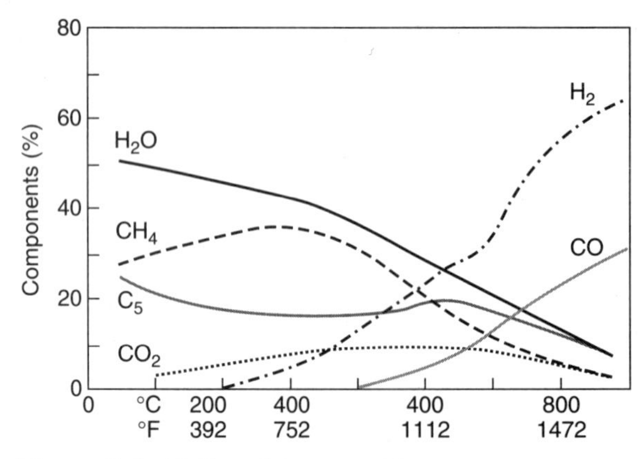

Figure 9-1. Methanol decomposition.

carbon dioxide normally serve as oxidants to most metals and metal oxides, and hydrogen and carbon monoxide serve as reducing agents. The ratio of oxidant content to reducing agent content in relation to the same ratio at equilibrium indicates that the resulting atmosphere will either be oxidizing or reducing.

9.2.2 Important characteristics of gaseous atmosphere

During soldering, two of the most important characteristics of the atmosphere under which the soldering is occurring are

- Reduction-oxidation potential
- Heat transfer efficiency

For a reactive atmosphere, whether it is composed of organic vapor pressure or a blend of gases, the capability to reduce metal oxide is primarily controlled by the reduction-oxidation thermodynamics of the specific metal in relation to the atmosphere as discussed in Sec. 9.4 and by the operating temperature. Soldering is a nonequilibrium heating process. Heat transfer efficiency thus is influential to the overall process under both reactive and protective atmospheres. Table 9-2 lists the thermal conductivity of common gases in ascending order.

Table 9-2. Thermal Conductivity of Common Gases

Gas	Thermal conductivity cal/(s)(cm^2)(C°/cm) $\times 10^{-6}$
Carbon dioxide	44
Argon	46
Water	47
Carbon monoxide	64
Ammonia	65
Nitrogen	66
Air	66
Helium	376
Hydrogen	471

9.3 Key Process Parameters

9.3.1 Gas Flow Rate

The gas flow rate required to achieve a specific level of oxygen in the dynamic state of the reflow oven is largely controlled by the type of oven, categorized as closed-system, semiclosed system, and open system. The relationship of the flow rate versus the oxygen level within one type of oven and the relationship among the different types of ovens are summarized in Fig. 9-2. For a given oven, the required flow rate increases when the allowable oxygen level is lowered. At a given flow rate, when the air tightness in oven construction is reduced, the achievable oxygen level will increase. One practical example is that at a nitrogen flow rate of 1000 to 2500 ft^3/h, the closed system may reach 10 ppm of oxygen and the semiclosed system will stabilize to 500 to 1000 ppm.

As expected, for a given reflow system, the oxygen level is inversely related to the gas flow rate as shown in Fig. 9-3.[2] The gas flow rate also affects the temperature distribution and temperature uniformity of the assembly. Figure 9-4 exhibits the temperature gradient between the component PLCC-84 and the board surface, indicating that a higher gas flow rate reduces the temperature gradient of an assembly.[2] However, the down side of using a high flow rate is that there is a higher gas consumption and energy consumption and, therefore, an increase in cost. The cost impact may be mitigated when the oven is designed to recirculate the internal gas in an efficient fashion.

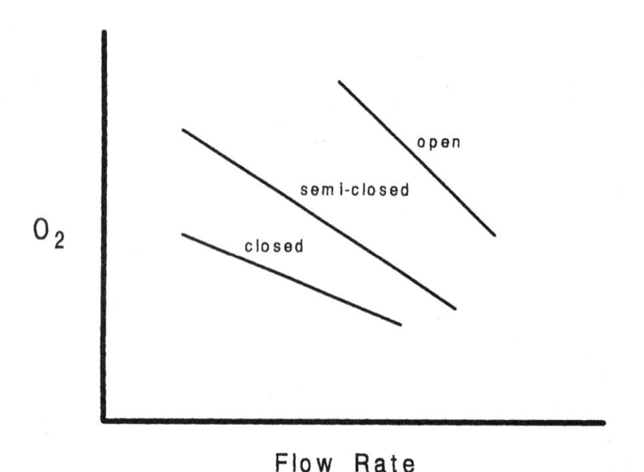

Figure 9-2. Gas flow rate versus oxygen level for open, semiclosed, and closed oven systems.

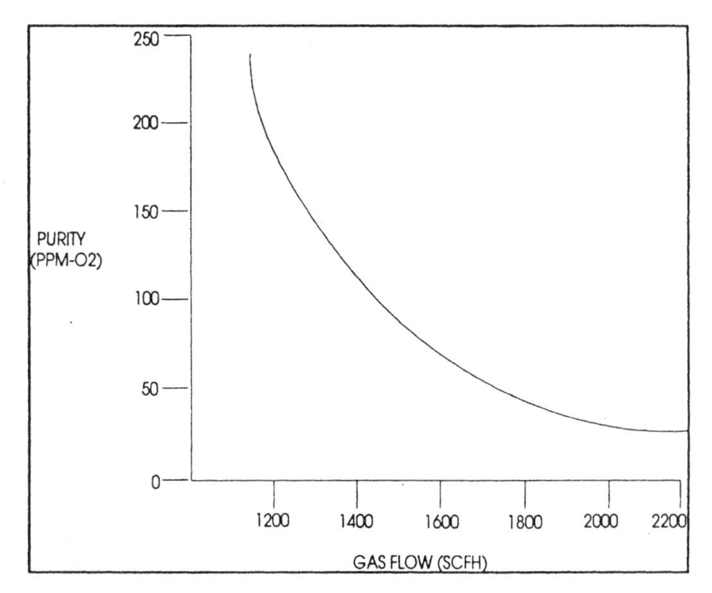

Figure 9-3. Gas flow rate versus oxygen level.

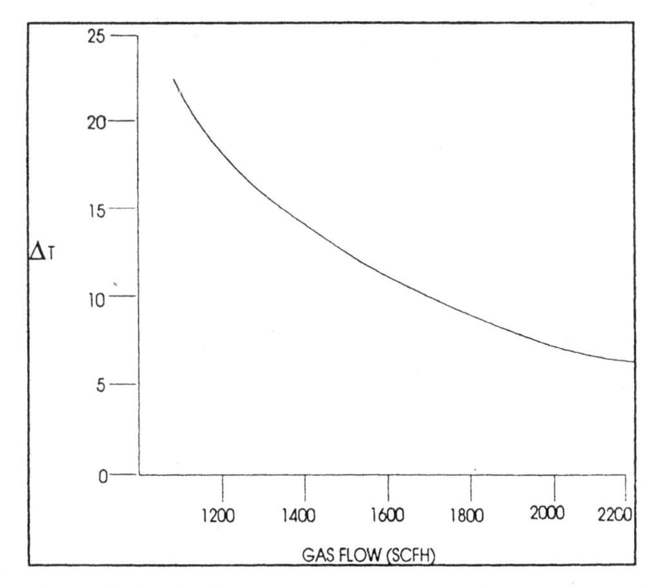

Figure 9-4. Gas flow rate versus temperature gradient of board.

9.3.2 Humidity and water vapor pressure

Factors affecting the water vapor pressure inside the soldering oven are

- The composition and purity of the atmosphere
- The reaction product of the flux/vehicle chemical system with metal substrates
- The moisture release from the assembly including components and board
- The ambient humidity

Since water vapor is essentially oxidizing to metal substrates that are to be joined by soldering, its partial pressure in the oven affects the overall function of the atmosphere.

The partial pressure of water vapor in an atmospheric gas is conveniently expressed as the *dew point,* that is, the temperature at which condensation of water vapor in air takes place. The dew point can be measured by a hygrometer or a dewpointer by means of a fog chamber, chilled mirror, or aluminum oxide technique. The relationship of the dew point to the vapor pressure is shown in Figs. 9-5 and 9-6. The relative humidity (RH) is related to the actual vapor pressure of water

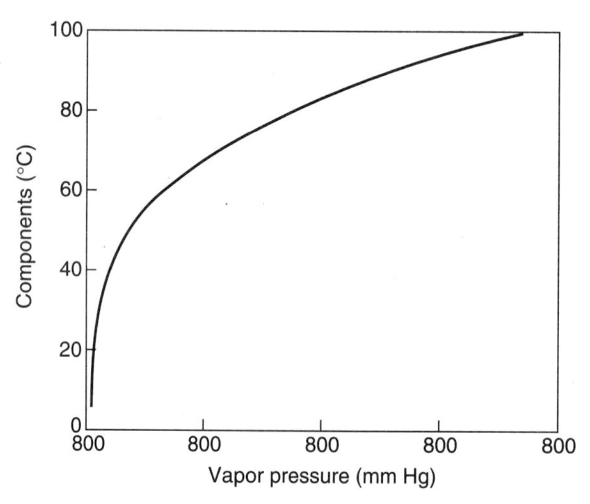

Figure 9-5. Water vapor pressure versus dew point (0 → 100°C).

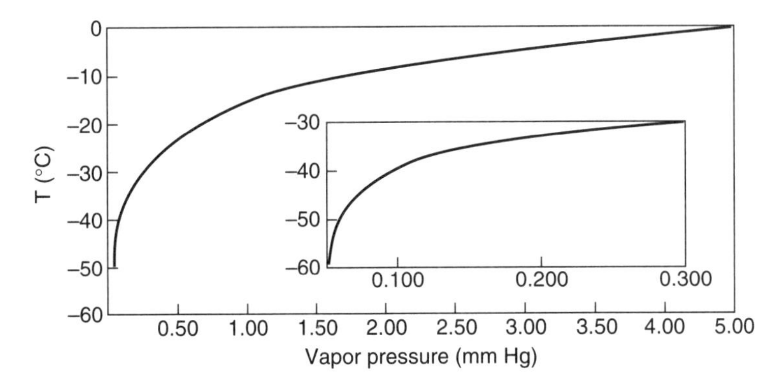

Figure 9-6. Water vapor pressure versus dew point ($-60 \rightarrow 0$°C).

(or represented by the dew point) P_w and the saturated vapor pressure at the prevailing ambient temperature P_s as follows:

$$\mathrm{RH} = \frac{P_w}{P_s}$$

The purity of incoming gas in terms of moisture can be monitored by measuring the dew point. The water vapor pressure in parts per million by volume is then calculated from the measured dew point as shown by the following example.

Example Find the water vapor pressure in a nitrogen gas stream where the total pressure is 5.1 atm and the measured dew point is −20°C.

solution The dew point of −20°C corresponds to a partial water vapor pressure of 0.776 mm. Therefore,

$$\text{Total pressure} = 5.1 \text{ atm} = 3876 \text{ mm}$$

The water vapor pressure in parts per million by volume is the ratio of the partial pressure of water vapor to the total pressure of the carrier gas. Thus

$$\text{Water vapor pressure} = \frac{0.776}{3876} = 200 \text{ ppm}$$

9.3.3 Belt speed

For an evenly spaced loading on the belt, the belt speed not only determines the throughput but also affects other operating parameters that can alter the soldering results. As examples, the parameters that are affected by the change of belt speed include:

- *Peak temperature.* At fixed temperature settings, an increasing belt speed results in the decrease of peak temperature as shown in Fig. 12-12.

- *Atmospheric composition.* While other conditions are equal, the change of belt speed may alter the oxygen level (including moisture content).

9.3.4 Temperature

The operating temperature or temperature profile as discussed in Sec. 12.6 is an integral part of the soldering process. It affects the physical activity and chemical reaction of the organic system in solder paste or flux. The operating temperature, particularly peak temperature, changes the wetting ability of molten solder on the metal substrate; wetting ability generally increases with increasing temperature. Chemical reactions and thermal decompositions respond to the rising temperature and the temperature profile.

9.3.5 Oxygen level

Various studies have focused on the application of no-clean processes and on the determination of the maximum oxygen level allowed for using the nitrogen-based no-clean soldering process in solder paste reflow and in wave soldering.[3–13] Each study was performed with a specific solder paste and flux or with a selected series of paste and flux. Tests were conducted with specific equipment and under a designated process. In view of the continued introduction of new equipment and the diversity of processes coupled with the versatility of solder paste and flux compositions, the test results are expected to represent the specific system tested (paste, oven, process, assembly) and at best to provide a guideline reference point. For example, a solder paste from vendor 1 to be used with process *A* may require a maximum oxygen level of 20 ppm in order to obtain good solderability, to be grossly solder-ball free, and to have acceptable after-soldering residue. To achieve similar results with process *B*, the same paste may only need a maximum of 300 ppm oxygen. The same could be true for a different paste used with process *A*.

The precise oxygen level requirement for no-clean soldering is impractical to pin down. Instead, the general principle and trends in the relationship between the performance feature and the allowable maximum oxygen level can be derived. Figure 9-7 presented the trend of performance feature merit in relation to oxygen level for a series of solder paste containing various levels of solid contents. For a given performance feature, Fig. 9-8 offers the trend of the effect of solid con-

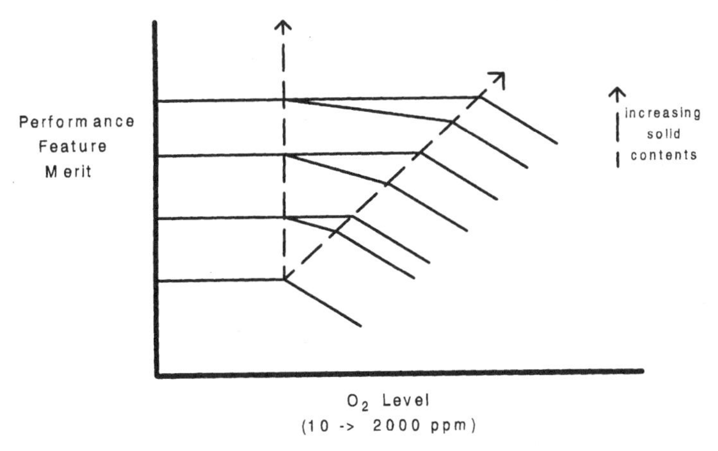

Figure 9-7. Performance failure versus oxygen content for solder paste containing increasing level of solids.

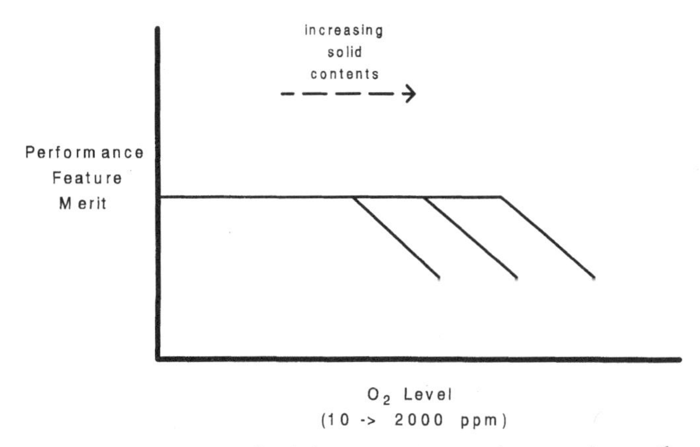

Figure 9-8. Effect of solid content in no-clean paste on the required oxygen content.

tent in no-clean paste on oxygen level requirement during reflow soldering. The performance feature denotes the solderability or the reduction in solder balling.

For convenience, solderability may be monitored by measuring wetting time, wetting force, meniscus rise or wetting angle, or by visual wetting quality. The series of the curves represents the generic groupings of no-clean solder paste or flux by the level of solid contents. This is, however, based on the fact that the solid content possesses a good flux system. As shown in Fig. 9-7, solderability is enhanced and solder balling is decreased when the solid content increases, until at a given solid content, beyond a threshold of oxygen level, the performance significantly drops. It should be noted that increasing solid content delivers an increasing amount of after-soldering residue. The required oxygen level may fall in any place within the region, depending on other factors as discussed above. For a given level of performance, the allowable oxygen level will be relieved as the solid content increases, as depicted in Fig. 9-8.

9.3.6 Internal gas flow pattern

The gas flow pattern which directs the exhausting fume in the oven is equally important to the soldering performance as is the gas flow rate. A typical flow pattern in an oven chamber is illustrated in the schematic of Fig. 9-9. The gas flow rate should be high enough to avoid localized atmosphere buildup as a result of local reactions. In order to achieve the best performance and cost results, the required flow rate is determined by the characteristics of solder paste being used, furnace belt speed, loading pattern, belt width, and other furnace parameters. The exhaust efficiency and its flow pattern in combination with the flow rate dominate the removal of volatile components generated from the pyrolysis and evaporation of chemicals in the solder material, which in turn affects solderability.

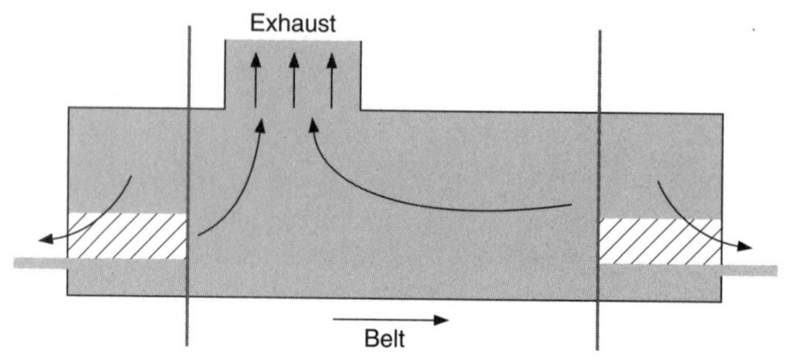

Figure 9-9. Schematic of flow pattern in oven chamber.

9.3.7 Interrelation of process parameters

In addition to the individual function of process parameters, their interdependency further complicates the operation of atmosphere controlled soldering. Some example of one-to-one correlations are

- Belt speed versus oxygen level
- Belt speed versus operating temperature
- Operating temperature versus oxygen level
- Ambient humidity versus oxygen level

Increasing the belt speed may cause an increase in the oxygen level as a result of the drag-in (moisture and air) factor; increasing operating temperature could decrease the total oxygen level in the oven chamber; and changing ambient humidity may alter the in-chamber oxygen level while all other factors remain the same.

9.4 Thermodynamic and Kinetic Considerations

During soldering the reactions and interactions of chemical in solder paste and those between chemicals and metal surfaces can be quite complex. In simple terms, the mechanisms may include evaporation, pyrolysis, oxidation, and reduction. The generalized oxidation and reduction reactions can be expressed as

$$xM + yO_2 \leftrightarrows Mx\,O_{2y}$$

$$MxOy + yH_2 \leftrightarrows xM + yH_2O$$

$$xM + yCO_2 \leftrightarrows MxOy + yCO$$

To obtain the thermodynamic equilibrium constant K for each of the above reactions,

$$xM\,(s) + yO_2\,(g) \rightarrow M_xO_{2y}\,(s) \tag{9.1}$$

$$K_1 = \frac{a_{M_xO_{2y}}}{a_M{}^x\, a_{O_2}{}^y}$$

$$M_xO_y\,(s) + y\,H_2\,(g) \rightarrow x\,M\,(s) + y\,H_2O\,(g) \tag{9.2}$$

$$K_2 = \frac{a_M{}^x \, a_{H_2O}{}^y}{a_{MxOy} \, a_H{}^{2y}}$$

$$x\,M\,(s) + y\,CO_2\,(g) \rightarrow M_xO_y\,(s) + y\,CO\,(g) \tag{9.3}$$

$$K_3 = \frac{a_{MxOy} \, a_{CO}{}^y}{a_M{}^x \, a_{CO}{}^y}$$

Assuming the compositions of solids remain constant and the gases behave ideally, we have

$$K_1 = \frac{1}{P_{O_2}{}^y}$$

$$K_2 = \frac{P_{H_2O}{}^y}{P_{H_2}{}^y}$$

$$K_3 = \frac{P_{CO}{}^y}{P_{CO_2}{}^y}$$

where a represents the activity of each reactant and product of reactions (9.1) through (9.3), and P_{H_2O}, P_{H_2}, P_{CO}, and P_{CO_2} represent the partial pressure of H_2O, H_2, CO, and CO_2, respectively.

By introducing the relationship between free energy $\Delta G°$ and the equilibrium constant K

$$\Delta G° = -RT \ln K$$

with R the gas constant and T the temperature, it is seen that the reactions (9.1) through (9.3) can proceed in a forward or reverse direction depending on the temperature and the ratio of P_{CO_2}/P_{CO}, P_{H_2O}/P_H, and P_{O_2}.

Figure 9-10 shows the standard free energy of formation for some metal/metal oxide systems and CO_2/CO, H_2O/H_2, CO_2/C atmospheres as a function of temperature.[5] Assuming they are under equilibrium conditions and at a soldering temperature of 250°C, lead oxide and copper oxide can be reduced by hydrogen; however, hydrogen is not effective in reducing tin oxide until the temperature reaches 600°C. Equation (9.2) also indicates that the presence of too much water vapor in the furnace atmosphere will cause oxidation of certain metals. The partial pressure of water vapor should therefore be maintained at a constant and defined value.

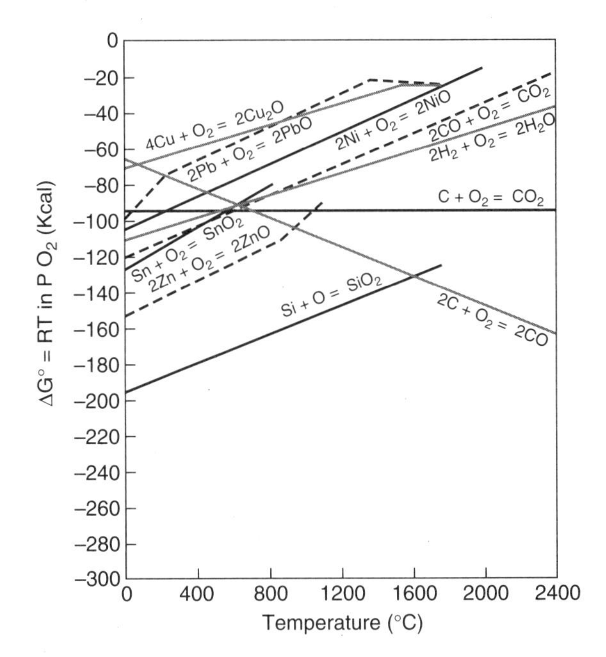

Figure 9-10. Standard free energy of formation of selected metal/metal oxides and common gases.

9.5 Reactive Atmosphere Soldering

Hydrogen and low-molecular-weight organic acids that have a high vapor pressure at the soldering temperature in an inert carrier gas such as nitrogen are a common reactive atmosphere. The proof-of-principle study was conducted using diluted acetic acid or formic acid.

Plasma-assisted solder has been studied in the literature.[14] One process, known as plasma-assisted dry (PAD) soldering[16] utilizes the reactive free radicals, atomic fluorine (F), to remove surface oxides of the parts to be soldered. The fluorine free radicals are generated from gases like CF_4 or SF_6 with plasma energy. The fluorine free radicals then react with the oxides, forming a passivation film in oxyfluoro-metal compositions (e.g., SnO_xF_y). As solder melts, this oxyfluoro-metal film breaks apart, thus exposing the clean metal substrate ready for wetting. This process may be expressed by the following chemical processes:

$$SF_6 \text{ energy} \rightarrow 4F + SF_2 + \text{electrons}$$

$$yF + SnO_x \rightarrow SnO_xF_y$$

$$SnO_xF_y \text{ (molten solder)} \rightarrow Sn \text{ (wetted with solder)} + \text{oxyfluoride residues}$$

Through wetting balance tests, the PAD-treated surface has been shown to have a wetting force that is essentially comparable to that fluxed with RA-type flux and to have a wetting time similar to that treated with RMA-type flux. The treated surface reportedly can maintain its freshness under storage up to 1 week in ambient air.

The results indicate that the PAD treatment, in principle, can be used in lieu of conventional flux for soldering (reflow) under either nitrogen or ambient air conditions.

9.6 Examples of Test Results

Figures 9-11 through 9-16 depict the solderability performance of a solder paste under the atmospheric composition of N_2, 95 N_2/5 H_2, 85 N_2/15 H_2, 70 N_2/30 H_2, H_2, and ambient air, respectively, with other conditions being equal. It is apparent that the solderability under N_2 atmosphere is significantly improved as represented by the elimination of solder balls formed under ambient air (compare Figs. 9-11 and 9-16). By examining the results as shown in Figs. 9-11 through 9-15, it is seen that the increasing H_2 content in the atmosphere does not benefit the solderability at any detectable level; this is in conformance with

Figure 9-11. A paste reflowed under N_2.

Figure 9-12. A paste reflowed under 95 N_2/5 H_2.

Figure 9-13. A paste reflowed under 85 N_2/15 H_2.

Figure 9-14. A paste reflowed under 70 N_2/30 H_2.

Figure 9-15. A paste reflowed under H_2.

the thermodynamic data. With respect to residue characteristics, the controlled atmospheres, such as N_2, H_2, and N_2/H_2 blends, provide more intact residue and less charring tendency. Furthermore, the controlled atmosphere soldering under significantly different temperature profiles does not exhibit a detectable difference in residue appearance, board material intactness, and solderability. The results indicate that

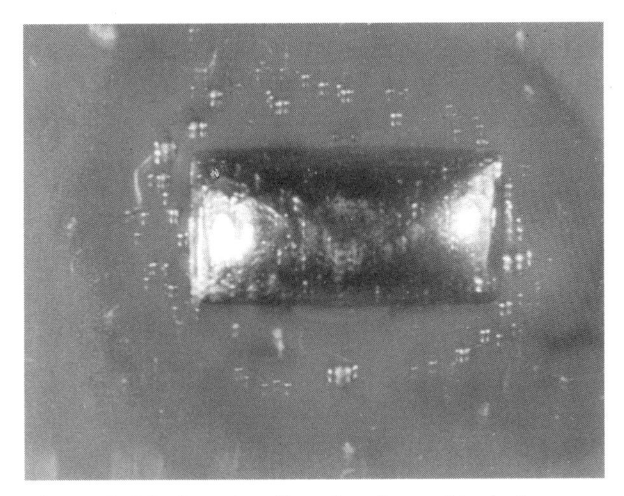

Figure 9-16. A paste reflowed under ambient air.

the system with the specific combination of the rosin-based solder paste and the process worked better in nitrogen than in air, but the results do not necessarily conclude that a nitrogen atmosphere will demonstrate superior results for all systems or conditions.

Figures 9-17 and 9-18 exhibit the performance of a no-clean solder paste under N_2 and ambient air soldering. Under the N_2 atmosphere, the amount of residue appears negligible, and solderability is enhanced without the formation of detectable solder balls.

A study of two kinds of solder paste (RMA and low-residue) used on two groups of FR-4 boards and reflowed with two types of ovens containing nitrogen gas with 10, 100, and 1000 ppm oxygen, respec-

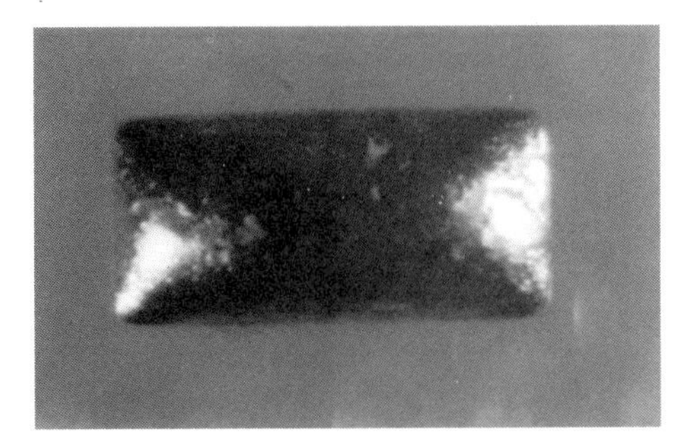

Figure 9-17. No-clean paste reflowed under N_2.

Figure 9-18. No-clean paste reflowed under ambient air.

tively, reported the following results in comparison with the tests conducted in air:[12]

- Soldering performance was improved by nitrogen atmosphere with 1000 ppm oxygen.
- Below 1000 ppm oxygen, there was no further beneficial effect detected.

This conclusion, however, may not apply to other systems which use a different solder paste or a different reflow process. Other paste compositions may call for an atmosphere containing much less than 1000 ppm oxygen.

9.7 Cost Considerations

For the purpose of an exercise in cost assessment, two scenarios covering the likely options are considered.

9.7.1 Nitrogen-based closed system capable of maintaining 10 to 50 ppm oxygen

- Capital equipment expenditure = \$120,000 to \$200,000
- Assume unit price of nitrogen = \$0.60/100 ft^3
- Gas flow rate = 2000 ft^3/h
- Two shifts per day
- 230 operation days per year

$$\text{Gas consumption without recirculation} = \frac{\$0.60}{100\ \text{ft}^3} \times 2000\ \text{ft}^3/\text{h} \times 16\ \text{h/day} \times 230\ \text{day/yr} = \$44{,}160\ \text{per year}$$

9.7.2 Nitrogen-based semiclosed system capable of maintaining 500 to 1000 ppm oxygen

- Capital equipment expenditure = \$30,000 to \$80,000
- Gas flow rate = 1500 ft^3/h

$$\text{Gas consumption without recirculation} = \frac{\$0.60}{100\ \text{ft}^3} \times 1500\ \text{ft}^3/\text{h} \times 16\ \text{h/day} \times 230\ \text{day/h} = \$33{,}120\ \text{per year}$$

The gas consumption, equipment expenditure and energy consumption vary with the use of different oven systems.

Overall, as shown in Fig. 9-19, the cost as the result of gas consumption increases as the flow rate increases. The cost is also higher when the allowable oxygen level is lowered. To combat the gas consumption cost, internal recirculation capability that is built in to the oven design and construction continues to be desired.

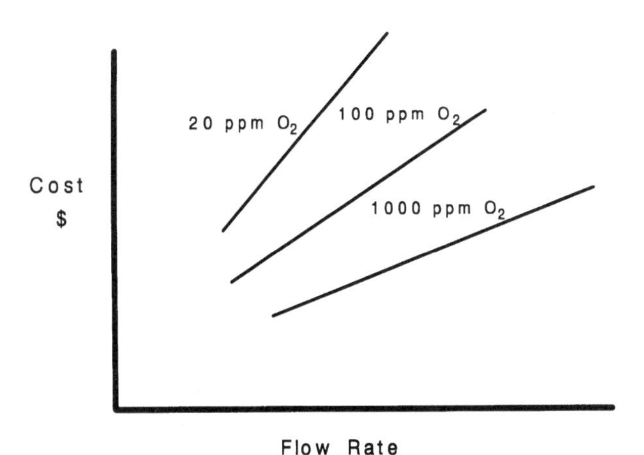

Figure 9-19. Cost versus gas flow rate for three levels of oxygen content.

9.8 Performance versus Oxygen Level versus Cost

When cost is not a factor, a nitrogen-based inert atmosphere is expected to deliver potential benefits in solderability enhancement, minimizing solder balling, protecting the board from discoloration, and providing a wider process window compared with ambient air soldering. Under compatible conditions, a nitrogen atmosphere may facilitate the carrying away or volatilization of decomposed organic chemicals during the no-clean process, thus likely resulting in less residue.

With today's competitive manufacturing, continued cost reduction is a sustained effort. The increased cost of using a controlled atmosphere needs to be justified by improved quality and increased yield. It should be noted that the enhanced performance may or may not be detected in a manufacturing environment, depending on the existing level of quality, the control of the process, and the materials and components involved. For an assembly which is composed of the components and boards with good solderability and the solder paste with a wide performance latitude, the incorporation of a nitrogen atmosphere may not demonstrate any detectable improvement. It is analogous to the difficulty in improving an already six-sigma production.

As to what is the best oxygen level for nitrogen-based soldering, the answer to the question is, It depends. It varies with the chemistry and makeup of the solder paste and flux, the process parameters, the equipment, and the condition of components and boards. Looking at solder paste alone, the paste can be designed to suit different levels of oxygen content in the reflow oven. Considering the real-world applications, Figs. 9-7 and 9-8 provide the answer to the requirement trend of oxygen level for nitrogen atmosphere soldering. One can derive the oxygen requirement from the experimentation for a specific system, but a universal criterion cannot be made.

9.9 References

1. J. McGinn, "Effects of Furnace Atmosphere Chemistry in Sintering," *Industrial Heating,* May 1985, p. 24.
2. Normal R. Cox, "The Influence of Varying Input Gas Flow on the Performance of a Nitrogen/Convection Oven," *Proceedings NEPCON East,* 1994, p. 323.
3. Mark S. Nowotarski and Michael J. Mead, "Effects of Nitrogen for IR Reflow Soldering," *SMART Conference Proceedings,* 1988.
4. Bandypadhyay and M. Marczi, "Development of a Fluxless Soldering

Process for Surface Mount Technology," *Journal of SMT,* October 1989, pp. 10–18.

5. Jennie S. Hwang, "Controlled Atmosphere Soldering—Principle and Practice," *Proceedings NEPCON West,* 1990.
6. Kermit Agnayo, "Case History/Utilization of Nitrogen in IR Reflow Soldering," *Proceedings NEPCON West,* 1990.
7. Ahmet Arslancan, "IR Solder Reflow in Controlled Atmosphere of Air and Nitrogen," *Proceedings NEPCON West,* 1990.
8. Jay Leonard, "Nitrogen Versus Air in IR Reflow Soldering," *Proceedings NEPCON West,* 1990.
9. Jennie S. Hwang, "Considerations for Soldering and Solder Paste in Electronic Packaging," *ISHM-Chicago/Milwaukee Annual Symposium,* February 1989.
10. Jennie S. Hwang, "Solderpaste—Present and Future," *First International Conference on Solder Flux Technology,* April 1989.
11. P. F. Stratton et al., "The Effect of Adventitious Oxygen on Nitrogen Inerted IR Reflow Soldering with Low Residue Pastes," *Soldering and Surface Mount Technology, no. 13,* February 1993.
12. Klein Wassink and M. M. F. Verguld, "Use of Nitrogen in Reflow Soldering," *Proceedings SMI,* 193, p. 1.
13. N. Potier et al., "Comparative Study of Low Residue Solder Pastes: The Influence of N_2-Based Atmospheres Containing Various O_2 Levels," *Proceedings NEPCON West,* 1994, p. 293.
14. G. Dishon et al., "Plasma Assisted Fluxless Soldering," *Proceedings NEP-CON-West,* 1992.

9.10 Suggested Readings

Brammer, Dieter, "Maximize Reflow Oven Performance," *Surface Mount Technology,* June 1994, p. 31.

Cole, Hugh, and Eli Westerlaken, "The Control of Solder Ball Formation During Wave Soldering Using No Clean Fluxes and Inert Gas Atmospheres," *Proceedings NEPCON-West,* 1993, p. 971.

Hall, James, "Implication of Inerting in a Convection Reflow Oven," *Proceedings NEPCON-West,* 1994, p. 313.

Howard, R., "Fluxless Soldering Process Using a Silane Atmosphere," U.S. Patent # 4,646,958, 1987.

Lea, C., "Relation Between Oxygen Level in an Infrared Oven & Soldering Quality," *Controlled Atmosphere Soldering Conference,* London, November 1991.

Lechok, Cheryl A., "Improved Soldering via Membrane Nitrogen: 5 Case Studies," *Surface Mount Technology,* April 1993, p, 27.

McKean, Kevin P., "An Atmosphere-Purity Selection Guideline," *Surface Mount Technology,* November 1994, p. 32.
Moskowitz, P., and A. Davidson, "Laser Assisted Dry Process Soldering," *Vacuum Science Technology,* vol. 3, no. 3, May/June 1985.
Solberg, Vern, et. al., "No Clean Reflow Process for 0.4 mm Pitch, 256-Pin Fine Pitch Quad Flat Pack," *Proceedings NEPCON-West,* 1993, p. 1090.
Warwick, M., and P. Hedges, "Design Consideration for Solder Paste Used in Fine Pitch Inert Gas Soldering," *Proceedings NEPCON-West,* 1993, p. 995.
Yeh, H. et al., "Palladium Enhanced Fluxless Soldering and Bonding of Semiconductor Device Contacts," U.S. Patent # 5,048,744.

10 Surface Mount Fine-Pitch Technology

10.1 Introduction

In 1985, the industry primarily engaged in the use of IC packages of 68 pins or lower with a 50-mil (1.26-mm) pitch or wider design. During the period of 1985 through 1990, the 25-mil- (0.63-mm-) pitch packages with a pin count of 70 to 100 progressively replaced some 50-mil- (1.26-mm-) pitch components. In 1993, it was estimated that there were 4.3 million units of IC packages designed with an over 100-pin count, which accounts for approximately 5 percent of total IC packages.[1] The packages having a finer-pitch design than 25 mil (0.63 mm) will continue to gain the share of overall IC packages, complementing other design architectures, such as array packages which will be discussed in Chap. 11.

This chapter addresses all aspects of soldering and solder paste for fine-pitch applications with respect to materials and processes; in particular, this chapter focuses on the producibility and reliability for making surface mount fine-pitch solder joints from the application point of view. From a manufacturing perspective the important aspects in fluxing, solderability, solder paste characteristics, solder powder requirements, paste printing parameters, alternate soldering techniques, and fine-pitch solder defects are presented in a guideline format to facilitate the fine-pitch implementation. For the convenience of technology assessment, the author defines *fine-pitch components* to be those with less than a 25-mil-(0.64-mm-) pitch design.

10.2 Factors to Be Considered

When using a soldering process to connect fine-pitch components onto PCBs with conventional solder paste or other alternatives, some additional factors which are either inherent or as a result of real-world practice must be considered:

Thinner and fragile leads

Smaller solder pads

Narrower spacing between pads

Varying solderability

Degraded solderability

Nonstandardized lead configuration

Lack of lead coplanarity

These factors significantly contribute to the final yield of the assembly process and potentially aggravate the common soldering ailments and defects. With respect to commercial availability, lead coplanarity is generally no better than 0.002 in (0.05 mm), the lead foot angle of plastic QFPs could vary from 2 to 10°, and the lead tip-to-tip dimension ranges from 30 to 32 mm (1.2 to 1.28 mils). These variations and lack of consistency in components are one part of the main hurdles for producing solder joints with a high production yield.

It should also be recognized that the fine-pitch component leads may impart various levels of solderability, depending on the subsequent steps conducted after the lead coating process, the conditions under which the subsequent steps are conducted, and the storage time and conditions of the components. The subsequent steps after lead coating could involve molding, deflashing, burn-in, and lead trimming and forming. The degraded solderability may in turn jeopardize wetting and therefore the integrity of the resulting solder joints. When the poor solderability of component leads cannot be conveniently corrected through other means on the production floor, wetting problems are often expected to be remedied by the solder paste or the soldering process. Therefore, understanding the properties and application of solder paste and the soldering process becomes increasingly important.

The smaller pads and narrower spacing between pads require a higher level of precision in solder paste deposition and in component placement and require higher quality and consistency in solder paste. Overall, fine-pitch components put higher demands on the material, equipment, and process, leaving less tolerance of error in every step of the assembly.

10.3 Solder Material Deposition

10.3.1 General techniques

For depositing solder material in making solder interconnections, techniques include

1. Solder paste deposition
2. Electroplating
3. Hot dipping
4. Using preform
5. Solder precoating via in situ chemical reaction

10.3.2 Paste rheology—general[2,3]

Rheology is the science of studying deformation and flow of materials. The goal of studying rheology is to understand and anticipate the force needed to cause a given deformation or flow in a material under a set of specified conditions. Rheology of solder paste is one of the important characteristics with respect to application, as the flow and deformation behavior of solder paste directly affect the quality of paste deposition on the intended solder pads. Therefore, the correlation of paste rheology to the applications and its specific parameters is crucial to the control of rheological quality when manufacturing paste, as well as to the setting and control of application parameters when using paste.

In many rheological aspects, there are similarities between coating products where fine oxide particles are dispersed in an organic carrier phase and in solder pastes. Yet, different inherent factors are involved in solder pastes, making the solder paste rheology more of a challenge. Factors such as the following are influential:

- Drastic difference in density between organic matrix and metallic particles
- Suspended particle of metallic nature
- Relatively large particle size
- Spherical particles
- High load of particles

The drastic difference in density between the organic matrix (0.7 to 1.2 g/cm^3) and the metallic particles (7 to 12 g/cm^3) facilitates the settling of particles, and the large size (up to 75 μm) of particles with small surface area reduces the interactions between the particle and matrix, which further facilitates settling. The metallic nature of particles renders the mixing or dispersion technique more delicate. The high load of particles, commonly more than 40 volume percentage (vol %), contributes to the complexity of the rheological behavior.

The ideal rheology of solder paste for both dispensing and printing applications from an application point of view would deliver the following properties:

- Specified shelf stability
- No separation or settling
- No dripping
- No stringiness, but enough tackiness to hold components
- No slumping or sagging, but capable of leveling
- Perfect transfer through printing pattern
- Clean release from the screen
- Clean breakoff through fine needles
- Nearly instant recovery after transfer
- Temperature insensitive

In practice, true instant recovery after paste transfer and temperature insensitivity are difficult to come by. It is thus obvious that tradeoffs and balance are essential to the design of pastes having a rheology able to accommodate the application processing parameters.

The primary driving forces underlying the rheology of solder paste are kinetic and thermodynamic. In principle, these cover thermal motion, with the magnitude being determined by the mean free path of molecules/particles and by temperature, in conjunction with intermolecular/interparticle forces. The forces can be viewed as electrostatic repulsion and attraction between particles, and the affinity between the chemical matrix and the particle surface. The particles will not aggregate when the affinity of the particle surface for the chemical matrix exceeds the attractive forces between particles, or when steric effects through absorption hinder the aggregation. Therefore, the rheology of solder paste may be affected by the following factors:

- The composition, shape, and size of suspended particles
- The chemical composition of suspending the matrix
- The relative concentration of effective ingredients in the matrix
- The structure of ingredients in the matrix
- The interactions between the matrix and suspended particles either of a physical or chemical nature, including wetting and solvation
- The volume fraction occupied by the suspended particles, usually, the higher the amount of particles, the more deviation from viscous flow
- The internal structure and its response to external forces
- The interactions among particles and resulting aggregates and flocculants.
- Temperature

The difficulty of predicting the rheology of such a system is apparent, due to the lack of knowledge of the structure and the nature of forces exerted by molecules/particles. However, rheological behavior can be characterized, and the characterization can then be correlated to performance. It is observed that solder paste is not an elastic material, nor a pure viscous material. Viscoelasticity normally describes the behavior of solder paste. *Viscoelasticity* is a rheological property of a material which is viscous but also exhibits certain elastic qualities such as the ability to store energy of deformation, and in which the application of a stress to the material gives rise to a strain that approaches its equilibrium strain value slowly. Sections 10.3.2.1 and 10.3.2.2 introduce the two types of basic rheological behavior: flow behavior and elastic behavior. A solder paste can be adequately characterized by its flow behavior and elastic behavior. The characterization of the paste has to be correlated with the processing parameters as well as with the paste performance under a specific set of conditions.

10.3.2.1 Paste rheology—flow behavior. For viscous materials, Newton's Law defines

$$\tau = \eta\dot{\gamma}$$

where τ = shear stress
$\dot{\gamma}$ = shear rate
η = apparent viscosity

The viscous flow described by this relationship is called newtonian flow, as shown in Fig. 10-1.

In addition to idealized newtonian flow, nonnewtonian types are exemplified by plastic, pseudoplastic, dilatant, thixotropic, and rheopectic flow. As illustrated in Figs. 10-2 through 10-6, the viscosity of pseudoplastic material decreases with increasing shear rate due primarily to the alignment of molecules/particles along with the direction of shearing. The dilatant material displays increased viscosity as the shear rate increases. It is believed that the flow resistance is attributed to the formation of close-packed structure. It often occurs in the systems containing very small solid particles being suspended in a highly deflocculated manner. The plasticity of materials is characterized by the presence of a yield value which is the threshold shear stress needed to overcome the forces of internal structure to start the flow. In addition to shear thinning effect, the thixotropic material exhibiting a time-dependent reestablishment in viscosity results in a hysteresis loop in the ramping up and down cycle. The rheopectic flow is not well understood. It is postulated that a perfect structural orientation is associated with this phenomenon, assuming an in situ chemical reaction and evaporation are absent. Flow behavior can be characterized by several ready-to-measure parameters: shear stress versus shear rate, yield point, viscosity versus shear rate, viscosity versus time, and shear stress versus time.

The relationship of shear stress versus shear rate indicates the type of flow as shown in Figs. 10-2 through 10-6. There is more than one way to determine the yield point. For convenience, the yield point may be

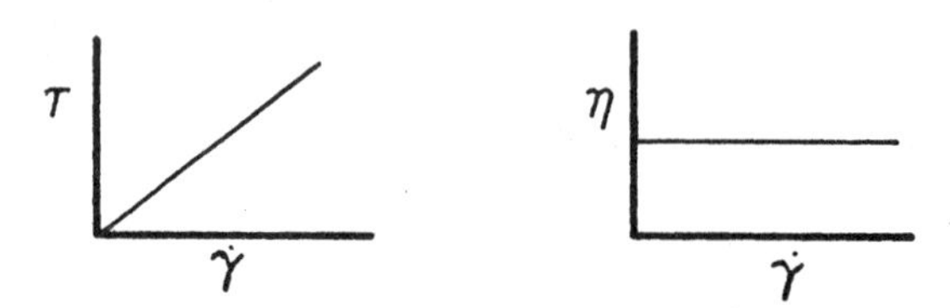

Figure 10-1. Newtonian flow curve.

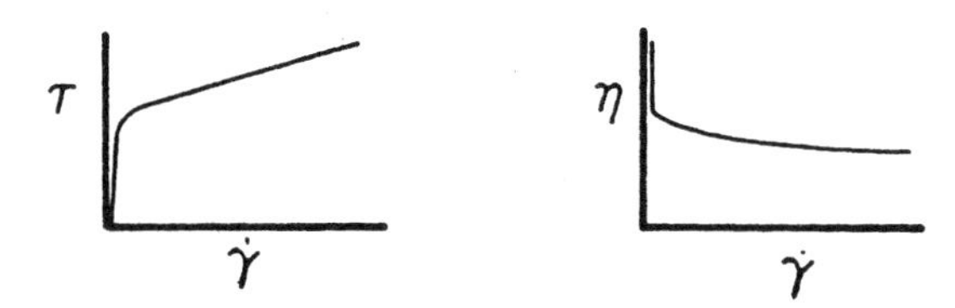

Figure 10-2. Plastic flow curve.

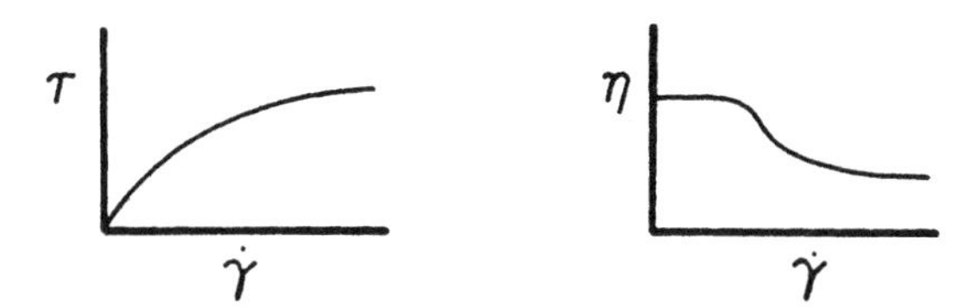

Figure 10-3. Pseudoplastic flow curve.

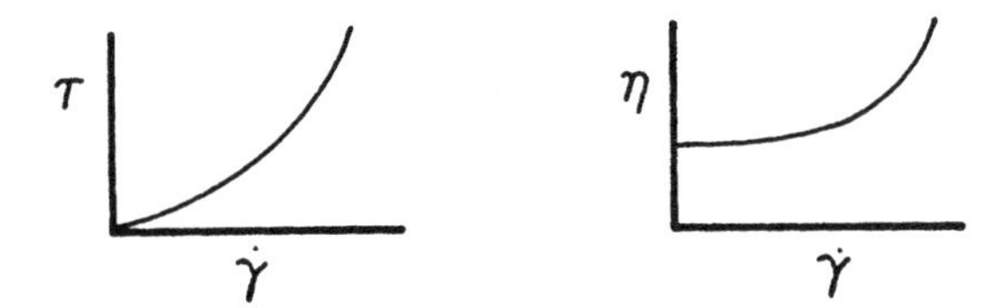

Figure 10-4. Dilatant flow curve.

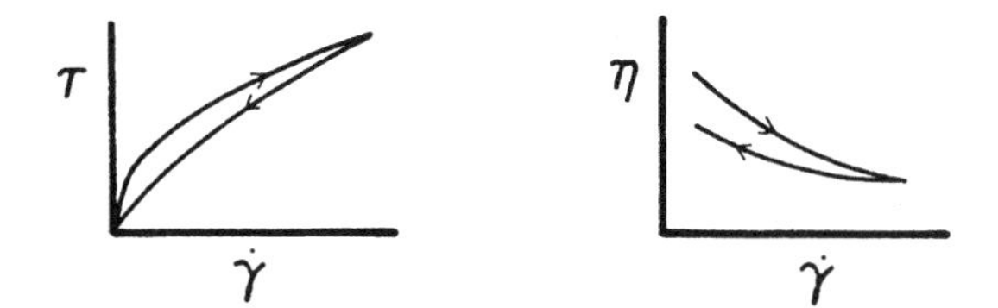

Figure 10-5. Thixotropic flow curve.

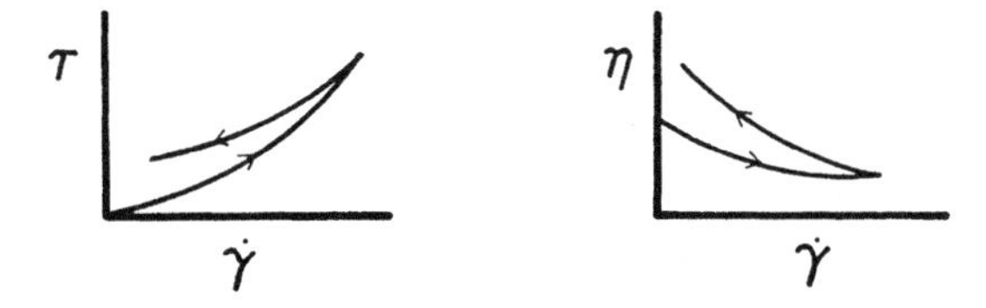

Figure 10-6. Rheopectic flow curve.

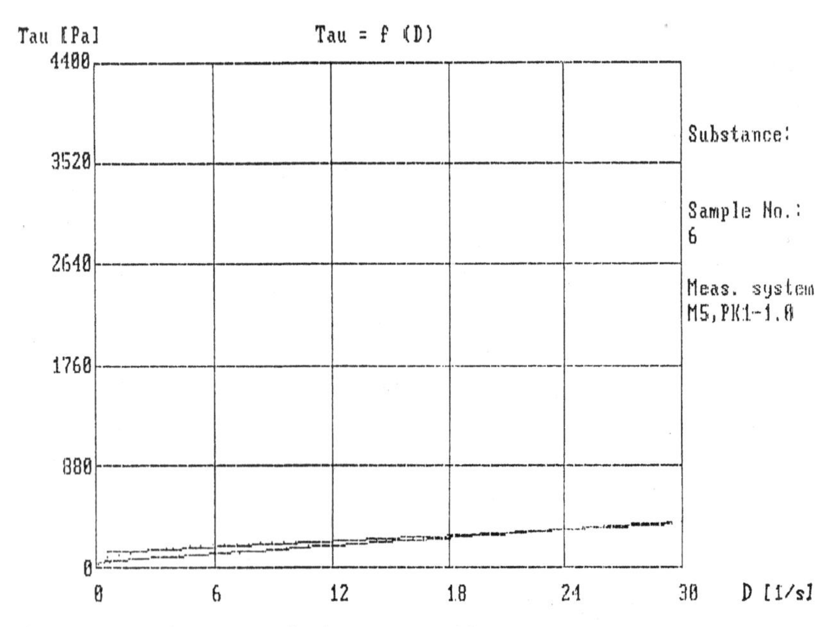

Figure 10-7. Flow curve of a dispensing solder paste.

estimated by the extrapolation of the flow curve to intercept the shear stress axis, or be taken as the shoulder point (inflection) of the curve, if present. These techniques are appropriate only for comparison purposes. When using these techniques to derive the yield point, it is more reliable to measure flow at low shear rate conditions ($< 50\ s^{-1}$).

The viscosity versus shear rate relation is equivalent to shear stress versus shear rate. However, it provides direct readings on viscosity change in response to shear rate change. The rheometers manufactured by Haake Buehler Instruments, Inc. (Rotovisco RV 20) and Rheometrics, Inc. (RMS-800) or their equivalent employing cone-and-plate rotational techniques can be used for measurements.

Figure 10-7 shows the flow profiles of a dispensing solder paste, indicating very slight shear rate dependency. The shear rate ramping up and down curves indicate a small amount of thixotropy. They also show a very low yield point. In comparison, Fig. 10-8 is the flow profile of a printing paste, showing a relatively large yield point, moderate thixotropy, and high shear-rate dependency. The change of viscosity with time of a dispensing paste after a constant shear rate is applied is shown in Fig. 10-9. The stress relaxation curve of a printing paste is shown in Fig. 10-10. The profile of the curve and the area under the

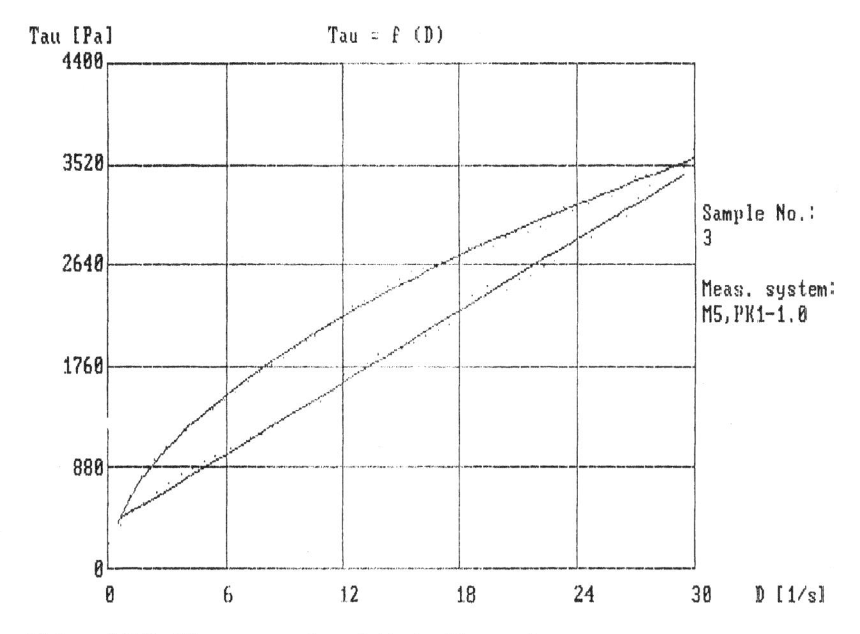

Figure 10-8. Flow curve of a printing solder paste.

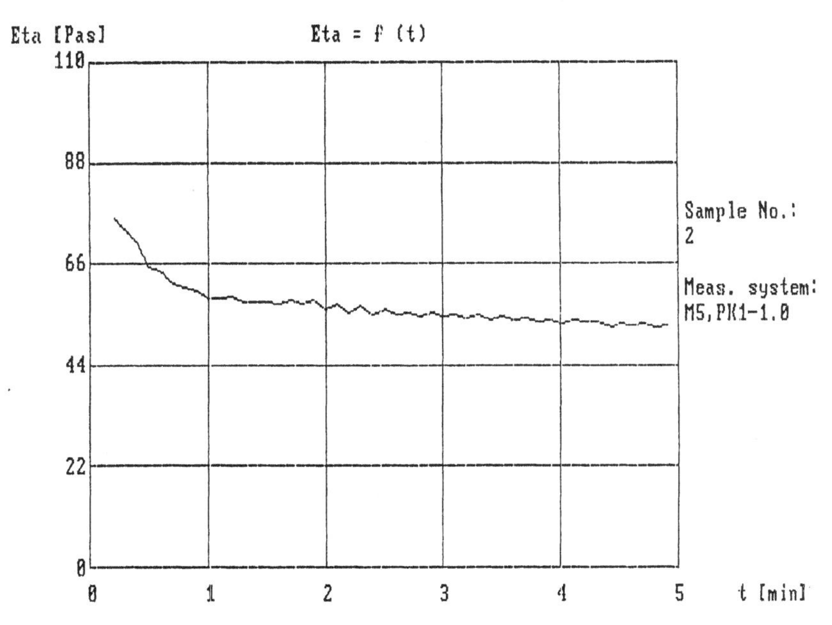

Figure 10-9. Viscosity versus time of a dispensing paste.

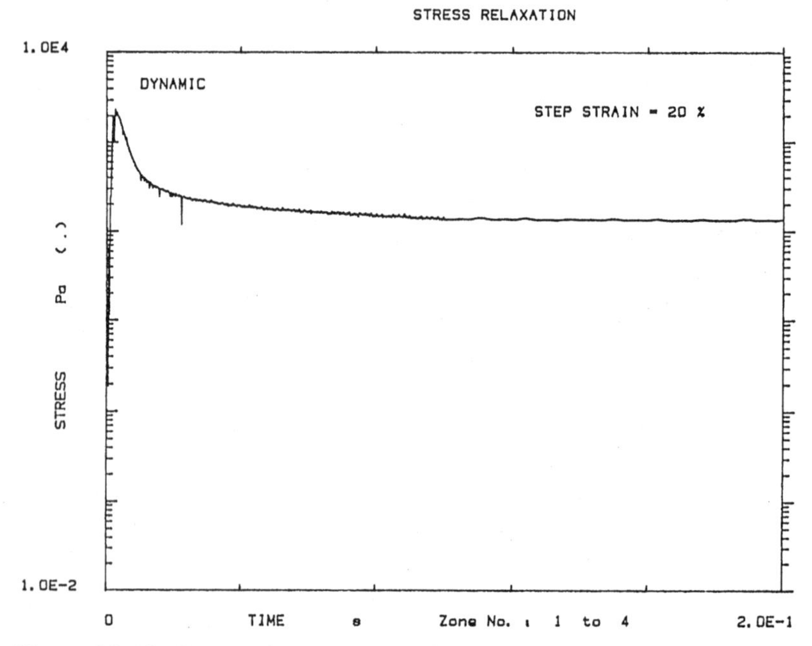

Figure 10-10. Stress relaxation curve of a printing paste.

curve reflect the characteristics of a solder paste such as yield value, leveling viscosity, and the internal structure change of the paste.

10.3.2.2 Paste rheology—elastic behavior. Parallel to Newton's Law for viscous fluids is Hooke's Law for solids. Hooke's Law relates shear stress τ and shear strain r in the following equation:

$$\tau = Gr$$

where G is the shear modulus. Under stress, the deformation of most materials differs substantially from the idealized case as described by Hooke's Law.

Solder paste is neither a purely viscous nor a purely elastic material; it exhibits viscoelastic properties. Therefore, solder paste is characterized by its flow behavior as well as its elastic properties. Because of the time factor involved in the viscoelastic phenomenon, dynamic mechanical measurements are needed for characterizing the viscoelasticity. The theoretical and mathematical treatment of viscoelastic properties are thoroughly covered by J. D. Ferry.[3]

At a constant temperature, the dynamic mechanical measurements that characterize the elastic properties include three parameters—storage modulus, loss modulus, and complex viscosity—in response to the operating parameters of frequency, strain amplitude, and time.

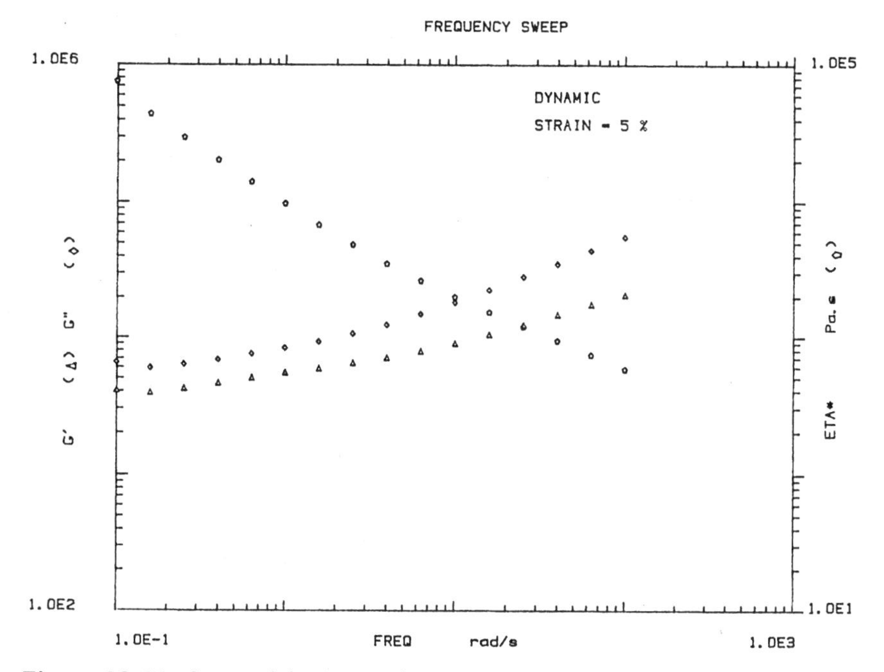

Figure 10-11. Sinusoidal stress and strain.

Under dynamic measurements, the stress is applied sinusoidally at a given frequency $\overline{\omega}$ as represented by Fig. 10-11. The strain also alternates sinusoidally but is out of phase with the stress at phase angle δ. Therefore, the mathematical equations for stress and strain, respectively, are

$$\tau = \tau_0 \sin \overline{\omega}\tau$$

$$r = r_0 \sin (\overline{\omega}\tau + \delta)$$

The dynamic shear modulus may be written as a complex modulus G^*, with the real and imaginary parts, as shown in the vectorial components.

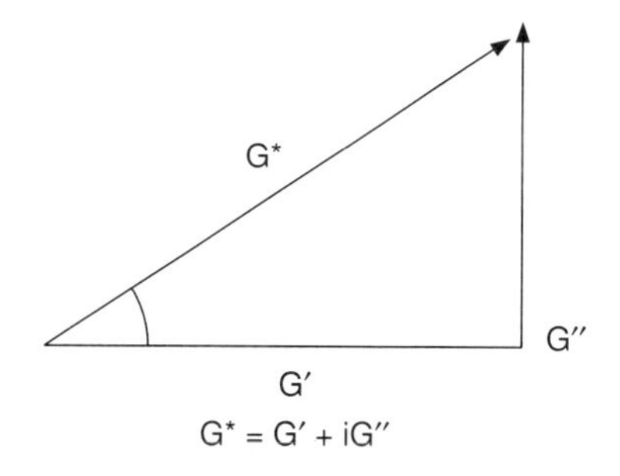

The imaginary part of the complex modulus is the damping term which determines the dissipation of energy when material is deformed and is called the *loss modulus* G''. The real part of the complex modulus represents the amount of recoverable energy stored as elastic energy and is called the *storage modulus* G'. The ratio of the loss modulus to the storage modulus is called the *loss tangent*, tan δ:

$$\frac{G''}{G'} = \tan \delta$$

The above equation is a convenient parameter in practical measurements, as it indicates the relative portion of viscous and elastic properties of a material.

The characteristics may be defined by analyzing the following features:

- Magnitude of the three properties
- Relative magnitude of loss and storage modulus
- Relative changes of loss and storage modulus in relation to frequency of change, strain amplitude, or time
- Relative rate of change in complex viscosity versus frequency change, or strain amplitude, or time

The storage modulus, loss modulus, and complex viscosity can be monitored in relation to frequency sweep under constant strain. These three properties can also be characterized in relation to strain sweep under constant frequency. Figure 10-12 is a measurement of storage modulus, loss modulus, and complex viscosity of a printing paste in relation to frequency under the constant strain of 5 percent. The relative weight of viscous and elastic portions in solder paste can be directly measured as the loss tangent value, as shown in Fig. 10-13.

10.3.2.3 Viscosity measurement. Under two main paste application methods —fine dot dispensing and pattern printing—a paste possessing a low yield point and slight plastic behavior is found most suitable for dispensing applications. This is consistent with the formulation requirement that a dispensing paste normally has a lower upper limit in volume fraction of metal load as well as a lower viscosity range, in comparison with pastes designed for stencil/screen printing applications. Table 10-1 lists typical viscosities and metal load percentages for the dispensing and printing application techniques, although some exceptions may exist.

A low yield value physically signifies that a paste will start to flow under low shear stress. Too high a yield point may result in poor level-

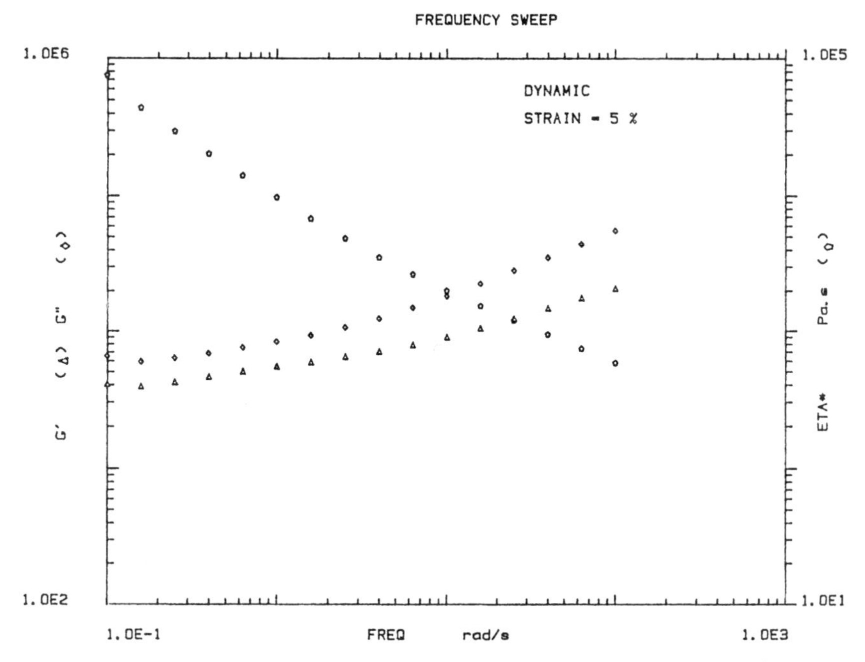

Figure 10-12. Storage modules, loss modules, and complex viscosity of a printing paste.

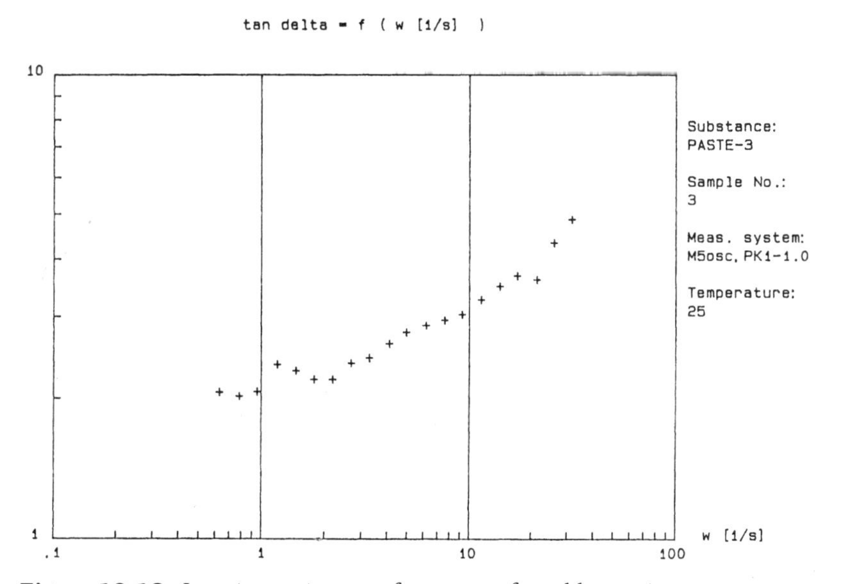

Figure 10-13. Loss tangent versus frequency of a solder paste.

Table 10-1. Viscosity and Metal Load of Dispensing and Printing Paste

Solder paste type	Viscosity, centipoise (cP)	Metal load, wt %
Fine dot dispensing	200,000–450,000	Up to 88
Screen printing	450,000–750,000	Up to 92
Stencil printing	650,000–1,200,000	Up to 92

ing which leaves an unsmooth surface, and too low a yield point may cause running and slumping.

A printing paste normally demands a higher viscosity range. A moderate yield point, slight thixotropy, and shear thinning are found to be suitable for printing mechanics. Too high a degree of thixotropy may cause slumping and sagging. A proper shear thinning is desirable, especially at a high shear rate, as is encountered when a paste is being transferred through the opening during a printing process.

A single-point viscosity measurement as a criterion to represent the rheology of a solder paste is prevalent in the industry. It is thus appropriate to discuss some practical aspects of single-point viscosity measurement. One of the common techniques adopted is to use the Brookfield viscometer with a helipath spindle TF at 5 r/min under a specified condition. Measurements obtained by this technique are simple and convenient. The new revision of ANSI/J-STD-005 "Requirements and Test Methods for Solder Paste" provides alternate test methods: T-bar spindle (Brookfield RVT or equivalent) or spiral pump (Malcom, Brookfield or rheometer) for viscosity measurement. However, single-point viscosity is valid only when the following conditions are met:

- The rheology of the paste is known.
- The paste handling technique is standardized immediately prior to measurement.
- The measurement technique of the paste is standardized.
- The paste container is specified in size and dimensions.

To illustrate the point, Table 10-2 is a summary of viscosity data obtained by the use of the Brookfield Model RTV viscometer with helipath spindle TF at 5 r/min for two pastes, paste *A* and paste *B*. The variables in this table are paste handling prior to descending spindle as expressed by the mixing time, and the measuring time (time-to-read), i.e, the time interval from the start of the spindle descending to

Table 10-2. Single-Point Viscosity of Pastes *A* and *B* in Relation to Premixing Time and Time-to-Read

	Viscosity, at time to read, kilocentipoise (kcP)						
Mixing time,min	30 s	45 s	60 s	90 s	120 s	100 s	180 s
Paste *B*							
0	—	—	1440	1550	1700		
1	—	—	840	950	110		
2	—	600	660	720	800	900	
3	—	580	640	700	800	880	900
Paste *A*							
0	200	220	220	280	400	430	
1	190	210	220	240	270	290	
2	190	200	210	220	230	250	
3	170	180	200	200	220	230	

obtaining the dial reading. This time interval also indicates the spindle position in the paste container.

Figure 10-14 plots the viscosity versus time-to-read dial for four different premixing times of paste *B*—no mixing, 1-min mixing, 2-min mixing, and 3-min mixing. The mixing technique for all measurements is kept the same with a 1¼-in diameter, three-blade propeller at a speed of 240 r/min. All measurements are made in the same size container (4-fluid-oz jar with 2-in diameter and 4-in height).

The data clearly indicate the effect of premixing. Longer mixing generates a lower viscosity number across the full range of reading time from 0.5 to 3 min. Under the test conditions, after the paste is mixed thoroughly for 2 min, further mixing does not significantly affect viscosity. The data also indicate that the viscosity reading increases with increasing time-to-read, due to increased friction as the spindle is descending.

An equivalent test is carried out for paste *A*. The comparison between paste *A* and paste *B* indicates that paste *A* has much less variation in response to mixing time and time-to-read. Again, as paste is mixed for 2 min, further mixing has no significant effect.

The comparative results demonstrate the relative sensitivity of pastes to handling (mixing) and measuring techniques. With these variations in mind, it is not surprising to see the inconsistency and lack of agreement from laboratory to laboratory and operator to operator. Nonetheless, with the proper technique and control, viscosity data can be consistently obtained. Figure 10-14 shows the good reproducibility of measurements conducted under the specified technique.

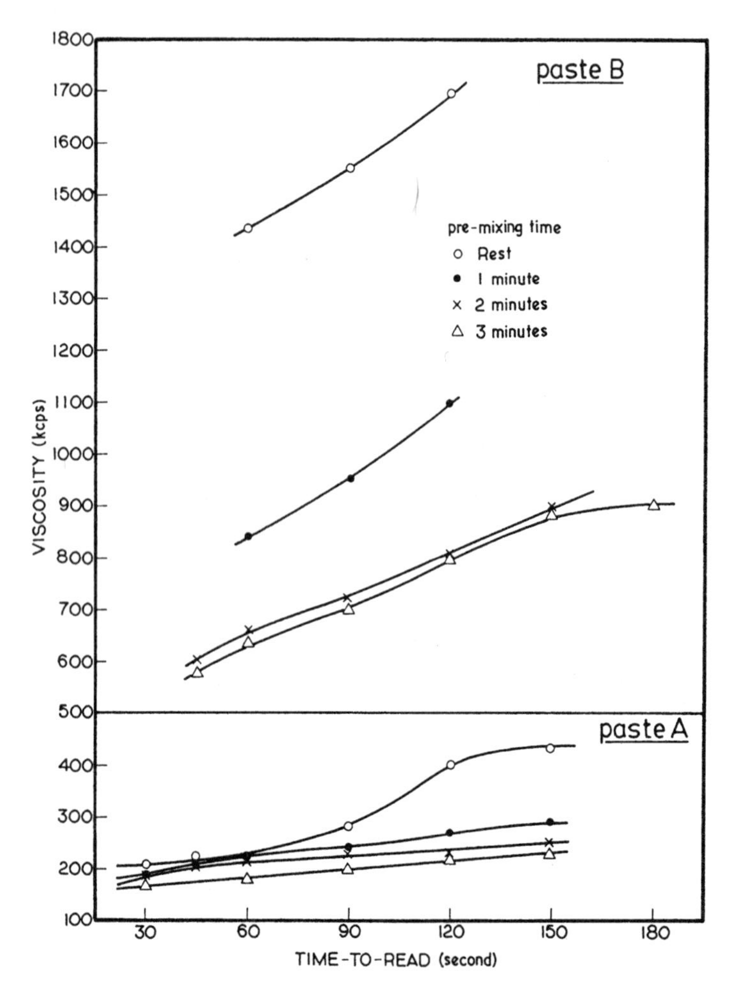

Figure 10-14. Viscosity versus time-to-read at four premixing times.

In practice, realizing such variations as well as being able to control them is important to the establishment of specifications and communication between users and vendors, which in turn are important to overall quality and production yield.

Single-point viscosity can also be utilized to develop an indicator for the shear thinning factor by measuring viscosities at two significantly different shear rates in revolutions per minute, such as viscosities at 1 r/min and 20 r/min. The ratio of these two viscosities is the shear thinning factor. Again, the paste handling has to be kept consistent in order to obtain meaningful results.

10.3.2.4 Paste transfer technique. Three major techniques for applying paste onto the designated solder pads are

1. Printing
2. Dispensing
3. Pin transfer

Printing. The printing parameters and factors affecting printing results are covered in Sec. 10.4.

Dispensing.[2] Basically, a pneumatic dispenser delivers paste by means of air pressure that pushes the paste through a specified channel. The size of the paste deposit for attaching components or leads of electronic assemblies is usually very small. Table 10-3 provides the common needle gauge and corresponding diameters. The paste reservoir is a syringe or equivalent container in different sizes, from 10 to 600 cm^3 (20 fl oz). The amount of paste packed in different sizes of syringes depends on the type of alloys and metal content. The common ranges are listed in Table 10-4. It is advisable to select a proper size of dispensing container (syringe) to accommodate a specific production flow. Too big a syringe may jeopardize the shelf stability of the unused paste; and too small a tube requires frequent tube changes, reducing production throughput. Air pressure applied on the top surface of the paste presses the paste out of the syringe reservoir, through the fine needle, and deposits it on the desired locations of the substrate. For discrete dot dispensing, pulsed air pressure regulated by a controller is used. For a continuous line or bead of paste, continuous pressure is used. Operation can be manual or automated. Both types of equipment are available. The microprocessor, with numerically controlled *xyz* table, directs the precise location of paste deposit on the designated locations.

Table 10-3. Dispensing Needle Gauges and Inner Diameters

	Inner Diameter	
Needle gauge	in	mm
15	0.054	1.37
18	0.033	0.84
20	0.023	0.58
21	0.020	0.51
22	0.016	0.41
23	0.013	0.33
25	0.010	0.25
30	0.006	0.15

Table 10-4. Common Dispensing Paste Reservoir Container Size and Amount of Paste

Paste reservoir container, cm^3	Amount of paste, *g*
10	35–45
30	95–105
35	100–110
60	190–210
180 (6 fl oz)	450–650
360 (12 fl oz)	1000–1200
600 (20 fl oz)	1900–2100

For a reliable process, the key requirements are to maintain constant pressure and flow rate in the syringe and to have a pressure-stable paste. The pressure required to produce a constant flow rate depends on several parameters. Considering that the system is equivalent to pumping viscous fluid through a pipe, the flow rate increases approximately with the square root of the applied pressure, and the pressure P, is related to the viscosity of the fluid η, the length of the tube L, the volume of the fluid V, the radius of the tube R, and the time t, by the following equation:[4]

$$P = \frac{\eta L\, V}{R^4 t}$$

Thus, the combination of smaller volume, short passing distance, less viscous fluid, and the large radius of the tube demands less pressure for dispensing.

The pressure decay in the pipelines depends on the viscous resistance of the material and the pipeline transitions such as valves, elbows, bends, and pipe expansion and contraction.[5] In view of this, straight-line dispensing with minimum pipeline transitions facilitates smooth dispensing. As indicated in the equation, the pressure is inversely proportional to the fourth power of the radius, explaining that one increment of the needle gauge can affect the paste dispensability drastically.

The dispensing time depends on the size of the syringe and the size of the dot placed. For an application using a 30-cm^3 syringe containing 100 g of paste to dispense a single dot through a 20 gauge needle by 1 s on/1 s off pulsing, the dispensing time is approximately 15 h. Therefore, the paste for this application must be able to withstand the applied pressure at the ambient temperature for 15 h. Equally important are the quality of the equipment and the setup of the equipment.

Many practical points are to be considered for obtaining consistent and smooth dispensing, as the following indicates.

- Carefully select a proper dispensing system whether for manual or robotic application.
- Thoroughly check the whole system including the needle and connection tubes (if any) to assure the system is clean and clear.
- Always flush the system with a compatible lubricant to wet the wall of the channel so that the paste will experience a minimum of friction to start.
- The paste syringes should be kept under cold storage (or under conditions paste manufacturers recommend) to assure stability and freedom from physical separation.
- The paste syringe should be brought to a temperature in equilibrium with the ambient before starting dispensing. The flow rate depends on viscosity, which can be altered by temperature change. Even a small temperature difference may cause a wrong start, resulting in erratic results, particularly for very fine dots (i.e., 20 gauge or finer needle).
- Make sure that no foreign particles are introduced and that no solder particles are flaked, since solder alloys are relatively soft and easily deformed.
- The pressure applied in the syringe should be kept at a minimum, and the proper head pressure kept in the range of 15 to 25 lb/in^2. In cases where a paste requires much higher pressure (more than 40 lb/in^2) to dispense, the probability of having a consistent and continuous dispensing is very low.
- The external air pressure supply should be maintained constant. An air regulator to assure constant pressure input is needed.
- The clearance (gap) between the needle and the substrate affects the shape and quality of the dot dispensed. If the clearance is too little, the dot tends to be flattened out, and if too large, the dot tends to have long tailing. The Hershey Kiss shape is considered desirable.
- Select the right paste for the dispensing system. The required characteristics of a dispensing paste include: a viscosity of 350,000 ± 100,000 cP (Brookfield RVT, helipath TF/5 r/min, 3 min mixing/2 min reading); rheology displaying low thixotropy and low yield point, yet no drooping during dispensing and idle time; clean breakup without trailing tail (or stringiness); and physical stability in ambient temperature under applied pressure during the entire

dispensing time. A viable dispensing paste should provide consistent and reproducible dot placement without any disruption until the whole syringe is completely depleted. The powder used in the paste is another important factor to the performance and stability of the paste. A general guideline in selecting powder size for various needle gauges is provided in Table 10-5.

- Chemical compatibility is essential. The materials used in a dispensing system, metal or polymer in nature, with which the paste is in direct contact must be checked for chemical compatibility with the paste to be used.
- For extensive downtime, the whole dispensing system should be flushed without leaving any paste in any part of the system. This step may appear to be redundant, but in the long run it is time-saving and a quality assurance.

Pin transfer. The potential advantage of this technique is being able to deposit paste in small and multiple dot form to already populated boards where a printing process is not suitable and where a large number of dots can be placed simultaneously. The process requires the design of a pin pattern which exactly corresponds to the board design and to the uniform wetting of the tip of the pin. The basic parameters to be determined are the shape, the diameter, and the length of the pin, the technique of making a reservoir layer and its control, and the characteristics of the paste.

A suitable paste for this process should possess the proper fluidity and wetting ability for the pins at the moment of dipping but not be too fluid to stay on the pin so that a complete transfer to the designated location can be achieved. The viscosity for a well-designed paste for pin transfer is in the range of 400,000 ± 100,000 cps (Brookfield RVT,

Table 10-5. Selection Guideline of Solder Powder for Dispensing Application

Needle gauge	Needle inner diameter in	Needle inner diameter μm	Applicable powder (mesh cut)
20	0.023	575	−200/+325 or −325/+500
21	0.020	500	−325/+500 or −325
22	0.016	400	−325/+500 or −325
23	0.13	325	−325/+500 or −325
24	0.12	300	−325/+500 or −325

helipath TF/rpm, 3 min mixing/2 min reading). In addition, the paste requires good environmental stability, since the paste reservoir is essentially an open system exposed to the ambient environment.

10.3.3 Solder precoating via in situ chemical reaction[6]

A pasty mixture, that is composed of a Pb salt of organic acid and Sn powder in a predesigned stoichiometric ratio, is deposited in lines (beads) across fine-pitch pads. At the heating temperature of above 183°C, the Pb element is released and it combines with Sn to form Sn-Pb eutectic solder, which concurrently wets the substrate solder pads, forming a solder precoat with a composition of 63 Sn/37 Pb. The reaction is expressed as

$$(RCOO)_2^- Pb + Sn \rightarrow SnPb + RCOO^-$$

The solder deposit made by this technique reportedly offered the lowest oxygen content as compared with electroplated solder and as-reflowed solder paste. Figure 10-15 illustrates the formation of precoat. The thickness of precoat solder ranges from 10 to 100 μm. It is found that the bond strength becomes stable and consistent as the solder thickness reaches 18 μm or greater. For hot bar soldering, it is suggested that a solder precoat thickness of 25 to 30 μm is desirable. Figure 10-16 depicts the lead bond strength in relation to solder precoat thickness for TAB lead [lead width 100 μm (0.004 in); lead pitch 0.3 mm (0.12 in)] after hot bar solder.[7]

It is also found that interpad migration resistance after precoating with the above process, as measured by insulation resistance, was superior to that made with other conditions. The order of migration resistance was ranked as in situ reaction precoat solder > electroplating > as-reflowed solder paste > bare Cu pad.

10.4 Printing Fine-Pitch Solder Paste

10.4.1 Major factors

Major factors contributing to the printing results of solder paste in addition to the selection of solder paste are

1. Printing parameters
2. Characteristics and selection of solder powder

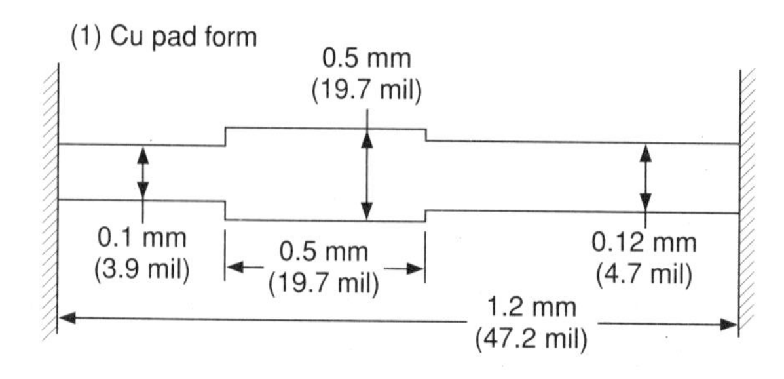

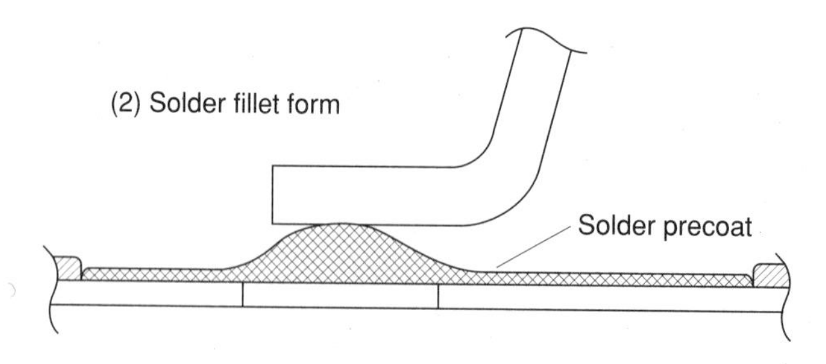

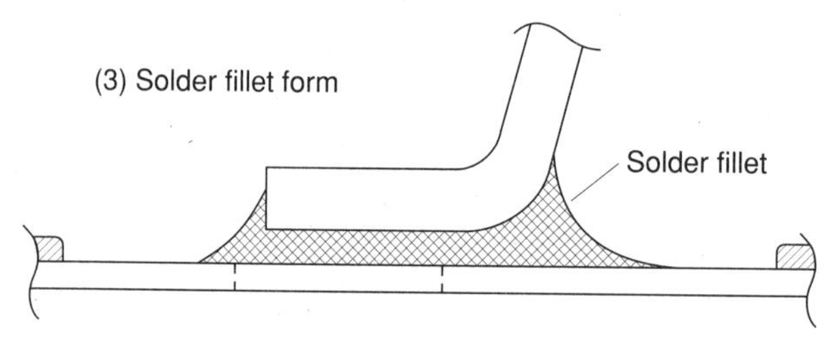

Figure 10-15. Formation of precoat via in situ chemical reaction.

3. Stencil thickness versus aperture design
4. Stencil aperture versus land pattern
5. Stencil selection

10.4.2 General printing principles and parameters

Several forces contribute to the paste transfer process:

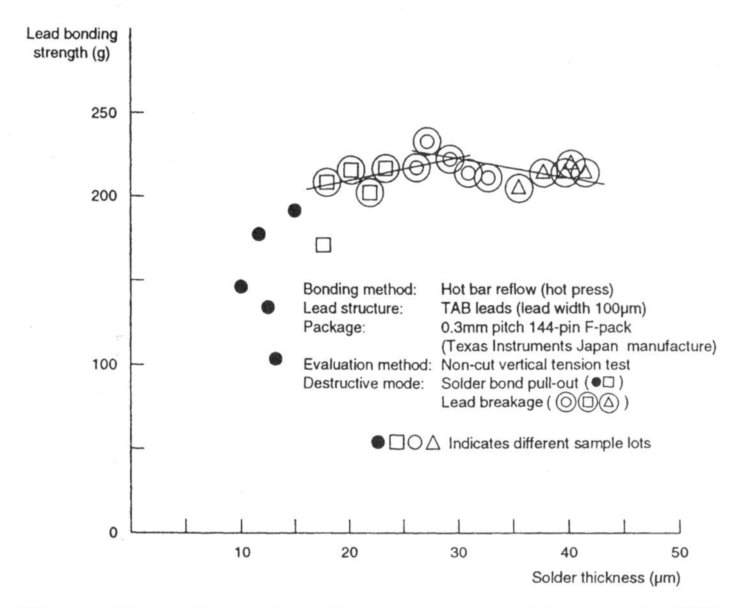

Figure 10-16. Bond strength versus precoat thickness for TAB leads.

1. The relative adhesion between the paste/stencil interface and the paste/substrate interface
2. The pushing force by the squeegee along the y direction
3. The pressure (force per unit area) exerted by the squeegee along the z direction
4. The hydraulic pressure created in the paste when the paste is compressed between the squeegee and the stencil
5. The gravity force.

The squeegee makes a line contact with the stencil and a close contact with a given amount of paste. The paste thus moves in front of the squeegee while the squeegee is in motion and is released behind the squeegee through the openings to the substrate. During squeegee motion, the hydraulic pressure developed in the paste is believed to press against the stencil and the squeegee, resulting in the lifting of the squeegee edge and the extrusion of paste through the stencil opening. General printing variables include

Squeegee system

- Blade material type

- Blade hardness
- Blade edge angles
- Blade edge flatness
- Blade configuration
- Squeegee size
- Squeegee speed
- Downstop pressure
- External pressure

Substrate

- Material type
- Surface condition
- Surface flatness
- Substrate size

Screen system

- Mesh screen

  Wire material type
  Wire diameter
  Mesh tension
  Mesh count
  Mesh orientation
  Emulsion type
  Emulsion thickness
  Print pattern
  Screen size

- Metal mask stencil

  Material type
  Flatness
  Thickness
  Print pattern
  Stencil size

Paste

- Paste consistency
- Paste rheology
- Paste stability
- Paste tackiness
- Chemical compatibility with screen and squeegee
- Compatible thickness with printing pattern

Snap off

With a good fine-pitch paste available, fine-pitch printing requires a different set of printing parameters from those for non-fine-pitch printing. The general direction is outlined below as a starting point.

1. *Registration.* Desirably 0.0005 in (0.013 mm) accuracy.
2. *Printing mode.* Preferably one-direction printing to retain precise alignment.
3. *Board fixation.* High vacuum volume to secure the board.
4. *Squeegee hardness.* Greater hardness, 80 to 90 durometer or flexible metal blade. However, when a step-stencil is used or when the stencil is thick, a lower durometer squeegee may be more forgiving.
5. *Squeegee speed.*
   - More critical as stencil thickness increases. When the stencil is less than 0.004 in (0.10 mm) thick, the effect of squeegee speed diminishes.[8]
   - Slower speed imparts a more consistent print height and the height is closer to the stencil thickness.
   - General speed range of 0.1 to 1.0 in/s (2.5 to 25 mm/s).
   - Squeegee speed should be set slower as stencil is thicker, e.g., 850.000 cP paste, 20-mil (0.50-mm) pitch.

Stencil thickness, in (mm)	Squeegee speed, in/s (mm/s)
0.008 (0.20)	0.1–0.4 (2.54–10.2)
0.006 (0.15)	0.3–0.8 (7.62–20.3)
0.004 (0.10)	0.5–1.0 (12.70–25.4)

6. *Squeegee pressure.* Always avoid excessive pressure. External pressure depends on equipment and other setup parameters. Start with low pressure (15 to 20 lb) and gradually increase the pressure until the squeegee obtains a clean sweep on the stencil.
7. *Component consideration.* Possible variations in component lead dimension and configuration should be considered when depositing solder paste.[9,10]

10.4.3 Characteristics of solder powder

As the quality of solder paste is raised for fine-pitch applications, the requirements of solder are also more stringent. The following outlines the properties and characteristics of solder powder which are important to the quality of fine-pitch solder paste and to the printing results.

1. *Appearance.* The solder powder shall have a clean, dry-sand appearance without agglomeration and have the inherent color of the specific alloy.

2. *Flow rate.* According to the ASTM B-213, "Test for Flow Rate of Metal Powders," the fine solder powder shall be free flowing through the orifice of a Hall flowmeter without stoppage or clogging.

3. *Density.* High density is required. Desirably, each individual particle should have a density as close to the theoretical density of the alloy as possible with no internal voids. Precision sectioning of one particle coupled with microscopic examination is one technique for confirmation.

4. *Particle size and size distribution.* The powder has been customarily represented by mesh designation according to ASTM B-214, *Test Method for Sieve Analysis of Granular Metal Powders.* Table 10-6 summarizes the typical range of particle sizes, average diameter, and distribution, for two types of powder cuts: type *A* (−200 to +325) and type *B* (−325 to +500). It should be noted that the sieving parameters and the material of the sieve mesh may affect the actual measurements. The measurements may also change with the instrument and technique being employed. The ratio for the surface area per unit volume is estimated based on the assumptions that the powder particles are perfectly spherical and the volume is occupied by a close-packed structure with more than 95 percent efficiency.

5. *Particle shape and morphology.* To balance the flux demand, rheology, and metal load acceptance in paste form, solder particles should have a smooth surface and be spherical in shape. Figures 10-17 through 10-20 exhibit the general surface characteristics of four 63 Sn/37 Pb powders classified in −200/+325 mesh, −325/+500 mesh, −400/+500 mesh, and −500 mesh. Figure 10-21 further reveals the morphology and metallurgy of good-quality solder powder with well-distributed two-phase microstructure.

10.4.4 Selection of solder powder

The required particle size of solder powder incorporated in solder paste may vary with the specific flux/vehicle system designed by a manufacturer; nonetheless, Table 10-7 serves as a guideline of relationships among the component pitch, the selected stencil aperture design, the maximum powder size allowable, and the corresponding powder mesh cut. To obtain good printability for 25-mil (0.63-mm) pitch com-

Table 10-6A. Particle Size Distribution (−200/+325)

Size interval through	Percent by count	Percent by volume (weight)
71	99.90	99.77
70	99.90	99.77
69	99.71	99.35
68	99.71	99.35
67	98.95	97.79
66	98.95	97.79
65	97.99	95.98
64	96.08	92.64
63	96.08	92.64
62	92.26	86.46
61	92.26	86.46
60	87.95	80.06
59	81.55	71.31
58	81.55	71.31
57	74.28	62.20
56	74.28	62.20
55	63.29	49.60
54	51.91	37.72
53	51.91	37.72
52	43.02	29.27
51	43.02	29.27
50	30.88	18.82
49	21.99	11.90
48	21.99	11.90
47	14.15	6.42
46	14.15	6.42
45	9.94	3.78
44	9.94	3.78
43	7.07	2.17
42	5.16	1.22
41	5.16	1.22
40	4.68	1.01
39	4.69	1.01
38	3.73	0.63
37	3.35	0.50
36	3.35	0.50
35	2.58	0.28
34	2.58	0.28
33	2.39	0.23
32	2.20	0.19
31	2.20	0.19
30	1.82	0.12
29	1.82	0.12
28	1.53	0.07
27	1.24	0.03
26	1.24	0.03
25	1.24	0.03

Table 10-6B. Particle Size Distribution (−325/+500)

Size interval through	Percent by count	Percent by volume (weight)
43	99.26	98.20
42	99.26	98.20
41	98.51	96.65
40	96.65	93.01
39	95.35	90.62
38	92.94	86.46
37	84.94	73.98
36	80.67	67.96
35	72.49	57.23
34	68.03	51.79
33	59.29	41.89
32	47.96	30.44
31	40.15	23.44
30	33.83	18.22
29	29.55	14.97
28	23.61	10.83
27	18.59	7.64
26	11.52	3.72
25	10.22	3.10
24	8.36	2.30
23	5.76	1.29
22	4.09	0.70
21	2.60	0.25
20	2.60	0.25
19	2.23	0.17
18	1.86	0.10
17	1.67	0.07
16	1.49	0.05
15	1.49	0.05
14	1.30	0.03
13	1.12	0.02
12	1.12	0.02
11	0.93	0.01
10	0.74	0.00
9	0.74	0.00
8	0.74	0.00
7	0.74	0.00
6	0.37	0.00
5	0.19	0.00
4	0.19	0.00
3	0.00	0.00
2	0.00	0.00
1	0.00	0.00
0	0.00	0.00

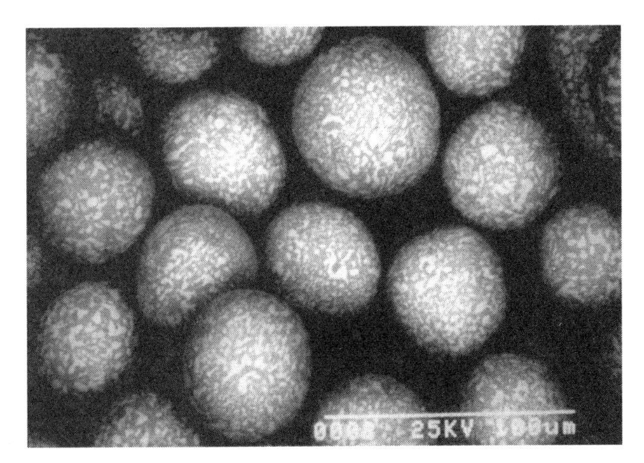

Figure 10-17. SEM micrograph of 63 Sn/37 Pb solder powder in −200/+325 mesh.

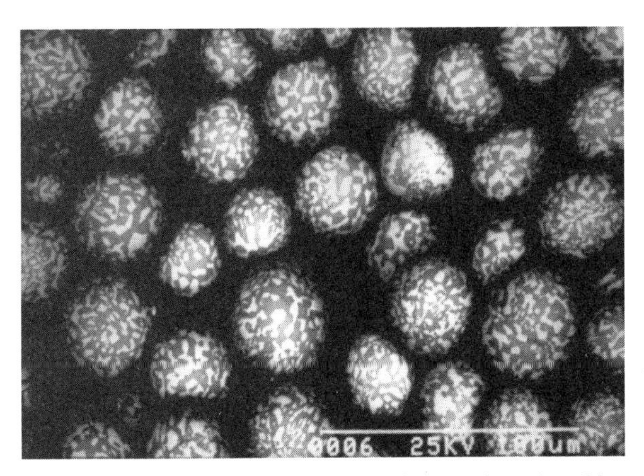

Figure 10-18. SEM micrograph of 63 Sn/37 Pb solder powder in −325/+500 mesh.

ponent pads, the paste containing either −200/+325 or −325/+500 powder can be used. In this case, the powder selection depends on the chemistry and rheology of the flux/vehicle system. For printing a 15-mil (0.38-mm) pitch pattern, the mesh cuts of −325/+500 or −400/+500 or finer can be used and the choice among −325/+500 and −400/+500 or finer powders also depends on the specific flux/vehicle system. It has been demonstrated that a well-designed paste can readily print a 15-mil (0.38-mm) pitch pattern if a −325/+500 powder is used.

Figure 10-19. SEM micrograph of 63 Sn/37 Pb solder powder in −400/+500 mesh.

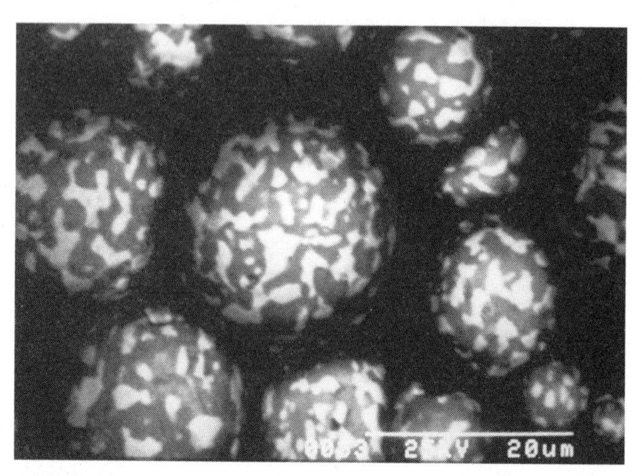

Figure 10-20. SEM micrograph of 63 Sn/37 Pb solder powder in a range of 10 to 25 μm.

It is always advantageous to use the coarsest powder which is allowable by a flux/vehicle system for achieving the printability, so that the reduced cost and proper fluxing activity can be obtained.

10.4.5 Stencil thickness versus aperture design

When printing solder paste, the design of the relative dimensions of stencil thickness and stencil aperture is to achieve a balance between

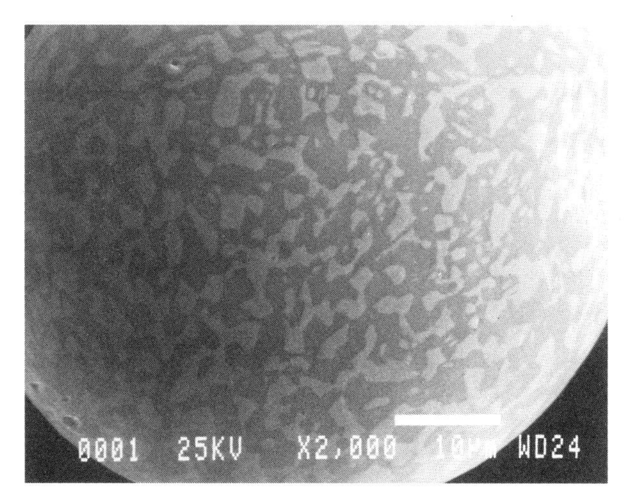

Figure 10-21. SEM micrograph of two-phase microstructure on a 63 Sn/37 Pb particle surface.

Table 10-7. Selection Guideline of Solder Powder for Printing Applications

Component pitch		Stencil aperture width		Maximum powder size, μm	Applicable mesh cut
mil	mm	mil	mm		
50	1.26	26	0.66	90 ± 5	−200/+325
25	0.63	14	0.36	65 ± 5	−200/+325 or −325/+500
20	0.50	10	0.25	53 ± 5	−325/+500
15	0.38	8	0.20	41 ± 5	−325/+500 or −400/+500 or finer
10	0.25	6	0.12	25 ± 5	−500 or 5–15 μm

the printing resolution and the proper amount of solder deposit in order to avoid starved solder joints or pad bridging. For a selected stencil thickness, too small a stencil aperture width leads to open joints or starved joints and too large an aperture width causes pad bridging. As a guideline, as shown in Table 10-8, the ratio of stencil aperture width to stencil thickness should be greater or equal to 1.6.

Table 10-8. Guideline of Stencil Thickness versus Aperture Width

Component lead pitch		Aperture width		Maximum stencil thickness	
in	mm	in	mm	in	mm
0.050	1.26	0.023	0.58	0.014	0.35
0.025	0.63	0.012	0.30	0.0075	0.19
0.020	0.50	0.010	0.25	0.0063	0.16
0.015	0.38	0.007	0.18	0.0043	0.11
0.008	0.20	0.004	0.10	0.0025	0.06

10.4.6 Stencil aperture design versus land pattern

In order to make solder joints by a one-pass printing process, the stencil thickness must be selected for transferring a sufficient amount of paste onto the non-fine-pitch solder pads, while avoiding depositing an excessive amount of paste onto the fine-pitch pads. There are several options to achieve the deposition of a proper amount of solder paste on the land pattern to accommodate a mix of solder pads sizes.

1. *Step-down stencil.* This is commonly achieved by chemically etching the non-fine-pitch pattern area from one side of the stencil while etching the step-down area with the fine-pitch pattern to the other side during a double-sided etching process. Alternatively, the step-down area, is etched in one foil and the non-fine-pitch pattern is etched in the other foil. Then the two foils are registered and glued together. The practical step gradient is 0.002 in (0.05 mm). Some common combinations are (1) 0.008 in (0.20 mm) for non-fine-pitch and 0.006 in (0.10 mm) for fine-pitch and (2) 0.006 in (0.15 mm) for non-fine-pitch and 0.004 in (0.15 mm) for fine-pitch.

2. *Uniform reduction on four sides of apertures.* The dimensions of the fine-pitch aperture on the stencil are reduced by 10 to 30 percent in relation to those of the land pattern. This reduces the amount of paste deposition on the fine-pitch land pattern and also provides some room for printing misregistration and paste slump, if any.

3. *Staggered print.* The opening of the stencil is only one-half the length of the solder pad and is arranged in an alternate manner as shown in Fig. 10-22. For tin-lead–coated solder pads, when the paste starts to melt during reflow, the molten solder is expected to flow to the other half of the pad, making the complete coverage. With a bare

Figure 10-22. Staggered print pattern.

copper or nickel surface, the molten solder may not flow out to cover the area that the paste is not printed on.

4. *Length or width reduction.* The dimensions of the stencil opening are reduced along the length or width by 10 to 30 percent in relation to that of the solder pads, achieving the reduction of the amount of paste deposited.

5. *Other shapes.* The stencil openings are made with selected shapes such as a triangle or a teardrop in order to achieve the reduced solder paste deposition on the fine-pitch pattern.

6. *Compromised stencil thickness.* Instead of using the specific thickness which is considered as the most suitable for a specific land pattern, select a thickness which is practical to both fine-pitch and non-fine-pitch patterns. The following table lists some examples.

Land pattern pitch combinations	Compromised stencil thickness	
	in	mm
50 mil (1.26 mm) and 25 mil (0.63 mm)	0.007–0.008	0.18–0.20
50 mil (1.26 mm) and 20 mil (0.50 mm)	0.006–0.007	0.15–0.18

10.4.7 Printability test

One of the test designs is outlined in Table 10-9. It illustrates the printability of various stencil aperture designs and land patterns for components with various lead pitches by using a fine-pitch water-soluble paste.

Table 10-9. Fine-Pitch Water-Soluble Paste Printability Test Design

Component pitch, mil (mm)	Pad dimensions, in (mm)	Stencil aperture design			
		Area 1, in (mm)	Area 2, in (mm)	Area 3, in (mm)	Area 4, in (mm)
50 (1.26)	0.022 × 0.078 (0.55 × 1.95)	0.032 × 0.078 (0.80 × 1.95)	0.030 × 0.078 (0.75 × 1.95)	0.027 × 0 .078 (0.68 × 1.95)	0.022 × 0.078 0.055 × 1.95)
25 (0.63)	0.014 × 0.095 (0.36 × 2.41)	0.018 × 0.050 (0.45 × 1.27)	0.0112 × 0.095 (0.28 × 2.41)	0.0126 × 0.095 (0.32 × 2.41)	0.014 × 0.095 (0.36 × 2.41)
20 (0.50)	0.010 × 0.095 (0.25 × 2.44)	0.013 × 0.050 (0.33 × 1.27)	0.008 × 0.095 (0.20 × 2.41)	0.009 × 0 .095 (0.22 × 2.41)	0.10 × 0.095 (0.025 × 2.41)
15 (0.38)	0.008 × 0.095 (0.20 × 2.41)	0.010 × 0.050 (0.25 × 1.27)	0.0064 × 0.095 (0.16 × 2.41)	0.0072 × 0.095 (0.18 × 2.41)	0.008 × 0.095 (0.20 × 2.41)

SOURCE: Test design courtesy of Storagetek Corp. and solder paste courtesy of IEM Corp.

Test parameters

Board design. Land pattern containing 50-mil (1.26-mm) pitch, 25-mil (0.63-mm) pitch, 20-mil (0.50-mm) pitch, 15-mil (0.38-mm) pitch, passive components, and discretes.

Stencil aperture design. Dimensions are specified in Table 10-9.

- *Area 1.* Staggered pattern with width increase of 25 to 30 percent.
- *Area 2.* Twenty percent width reduction for 25-mil-, 20-mil-, and 15-mil-pitch components. There is a 0.008-in (0.20-mm) width increase for 50-mil components.
- *Area 3.* Ten percent width reduction for 25-mil-, 20-mil-, and 15-mil-pitch components. There is a 0.005-in (0.13-mm) width increase for 50-mil components.
- *Area 4.* One hundred percent pad dimensions.

Stencil. Made of stainless steel. The step down is 6 mil and 4 mil (0.15 mm and 0.10 mm). Step down of 6 mil (0.15 mm) for 50-mil components, and step down of 4 mil (0.10 mm) for 25-mil, 20-mil, and 15-mil components.

Squeegee speed. 0.39 to 0.78 in/s (10 to 20 mm/s)

Squeegee pressure. 1.5 to 2.5 lb/in.

Test results

All four areas with four different stencil aperture designs showed good printability—no wet bridging, no smudging, no tailing as shown in Fig. 10-23—and there was good solderability without pad bridging after reflow as shown in Fig. 10-24. Further examination, under an optical microscopic, revealed that the stencil aperture design area (3) delivered the best print resolution for all fine-pitch patterns as shown in Fig. 10-25*a*, *b*, and *c*. This design area with 10 percent width reduction in stencil aperture provides a good balance of various tradeoffs: solder volume, printability, placement variation, and paste characteristics.

10.4.8 Stencil selection

The performance of stencils is primarily driven by the foil metal and the process being used to create the printing pattern. Currently five types of stencil materials are commercially available—brass, stainless steel, molybdenum, Alloy 42, and electroformed nickel. The processes making the stencils may involve chemical etching, laser cutting, elec-

Figure 10-23. Fine-pitch prints.

Figure 10-24. Reflowed solder joints of fine-pitch prints.

tropolishing, electroplating, and electroforming. Each type of foil or fabricating process possesses inherent merits and limitations.[11,12]

The key performance of a stencil is assessed by the straight vertical wall, wall smoothness, and dimensional precision. In addition, durability, chemical resistance, fine opening capability, and cost are also important factors.

Brass foil has the lowest cost of all, is easy to etch, and provides a uniform wall finish. However, it is less durable and less chemical resistant. Stainless steel is harder to etch than brass or Alloy 42, and it is

Figure 10-25. Fine-pitch prints of stencil aperture with 10 percent reduction of opening dimension.

difficult to obtain a smooth uniform surface, although it is more durable. Thus stainless steel with a polished surface wall is particularly appropriate for applications requiring very thin foils such as filling via holes or depositing ultra-fine-pitch solder paste. Alloy 42 is also easy to etch as well as being compatible with laser cutting. It is particularly suitable for thicker stencils, 0.004 in (0.1 mm) or larger. Molybdenum is the most expensive of all, but provides a straight vertical and smooth wall as well as easy paste releasing. Table 10-10 summarizes the relative performance of stencil materials in key areas.

The selection of stencil thickness varies with the component type, since the amount of paste deposit determined by the stencil thickness depends on the land pattern for a specific type of component. Passive components, large-pitch chip carriers, or solder ball bumping for BGA need a larger amount of paste, and thus a thicker stencil, while fine-pitch chip carriers require a smaller amount of paste, and thus a thinner stencil. In order to achieve the selected amount of solder paste deposits required for various components to make good solder interconnections, several techniques can be considered as discussed in Sec. 10.4.6.

Chemical etching has been successful for making stencil apertures for many years. The process involves mounting and developing photoresist onto both sides of a metal foil and then etching from both sides. The etching action works in both vertical and horizontal directions. The lateral etching effect demands a design-in compensation factor to produce the desired dimensions and to minimize the formation

Table 10-10. Comparison of Stencil Materials in Key Performance Areas

Performance	Brass	Stainless steel	Molybdenum	Alloy 42	Ni (electroforming)
Mechanical strength	Unfavorable	Favorable	Favorable	Favorable	Favorable
Chemical resistance	Unfavorable	Favorable	Unfavorable	Favorable	Favorable
Etchability	Favorable	Less favorable	Favorable	Favorable	N/A
Sheet stock availability	Favorable	Favorable	Unfavorable	Favorable	N/A
Cost	Favorable	Less favorable	Unfavorable	Less favorable	N/A
Fine-pitch (openings) capability	Favorable	May need electropolishing	Favorable	May need electropolishing	Most favorable
Unique feature	Lowest cost	Durable	Self-lubricating; smooth wall	—	Finest opening

NOTE: N/A = not applicable.

Figure 10-26. Alloy 42 stencil with 1 mil under etch. (*Courtesy of Photo Stencil, Inc.*)

of a knife edge, as shown in Fig. 10-26 for an Alloy 42 stencil with a 6-mil (0.15-mm) thickness and a 10-mil (0.25-mm) nominal aperture.[11] To design the compensation factor, the rule of thumb is to reduce the openings by approximately one-half the thickness of the stencil foil. The extent of the knife edge and the occurrence of underetched or overetched openings depend on the etching parameters and their relation to the foil material. Figure 10-26 illustrates the 1-mil underetched aperture, and Fig. 10-27 shows the 1-mil overetched apertures for an Alloy 42 stencil with 6-mil (0.15-mm) thickness and 10-mil (0.25-mm) nominal aperture. It was also found that the efficiency and accuracy of chemical etching decreases as the foil thickness approaches 4 mil (0.1 mm) or below. It should be noted that the required compensation factors for various apertures may differ. This is due to the fact that the

Figure 10-27. Alloy 42 stencil with 1 mil over etch. (*Courtesy of Photo Stencil, Inc.*)

etch rate increases as the aperture size increases. A quality stencil comes from the careful design of the compensation factors accordingly.

The laser-quality cutting technique utilizes a YAG-laser energy beam with a diameter of 0.002 to 0.004 in (0.05 to 0.10 mm) to produce apertures. Its inherent repeatability and fine feature capability are its main merits. However, the surface of the aperture wall generally appears rough and grainy. For a stencil with a large number of apertures, it may take a longer time to form the stencil because of the sequential cutting operation.

The electroforming technique is based on the electrodeposition of nickel, atom by atom, commercially described as the E-FAB process.[13] It is characterized by precision, gasket effect, and sharp images. Figure 10-28 illustrates the stencil apertures made by electroforming, chemical etch, and laser cut.[13] The resulting aperture size has tighter tolerance limits compared with the chemical etch and laser cut processes. Figure 10-29 shows histograms of aperture width distribution with a target value of 0.0045 in (0.114 mm), lower tolerance limits of 0.0040 in (0.10 mm), and upper tolerance limits of 0.0050 in (0.13 mm) for the three processes. Ultrafine-pitch deposits of solder paste produced by an electroformed stencil with an 0.0045-in (0.114-mm) aperture width are represented in Fig. 10-30. The manufacturing process of electroformed stencils is represented in Fig. 10-31.[13]

Other techniques have been found to be beneficial to the quality and performance of stencils in transferring solder paste. Electropolishing is a technique that has demonstrated improvement to the smoothness of the aperture wall. Figures 10-32 and 10-33 compare the surfaces of an etched stainless-steel stencil and one that was etched and then electropolished. Figures 10-34 and 10-35 compare the wall surfaces of a laser-cut Alloy 42 stencil with one that was laser cut and the electropolished.[11] It should be noted that electropolishing can in the meantime increase the aperture size—a factor that needs to be considered at the design stage. It has been reported that nickel plating further improves the smoothness of the aperture wall of an electropolished stainless-steel stencil.[11] In addition, nickel plating results in a reduction of the opening, thus achieving the smaller size of aperture for fine pitch. For example, a coating with a 0.0005-in (0.013-mm) thickness per side reduces the opening by 0.001 in (0.025 mm).

It is also reported that a trapezoidal aperture as shown in Fig. 10-36 facilitates paste release. General recommendations on the taper parameters for various applications are listed in Table 10-11.[11]

Each technique has its inherent characteristics.[12–14] Generally, electropolishing increases the latitude of the printing process, particularly

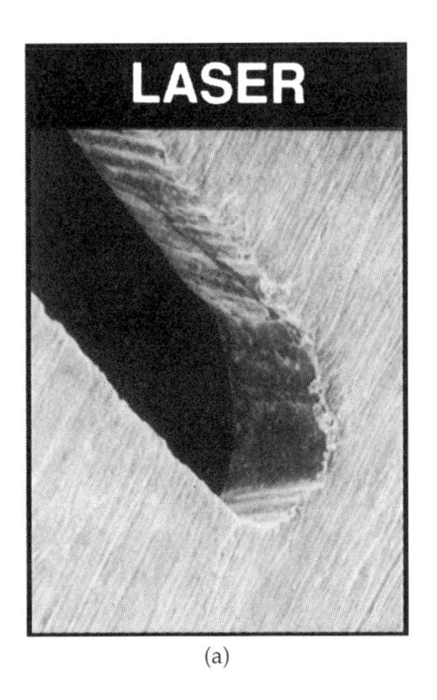

(a)

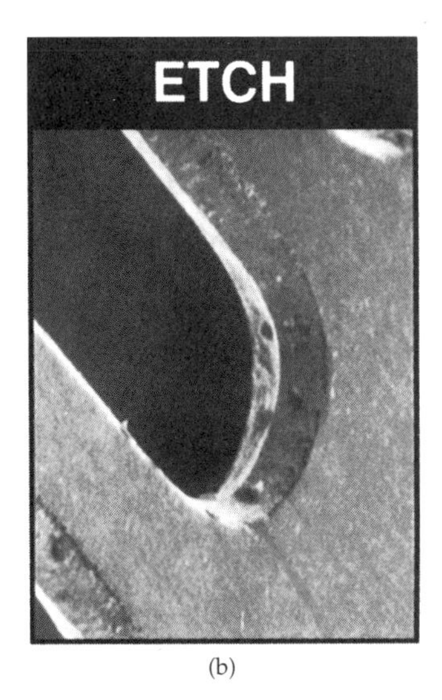

(b)

(c)

Figure 10-28. Stencil apertures made by (*a*) laser cut; (*b*) chemical etch; (*c*) electroforming. (*Courtesy of Photo Stencil, Inc.*)

for 20-mil (0.5-mm) pitch or finer patterns; the trapezoidal aperture contributes further to accommodating paste release when the ratio of aperture to stencil thickness decreases (approaching 1.6 or below).

A *hybrid stencil* combining laser cutting and chemical etching is recommended.[11] This approach exploits the accuracy that the laser offers

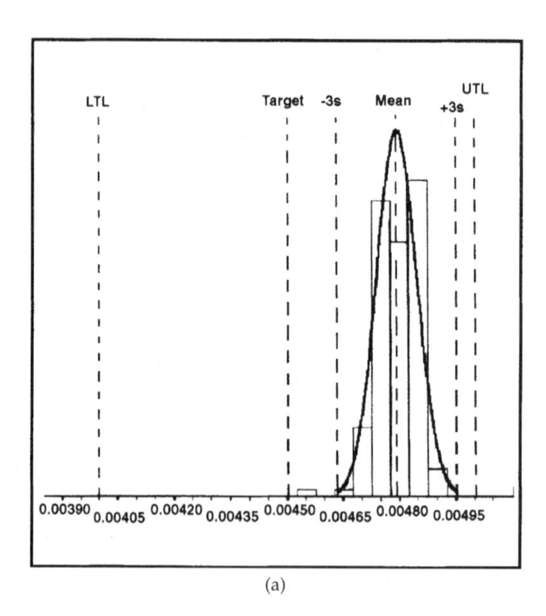

(a)

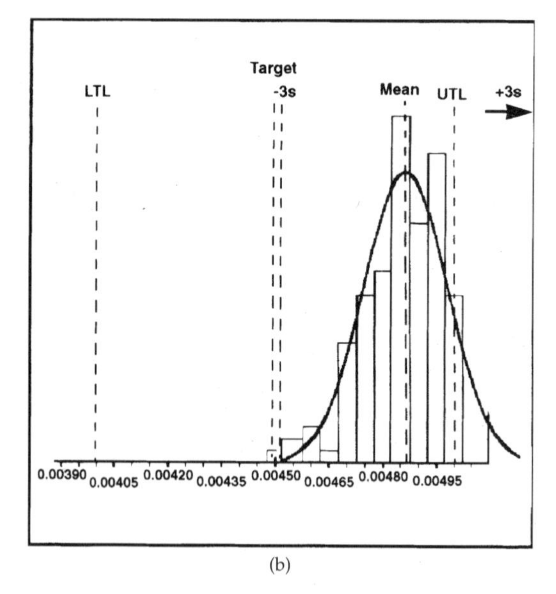

(b)

Figure 10-29. Histograms of aperture width distribution for (*a*) electroforming; (*b*) chemical etch; (*c*) laser cut. (*Courtesy of AMTX, Inc.*)

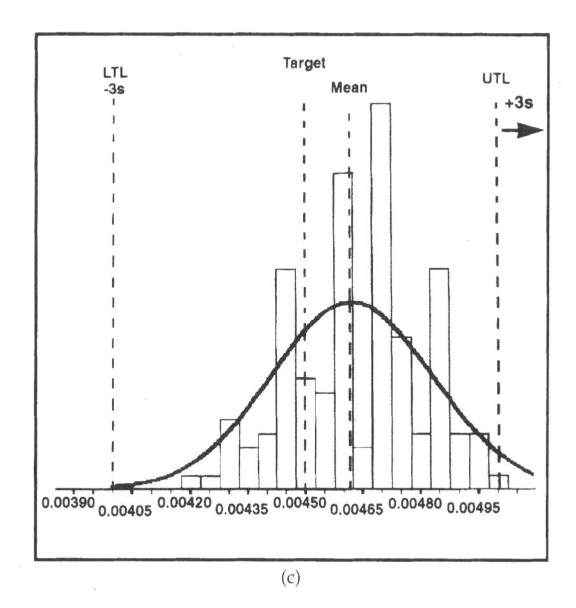

(c)

Figure 10-29. (*Continued*) istograms of aperture width distribution for (*a*) electroforming; (*b*) chemical etch; (*c*) laser cut. (*Courtesy of AMTX, Inc.*)

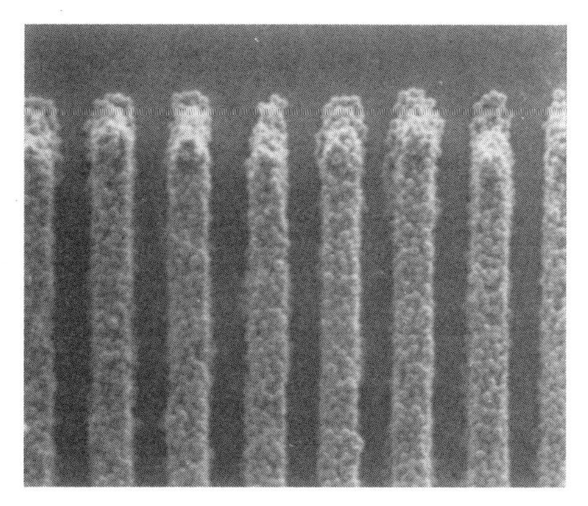

Figure 10-30. Ultra-fine pitch paste deposit with 0.0045-in aperture (30X). (*Courtesy of AMTX, Inc.*)

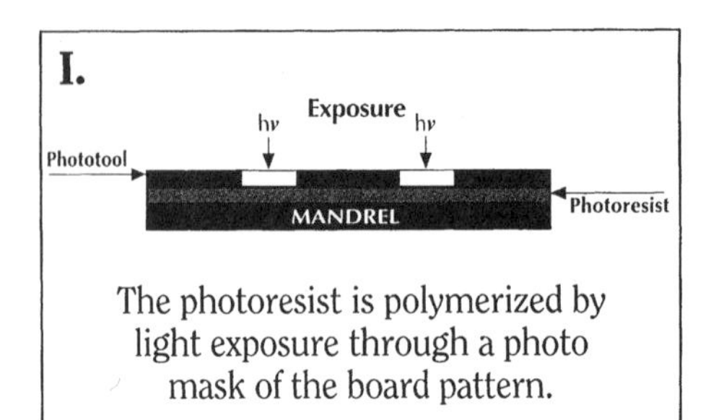

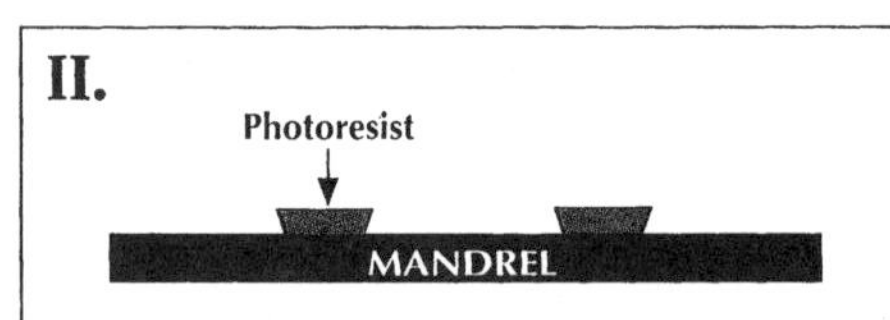

After developing, a negative image is created on the mandrel where only the apertures on the stencil remain covered by the photoresist.

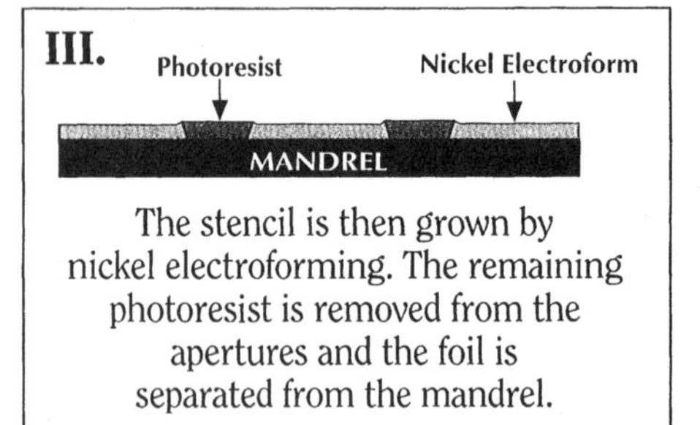

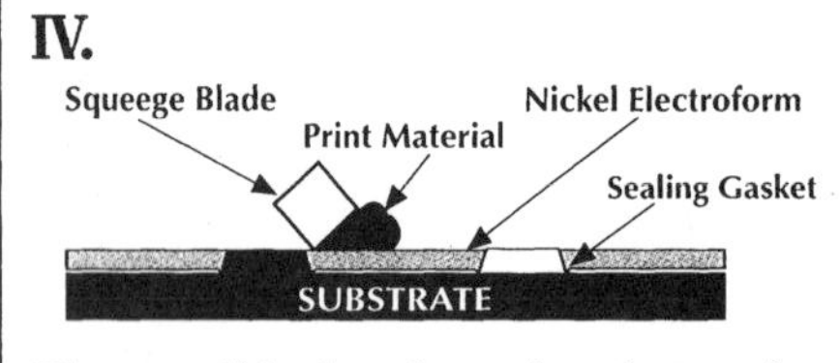

The stencil is then framed so that as the print material is spread across the stencil, the patented* sealing gasket helps to minimize print material bleeding, and the tapered side walls maximize material release from the stencil.

Figure 10-31. Schematic of electroforming process in manufacture of stencil. (*Courtesy of AMTX, Inc.*)

Figure 10-32. Aperture surface of chemical-etched stencil. (*Courtesy of Photo Stencil, Inc.*)

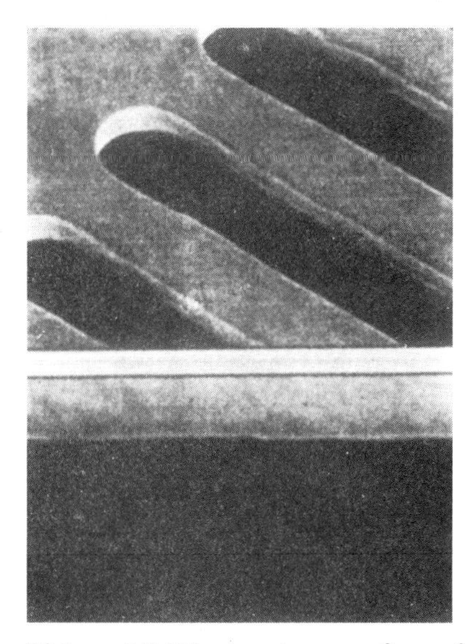

Figure 10-33. Aperture surface of stencil chemical-etched followed by electropolishing. (*Courtesy of Photo Stencil, Inc.*)

Figure 10-34. Aperture surface of laser-cut Alloy 42 stencil. (*Courtesy of Photo Stencil, Inc.*)

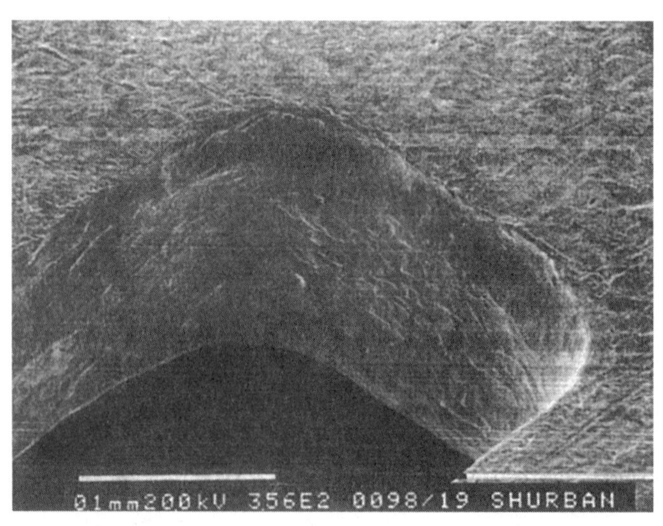

Figure 10-35. Aperture surface of laser-cut stencil followed by electropolishing. (*Courtesy of Photo Stencil, Inc.*)

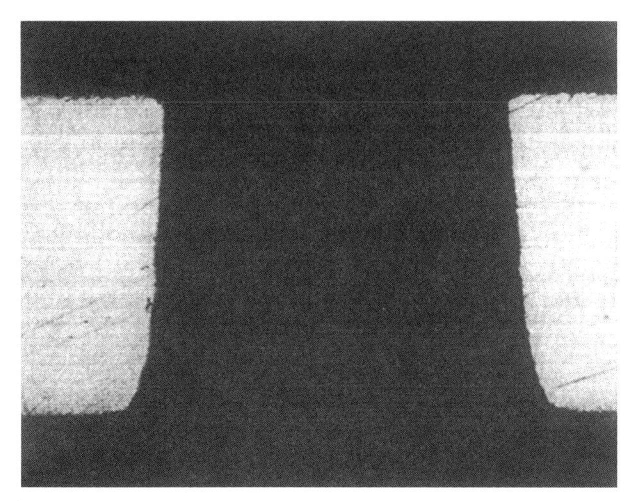

Figure 10-36. Trapezoidal aperture of electropolished laser-cut stencil. (*Courtesy of Photo Stencil, Inc.*)

Table 10-11. Trapezoidal Aperture Recommendations

	Stencil thickness		Taper size	
Application	in	mm	in	mm
Fine pitch	0.002–0.007	0.05–0.175	0.0005–0.00075	0.013–0.019
LGA, BGA	0.006–0.012	0.15–0.30	0.001–0.002	0. 025–0.050
Through-hole	0.008–0.030	0.20–0.75	0.001–0.003	0 .025–0.075
Solder bump	0.020–0.030	0.50–0.75	0.002–0.003	0 .050–0.075

to fulfill fine-pitch demands and the cost and time efficiency that the chemical etch provides in making apertures for larger components so that the performance/cost ratio can be maximized.

When the step printing is employed, both laser cutting and electroplating may encounter some limitations. Table 10-12 summarizes the characteristics and unique features of the techniques in fabricating stencils.

Another tip regarding stencil fabrication is that using a large-size stencil with a small image area always facilitates fine-pitch printing.

Table 10-12. Comparison of Techniques in Building Stencils

Techniques	Characteristics	Superior capabilities or features
Chemical etching	Most established process; sensitivity of fine-pitch capability to process and control; sensitivity of aperture size and vertical wall control	Versatile, economic
Laser cutting	Grainy wall surface; sequential cut; not concurrent formation of openings; higher cost; difficulty in making step stencil	Fine-pitch capability; no photo tools and resist needed
Electropolishing	Complementary step to produce smooth wall surface	Smooth wall surface
Ni-plating on aperture wall	Reducing aperture opening; smooth surface	Finer opening
Electroforming	Additive process via electrodeposition; concern about foil strength; difficulty in making step stemcil; suitable for stencil thicknesses of 0.001 to 0.012 in	Gasket effect minimizing bleed out; capability of producing very fine opening; no need of electropolishing

10.5 Additional Requirements of Fine-Pitch Solder Paste

10.5.1 Flux and fluxing

Solder powder for fine-pitch solder paste is generally smaller in size. As the powder becomes finer, higher fluxing activity is required to achieve proper solderability; this is because there is a larger surface area to be fluxed as shown in Table 10-13. Another factor is that as the component pitch becomes smaller, the solderability of component leads has been found to be less consistent. This also leads to the raised demands on flux performance. Pad bridging and open joints, which are two prevalent problems for fine-pitch soldering, are directly related to inadequate fluxing. Table 10-13 displays the particle size range, particle size average diameter, and the relative surface area for four mesh cuts of solder powder: −200/+325, −325/+500, −400/+500, −500.

Table 10-13. Solder Powder Mesh Designations versus Particle Size and Relative Surface Area

Mesh designation	Particle size range, μm	Particle size average, μm	Surface area ratio
−200/+325	74–44	60	1.0
−325/+500	44–25	35	1.71
−400/+500	37–25	31	1.93
−500	25–10	18	3.33

10.5.2 Tack time and open time

Tack time represents a paste's ability to hold components. It is the time elapsed from the time that the fresh paste was deposited. The *open time* reflects the paste's ability to withstand ambient environment. It is the time elapsed from the time that the fresh paste is employed.

For fine-pitch components, a relatively thinner solder paste deposition on fine-pitch solder pads is needed regardless of the stencil aperture design. For example, the 20-mil- (0.50-mm-) pitch pad will not be able to take a 10-mil (0.25-mm) thickness of paste. The thinner and smaller volume of paste requires better tackiness to hold the components in place before the final joints are made. However, too high a level of tackiness may be detrimental to the printing resolution. A delicate balance among the absolute tackiness, the tack time, and the paste intrinsic cohesiveness is required for designing a fine-pitch solder paste.

The open time and tack time, as the performance and characteristics of solder paste, must match the actual requirements on the production floor.

10.5.3 Rheology and viscosity

Qualitatively, the paste possessing a plastic flow property and moderate yield point has the most suitable rheology for fine-pitch applications as discussed in Sec. 10.3.2.1.

A stable rheological structure in solder paste in relation to external forces (squeegee motion) and the open time is important to the printing results. The phenomenon of *paste dry-up* on the printer at the production floor has been frequently reported. The occurrence of dry-up is often attributed to solvent evaporation. It should be noted that the dry-up may be a direct reflection of the change in rheological structure in paste after continuous printing, causing the thickening or stiffening appearance as is often observed. It may not be a simple process of solvent evaporation.

10.6 Solderability of Fine-Pitch QFP Leads

The principles and parameters of the solderability of component leads are covered in Sec. 5.3. It is expected and is often encountered that fine-pitch leads are more demanding in order to achieve good wetting.

10.7 Multiple-Step Soldering

10.7.1 Mass printing and mass reflow

Instead of using mass printing and mass reflow technique, solder deposition for all solder joints (fine pitch or non-fine pitch) is accomplished in one step during paste printing and is then followed by mass reflow as shown below, fine-pitch solder joints are made by a separate step (FP = fine-pitch, FPC = fine-pitch component).

Mass printing solder paste for all land pattern	→	placing conventional components	→	lead forming and mounting of FPC	→	mass→ reflow

Five alternative routes for placing solder material and making solder interconnections for board-level assemblies are included in the following flow diagrams. The term *reflow* used in the flow diagram denotes the furnace-type conveyorized mass soldering process; and the term *soldering* refers to using other heating processes to melt solder in a localized area using a localized heat source.

10.7.2 Printing, fine-pitch dispensing, two-step reflow (process 1)

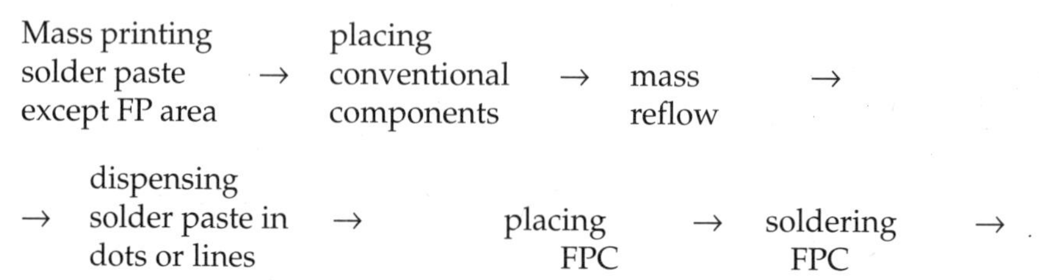

10.7.3 Mass printing, fine-pitch bumping, two-step reflow (process 2)

Mass printing solder paste for all land patterns → placing conventional components → mass reflow while forming solder bumps in FP area →

→ fluxing solder bump → placing FPC → soldering FPC →

10.7.4 Presolder-coated board, printing, mass reflow (process 3)

Presolder-coated board → mass printing solder paste except FP area → dispensing adhesive to FPC → dispensing flux to FPC

→ lead forming and mounting of FPC → placing conventional components → mass reflow

10.7.5 Presolder-coated board, printing, two-step reflow (process 4)

Presolder-coated board → mass printing solder paste except FP area → placing conventional components → mass reflow →

→ dispensing flux to FPC → lead forming and mounting of FPC → soldering FPC →

10.7.6 Presolder-coated board and leads, printing, two-step reflow (process 5)

Presolder-coated board → mass printing solder paste except FP area → placing conventional components → mass reflow

→ dispensing flux to FPC → placing FPC → soldering FPC →

↑ presolder-coated FP leads (into placing FPC)

Process 1 involves both printing and dispensing paste, using a dispensing technique to deposit paste on fine-pitch solder pads. The process also uses both mass reflow and special soldering. Process 2 is able to form solder bumps on fine-pitch solder pads while performing mass reflow. Process 3 is featured with precoated solder on board and with the use of adhesive to hold fine-pitch components, and then goes through one-pass mass reflow to make solder interconnections. Process 4, however, does two-step soldering: local soldering of fine-pitch components and mass reflow for the rest. Process 5 starts with both precoated board and precoated leads of fine-pitch components; it uses two-step soldering: local soldering of fine-pitch components and mass reflow for the rest. The precoating on both the board and fine-pitch leads provides additional solder volume for final solder joints, compared with precoating on either the board or leads. In most cases, the added volume is beneficial since the coating technique, such as electroplating, can only deliver a limited thickness of solder deposition within a reasonable level of cost and plating time. In contrast to process 2, processes 3 through 5 do not require a fine-pitch printing. Comparing process 3 and 4, process 3 is advantaged by the one-step mass reflow and process 4 is merited with the flexibility of local fine-pitch soldering.

10.7.7 Heating sources for soldering fine-pitch components

Other heating sources for attaching individual fine-pitch components include

1. Laser
2. Hot bar
3. Hot gas
4. Focused IR
5. Soft beam

Laser soldering. A heating process with a laser capable of focusing a 0.002-in (0.05-mm) diameter beam in the wavelength of 1.05 μm (Nd: YAG).

Hot-bar thermode. A contact process imposing localized heat and load to the component leads. It utilizes electrically resistive elements heated by the passage of a heavy current. The thermode holds the component down to the pads while it transfers heat to form solder joints.[15,16] It demands specific tooling for specific components. Critical requirements to form good solder joints are uniform force, proper amount of force, and uniform heat.

Hot gas. A contact process with good flexibility to accommodate various sizes and shapes of nozzles to deliver different heat flow patterns, which can be supplied in air or inert gas. It may not deliver a focused heat or a controlled heating profile.

Hot gas/tooling for lead-to-pad contact.[17–19] A noncontact heating process by hot gas, in combination with the various tooling to ensure lead-to-pad contact for a variety of devices.

Focused IR. A noncontact heating with controlled heating profile. It needs a fixturing design for holding components.

Soft beam. A noncontact heating process using energy in the wavelength of white light.

10.8 Liquid Solder Jetting

The liquid solder jetting technique which deposits solder in a very small size and volume recently has been developed. It employs the same principle involved in ink jet printing. Using the Maxwell theory, a magnetic field and an electric current applied at right angles to a conductor will result in a force directed at right angles to both the current and the magnetic field. Electric current is passed through molten solder in the presence of a magnetic field. The resulting force acts on the solder column, driving a pressure pulse to the nozzle tip through the solder. By controlling the applied voltage, first positive and then negative, a small droplet of solder can be formed at the tip of the nozzle located at the base of the solder column.[20]

The technique and tooling are reportedly superior to other methods based on the piezoelectric crystal driver which is sensitive to the temperature required in this process (> 183°C).

The molten solder jetting is capable of dispensing a 63 Sn/37 Pb solder ball with a diameter down to 0.004 in (0.1 mm). Since the technique and the device involve handling the molten solder and fine orifice, issues such as oxidation, solder dispensability, and material compatibility need to be addressed.

10.9 Soldering-related Defects

The defects that are most likely to occur or are prone to being magnified by fine-pitch soldering are

1. Bridging
2. Open and insufficient solder joint
3. Inadequate wetting
4. Solder balling
5. Voids
6. Foreign contamination
7. Package cracking
8. Component misalignment
9. Wicking
10. Tack time versus printability

10.9.1 Pad bridging

As spacing between pads decreases, bridging problems are more probable. Avoiding this problem requires

1. *Quality solder paste with superior printability and rheology.* Slump resistance is the most important characteristic. When using solder paste as the board level interconnecting material, the rheological characteristics of the paste, at the ambient temperature and as the temperature rises, is critical to the paste slump phenomenon during printing and at reflow. The slump resistance property is largely attributed to the intrinsic thermodynamic property of the flux/vehicle system (chemical portion of the paste without the solder powder). The objective is to design a flux/vehicle system capable of intimately wetting the surface of solder particles while offering the adequate surface tension so that solder particles are imbedded in the matrix of the flux/vehicle system with a high cohesive force F_c. As illustrated in Fig. 7-4, when the cohesive force of the system is not overcome by the combination of the gravity force F_g and thermal energy F_t, the paste exhibits the slump resistance and is able to retain its shape without collapsing as the temperature rises.

2. *Proper amount of solder paste.* Section 10.4.6 provides a guideline as to stencil thickness versus various fine-pitch components.

3. *Component placement precision.* The accuracy and precision of picking and placing the component are increasingly demanding as the pitch of components gets finer. A variation within ± 0.001 in (0.025 mm) is needed for consistent placement; this can be a challenging performance for most equipment and pick-and-place processes.

10.9.2 Open joints and insufficient solder joints

Open joints in fine-pitch soldering are primarily the result of (1) component leads lacking coplanarity, (2) insufficient fluxing activity, (3) insufficient solder volume, (4) excessive misregistration in component placement, and (5) solder wicking caused by a preferential temperature rising in component leads. Remedies to overcome this problem must embrace all the above causes with counteracting solutions: component coplanarity, good solder paste, proper stencil aperture design, and adequate pick-and-place equipment.

10.9.2.1 Case study. Solder joint failure (open joint) on production floor.

Description

An original equipment manufacturer (OEM) manufacturing 12-layer PCBs for super-minicomputers (four coprocessors) encountered functional failure (reject) of the boards during the electrical function tests. The boards were two-sided, epoxy-fiber–based and contained a mix of surface mount components and through-hole components. Each board had many via holes of different sizes. The finest-pitch surface mount component was a 20-mil (0.50-mm) P-QFP. The solder pads on the boards were typically Sn-Pb electroplated Cu. The manufacturing system of the boards consisted of four primary processes:

1. Dry the board overnight at around 110°C.
2. Assemble the bottom side of the board by the steps of printing solder paste, applying adhesive, placing chip components, and curing/reflow. The preheat stage around 150°C during the reflow process also cures the adhesive.
3. Assemble the top side of the board by typical surface mount steps of printing solder paste, placing components, and IR reflow with peak temperature around 200°C as measured by thermocouple.
4. Drag solder by floating the board with the bottom side in contact with the molten solder at a peak temperature around 230°C for 10 to 15 s. At the completion of the drag solder, all via holes are plugged so that the probe needle will not be caught and all through-hole components are soldered.

Observations

After process 3, all solder joints were examined and found to be shiny, smooth, and have good wetting and solder fillet. After process 4, the board underwent an electrical function test and failed. The failure was attributed to open solder joints which were found to be associated with 20-mil (0.50-mm) pitch surface mount components. It was further detected that the open solder joints occurred only with those that were parallel to the travel direction of the board. The examination also showed that the open joints having the solder fillet primarily remained on the leads of the component and appeared to be shiny and intact. The portion of the solder fillet being sandwiched between the lead and pad looked dull, rough, and soft.

Analysis

It was first thought that the repeated heat excursion may have promoted excessive intermetallics, therefore causing degraded wettability of the pads. It was also thought that the electroplated solder pads might have a poor solder surface. Attempts were made to shorten the heating cycles and to replace electroplated boards with solder-dipped/air knife boards. The problem persisted, indicating that the solderability versus intermetallics would not be the main source of the open solder joints.

Excess heat exposure during process 4 was suspected. High temperature tape was used to cover the bottom side of the 20-mil (0.50-mm) pitch component before going through the drag solder pot. The problem was alleviated. It was found that the relatively large size (0.40-in diameter) and large number of via holes (240) next to the four sides of the component allowed sufficient heat to be transferred to the solder joints which approached melting temperature. The heat transfer was facilitated by the travel direction and the relative location of via holes. This accounts for the occurrence of open joints aligned longitudinally with the travel direction, not perpendicular to the direction.

Moral

Any step of the process could contribute to the overall manufacturing quality and reliability. The initial circuitry design is often incongruent with the manufacturing process being selected.

10.9.3 Inadequate wetting

As the component pitch becomes finer, its solderability has been found to be less consistent, leading to the raised demands on flux performance. Furthermore, the solder powder for fine-pitch solder paste is generally smaller in size. As the powder becomes finer, higher fluxing activity is required in order to achieve the proper solderability because of the large surface area to be fluxed. The estimated ratios in surface area per unit volume for commonly used powder mesh cuts per ASTM B-214 "Test for Sieve Analysis of Granular Metal Powders: −200/+325, −325/+500, −400/+500, −500" are included in Table 10-13. Using a good solder paste with a large latitude of fluxing activity coupled with a good storage condition for fine-pitch components to minimize any solderability degradation is the practical solution.

10.9.4 Solder balling

For finer-pitch packages, the solder paste used contains finer powder, which promotes solder balling during reflow. The formation of excess solder balls may result in a potential electrical short during service if the solder balls are not removed. It can also deprive the solder volume intended for making good interconnections from the solder fillet. The solder balling phenomenon can be defined as the situation that occurs when small spherical particles with various diameters are formed away from the main solder pool during reflow. They do not coalesce with the solder pool after solidification. Likely causes and remedies are discussed in Sec. 13.3.

10.9.5 Voids

The generation of voids and its possible causes and remedies are discussed in Sec. 13.4.

10.9.6 Foreign contamination

Since the solder joint is much smaller for fine-pitch interconnections, the effect of the foreign contamination is much more pronounced. It is apparent that the concentration and distribution of element dissolution in solder joints should be estimated and examined based on the total volume of the solder fillet. Two types of contaminations should be considered, one of a chemical nature and the other of a metallurgical nature (see Sec. 13.5).

10.9.7 Package cracking

Plastic molded fine-pitch packages are vulnerable to cracks during reflow soldering (see Sec. 13.6).

10.9.8 Component misalignment

The tolerance for component misalignment is much less than that for array packages because of the less self-aligning ability during reflow and smaller pitch spacing. Misalignment is a contributor to the problems of both bridging and open or insufficient solder joints.

10.9.9 Wicking

This phenomenon is mainly driven by the temperature differential and relative metallurgical wetting rate. Preferential wetting between the leads of components and the solder pads on the board or between the different areas of wettable circuit traces may occur when there is a transient temperature differential. The potential remedies are to remove the temperature difference or block the path of wicking.

10.9.10 Tack time versus printability

For fine-pitch components, thinner solder paste is deposited. The thinner and smaller volume of paste is more prone to the degradation of paste characteristics under ambient conditions. Inadequate tack time may cause the component alignment problem during component placement or after the placement before the reflow, causing bridging problems. A poor open time is often associated with excessive solder balling and wetting problems. The tack time and open time of a solder paste must match the time and conditions that are required for a specific production environment.

10.10 References

1. Printed Circuits Fabrication, November 1990, p. 36.
2. Jennie S. Hwang, *Solder Paste in Electronics Packaging—Technology and Applications for Surface Mount, Hybrid Circuits, and Component Assembly*, Van Nostrand Reinhold, New York, 1989, Chapters 5 and 6.
3. John D. Ferry, *Viscoelastic Properties of Polymers*, Wiley, New York, 1970.
4. W. H. Herschel, "Experimental Investigations Upon the Flow of Liquid in Tubes of Very Small Diameter by J. L. M. Poiseuille," *Rheol. Met. Soc. of Rheology*, 1940, p. 47.
5. W. H. McAdams, *Heat Transmission*, McGraw-Hill, New York, 1954.
6. O. Obra et al., "New Soldering Technology for Fine Pitch (0.15 mm) Applications," *ISHM 1992 Proceedings*.

7. Yuichi Obara, Toshiaki Amano, and Seiji Kumamoto, "Mounting of Fine Pitch Components with Precoat Soldering," *ISHM 1993 Proceedings*, p. 355.
8. K. L. Kent, "Solder Paste Printing for Fine Pitch Devices," *Proc. ISHM Annual Symposium*, 1990, p. 136.
9. G. W. Peek and J. M. Alexander, "Fine Pitch Yield Optimization," *Proceedings NEPCON West*, 1990, p. 741.
10. T. Garrison, "Mechanical Dimensional Commonality for Fine Pitch Components," *Proc. NEPCON West*, 1990, p. 746.
11. E. Coleman Williams, "Photochemically Etched Stencils for Ultra-Fine-Pitch Printing," *Surface Mount Technology*, June 1993, p. 18.
12. Mark D. Herbst, "Metal Mask Stencils: The Key Components," *Surface Mount Technology*, October 1992, p. 21.
13. R. S. Clouthier, AMTX, Inc., Product Brochure.
14. William E. Coleman, Photo Stencil, Inc., Product Brochure.
15. L. Roberts, "Development in Hot-Bar Reflow," *Circuits Assembly*, 1991, p. 56.
16. G. Zinner, "Using Advanced Pulsed Hot-Bar Technology for Reliable Positioning and Mounting of High-Lead-Count Flat Packs and TAB Devices," *Proceedings 2nd International TAB Symposium*, 1990, p. 160.
17. D. J. Spigarelli, "New Methods of Outer Lead Bonding Enhance Placement," *Electronic Packaging and Production*, July 1991, p. 69.
18. D. J. Spigarelli, "A New Approach to Precision Placement of Fine Pitch Devices," *Electronic Packing and Production*, July 1991, p. 69.
19. D. Peck, "Tab and Ultra-Fine Pitch Assembly: Controlling Process Problems," *Surface Mount Technology*, November 1991, p. 49.
20. Tom Schiesser et al., "Micro Dynamic Solder Pump: Drop on Demand Eutectic Solder Dispensing Device," *Proceedings SMI*, 1994, p. 501.

10.11 Recommended Readings

Hutchins, Charles, "Understanding & Using SM & Fine Pitch Technology," Hutchins Associates, 1993.

Prasad, Ray, *Surface Mount Technology—Principles & Practice*, VNR, New York, 1989.

11 Ball Grid Array Technology

11.1 Characteristics and Benefits

The BGA package is able to provide a higher density of I/O interconnections than the QFP for a given board real estate without requiring impractically narrow spacing between I/O leads. For example, a BGA package with a 1.5-mm (0.050-in) pitch and 225 I/Os occupies significantly less area on a mother board than a QFP with a 0.5-mm (0.020-in) pitch and 208 leads. Figures 11-1 and 11-2 exhibit the size advantage of the BGA package over the QFP for various lead pitches and pin counts.[1,2]

The BGA package is also featured with a low profile and shorter length of interconnections, providing superior electrical performance and reduced inductance as shown in Table 11-1.

The improved electrical performance in the system level signifies a high-speed, low-noise signal transmission. In most cases, superior thermal performance in drawing heat from the device to the mother board is also expected.

The absence of fine, fragile leads in a BGA eliminates all leads-related problems such as lead coplanarity, skewing, and bridging due to narrow interlead spacing. Since the spacing between the interconnecting sites of an array package is much larger, a BGA-type package provides more component placement tolerance as well as printing toler-

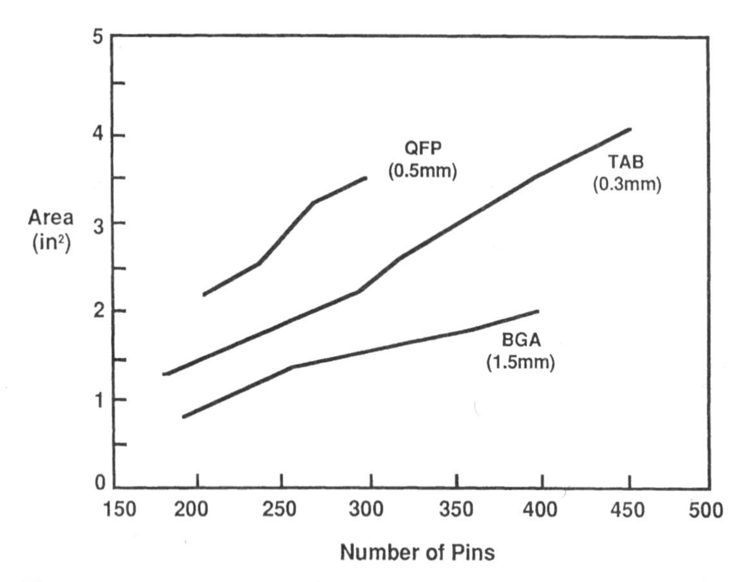

Figure 11-1. Comparison of board area occupied by BGA, QFP, and TAB.

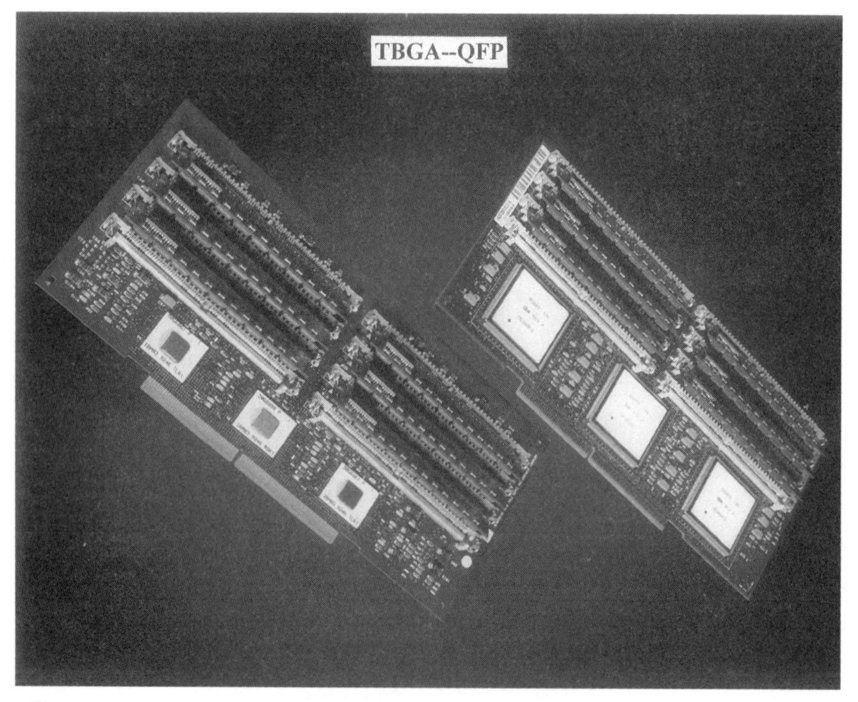

Figure 11-2. Comparison of QFP (*upper*) and TBGA (*lower*) on a similar assembly. (*Courtesy of IBM Corp.*)

Table 11-1. Electrical Properties of the BGA versus the QFP[2]

Package	Inductance, nH	Capacitance, pF	DC resistance, mΩ
225 BGA/2 layer	5.02–9.07	1.28–1.31	20–24
208 QFP/Cu lead frame	9.9–14.5	< 2.3	70–80

NOTE: nH = nanohenrys, pF = picofarads.

ance. On the negative side, all solder interconnections are hidden under the carrier substrate, making repair of a single solder joint impossible. This also makes visual inspection of inner rows of solder joints impossible. In this case, techniques based on x-ray emission are required in order to conduct inspections. The quality and integrity of solder joints are critically important. The manufacturing process to assure the quality and consistency is essential. With the established quality and process, inspection and repair can thus be spared.

Another area to be considered is the board layer count and routing. Because of the reduction in board surface area and the increase of pin density on the mother board, the increase in board layer count may be required. However, it is reported that the BGA is not expected to drive up the layer count for the applications which replace 0.50-mm (0.020-in) QFPs with 208 leads. The relative comparison of routing difficulty in the mother board between the QFP and BGA at various pin counts is shown in Fig. 11-3. In some cases, the results show that the BGA is

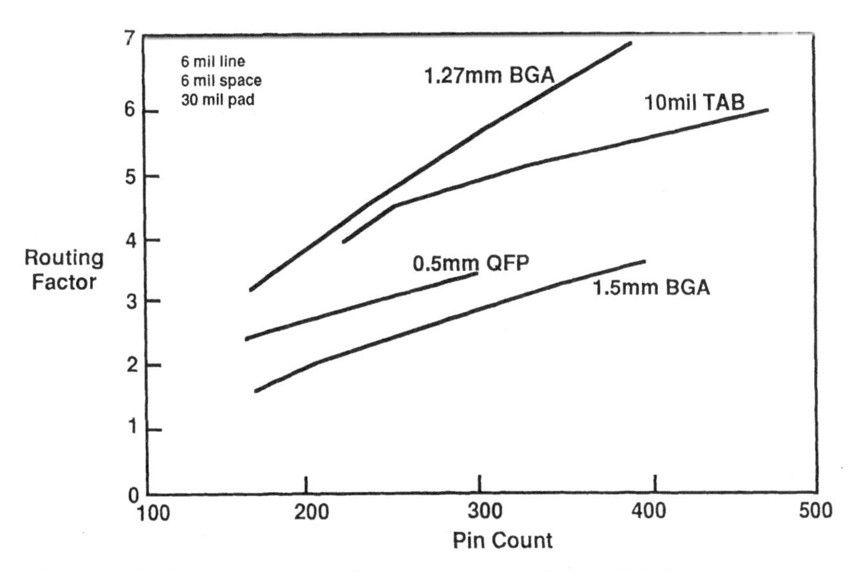

Figure 11-3. Relative routing efficiency of BGA, TAB, and QFP.

indeed more difficult to route than a QFP with a comparable pin count.

11.2 Types of Packages

11.2.1 General

Under the concept of the array pattern, one convenient way to classify the BGA is based on the type of material the carrier substrate is made of: (1) plastic substrate (P-BGA), for example, over-molded pad array carrier (OMPAC); (2) ceramic substrate (C-BGA), primarily the extension of collapse controlled chip connection (C-4) technology; (3) metal substrate (M-BGA) using anodized aluminum; (4) metallized tape substrate (TBGA); (5) flex circuit material for μ-BGA. The packages can house either a single chip or multiple chips known as a multiple-chip module. At the time of this writing, the JEDEC has registered ceramic ball grid array (C-BGA), plastic ball grid array (P-BGA), and tape ball grid array (T-BGA) packages.

11.2.2 P-BGA[3-5]

A die is mounted on a gold-plated die attach pad using a silver-filled epoxy. Typically, the die is thermomechanically wire-bonded to a circuit pattern on a bismaleimide-triazine (BT) epoxy substrate. It is then overmolded by conventional plastic transfer molding to encapsulate the package. The carrier substrate is 0.5-oz, copper-clad, BT epoxy-glass laminate with a 0.010-in thickness. Circuit patterns are constructed of die attach pads for back-side grounding, wire bond pads, plated through-holes, and conductors to the solder bump pad. Copper traces are routed to an array of metal pads on the bottom side of the carrier substrate to which solder balls are attached. Figure 11-4 depicts the construction of a P-BGA. A series of P-BGAs with 1.5-mm pitch in an

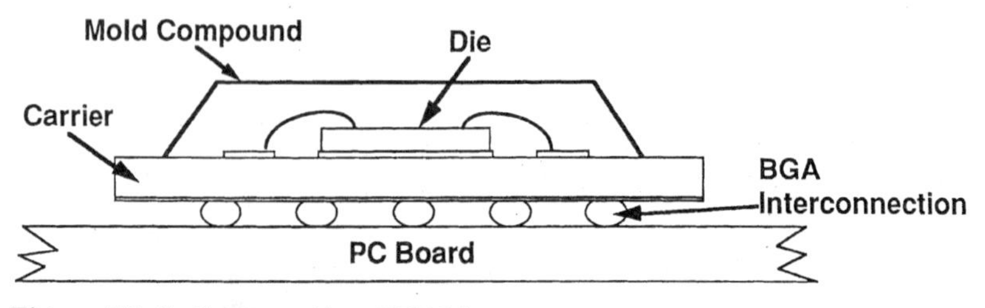

Figure 11-4. Cross section of PBGA.

Figure 11-5. PBGA in OMPAC-225 leads. (*Courtesy of Motorola, Inc., ASIC Division*).

ascending number of pin counts will become more readily available. They range in 86 pins, 119 pins, 169 pins and 225 pins which is shown in Fig. 11-5. Packages with higher numbers of pins, such as 313 pins, 479 pins, 503 pins, and higher are expected to be introduced in the near future. As the pin count increases, the lead pitch has to be reduced to maintain the package size. The lead pitch of these packages is expected to go down to a 1.2-mm pitch or lower. For finer pitches, new demands will then set in.

11.2.3 C-BGA

The carrier substrate is a ceramic instead of a BT resin epoxy-glass laminate. Superior thermal conductivity and hermeticity are considered general advantages of ceramic packages.[6–10] The multilayer cofired ceramic substrate has tungsten or molybdenum metallization paste screened on the desired circuit pattern, including the die attach pads, wire bonding, C-4 pads, and back-side solder bump pads. All patterns are photodefined, etched, and electrolytically plated with copper, nickel, and gold. Preferably, a thin layer of gold is deposited on the solder bumps and C-4 pads, and a thicker layer of gold is deposited on the

Figure 11-6. Size comparison of CBGA and PGA.

wire bond pads. Solder paste is screened on the back side of the carrier and is reflowed. At this point, the solder bump pads are ready for the solder ball attachment. Solder balls are attached either by a pick-and-place automated process or by the use of a stainless-steel template.

To compare C-BGA with the conventional PGA package, C-BGA offers a 60 percent area reduction for a lead count in the range of 600, as shown in Fig. 11-6. There is also a significant decrease in the signal line parasitic capacitance in the C-BGA package. Furthermore, the shorter path and tighter signal-to-power coupling in the C-BGA provides an effective inductance of the solder balls, generally one-half to one-third that of equivalent PGA modules.

Comparing C-BGA with the QFP, Fig. 11-7 exhibits the size difference between C-BGA and QFP at a comparable number of leads in the range of 304 to 361. With respect to package features, Table 11-2 summarizes the relative comparison between C-BGAs and QFPs, and Table 11-3 summarizes that between C-BGAs and PGAs.[7]

Table 11-4 compares the electrical characteristics among a standard 196-pin PGA, a 324-I/O BGA, and a 256-lead QFP.[11]

Figure 11-7. Size comparison of CBGA with QFP.

Table 11-2. Example of Package Features of QFPs versus C-BGAs[7]

		Lead pitch		Size	
Package	Lead count	mm	in	mm	in
QFP	304	0.50	0.020	$(42.6)^2$	$(1.7)^2$
C-BGA	361	1.27	1.27	$(25)^2$	$(1)^2$

Table 11-3. Example of Package Features of PGAs versus C-BGAs[7]

		Lead pitch		Size	
Package	Lead count	mm	in	mm	in
PGA	685	2.54	0.100	$(50)^2$	$(2.0)^2$
C-BGA	625	1.27	0.050	$(32)^2$	$(1.27)^2$

One IBM design illustrates that the carrier solder is formed in solder ball connectors and solder column connectors[11] as shown in Fig. 11-8*a* and *b*, respectively. Specific design parameters employed are

Solder alloy (ball or column)	10 Sn/90 Pb
Solder ball diameter	0.035 in (0.88 mm)
Solder column diameter	0.020 in (0.50 mm)
Solder column height	0.087 in (2.18 mm) or 0.050 in (1.25 mm)

Table 11-4. Electrical Characteristics of PGAs, BGAs, and QFPs[11]

Characteristic	PGA	BGA	QFP
Size	43.5×43.5 mm	21×21 mm	26×26 mm
I/O	196	324	256
Type of I/O	Pin	Pad	Lead
Pitch	3.94 mm (0.157 in)	1.0 mm (0.040 in)	0.4 mm (0.016 in)
Signal-to-ground capacitance	6.76 pF	1.85 pF	1.78 pF
Signal inductance	8.57 nH	8.26 nH	20.31 nH

Solder pad pitch	0.050 in (1.25 mm)
Solder for attaching balls or columns	63 Sn/37 Pb

Both structures are recommended as cost-effective packaging solutions for applications requiring more than 200 I/Os and/or demanding good thermal dissipation and requiring the incorporation of a ground plane. In addition, the column structure provides more stress endurance when the package contains more than 500 I/Os. For performing under especially harsh service conditions and when the solder joints must endure a prolonged cyclic lifetime, the column solder joints demonstrated superior performance.[11] Further discussions will be covered in Sec. 11.4.

11.2.4 T-BGA[12,13]

The T-BGA is characterized by a flexible polyimide tape, flip-chip stiffener, and high-temperature solder balls. The tape is a 0.002-in-(0.05-mm-) thick polyimide layer with copper metallization on each side. Fine-line copper traces carry the signals on one side, and the other side serves as a ground plane. The close proximity of signal traces to the ground plane (0.002 in; 0.05 mm) provides a very low signal noise. Plated vias provide an electrical path from one side of the tape to the other. Copper stiffener plates are bonded to the tape by an adhesive in order to give mechanical rigidity to the package. Thermal adhesive may be used to attach a cover plate or heat sink on the back of the flip chip. Figure 11-9 is a schematic cross section of a T-BGA. Typical material and design parameters are

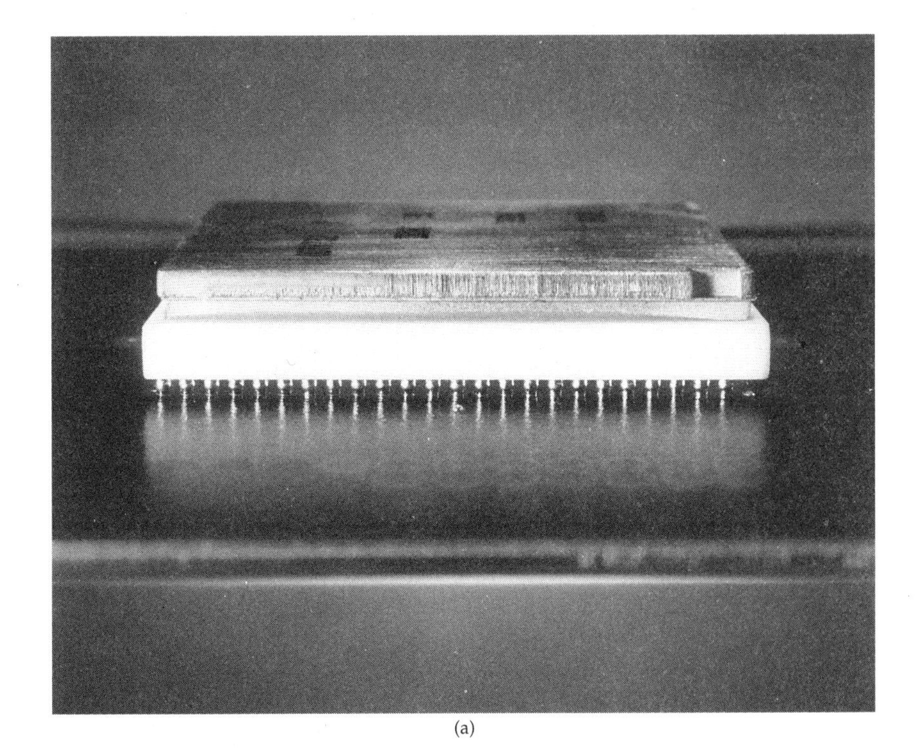

(a)

(b)

Figure 11-8. (*a*) CBGA ball connection (*Courtesy of IBM Microelectronics*); (*b*) CBGA column connection. (*Courtesy of IBM Microelectronics*).

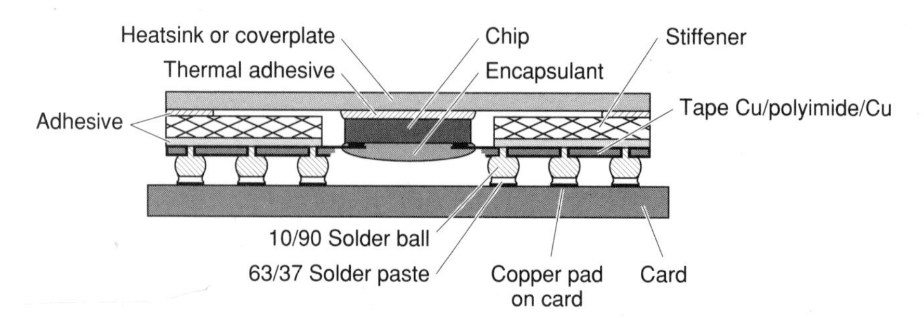

Figure 11-9. Cross section of a TBGA.

Solder alloy of balls	10 Sn/90 Pb
Solder ball diameter	0.025 in (0.63 mm)
Solder array pitch	0.040 in (1.0 mm) or 0.050 in (1.25 mm) or 0.060 in (1.5 mm)
Attaching balls to substrate	Weld to via pads
Solder for attaching T-BGA onto the board	63 Sn/37 Pb or equivalent

The T-BGA has a relatively low profile of 1.3 mm (0.052 in) without the overplate and 1.9 mm (0.076 in) with the overplate. Its body size ranges from 21 mm containing 193 solder interconnections to 40 mm with 736 solder interconnections. Solder interconnections are generally avoided in the area underneath the chip.

11.2.5 μ-BGA

The μ-BGA is designed as a single-chip package, combining the features of TAB, wire bonding, and flip chips. Figure 11-10 is a schematic of a μ-BGA.[14] The package consists of a flexible polyimide circuit that is attached to the face of the die with an elastomer. The elastomeric layer between the flex circuit and the chip acts like a cushion so that

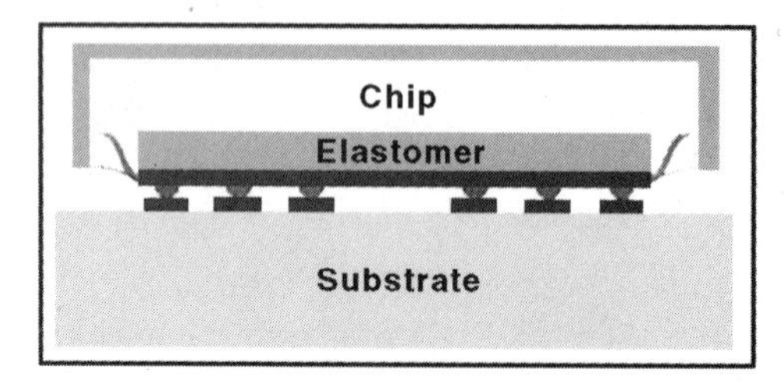

Figure 11-10. Schematic of a μBGA.

Figure 11-11. Interconnection sites of μBGA.

the package can be socketed for testing and burn-in. A thermosonic bonding technique is used to attach short, flexible gold leads to the die pad. The leads fan inwardly from the die pad to the bump array as shown in Fig. 11-10. The bottom view of a μ-BGA package with 188 gold bumps on a 0.5-mm (0.02-in) grid is shown in Fig. 11-11. Typical design and material parameters are

Bump material	Nickel with gold flash
Bump size	0.085 mm (0.0034 in) in height
Bump pitch	0.3 mm (0.012 in), 0.5 mm (0.020 in), 1.0 mm (0.040 in), 1.27 mm (0.050 in), 1.5 mm (0.060 in)
Elastomer thickness	0.12 ± 0.05 mm (0.005 ± 0.002 in)
Package height	0.8 mm (0.032 in)

Because of the small body size that can be achieved, the package is most suitable for a double-sized type 1 PCMCIA card. The μ-BGA is presented as a means of achieving the smallest package and the highest levels of performance at a commercially competitive cost.[13,14] The inherent merits are

- The package can be virtually the same size as the die; the traces fan in from the die pads (in contrast to the TAB's fan out pattern).
- The package can be burned in and tested.
- Electrical performance is excellent with low resistance, capacitance, and inductance of the short leads.
- The thermal path is through the back side of the die to a heat sink to provide efficient cooling.

Table 11-5 compares the key properties of various packaging technologies,[12] and Fig. 11-12 illustrates the package area as a function of I/O for QFPs, PGAs, BGAs, and μ-BGAs.[14]

11.2.6 μSMT

In this design, the chip is directly connected onto the silicon wafer substrate.[15] The I/O pads of the chip are rerouted from the conventional peripheral wire bonding pads to the array bonding pads by adding an aluminum conductor layer and a dielectric layer with I/O vias and solder wettable pad metallization. The rerouting also increases the pitch from as small as 5 mil (0.13 mm) or 16 mil (0.40 mm) or larger, depending on the chip size and the number of I/Os. Figure 11-13 represents a cross-sectional view of this type of package.

TABLE 11.5. Comparison of Key Features of Various Packages

Feature	PGA	P-QFP	BGA	T-BGA	μ-BGA
I/O	208	208	225	224	313
Pitch, mm (in)	2.5 (0.100)	0.5 (0.020)	1.27 (0.050)	1.27 (0.050)	0.5 (0.020)
Footprint, mm^2	1140	785	670	530	252
Height, mm	3.55	3.37	2.3	1.5	0.80
Mass, g	25	—	1	5	0.47
Package-to-die ratio	11	8	7	5	1
Inductance, nH	3–7	6–7	3–5	1.3–5.5	0.5–2.1
Capacitance, pF	4–10	0.5–1	1	0.4–2.4	0.05–0. 2
Thermal resistance, °C/W	2–3	0.5–0.6	10	1.5	0.2–2

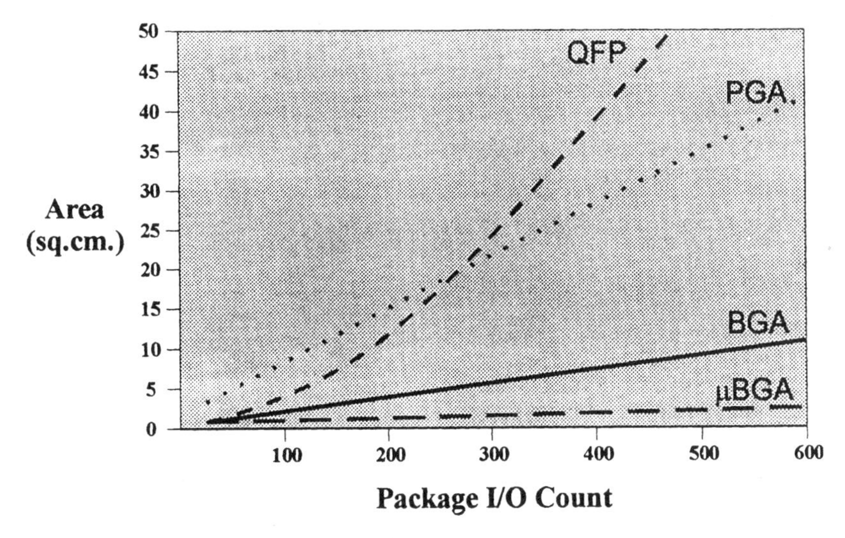

Figure 11-12. Package footprints for QFP, PGA, BGA, and μBGA as a function of I/O.

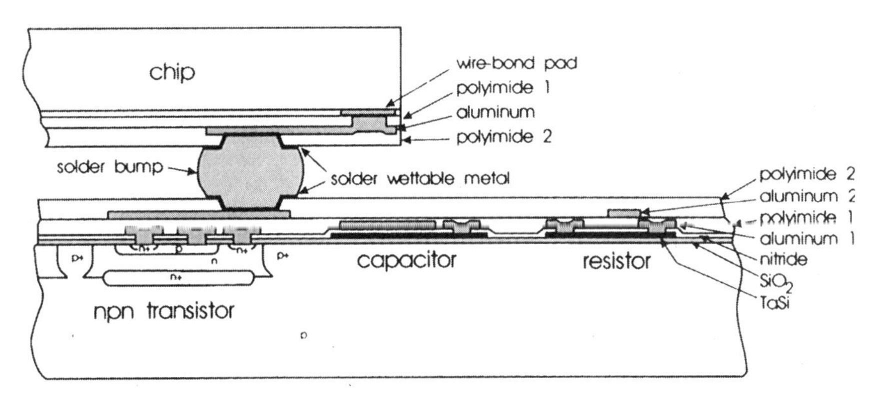

Figure 11-13. Cross-sectional view of μSMT.

11.2.7 mBGA

A die-sized package, known as m-BGA, is characterized by the addition of two layers of polyimide and two layers of metal to a conventional die.[16] The chip's peripheral pads are redistributed into a bumped array for a more forgiving pitch in 20 mil (0.5 mm) or 30 mil (0.75 mm) in comparison with the flip chip's 10-mil (0.25-mm) pitch.

11.2.8 M-BGA[17]

This package is featured with thin-film circuits deposited on the anodized aluminum substrate. The package consists of die attach, wire bonding, encapsulation, and normally 63 Sn/37 Pb solder balls. Reportedly, M-BGA packages can host the chip sizes from 5 mm to greater than 18 mm in packages of 27 mm, 31 mm, and 35 mm. A 256-lead, 27-mm package has a mounted height as low as 1.0 mm and a weight of 1.6 g.

Anodized aluminum substrate in direct contact with the die and circuitry acts as a heat sink as well as the wiring substrate, thus enhancing heat dissipation and reducing thermal resistance. Electrically, the direct contact between the circuitry and the anodized aluminum substrate in M-BGA packages makes the substrate behave as a ground plane, reducing inductance and capacitance. It is also expected that a multimetal layer design can further enhance electrical performance. In terms of routability, the number of lines per inch for M-BGAs can reach 500 which is higher than for other types of packages.[18] The routability is ranked as

M-BGA > T-BGA > P-BGA > C-BGA

The aluminum substrate is an effective shield against electromagnetic interferences. Currently, the cost of M-BGAs is considered compatible with that of T-BGAs or C-BGAs and is higher than that of P-BGAs.

11.2.9 Comparison with peripheral packages

The peripheral counterparts of BGAs in the QFP format are exemplified in the following tables.

	BGA-169	QFP-160
Pitch, in (mm)	0.060 (1.50)	0.025 (0.63)
Number of I/O (leads)	169	160

	BGA-225	QFP-208
Pitch, in (mm)	0.060 (1.50)	0.020 (0.50)
Number of I/O (leads)	225	208

Figure 11-14 exhibits the difference in carrier size and board area consumed by various array and peripheral packages: C-BGA, PGA, TAB-

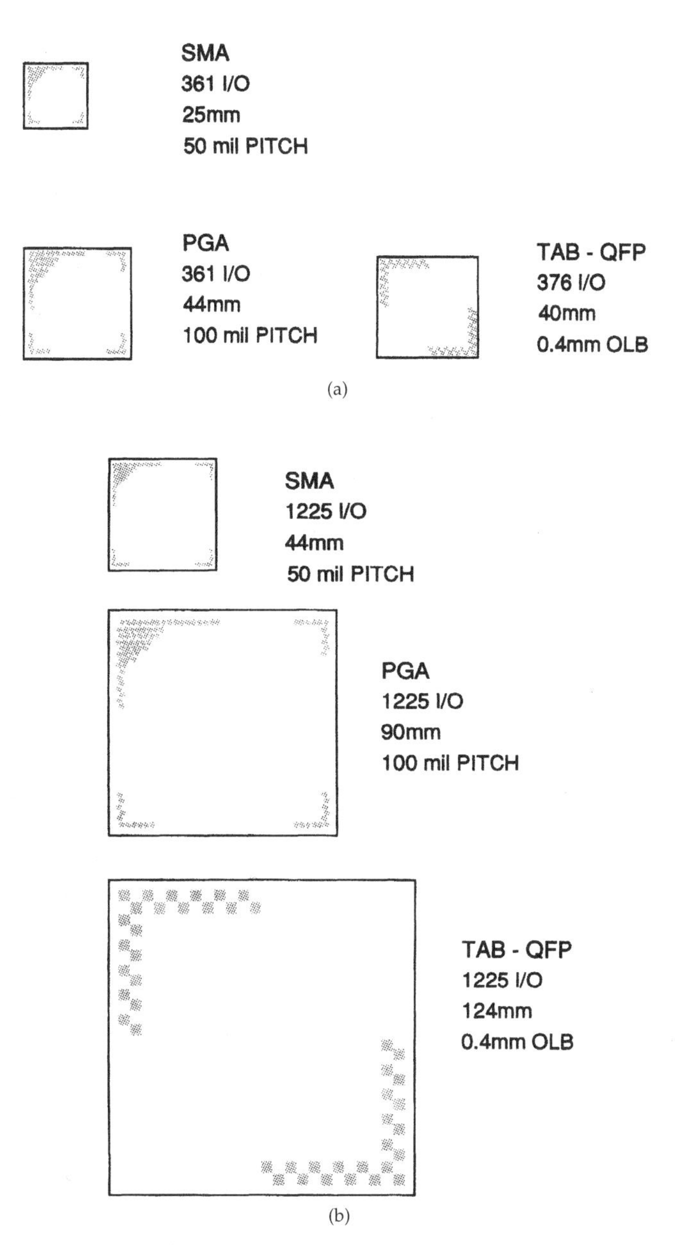

Figure 11-14. Relative carrier size and board area for C-BGA, PGA, TAB-QFP: (*a*) 361 I/O (*b*) 1225 I/O. (*Courtesy of IBM Corp.*)

QFP.[11] It is evident that the array packages occupy a smaller area as well as have a larger pitch to accommodate a given number of I/Os. Specific examples at a given number of I/O pins of 361 are given in the following table.

	BGA	PGA	QFP
Package dimension, mm	25	40–44	40–44
Pitch, mm	1.25	2.54	0.50

At a given area of 10 cm^2, the BGA package format can take a higher number of I/O pins as shown in the following table.

	BGA	PGA	QFP
Number of I/O leads	625	150	150

11.3 Materials and Processes

The basic alloy for the solder bump of die level, particularly for the flip-chip scheme, often contains high-temperature, high-lead solder compositions, such as 5 Sn/95 Pb, 10 Sn/90 Pb. Eutectic or near-eutectic alloys such as 60 Sn/40 Pb, 62 Sn/36 Pb/2 Ag, and 63 Sn/37 Pb have also been used successfully.

The solder ball on the underside of the carrier substrate can be made of either high-temperature, high-lead solder or eutectic or near-eutectic tin-lead or tin-lead-silver solders. Board-level solder for attaching the BGA carrier to the PCB is limited to eutectic or near-eutectic tin-lead and tin-lead-silver solders, due to the temperature tolerance level of conventional board material, namely FR-4 material. In some cases, low-temperature solder compositions containing bismuth (Bi) or indium (In) may be used.

Based on the current and foreseeable future design, BGAs and PACs are in 0.63- to 1.5-mm (0.025- to 0.060-in) pitch, more commonly of 1.0- to 1.5-mm (0.040- to 0.060-in) pitch. The established surface mount manufacturing process using mass printing and mass reflow can be directly applied to mounting array packages onto the mother board. However, some unique steps or alternate solder compositions may apply to each of the specific BGA modules, besides following the standard surface mount processes.

11.3.1 C-BGA

The standard manufacturing steps and the solder paste containing 63 Sn/37 Pb or 62 Sn/36 Pb/2 Ag are adopted. The C-BGA is considered to be equivalent to leadless ceramic chip carriers (LCCC) with respect to the attachment on the PCB. The reliability issues are discussed in Sec. 11.4.

Case 1: Replacing C-QFP by C-BGA

For the IBM RISC system 6000 Workstation, the C-BGA replaced the 304-I/O C-QFP (40 mm size/0.5 mm pitch) for the memory controller modules that are part of the subassembly consisting of banks of SIMMs and three controller modules for expanded random access memory.[19] The C-BGA package demonstrated an advantage in board real estate as shown in Fig. 11-15. Electrical performance including a shorter signal path, and capacitance and inductance improvement over the lead frame interconnection of C-QFP are manifest. In addition, there was a higher assembly yield and a lower component reject rate. Figure 11-16 illustrates the chip-to-board (card) signal path. Overall the comparison between the C-QFP and the C-BGA is summarized in Table 11-6

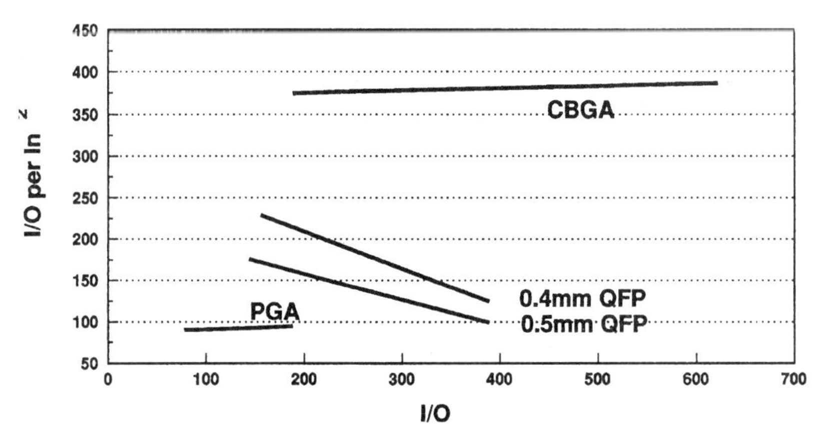

Figure 11-15. Board real estate for QFP, C-BGA, and PGA as a function of I/O.

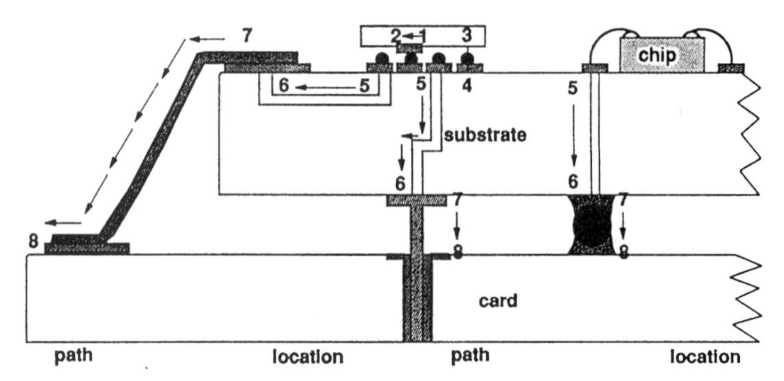

Figure 11-16. Signal path of chip-on-board card: 1 to 2 = on chip; 3 to 4 = c hip to substrate; 5 to 6 = substrate internal; 7 to 8 = substate to card.

Table 11-6. Comparison between the C-QFP and C-BGA of the Memory Controlled Module[19]

	C-QFP	C-BGA
Chip carrier size, cm^2	16	5.25
Pitch, mm (in)	0.50 (0.020)	1.27 (0.050)
Chip carrier cost	1 X	0.8 X
Signal noise	2.25 X	1 X
Assembly yield, ppm/lead	100	0.6
Component reject rate, % lead damage	7	0

1.3.2 P-BGA and OMPAC

Case 2: P-BGA[20]

The bare die is placed on an FR-4 carrier substrate (lower cost than BT-resin epoxy laminate) with a rim around the die. The carrier substrate provides the proper pads for wire bonding chips and for molded chip packages. Either flip-chip or TAB devices can be used. The key steps are

1. Die bonding.
2. Plasma cleaning.
3. Wire bonding.
4. Encapsulation.
5. Flip over the carrier substrate.
6. Print solder paste on the back side of the carrier substrate.
7. Reflow, forming semispherical solder bumps.

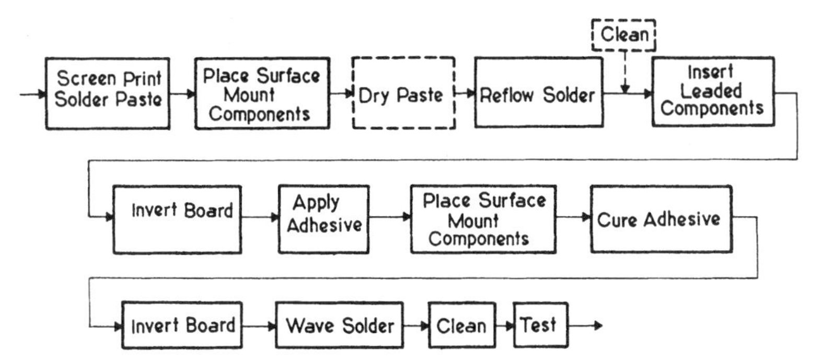

Figure 11-17. Standard SMT manufacturing steps.

8. Singulate into individual BGA packages.
9. Place individual BGA packages into the tube feeder.
10. Prepare for the next level (mother board level) of assembly using standard surface mount manufacturing steps as shown in Fig. 11-17.

The formation of solder bumps on the BGA carrier substrate is an important step. The solder volume and the uniformity among the solder bumps directly contribute to the quality and yield of board level assembly and to the resulting integrity of solder interconnections to the board. Printing equipment and printing parameters that have been developed for standard surface mount processes are applicable to attaching the P-BGA onto the board.[21–24] Yet some different demands to the process may be required, due to the higher volume of solder required for each interconnection. The higher solder volume is usually applied by using a thicker stencil, 14 to 18 mils in comparison with the standard surface mount stencil (6 to 8 mils). The thicker stencil may hinder the efficiency of paste transfer. To improve the efficiency of paste transfer, a few process and hardware-related tips can be considered:

1. *Stencil quality.* Smooth and straight openings become more important (see Sec. 10.4.8 for stencil selection).
2. *Stencil treatment.* A polymeric releasing agent applied on the stencil surface and the opening wall areas often improves the paste release.
3. *Squeegee speed.* Slow speeds facilitate paste transfer; less than 1.0 in/s (25 mm/s) is generally recommended.

4. *Delayed squeegee lifting.* After each squeegee stroke, the time delay in releasing the squeegee from contact with the stencil may provide the time for paste recovery and paste transfer.
5. *Heated squeegee.* A slight temperature elevation, 3 to 5°C, at the squeegee blade that is in contact with the paste facilitates paste transfer.

Case 3: OMPAC[25,26]

Key steps are

1. Die bonding.
2. Plasma cleaning.
3. Wire bonding.
4. Encapsulation.
5. Robotic work cell picks up solder balls.
6. Flux and glue solder spheres onto the backside of the carrier substrate.
7. Reflow and attach solder spheres on the carrier.
8. Prepare for the next level of assembly.

Rather than FR-4, BT-resin epoxy laminate is used as the carrier substrate. The top-side copper metallization consists of etched ICs, wire bond areas, and signal traces. The wire bond pads extend to plated through-holes are vias located around the periphery. The plated vias connect the top-side and bottom-side metallization patterns. This provides electrical continuity. Photolithographically defined copper traces provide the capability of finer pitches and traces. Encapsulation of the die is performed through a conventional plastic transfer molding technique. Electrical connections between the IC and the carrier are made through IC adhesives and thermosonic gold wire bonding. Solder mask is used to define pad arrays and to control the flow of solder. A schematic is shown in Fig. 11-18.[25] Copper traces are routed to an array of metal pads on the bottom side of the carrier to which solder balls of 62 Sn/36 Pb/2 Ag are reflowed and thus attached. Copper foil serves to distribute the heat to specific solder balls which are connected to the system PCB ground plane. A robotic work cell is designed to pick and place solder balls.[25] The work cell holds the solder ball in a pickup tool via a vacuum, passing over a consistent level of flux in the reservoir with a doctor blade. The flux should be viscous and tacky enough to act not only as a flux but also as an adhesive holding the solder balls in place before the permanent attachment is carried out.

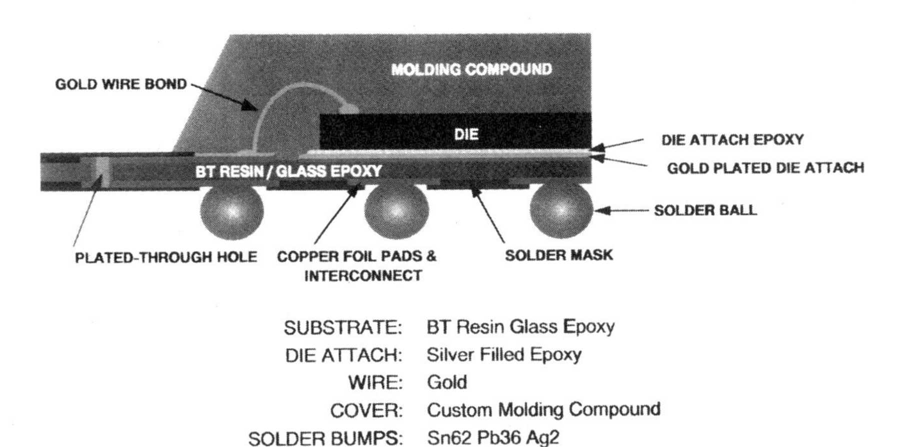

Figure 11-18. Schematic of OMPAC (*Courtesy of Motorola, Inc.*)

The resulting OMPAC packages possessing the following physical characteristics are ready for use in standard surface mount manufacturing.

Solder bump shape	Semisphere
Solder sphere diameter	0.76 mm (0.030 in)
Solder bump height	0.56 mm (0.022 in)
Pitch (225 leads or less)	1.5 mm (0.060 in)

Key steps for mounting the OMPAC onto the mother board are

1. Print solder paste on the solder pads of the mother board. The solder alloy is 62 Sn/36 Pb/2 Ag; the solder pad diameter is 0.76 mm (0.030 in); the paste thickness is 0.2 mm (0.008 in).
2. Place the OMPAC packages. Package edges can be used as a physical placement guide or use fiducial recognition with upward-looking vision. Because of the OMPACs strong self-alignment ability up to 0.3 mm (0.012 in) misplacement is allowable.
3. Reflow and form solder interconnections.

A typical final solder height is 0.41 mm (0.0216 in). It is reported that the defect level of the 169-pin OMPAC reached zero, compared with that of the 0.5-mm- (0.020-in-; 1.5 mm) pitch QFP of 120 to 700 ppm. A separate test showed that the defect level of the 80-pin OMPAC (0.060-

in) pitch was 0 to 7 ppm in comparison with that of the (0.025-in; 0.625-mm) pitch QFP of 20 to 35 ppm.

11.3.3 SLICC[27]

Case 4

The slightly larger than IC carrier (SLICC) package as shown in Fig. 11-19 integrates the features of both flip-chip technology and array technology. It is reported that the package offers the increased volumetric efficiency of IC packages. The solder bumped flip chip is attached to the PCB instead of to the traditionally used ceramic substrate. Because of the extreme CTE mismatch between silicon IC (≅ 2) and the PCB (≅ 16), solder interconnections in this package are expected to undergo adverse stress during the service life. In this case, the volume of solder joints is particularly small [0.1 to 0.13 mm (0.004 to 0 .005 in)]. This further compounds the concern about the integrity of solder joints. A polymeric underfill has been used in an attempt to distribute the stress and thus enhance the life of solder joints.[28]

Key process steps are

1. Flux solder pads on the carrier.
2. Place solder bumped flip chip using the vision system; solder bumping contains 60 Sn/40 Pb or high-lead solder.
3. Reflow and attach the flip chip on the carrier.
4. Clean.
5. Apply underfill and cure the polymer at 150°C for 45 to 50 min.
6. Solder bump the back side of the carrier using solder balls with a

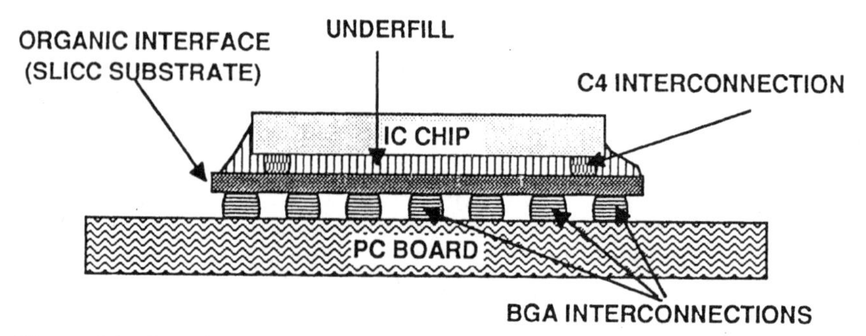

Figure 11-19. Schematic of SLICC.

0.50-mm (0.020-in) diameter and in a 0.88-mm (0.035-in) pitch array.

7. Singulate into the package unit.
8. Perform electrical tests.
9. Prepare for the next level of assembly using the standard surface mount manufacturing.

11.3.4 T-BGA

As covered in Sec. 11.2.4, T-BGA packages can be joined to the mother board with eutectic solder by following the surface mount process for 40-, 50-, and 60-mil pitch packages.

11.3.5 μ-BGA

Mounting a μ-BGA on the mother board follows the surface mount fine-pitch process. Key process steps are

1. Print solder paste.
2. Pick and place the fine-pitch μ-BGA.
3. Reflow.

Postsoldering underfill is not required.

11.3.6 μSMT

The assembly process involves stencil printing on an active silicon wafer using a proprietary solder paste, picking and placing the bare die face down in the solder paste, and then nitrogen reflow in an infrared oven. In this assembly, a typical surface mount process is employed except that the die is not prepackaged and the substrate is silicon wafer.[15]

11.3.7 M-BGA

Key process steps are

1. Deposit a thin-film circuit on an anodized aluminum substrate.
2. Die attach.
3. Wire bonding.
4. Encapsulation.

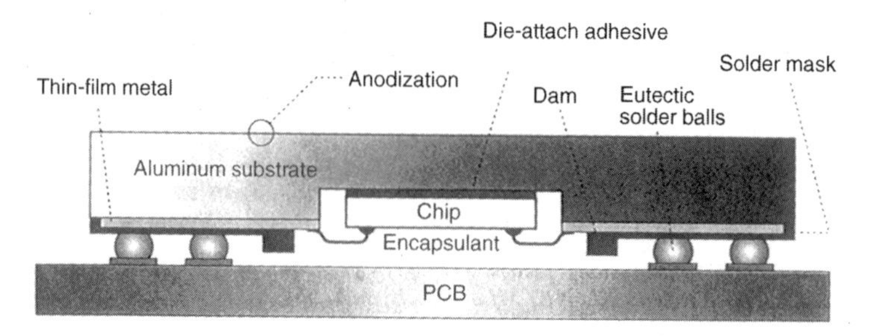

Figure 11-20. Cross-sectional view of M-BGA.

5. Eutectic solder attach.
6. Prepare for the next assembly using surface mount manufacturing.

A cross-sectional view of an M-BGA package is illustrated in Fig. 11-20.[17]

11.3.8 Flip-on-Flex[29]

This package is made of flip-chip mounting and flexible circuitry. The circuit material is featured with an adhesiveless polyimide copper laminate. Key process steps are

1. Print solder paste.
2. Mount solder bumped flip chip on the flexible circuit board.
3. Reflow and form permanent interconnections.
4. Clean.

11.3.9 PCMCIA

As discussed in Sec. 2.9, the Personal Computer Memory Card International Association (PCMCIA) requires that products possess high I/O, low profile, and high performance. Two package standards, type 1 with 3.3-mm thickness and type 2 with 5.0-mm thickness are evolving.[30] Flip-chip technology, MCM technology, and array packages are likely the integral part of PCMCIA cards. It has an expanding, high-volume market which is projected to reach more than $6 billion worldwide by the year 2000.

11.4 Reliability of Solder Interconnections

The topics of solder joint failure mode and solder joint reliability will be addressed in Chap. 16. This section, however, outlines the specific aspects that are important to BGA and other array solder interconnections. A summary of the state-of-the-art findings of solder joint reliability for BGA and other array solder is also provided.

The main interest and/or concern about the reliability of array solder interconnections stem from two areas:

- Array solder interconnections are less compliant in comparison with the conventional peripheral leaded solder joints. The decreased compliance is often linked with the reduced performance under fatigue environment due to thermal stress/strain imposed on the solder joints as a result of temperature cycling, ambient temperature fluctuation, and in-circuit power on-off.
- Array joints are relatively new and still in the infant stage of application for board-level assembly on a mass-production basis. There are less statistically substantiated data, and the field performance establishment is lacking.

The major influence on the reliability of array solder joints can be considered from seven areas:

- Component (die, substrate)
- Board material (CTE)
- Solder composition (eutectic, high-lead)
- Configuration and volume of solder joint
- Supplemental material (underfill polymer)
- CTE match between component, board, and solder
- Manufacturing process

11.4.1 Component

For a component package, the functional characteristics of the die, the die-to-package size ratio, the properties of the substrate, and power dissipation all contribute to the thermal profile of the array solder joints. A study was conducted on life tests of OMPAC solder joints that are grouped into five loops in relation to the location of the die. Figure 11-21 illustrates these five loops: corner, outer, middle, inner, and cen-

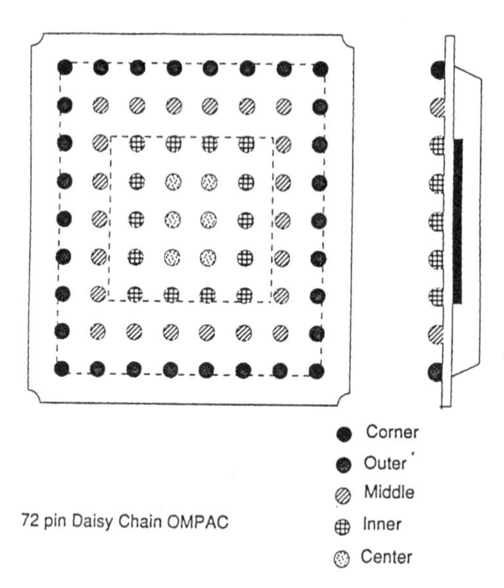

Figure 11-21. Five loops of solder joint in relation to the die location.

ter.[31] The test conditions included thermal shock (−65 to 150°C; −55 to 125°C). It was reported that the first solder joint failure occurred at the loop of solder joints near the edge of the die and that 75 to 80 percent of failures under temperature cycling occurred at the solder sphere to package interface. Finite-element analysis and cycle-to-failure calculation[32] were carried out on P-BGA under the thermal shock (−25 to 100°C), temperature cycling (−25 to 100°C) with a 15-min dwell time at each temperature extreme), and power cycling for various configurations of P-BGA as summarized in Table 11-7.

Under power cycling, the temperature distribution for the outermost solder joint reached 84.3°C, while the center joint was at 98.8°C (1.2°C

Table 11-7. Configuration of IC Devices

Number of pins	Die size, mil × mil) × mil	Bump pitch		Body size	
		mil	mm	inch	mm
81	270×270×15.5	60	1.5	0.60	15.0
165	437×437×21	60	1.5	0.78	19.7
225	389×389×18.7	60	1.5	0.91	22.7
421	400×400×20	50 (staggered)	1.27	1.60	40.0

Table 11-8. Thermal Shock Test Results of BGAs with Various Die Sizes

Number of pins	Die size, mil×mil ×mil	First failure	Cycles to 50% failure
72	270 × 270 × 11.5	1768	3403
165	437 × 437 × 21	966	2124
225	389 × 389 × 18.7	2350	2804

below the junction temperature) for an 81-pin package. However, for a 421-pin package which had a lower die-to-package size ratio, there was a large temperature differential between the outer joints of 56.2°C and the center joints of 98.5°C.[32] It was found that the 165-pin device which had the largest die size had the earliest failure, and its cycles-to-failure was lower than that of the 225-pin device, indicating that the size of the die makes a significant contribution to the solder joint performance. The solder joints proximate to the perimeters of the die failed first under temperature cycling. The results concluded that the cycles-to-failure does not follow the device size, rather it mostly depends on the die size. Another study[1] on P-BGA solder joints under thermal shock (−25 to 100°C) for various packages as shown in Table 11-8 revealed similar results indicating that the die size (or die thickness) has the most impact on the fatigue life of solder joints under temperature cycling and that the first temperature-cycling–induced failure occurred in the solder joint beneath the edge of the die.

11.4.2 Board material

The two characteristics of board material that have the most influence on the long-term performance of solder interconnections are planarity and the CTE. Poor board planarity adds to the coplanarity problem of the BGA package, contributing to the occurrence of solder joint distortion, which in turn may lead to early failure of the solder joint under cyclic stress. The 62 Sn/36 Pb/2 Ag solder joints of the pad array carrier (PAC-196) which is composed of alumina substrate and 0.76-mm-(0.030-in-) diameter solder balls on a 1.27-mm-(0.050-in-) pitch array was studied by continuously monitoring its in situ resistivity after being subjected to temperature cycling (−55 to 125°C, air-to-air, 15 min dwell time at each of the temperature extremes, and 15 min transition between upper and lower temperatures).[33] The criterion for an open solder joint was set at 100 Ω. The CTE of commonly used board material, such as FR-4, is around 15 ppm/°C which is much larger than that of alumina with its nominal CTE of 7 ppm/°C. The CTE of solder materials falls in the range of 21 to 30 $\times 10^{-6}$/°C, depending on the alloy composition. The differential in the CTE between the board material and the carrier substrate results in an additional driving force to cause plastic deformation in solder joints under fluctuating temperature conditions. It was found that the barrel solder joints between alumina substrate and aramide had no failure through 1000 temperature cyclings.[34] A closely matched CTE between the board and carrier substrate reduces thermally induced stresses, thus increasing the service life of solder joints.

11.4.3 Solder composition

While all other conditions are equal (alumina substrate, aramide board, barrel solder configuration), a special solder composition that displays a more ductile property than 62 Sn/36 Pb/2 Ag was reported to withstand temperature cycling (−55 to 125°C) better.[33] The solder composition used for the IC solder bumps (balls) affects the mechanical integrity of solder interconnections. Figure 11-22 compares the 60 Sn/40 Pb and 5 Sn/95 Pb used as IC bump solder material for interconnecting C-4 packages and the 60 Sn/40 Pb prebumped PCB in measuring their monotonic strength. Under both shear and tensile testing

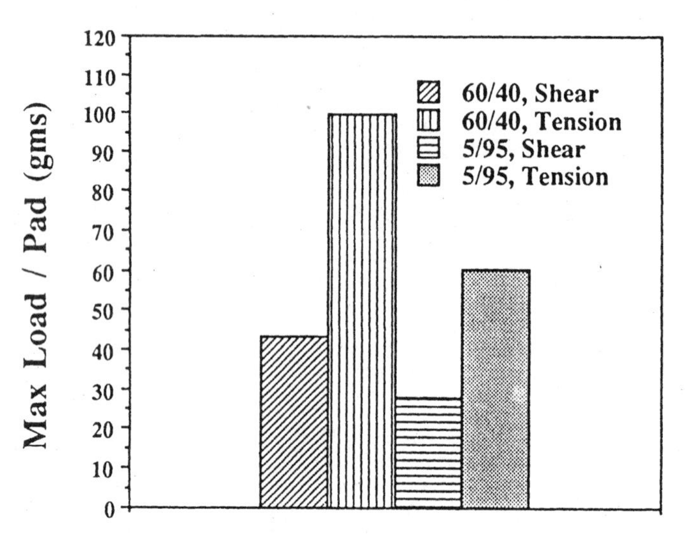

Figure 11-22. Monotonic strength of 60 Sn/40 Pb and 5 Sn/95 Pb as IC bump material.

modes, 60 Sn/40 Pb solder bumps delivered higher strength.[27] Furthermore, the failure mode differed with the change of the solder compositions. The failure in the case of 60 Sn/40 Pb solder bumps was through the solder adjacent to the component interface. The failure in the 5 Sn/95 Pb solder bumps, however, was through the reduced cross section where the board solder wicked up around the high-melting-point component solder bump.

11.4.4 Configuration and volume of solder joint

For a specific assembly (alumina substrate and aramide board), it was found that a more compliant solder joint configuration (hourglass) benefited the temperature cycling resistance when a more ductile solder composition was used.[33] The origin of the improved fatigue life of the hourglass solder joint relative to the barrel joint is attributed to the relative location of deformation. In the hourglass solder joint, the reduced cross section is at the middle between the PCB and the component; therefore, the deformation takes place in the ductile middle section of the solder joint, and the deformation is away from the brittle interface of the solder joint to the board or to the component. In contrast, the barrel joint has the reduced cross section at the interface with the board or the component side, where the mechanical properties are normally poor.

Another example involves C-BGA with solder ball [0.89 mm (0.030-in) diameter] or ceramic column grid arrays (C-CGA) with solder column [2.20-mm (0.087-in) height and 0.51-mm (0.020-in) diameter]. Both solder balls and columns are composed of 10 Sn/90 Pb solder composition and attached to substrate and bonded by 63 Sn/37 solder. After reflow of the packages on the board, the final standoff height of C-BGA is 0.94 mm (0.037 in) and that of C-CGA is 2.26 mm (0.089 in).[11,35]

Under temperature cycling, it was found that the fatigue life of C-CGA solder joints is 7 to 10 times longer than that of C-BGAs under the temperature cycling range of 0 to 100°C; however, under −40 to 125°C, the fatigue life of C-CGAs is about 5.5 times better than that of C-BGAs. The shift of failure mode was also observed in C-CGA solder interconnections; that is, from the crack through the column for the cycling temperature range of 0 to 100°C to the combined cracks through both the column and the eutectic solder used to attach the column to the ceramic substrate for the cycling temperature of −40 to 125°C.

11.4.5 Underfill material

Several studies demonstrated that an epoxy underfill which fills the air gap between the solder and the underside of the component is beneficial to the fatigue life of the solder joints.

Solder joints for OMPAC packages under temperature cycles (−40 to 125°C) were improved by nearly twofold with epoxy underfill.[31] The underfill around the chip and the substrate of the SLICC assembly was used to enhance the reliability of solder joints. The solder joint failure under thermal shock (−55 to 125°C, 10 min dwell at each extreme) was attributed to the separation of the underfill from the die surface. The loss of adhesion was further related to the foreign contamination that was not thoroughly removed during the cleaning procedure.[27]

A combination of FR-4 and PAC with alumina substrate was benefited by an epoxy underfill in its fatigue life by approximately a factor of 5.[33] It was evidenced that the underfill reduced the relative displacements of the PAC and FR-4 board by mechanically coupling them both. The results are indicative that the CTE of the underfill plays an important role in its performance; the closely matched CTE between the underfill and the solder composition minimized the vertical axis strain, therefore improving the fatigue life.

Figure 11-23 presents the fatigue life data for C-BGAs in relation to the predicted distribution.[35] The lower failure rate of C-BGAs with underfill than the predicted value is attributed to the presence of the underfill material. The C-CGA failure rate, as expected, is much lower

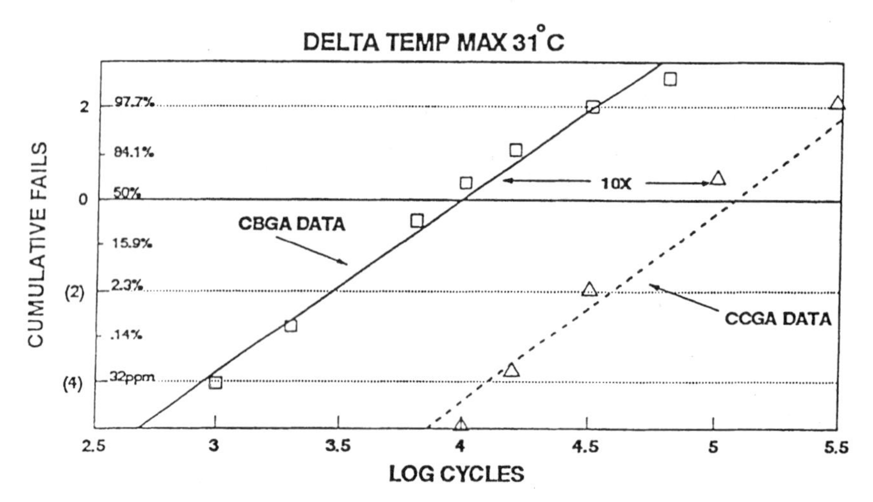

Figure 11-23. Fatigue life data of CBGA versus CCGA.

than that of the C-BGA. In general, the fatigue life is expected to follow the descending order of

C-CGA > C-BGA with underfill > C-BGA

In evaluating the performance of an encapsulant, the strain development as measured at the bump farthest from the center of the chip is 4.5 percent for FR-4 substrate and 1.5 percent for CTE-controlled PCB. With added encapsulant, the strain of 4.5 percent for FR-4 board drops to 1.0 percent.[36] The encapsulant is believed to be capable of converting the strain on the solder bump into deformation of the whole system of chip and substrate.

11.4.6 CTE match between component, board, and solder

When considering the sources of stress imposed on solder interconnections of electronics packages and assemblies, thermal stress, rather than mechanical load, plays a major part. Thermal stresses and strains generally result from

- Inherent temperature gradient over the board
- External temperature change
- In-circuit power on and off

The magnitude of the stress or strain as a result of the above sources is directly related to the difference in CTEs of the component, the board, and the solder materials, as represented by

$$\text{Effective strain} = \frac{b\, \Delta\alpha\, \Delta T\, L}{2t}$$

where b = geometry factor
$\Delta\alpha$ = difference in CTE
ΔT = temperature differential
L = length of component
t = thickness of solder joint

Under a given range of temperature fluctuation and for a specific assembly system, the effective strain is proportional to the difference in CTEs of the assembled parts. When their CTEs are closer to one another, the effective strain becomes smaller. For example, temperature cycling tests (−55 to 125°C) for assemblies with two different board materials—FR-4 (x axis CTE $\cong 15\times10^{-6}$/°C) and aramide paper

Table 11-9. Temperature Cycling Data for Two Types of Board Material

Board material	Number of solder joint failures			
	100 cycles	200 cycles	500 cycles	1000 cycles
FR-4	12	24	—	—
Aramide paper PCB	0	0	1	19

PCB (x axis CTE $\cong 7 \times 10^{-6}/°C$)—were conducted. The test results indicated that as the CTE of the board materials was brought closer to that of the component, the fatigue life of the solder joints was prolonged. Table 11-9 summarizes the test data for 62 Sn/36 Pb/2 Ag solder joints under 100, 200, 500, and 1000 temperature cycles for both board materials.[33]

11.4.7 Manufacturing process

In addition to the factors that contribute to the long-term performance of solder joints during their service life, the ability to produce high-quality solder joints is equally important. Although the installed surface mount operation in the industry can be directly applied to mounting BGAs on the mother board, the setup of process parameters, the control of the process, and the proper ambient environmental conditions are key to making quality solder joints. High humidity and high temperature which are generally detrimental to surface mount manufacturing should be avoided.

11.5 Coplanarity Tolerance

The required coplanarity of array solder balls is not as stringent as that for fine-pitch leads. Yet the better coplanarity is always beneficial to the yield of surface mount manufacturing. It is reported that the current achievable level of coplanarity for P-BGAs is around 200 μm (7.8 mils).[37]

The *coplanarity* is defined as the distance between the highest solder ball value and the lowest solder ball value as shown in Fig. 11-24. It was found that there is a direct relationship between the coplanarity and the carrier substrate warpage; this is not a surprise considering the fabrication process of and material used in P-BGA packages. It is also anticipated that the substrate warpage is linked with the size of the packages. Table 11-10 shows that the larger the package, the

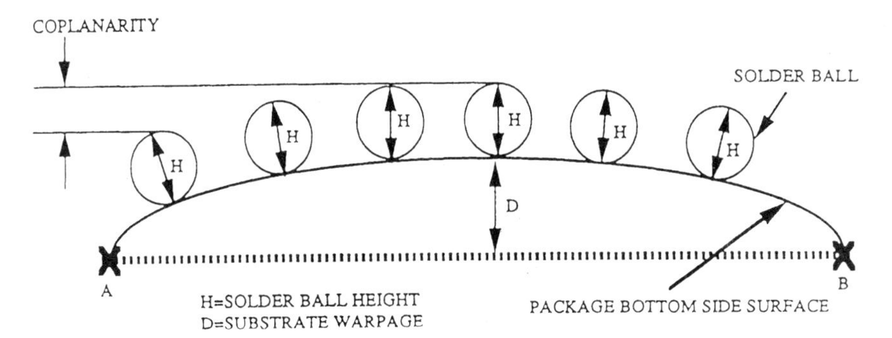

Figure 11-24. Coplanarity.

Table 11-10. BGA Package Size versus Substrate Warpage

	BGA Package		
	169L	225L	396L
Package size, mm^2	22.0	27.0	35.0
Substrate warpage (μm)			
Maximum	108.6	107.4	140.1
Standard deviation	8.6	4.7	9.7
Average	100.5	104.0	124.7

greater the substrate warpage. Consequently, as the package size increases, the coplanarity value increases (less coplanar) as shown in Table 11-11.

The thickness of the substrate core can be another factor affecting the coplanarity. Table 11-12 exemplifies the effect of the thickness of the substrate core on coplanarity, indicating that the increased thickness raises the level of coplanarity.

With respect to current specification, the JEDEC has standardized the coplanarity of 150 μm (5.9 mils).

Table 11-11. BGA Package Size versus Coplanarity

	BGA Package		
	169L	225L	396L
Package size, mm^2	22.0	27.0	35.0
Coplanarity value, μm			
Maximum	134.4	149.4	161.3
Standard deviation	12.0	10.5	10.4
Average	118.8	126.9	142.5

Table 11-12. Substrate Core Thickness versus Coplanarity

	BGA Package			
	225L		396L	
Substrate core thickness, mm	0.2	0.6	0.2	0.6
Coplanarity value, μm				
Maximum	149.4	120.1	161.3	122.3
Standard deviation	10.5	11.3	13.6	11.6
Average	126.9	104.0	141.1	112.0

11.6 Soldering-related Defects

Soldering-related defects may be construed as the product rejects during board manufacturing or create functional problems during service or contribute to the premature failure of circuitry function. These include

1. Bridging
2. Open or insufficient solder
3. Poor wetting
4. Solder balling
5. Voids
6. Foreign contamination—chemical induced, metallurgical dissolution
7. Plastic package cracking

11.6.1 Bridging

Array packages with 1.50-mm (0.060-in) pitch or 1.0-mm (0.050-in) pitch are less prone to bridging problems between two adjacent interconnection sites. As array density continues to increase to a finer pitch, bridging tendency also increases. In addition to dimensional spacing between solder interconnections, two other main causes are

- *Excessive amount of solder volume in relation to pad dimensions.* As a result of the mutual affinity of molten solder between two adjacent solder sites, bridging between two sites that carry an excessive amount of solder occurs. Each type of array package may behave differently. This depends on the alloy composition and the melting temperature of the carrier solder balls, the pad design in relation to the size of the carrier solder ball, and the weight of the carrier. For

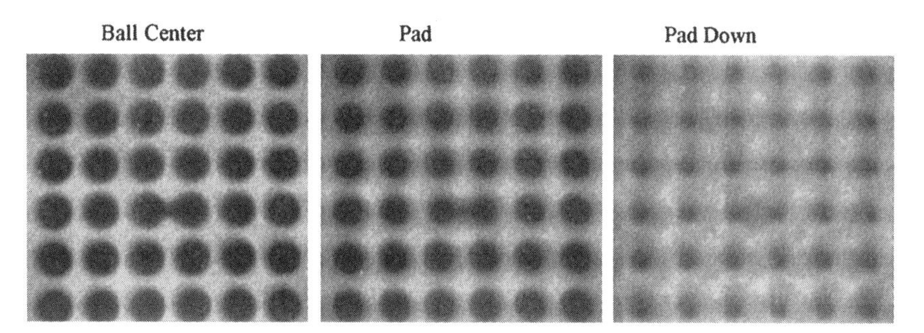

Figure 11-25. Bridging between array interconnections x-ray image. (*Courtesy of Four Pi Systems*)

example, the assembly with carrier solder balls containing high-temperature solder alloy which does not melt during the board assembly reflow has less tendency for bridging, when other conditions are equal. Figure 11-25 illustrates the bridging between array interconnections.

- *Paste slump.* When using solder paste to connect a BGA onto the mother board, the paste slump phenomenon during printing and reflow plays an important role to bridging. The desired property of slump resistance (ideally, zero slump) is largely attributed to the intrinsic thermodynamic behavior of the flux/vehicle system (chemical portion of the paste without the solder powder) as discussed in Sec. 10.3.

11.6.2 Open or insufficient solder

The major contributors to open or insufficient solder joints in attaching array packages to boards are

1. *Excessive board warpage.* Most BGA designs take into account local board warpage up to 0.13 mm (0.005 in) from the center to the edge of the package. However, when board warpage exceeds the designed tolerance level, open or insufficient solder joints may occur.
2. *Coplanarity tolerance.* As discussed in Sec. 11.5, the prevalent level of coplanarity is around 150 to 200 μm (5.9 to 7.8 mils) for P-BGA balls. It should be noted that the coplanarity is directly linked to the level of board warpage.

3. *Related to poor wetting.*
4. *Related to excessive solder balling during reflow.*

Sections 11.6.3 and 11.6.4 discuss wetting and solder balling, respectively.

11.6.3 Poor wetting

A wetting problem could occur at either a solder-to-pad interface or a solder-to-ball interface as indicated in Fig. 11-26. The lack of adequate wetting can be caused by external factors that produce a wettability variation among solder sites or among BGA packages and boards. Variations of the BGA fabrication process, board fabrication process, and subsequent handling, storage, and exposure conditions create inconsistencies in wettability. The wetting problem can also be caused by intrinsic interaction of the material system, relating to the nature of metal substrates or to the metallurgically limited stability between the solder alloy compositions and the metal substrates that are being used in electronics packaging and circuit boards or relating to the compatibility and effectiveness of fluxing chemistry. It is well recognized that for Sn-Pb–coated Cu leads or pads, metallurgical reactions between Sn and Cu proceed continuously at a rate that is determined by the ambient exposure conditions. The extent of the reaction affects the wettability of subsequent interconnections. The consistency and quality of the surface to be soldered, the control of reflow process, and the selection of solder paste and its consistency are obviously important factors in achieving good wetting.

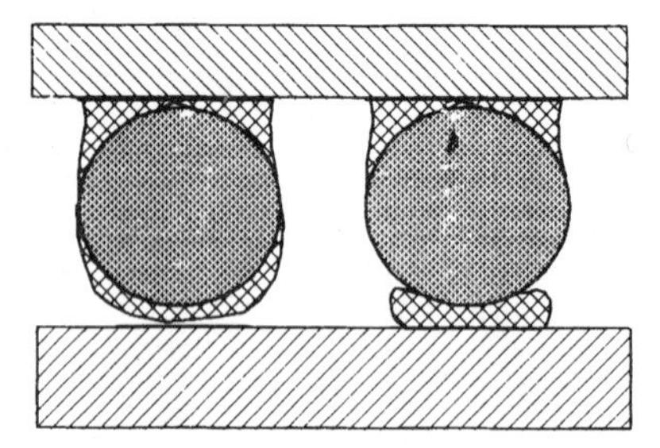

Figure 11-26. Wetting interface at solder-to-pad (left) and solder-to-ball (right).

11.6.4 Solder balling

When using solder paste to attach BGA packages to the board, the formation of excessive solder balls may result in a potential electrical short during service if solder balls are not removed. It can also deprive the solder volume necessary for making good interconnections from the solder filet. General causes and remedies follow the discussion in Sec. 13.3.

11.6.5 Voids

The voids in solder joints of BGA assembly should be examined in two areas: (1) solder balls (bumps) of carrier where voids may be introduced during package fabrication, and (2) solder interconnecting the carrier and the board where voids may be formed during board assembly. Possible causes and remedies are covered in Sec. 13.4.

11.6.6 Foreign contamination

The principle and data are discussed in Sec. 13.5.

11.6.7 Plastic package cracking

Among BGAs, the plastic molded packages such as P-BGA are prone to moisture absorption. The absorbed moisture induces package cracking during reflow (see Sec. 13.6).

11.7 Cost Consideration

Comparing the cost of BGAs with that of conventional, peripheral packages, today's plastic QFP packages generally enjoy a lower cost compared to plastic BGAs, particularly when the pin count is lower than 200. The higher cost of BGAs is largely attributed to one or a combination of factors as listed below:

- High cost of carrier substrate material (BT resin)
- Fine line circuitry
- More routing layers
- Infrastructure

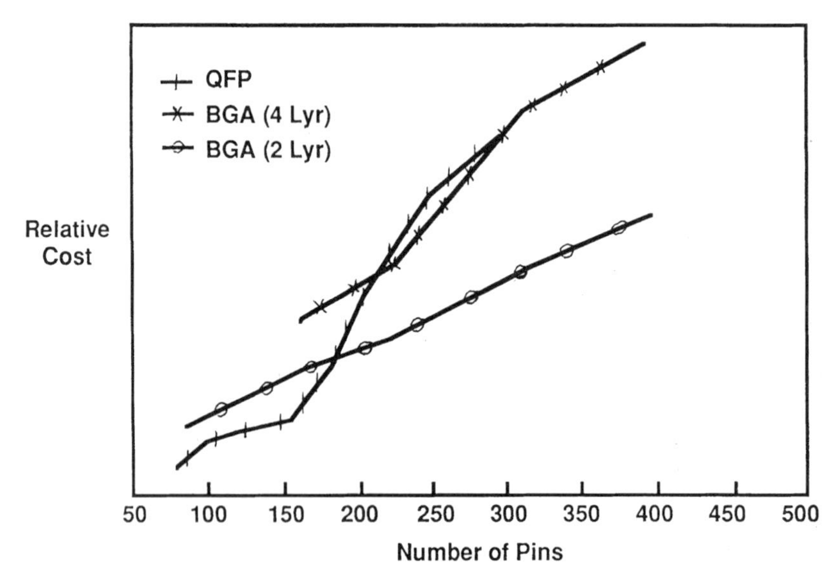

Figure 11-27. Comparison of package cost.

These higher cost factors are partially offset by the lower process cost on the board level. Figure 11-27 illustrates the comparison of package costs. It is shown that a BGA package with a two-layer carrier substrate has an equivalent cost to a QFP at about 200 pins. Above 200 pins, the two-layer plastic BGA package can offer a saving over the QFP, and the four-layer plastic BGA package essentially follows the QFP cost trend.[2]

It should be noted that assemblies with array packages requiring an underfill material in order to enhance the endurance of the solder interconnections under a thermal fatigue environment are normally nonrepairable. Such nonrepairable packages in multichip module design would be cost-prohibitive.

11.8 References

1. Bruce Freyman et al., "Surface Mount Process Technology for Ball Grid Array Packaging," *Proceedings SMI,* 1993, p. 81.
2. Randy Johnson et al., "A Feasibility Study of Ball Grid Array Package," *Proceedings NEPCON-East,* 1993, p. 413.
3. Jon Haughton, "Capturing Design Advances of BGA's," *Surface Mount Technology,* March 1994, p. 36.

4. Jan Vardaman, "Technology and Business Trends in BGA Packaging," *Proceedings NEPCON-East*, 1993, p. 431.
5. Jennie S. Hwang, "Innovation, Leadership and Competitiveness," *Surface Mount Technology*, May 1993, p. 88.
6. John Tuck, *Manufacturing Market Insider*, June 1992.
7. Jerry Bartley et al, "C-BGA: A Packaging Advantage," *Surface Mount Technology*, November 1993, p. 35.
8. F. F. Cappo, and J. C. Milliken, "Solving I/O Density Problems Using MLC Packages," *Surface Mount Technology*, November 1993, p. 35.
9. Panangathur N. Venkatachalam and David J. Perlman, "25 mm Solder Ball Connection Module for Improved Memory Performance," *Proceedings IEPS*, 1993, p. 709.
10. Richard E. Sigliana, "Using BGA packages," *Advanced Packaging*, March/April 1994, p. 36.
11. Thomas Caulfied, J. A. Benenati, and J. Acocella, "Cost-Effective Interconnections for High I/O MCM-C-to-Card Assemblies," *Surface Mount Technology*, July 1993, p. 18.
12. M. Coleand and T. Caufield, "Ball Grid Array Packaging," *Proceedings SMI*, 1994, p. 147.
13. P. Mescher and G. Phelan, "A Practical Comparison of Surface Mount Assembly for Ball Grid Array Components," *Proceedings SMI*, 1994, p. 164.
14. T. H. Ditefano et al., "μ-BGA for High Performance Applications," *Proceedings SMI*, 1994, p. 212.
15. T. D. Dudderar et al., "AT&T Surface Mount Assembly: A New Technology for the Large Volume Fabrication of Cost Effective Flip Chip MCMs," ISHM, *Journal of Microcircuit and Electronic Packaging*, vol. 17, no. 4, 1994, p. 361.
16. R. Treece, "mBGA Technology Overview," *Electrecon*, 1994, Indianapolis, Indiana.
17. Product Brochure, Olin Interconnect Technology, Manteca, California.
18. Semiconductor International, February 1995.
19. R. S. Malfattt et al., "Package Conversion from CQFP (40 mm/0.5 pitch) to CBGA in the IBM RISC System 6000 Workstation", *Proceedings NEPCON-East*, 1994, p. 281.
20. Dave Hattas, "BGAs Face Production Testing," *Advanced Packaging*, Summer 1993, p. 44.
21. Jennie S. Hwang, *Solder Paste in Electronics Packaging-Technology and Applications for Surface Mount, Hybrid Circuit and Component Assembly*, Van Nostrand Reinhold, 1989, Chap. 6.
22. Jennie S. Hwang, "Step-by-Step SMT: Solder Screen Printing," *Surface Mount Technology*, March, 1994, p. 44.

23. Chung H. Park, "Understanding Solder Paste Fundamentals," *Circuits Assembly,* August 1993, p. 29.
24. K. L. Kent, "Solder A Paste Printing for Fine Pitch Devices," *Proceedings ISHM, Annual Symposium,* 1990, p. 136.
25. Fonzell D. J. Martin, "C-5 Solder Sphere Robotic Placement Cell for Overloaded Pad Array Carrier," *Proceedings IEPS,* 1993, p. 740.
26. John Tuch, "Ball Grid Arrays Have Arrived," *Manufacturing Market,* January 1993.
27. Kingshuk Banerji, "Development of the Slightly Larger Than IC Carrier (SLICC)," *Proceedings NEPCON-West,* 1994, p. 1249.
28. Yutoka Tsukada and Yohko Mashimoto, "Low Temperature Flip Chip Attach Packaging on Epoxy Base Carrier," *Proceedings SMI,* 1992, p. 294.
29. K. Casson, "Flip-on-Flex: Solder Bumped IC's Based to Novaclad A New High Temperature, Adhesiveless, Flex Material," *Proceedings IEPS,* 1993, p. 321.
30. Portia Isaacson, "Dream This," *OEM Magazine,* May 1994, p. 100.
31. Paul T. Lin et al, "Manufacturability of Plastic Ball Grid Array versus QFP Packages," *Third International Microelectronics and Systems, 1993 Conference,* Kuala Lumpur, Malaysia, p. 77.
32. Andrew Mawer et al, "Calculation of Thermal Cycling and Application Fatigue Life of the Plastic Ball Grid Array (BGA) Package," *Proceedings IEPS,* 1993, p. 718.
33. Kevin Moore et al., "Solder Joint Reliability of Fine Pitch Solder Bumped Pad Array Carriers," *Proceedings NEPCON-West,* 1990, p. 264.
34. Tsunashima et al., "An Innovation to Multilayer Hybrid Circuit Laminate Composed of Sophisticated Resin Combinations of a New Aramid Paper and an Epoxy Impregnant," *Electronics Components Conference,* May 1989.
35. David Gerke, "Ceramic Solder-Grid-Array Interconnection Reliability Over a Wide Temperature Range," *Proceedings NEPCON-West,* 1994, p. 1087.
36. R. J. Lyn, "Encapsulation for PCMCIA Assemblies—An Overview," *Proceedings NEPCON-West,* 1993, p.743.
37. Junicho Shimizu, "Plastic Ball Grid Array Coplanarity," *Proceedings SMI,* 1993, p. 87.

12 Soldering Methodologies

12.1 Soldering Heat Source

Soldering refers to the process that produces solder interconnections from a solder material in a specific physical form by using a selected equipment and methodology as a heat source. Solder material can be preform, a predeposited coating layer, solder paste, or other forms. The methodologies and corresponding equipment are largely based on the following heat sources:

- Conduction
- Infrared
- Vapor phase (condensation heat)
- Hot gas
- Convection and forced-air convection
- Induction
- Laser
- Focused infrared lamp
- White beam lamp
- Hot bar (thermode)

Table 12-1. Reflow Methods—Feature Comparison

Reflow method	Benefits	Limitations
Conduction	Low equipment capital; rapid temperature changeover; visibility during reflow	Planar surface and single-side attachment requirement; limited surface area
Infrared	High thoughput; versatile temperature profiling and processing parameters; easier zone separation	Mass, geometry dependence
Vapor phase condensation	Uniform temperature; geometry independence; high throughput; consistent reflow profile	Difficult to change temperature; temperature limitation; relatively high operating cost
Hot gas	Low cost; fast heating rate; localized heating	Temperature control; low throughput
Convection	High throughput; versatility	Slower heating; higher demand for flux activity
Induction	Fast heating rate; high temperature capacity	Applicability to nonmagnetic metal parts only
Laser	Localized heating with high intensity; short reflow time; superior solder joint; package crack prevention	High equipment capital specialized paste requirement; limit in mass soldering
Focused infrared	Localized heating; suitable for rework and repair	Sequential heating; limit in mass soldering
White beam	Localized heating; suitable for rework and repair	Sequential heating; limit in mass soldering
Hot bar	Localized and selected heating	For individual components, e.g., QFP

Soldering is achieved with one of the heat sources listed above through different heat transfer methodologies. Each method has its unique features and merits in terms of cost, performance, or operational convenience. For localized and fast heating, laser heating surpasses all other methods, although it is a low-volume process, with hot-gas heating in second place. For uniform temperature, the vapor phase heating ranks first. The uniform temperature offered by vapor phase reflow is expected to be particularly beneficial for BGAs on mother board assemblies and other packages/assemblies that demand a higher level of temperature uniformity. For versatility, volume, and economy, infrared (with appropriate wavelength) and convection heating are favored. For concurrently soldering a large number of solder joints with a high throughput, as for high-density boards containing diverse components in various sizes and masses, forced-air convection is advantageous. Conduction heating, however, is convenient for hybrid assemblies and other assemblies that meet the requirement in surface planarity and substrate size. For conductive components requiring fast heating and high-temperature soldering, induction heating meets the requirements. Table 12-1 summarizes the strengths and limitations of each method.

12.2 Heat Transfer

Heat transfer from the heat source to the workpiece relies on one or a combination of the following mechanisms: conduction, radiation, and convection.

12.2.1 Thermal conduction

The heat of thermal conduction is transferred by molecular motion which occurs within the same body or between two bodies which are in physical contact with each other. As discussed in Sec. 3.2.3, heat in a solid is mainly conducted by elastic vibration of lattice movement in crystal transmitted in the form of waves (phonons) and/or by the movement of conduction electrons. The heat flow per unit area A is related to the thermal conductivity k and the temperature T gradient in a given distance x, as expressed by Fourier's Law:

$$A = -k\frac{\partial T}{\partial x}$$

Note that this is equivalent to Newton's Law of viscosity, relating shear stress between any two thin layers of fluid and a constant velocity gradient. The thermal conductivity of solids which conduct heat only by phonons is inversely proportional to the temperature above the Debye temperature. As the temperature is raised to a sufficiently high level, the mean free path decreases to a value near the lattice spacing, and the thermal conductivity is expected to be independent of temperature. At very low temperatures, the mean free path of phonons becomes of the same magnitude as the sample size and the conductivity decreases to 0 K. However, if conduction is electronic in nature, the thermal conductivity is expected to increase with increasing temperature.

12.2.2 Thermal radiation

The energy of thermal radiation is transferred by electromagnetic waves which originate from a substance at its elevated temperature. The thermal energy or emission power e is proportional to the fourth power of its absolute temperature T (K).

$$e = bT^4$$

where b is the Stefan-Boltzman constant. The maximum wavelength of the radiation γ_{max} is inversely proportional to the absolute temperature (K).

$$\gamma_{max} \propto \frac{1}{T}$$

These relations indicate that increasing temperature not only increases the emission power of an emitter but also shortens its maximum wavelength emitted. Figure 12-1 is the electromagnetic radiation spectrum and corresponding energy states. The radiation energy incident on the surface of a substance may be absorbed, reflected, or transmitted.

Metals normally absorb in the near-infrared region. Oxides and organics are, however, transparent in the near-infrared region and absorb in the middle- and far-infrared region. The infrared absorption spectra (absorbance versus wave number) of alumina, epoxy-glass, polyimide-glass, and solder paste are shown in Figs. 12-2, 12-3, 12-4, and 12-5, respectively.[1]

The spectra were obtained by using a Digilab FT-IR Spectrometer. Absorbance instead of transmittance was directly measured in relation to the wave number.

12.2.3 Thermal convection

The heat transfer by convection is normally expressed as a heat transfer coefficient h between two temperatures (heating source T_2, and workpiece T_1).

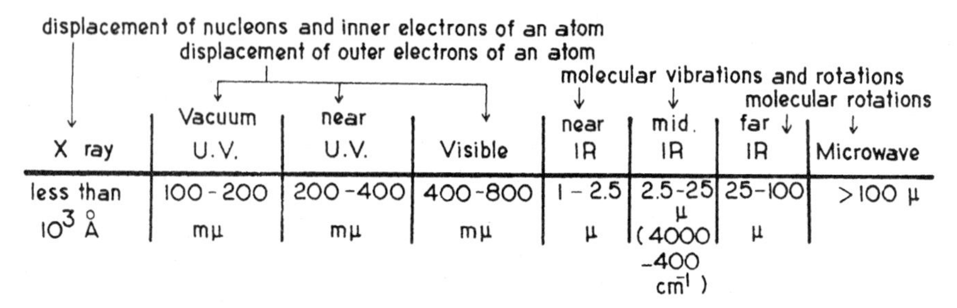

Figure 12-1. Electromagnetic radiation spectrum in energy state.

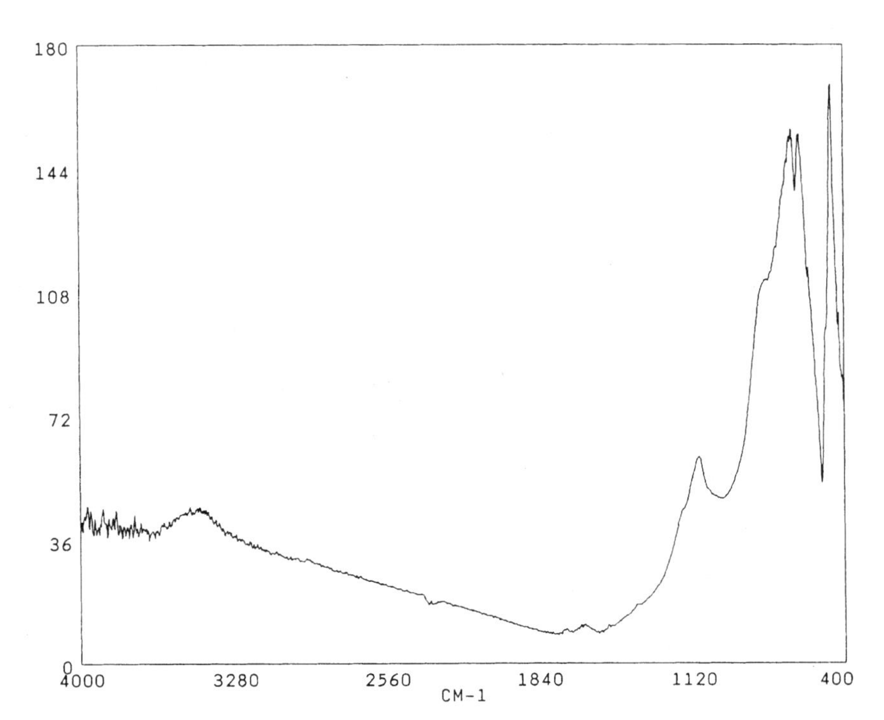

Figure 12-2. Infrared spectrum (absorbance versus wave number) of alumina.

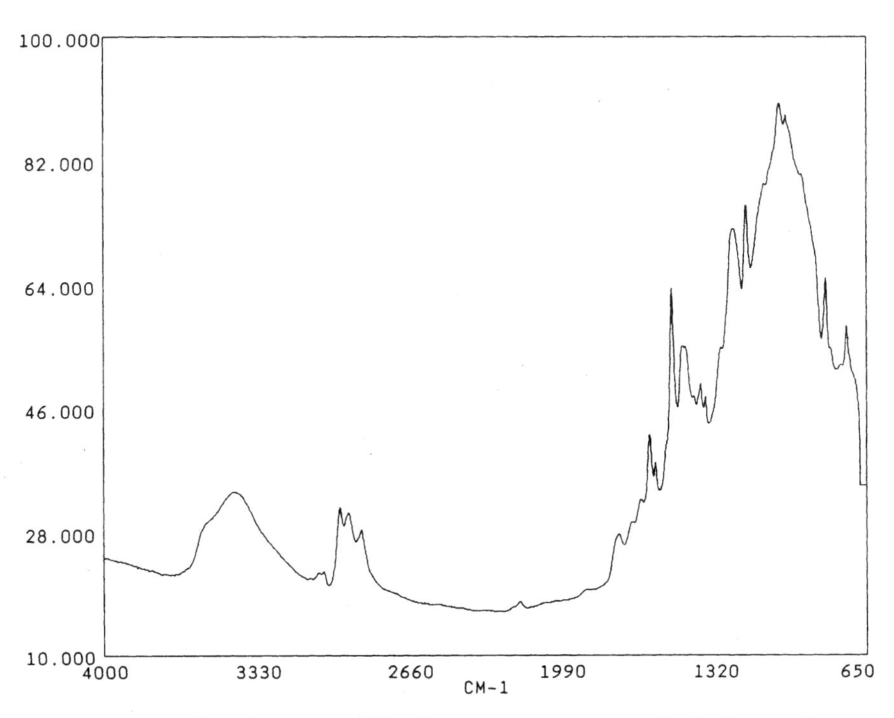

Figure 12-3. Infrared spectrum (absorbance versus wave number) of epoxy-glass.

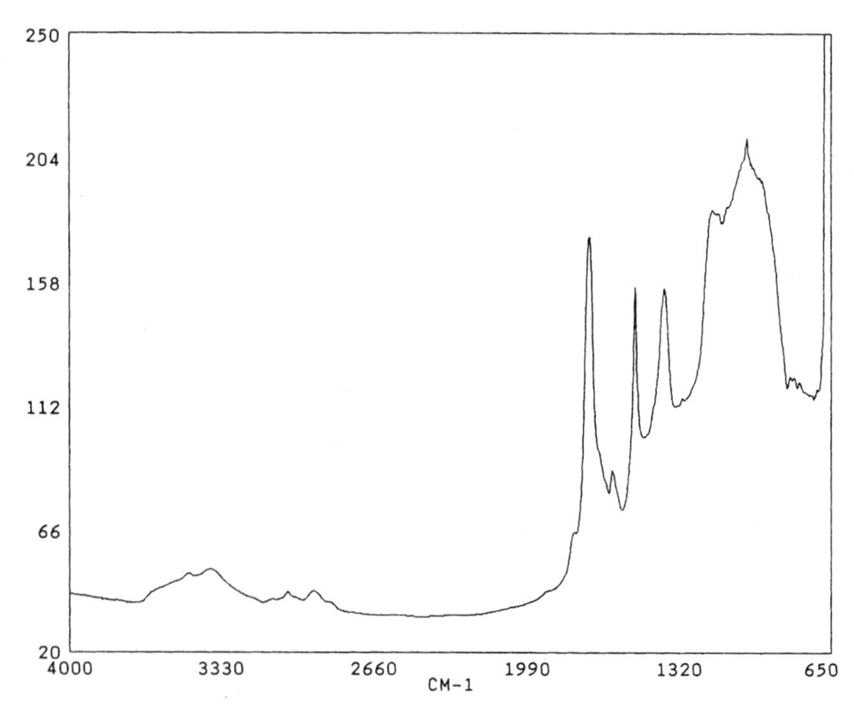

Figure 12-4. Infrared spectrum (absorbance versus wave number) of polyimide-glass.

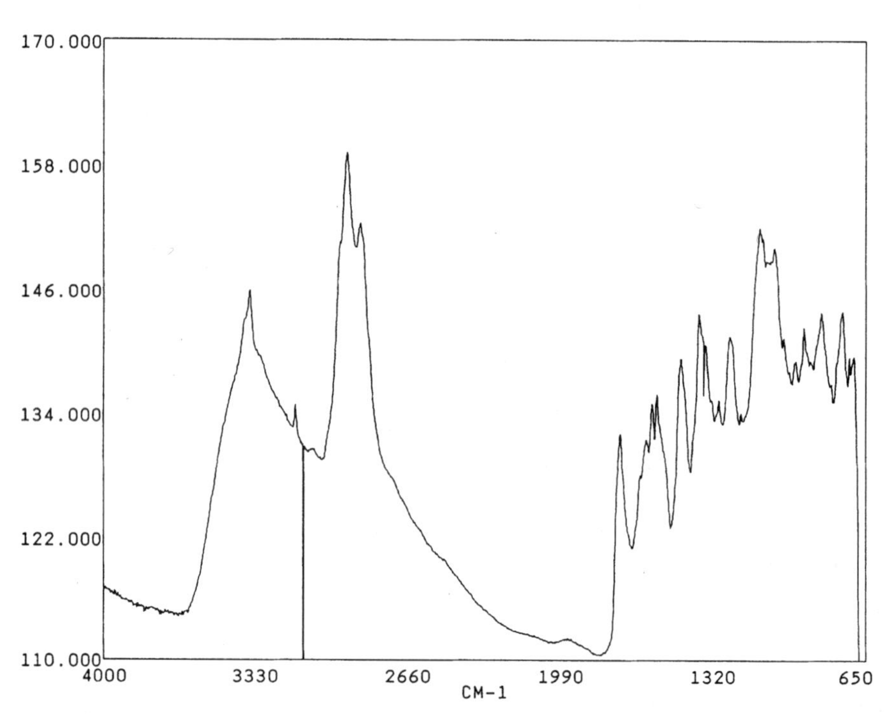

Figure 12-5. Infrared spectrum (absorbance versus wave number) of an RMA solder paste.

$$h = \frac{Q}{T_2 - T_1}$$

where Q is the heat transfer rate.

12.3 Soldering Principles

During soldering, a series of reactions and interactions occur in sequence or in parallel. These can be chemical or physical in nature in conjunction with heat transfer. The mechanism behind fluxing is often viewed as the reduction of metal oxides. Yet, in many situations, chemical erosion and dissolution of oxides and other foreign elements act as the primary fluxing mechanisms. For a process using solder paste as an example, the primary steps are represented by the following flowchart.

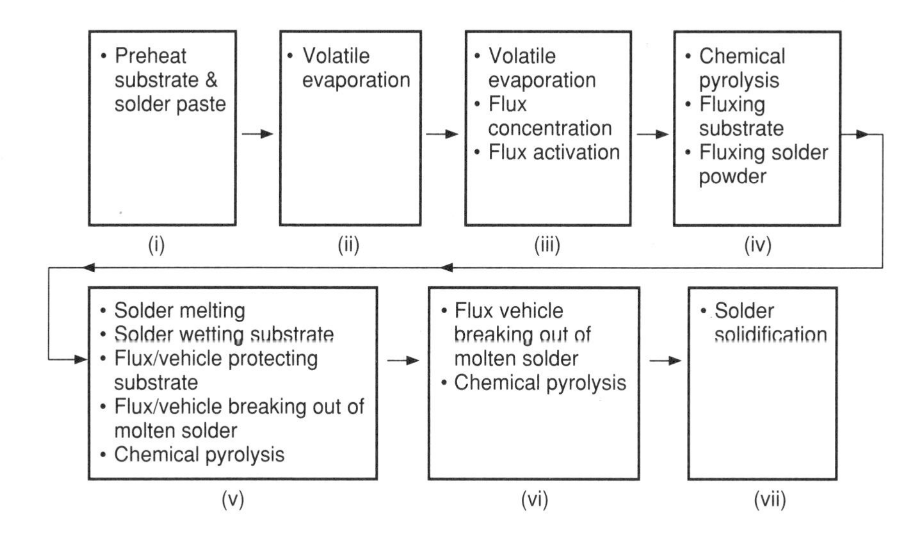

12.4 Mass Reflow Methodologies

Reflow used in this text refers to soldering on a mass production scale using primarily solder paste.

12.4.1 Conduction reflow

Because of the nature of heat transfer, this process is most suitable for assemblies with flat surfaces, composed of thermal conductive material as the substrate, and with single-side component/device populations. Therefore, hybrid assemblies having small, flat surfaces and relatively high thermally conductive alumina substrate meet all the criteria. Some limitations exist, yet this method provides a fast heating rate and operational simplicity. In addition, the visibility during the whole reflow process is a benefit and convenience in some cases. The ability to observe the reflow dynamics during the reflow process aids in the understanding of the response of solder paste to rising temperature. Problems such as component misalignment, tombstoning, and paste spattering can be visualized. This facilitates the fine tuning of the process and/or the solder paste. To solder less thermally conductive substrates such as FR-4 board or its equivalent, an infrared lamp on top of the conveyored heat platen supplements the heat transfer efficiency. FR-4 assemblies have been successfully produced by using the conduction/infrared lamp system. Equipment built with multiple platens which are independently controlled provides the capability of temperature profiling for various preheating conditions.

12.4.2 Infrared reflow

The performance of infrared furnaces has improved a lot in the 1980s, in order to accommodate the demanding requirements of surface mount technology and component manufacturing. The key design in furnaces is that of the heating source and heating control. Categorically, there are three types of heating design available: quartz-enveloped tungsten filament, nonfocused panel emitters, and honeycomb/downflow heating. The quartz-tungsten filament infrared source emits near- to middle-infrared radiation, with wavelengths ranging from 0.7 to 3.0 μm, as shown in Fig. 12-6.[2] The wavelength of emitted energy is controlled by the voltage applied to the filament. In this range of wavelength, organic materials are essentially transparent, and metals are absorptive or reflective depending on the surface condition. However, longer wavelengths in the range of 3 to 5 μm may also be present due to the remission from the back plane. For the nonfocused panel type, an emitter generates a wide range of infrared wavelength, 2.5 to 7.0 μm. Different designs and heating elements offer different ranges of infrared wavelength and different levels of radiation blended with convection heat. Equipment with multiple emitters which are independently controlled and with separate temperature settings can deliver more defined heating zones. Thus it is more efficient to separate top heating from the bottom as compared with other reflow techniques. The designed temperature differential between the top and bottom emitters is expected to facilitate the two-step reflow of double-sided boards and those with special temperature

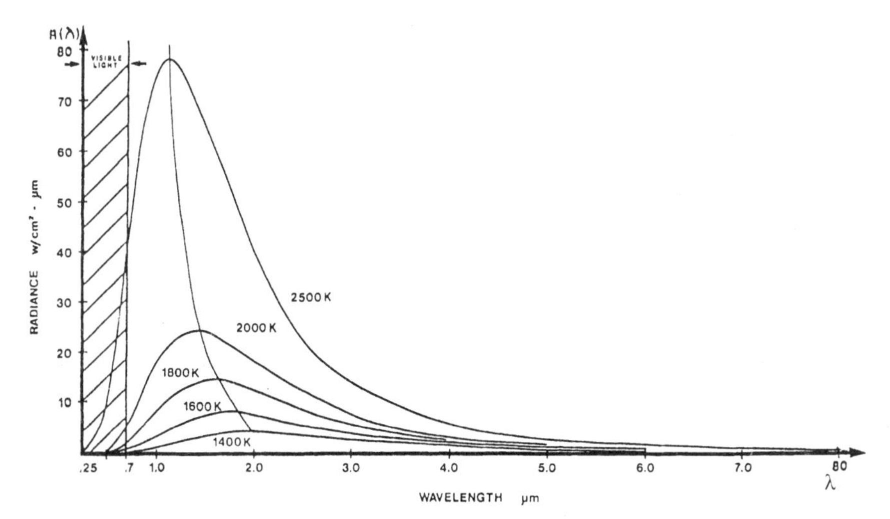

Figure 12-6. Infrared source of tungsten filament at various temperatures.

needs. By lowering the bottom heat supply, remelt of solder joints on the first side during the second reflow may be avoided. On the other hand, by lowering the top heat supply and increasing the bottom heat supply may facilitate the soldering of BGAs while minimizing the excessive heat exposure of other components.

12.4.3 Vapor phase reflow

Since vapor condensation soldering was invented by Western Electric in 1975, fluid development and equipment development have made vapor phase soldering commercially feasible. Chemicals such as perfluorocarbon, fluoropolyethers, perfluorotriamylamine, and perfluorophenanthrenes have been used as vapor phase fluids. The perfluorinated hydrocarbon has been the predominate fluid chemical.

It should be noted that perfluorinated fluids used in vapor phase reflow are completely fluorinated compounds without the presence of chlorine, bromine, or hydrogen. They do not contribute to ozone depletion and are not classified as volatile organic solvents.

The principle of the heat transfer in vapor phase reflow is straightforward. The release of latent heat ΔH, as a result of the phase transition from vapor to liquid, is the heat source. The latent heat stored in the vapor is transferred to the workpiece through the condensation process to make solder joints.

$$F\left(\begin{array}{cc} F & F \\ | & | \\ -C- & C- \\ | & | \\ F & F \end{array}\right)_n F \text{ (gas)} \rightarrow F\left(\begin{array}{cc} F & F \\ | & | \\ -C- & C- \\ | & | \\ F & F \end{array}\right)_n F \text{ (liquid)} + \Delta H$$

The whole workpiece is heated up to the temperature corresponding to the boiling point of the fluid used.

The selection of fluid is based on the solder alloy employed in the paste. A minimum of 25°C above the liquidus temperature of the alloy is required. For example, to reflow 63 Sn/37 Pb, the proper fluid has a boiling point of 215°C, and for 96.5 Sn/3.5 Ag, a fluid with a boiling point of 258 ± 7°C is needed, which is also the highest boiling point among the commercially available fluids.

The rate of heat transfer in vapor phase reflow is relatively high in comparison with that of infrared radiation or convection. The merits of the vapor phase process include uniform temperature distribution, precise temperature on workpiece, nonoxidizing atmosphere, and fast heat transfer. In addition, it is geometry-independent, providing uniform temperature to the workpiece. On the other hand, the fluid is costly, the equipment demands maintenance, and the operating temperature relies on the availability and quality of fluids.

Some practical points in using vapor phase reflow are as follows:

- The boiling point of the fluid used should be at least 25°C higher than the liquidus temperature of the solder alloy.
- The materials of components and boards should be compatible with the fluid.
- The predrying of the paste and preheating of the assembly will eliminate component skewing, tombstoning, and other problems that are commonly related to heating that is too fast and to moisture entrapment.
- The purity and cleanliness of the fluid in the sump should be maintained.

Recently introduced equipment includes preheating systems as shown in Fig. 12-7,[3] and multiple zones of preheating provide more versatile conditions for the process.

A case study demonstrated that the combined use of heating with a wide range of infrared wavelengths and of vapor phase reflow improved the reflow results of assembling solder bumped array packages on a reduced instruction set computing (RISC) board.[4] The temperature profile used is shown in Fig. 12-8. The process consisted of a

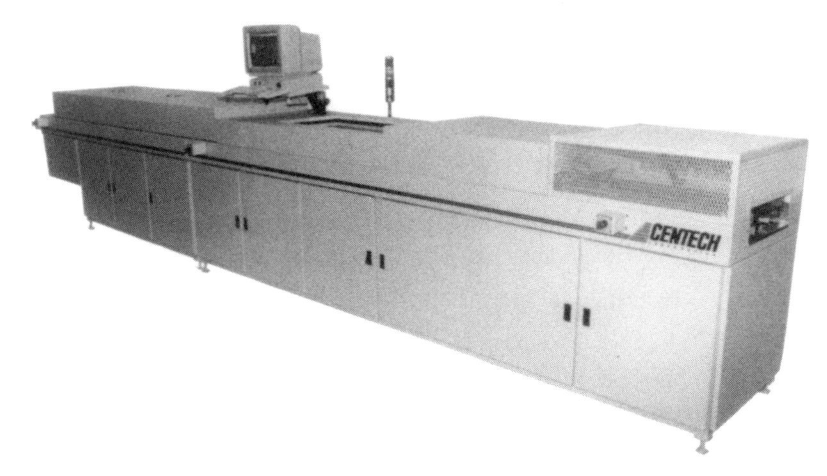

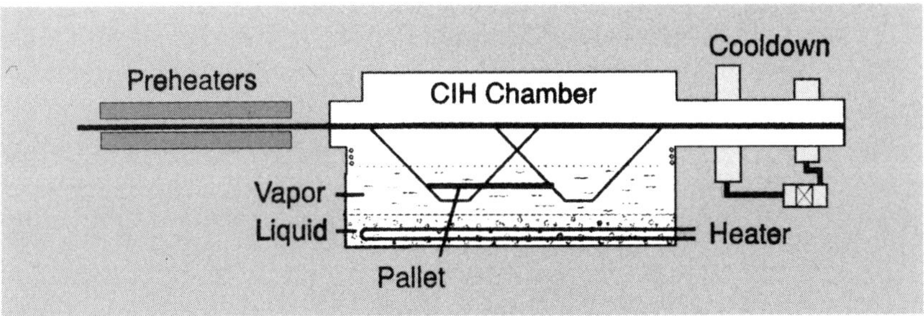

Figure 12-7. Vapor phase reflow equipment with preheating. (*Courtesy of Centech Corp.*)

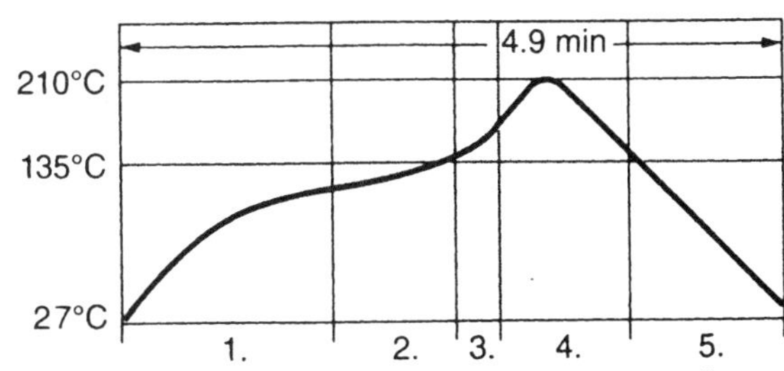

Figure 12-8. Temperature profile of vapor phase soldering.

preheating zone (using infrared heating) with the introduction of inert gas, a prevapor temperature stabilizing zone, and a vapor phase reflow zone. The process also was built with a control system to tune the process and a reflow design able to minimize the dragout of the reflow fluid. It was reported that the cost of the operation was $0.03 per board.

12.4.4 Convection reflow

Convection furnaces normally are capable of handling high-volume production. The heating process is relatively slower and needs more time to reach equilibrium conditions. Specific atmosphere can be easily introduced and controlled. A higher number of heating zones offer more flexibility in temperature profiling.

Forced-air (gas) convection enhances hot air circulation and thus facilitates heat transfer and improves the uniformity of heat distribution. The extent of air circulation is normally gauged by flow meters and is indicated as flow rate. A higher airflow rate, however, may demand higher flux activity due to the increased consumption of chemical contents. For the case of inert gas soldering, a higher flow rate represents higher cost if the gas is not fully recirculated.

12.4.5 Hot gas reflow

With proper adjustment in flow rate, gas temperature, and the distance between the nozzle and workpiece, soldering by means of heat transfer from hot gas to the workpiece has been found to be convenient in many cases. The hot gas can solder within a relatively short time and is able to handle high-temperature soldering with less vulnerability to residue charring.

12.4.6 Vertical reflow

Unless otherwise specified, traditionally conveyorized ovens are in horizontal configurations. Recently, a vertical oven has been introduced. The idea was prompted by the need to save floor space and/or to maintain the desired throughput, particularly for the processes that require a long operating time.

In a vertical oven, the fan blows air through heating elements, past a thermocouple, and then onto the assembly. The reflowed assembly is raised up to the top of the oven and conveyored to the separate vertical section, moving downward for unloading.

12.5 Process Parameters

Under mass reflow operation, both heating and cooling steps are important to the end results. It is generally understood that the heating and cooling rate of the reflow or soldering process essentially contributes the compositional fluctuation of the solder joint. This is particularly true when there are significant levels of metallurgical reactions occurring between the Sn-Pb solder and substrate metals. In the meantime, the cooling rate is expected to be responsible for the evolution of the microstructure.

The key process parameters that affect the production yield as well as the integrity of solder joints include

- Preheating temperature
- Preheating time
- Peak temperature
- Dwell time at peak temperature
- Cooling rate

It should be stressed that the reflow in a furnace (infrared, convection) is a dynamic heating process in that the conditions of the workpiece are constantly changing as it travels through the furnace in a relatively short reflow time. The momentary temperature that the workpiece experiences determines the reflow conditions and, therefore, the reflow result.

Figure 12-9 illustrates a simulated reflow profile (which is considered to be most desirable temperature versus time), comprising three stages of heating:

- Natural warm-up
- Predrying or preheating
- Spike and reflow

In the natural heating stage, the heating rate normally falls in the range of 1 to 2°C/s. The parameters of the predrying and preheating stage are very important to the reflow results. During soldering dynamics, the heating process contributes not only to the effectiveness of wetting but also to the extent of metallurgical reactions between solder and the substrates that the solder interfaces, particularly the peak temperature and the dwell time at the temperature above the liquidus of the solder.

Several events occur during this stage, as shown in steps (ii) and (iv) of the flowchart at the beginning of the chapter. These include temper-

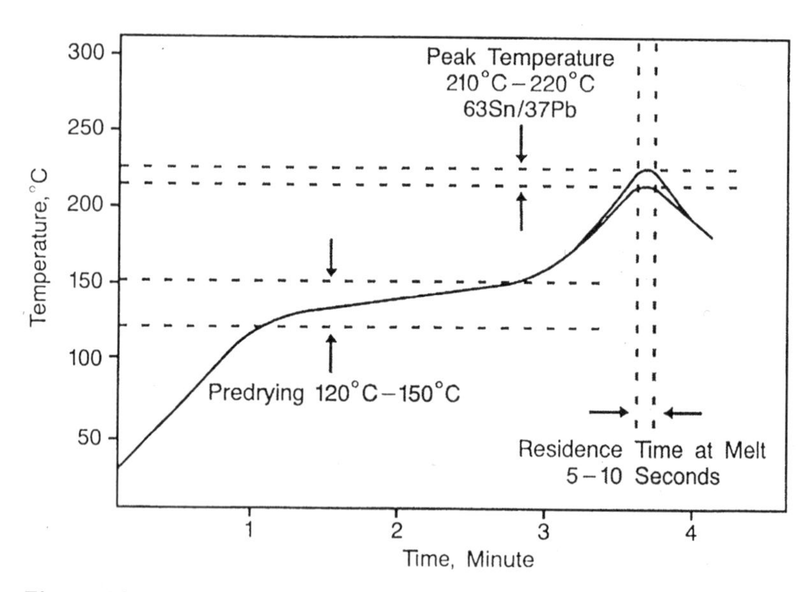

Figure 12-9. Simulated reflow temperature profile.

ature needs to be set to fit the specific flux activation temperature of the chemical system of the paste, and the time at heating to fit the constitutional makeup of the paste. Inadequate preheating often causes a spattering problem which manifests itself as discrete solder balls. Too high a temperature or too long a time at the elevated temperature results in insufficient fluxing and/or overdecomposition of organics, causing solder balling or hard-to-clean residue (if cleaning is chosen) or a change in residue appearance (if the no-clean route is adopted). The recommended general conditions for the second stage are 120 to 150°C (250 to 302°F) for a duration of 60 to 150 s. The third stage is to spike quickly to the peak reflow temperature at a rate of 1.2 to 2.0°C/s (2.18 to 3.6 °F/s). The purpose of temperature spiking is to minimize the exposure time of the organic system to high temperature, thus avoiding charring or overdrying. Another important characteristic is the dwell time at the peak temperature. The rule of thumb in setting the peak temperature is 25 to 50°C (77 to 122°F) above the liquidus or melting temperature; e.g., for the eutectic Sn-Pb composition, the range of peak temperatures is 208 to 233°C (354 to 451°F).

The wetting ability is directly related to the dwell time at the specific temperature in the proper temperature range and to the specific temperature being set. Other conditions being equal, the longer the dwell time, the more wetting is expected but only to a certain extent; the same trend applies at higher temperatures. However, as the peak temperature increases or the dwell time is prolonged, the extent of the formation of intermetallic compounds also increases as shown in Fig. 6-3. An excessive amount of intermetallics can be detrimental to long-term solder joint integrity. The peak temperature and dwell time should be set to reach a balance between good wetting and minimal intermetallic formation. The dwell time should be set as short as possible to avoid excessive formation of intermetallic compounds, yet long enough to achieve good wetting and to expel any nonsolder (organic) ingredients from the molten solder before it solidifies, thus minimizing void formation. For a given system, the cooling rate is directly associated with the resulting microstructure which in turn affects the mechanical behavior of solder joints (see Chaps. 6 and 16).

It was found that the microstructural variation and corresponding failure mechanisms of solder joints that were made under various reflow temperature profiles are extremely complex. Nonetheless, some correlation between the cooling rate and the basic properties can be obtained. The Cu-solder-Cu system is a good example, since it is still the most common material combination in electronics assemblies. In this system, 63 Sn/37 Pb solder joins Cu pads (coated or uncoated) on the PCB and the Sn-Pb–coated Cu leads of IC components.

For the tinned Cu–63 Sn/37 Pb–tinned Cu assembly, the reflowed solder joints are cooled in five different manners which deliver four cooling rates: 0.1, 1.0, 50, and 230°C/s, respectively, as measured above 100°C. The fifth cooling mode was conducted in a two-step cooling process, resulting in an uneven cooling with an average cooling rate of 12°C/s. Each of the five cooling modes produces a different development of microstructure of solder joint as discussed in Chap. 6.

12.6 Reflow Temperature Profile

The reflow temperature profile representing the relationship of temperature and time during the reflow process depends not only on the parameter settings but also on the capability and flexibility of equipment. Specifically, the instantaneous temperature conditions that a workpiece experiences is determined by

- Temperature settings to all zone controllers
- Ambient temperature
- Mass per board
- Total mass in the heating chamber (load)
- Efficiency of heat supply and heat transfer

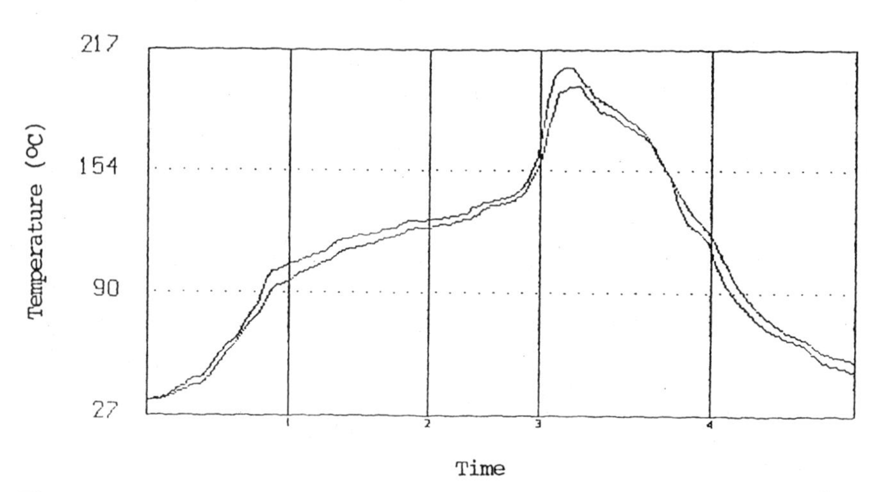

Figure 12-10. Reflow temperature profile of convection oven—lower preheat temperature.

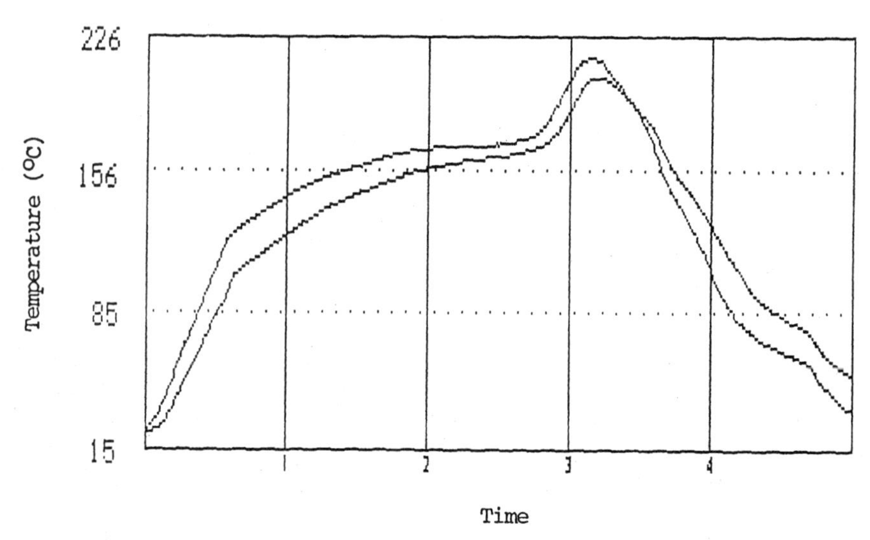

Figure 12-11. Reflow temperature profile of convection oven—higher preheat temperature.

For a furnace-type reflow process, two profiles are taken to illustrate the effect of temperature profile on the reflow results.

Figures 12-10 and 12-11 show the actual temperature profiles of a convection oven, with a relatively lower preheat temperature (Fig. 12-10) and a higher preheat temperature (Fig. 12-11). The importance of the compatibility of solder paste chemistry and the assembly system with the reflow temperature profile can be easily demonstrated. For instance, if the solder paste and the assembly require the temperature profile of Fig. 12-10, performing reflow under the temperature profile of Fig. 12-11 may give rise to the following phenomena:

- Deficiency of flux, resulting in solder balls
- Overheating of organics, resulting in cleaning difficulty for processes that are designed to include a cleaning step

On the other hand, if the paste is designed for the higher preheat temperature and/or the assembly requires additional heat, using the lower preheat temperature profile can produce the following phenomena:

- Uneven soldering, resulting in cold solder joints
- An excessive amount of residue remaining or nondry residue from no-clean paste

The temperature profile with boosted preheating conditions, as shown in Fig. 12-11, is most useful for the assembly that is densely populated with components that have a large disparity in their mass.

Depending on the type of conveyorized furnace, the mass of the assembly and the degree of loading, the major operating parameters to be monitored for effective reflow are the belt speed and the temperature settings of individual zones. The relationship between temperature settings and belt speed is reflected by the temperature profile, temperature versus time. At a given temperature profile, the peak temperature changes with belt speed; increasing belt speed decreases the resulting peak temperature, with other conditions being equal, as shown in Fig. 12-12. Since the required peak temperature is set at 25 to 50°C above the melting temperature of a solder alloy, the working range of peak temperature is always fixed. For every temperature profile, a relationship between peak temperature and belt speed can be established, and the usable range of belt speeds as depicted in Fig. 12-12 can be obtained.

12.7 Temperature Measurement and Uniformity

Two formidable uncertainties in the reflow process of circuit boards are the accuracy of temperature measurement and the level of temperature uniformity.

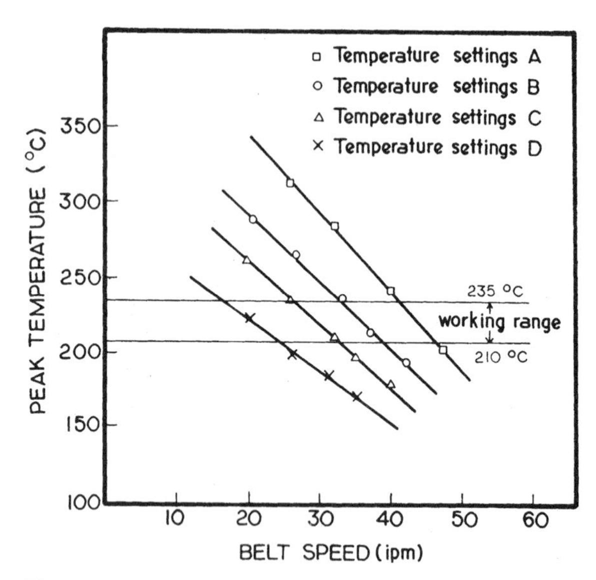

Figure 12-12. Peak temperature versus belt speed.

At a given location of the circuit board, the accuracy of temperatures measured during the reflow process is determined by

- Response efficiency of the thermocouple
- Consistency of contact between thermocouple and board

The response efficiency of a thermocouple is determined by the composition of the thermocouple wire and the diameter of the wire. Figure 12-13 indicates the relative temperature responses of two thermocouple compositions: type K (nickel-chromium/nickel-aluminum) and type E (nickel-chromium/copper-nickel). The relationship between the diameter of the thermocouple wire and the response time is shown in Fig. 12-14. The thinner wire provides a faster response time. At a steady state, the thermocouple is always able to read true temperature. However, in a dynamic environment, as in the reflow process, where the temperature of the board changes momentarily as the board travels through the heat chamber, the response efficiency of a thermocouple may affect the accuracy of the temperature measurement. Since the intimate physical contact between the thermocouple and the board contributes to the temperature measurement, the way in which the thermocouple is attached to the board should be consistent. Commonly, Kapton-type, high-temperature solder and high-temperature adhesive are the choices of bond material to attach the thermocou-

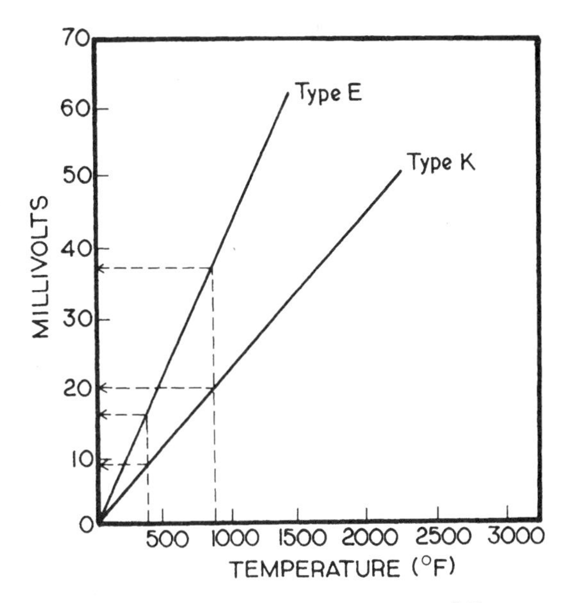

Figure 12-13. Temperature response of thermocouple types K and E.

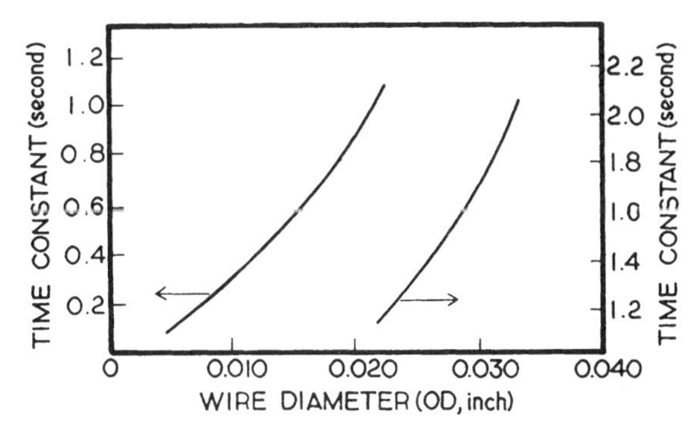

Figure 12-14. Thermocouple diameter versus response time.

ple at the specified location on the board. However, high-temperature solder, such as 10 Sn/88 Pb/2 Ag, is most favorable when performing the attachment. For best results, an excessive amount of solder material should be avoided.

It should be noted that the temperature gradient across the surface of the PCB in most operations could be significant. This would occur in peak temperature zones, or it may exist even after oven cooling

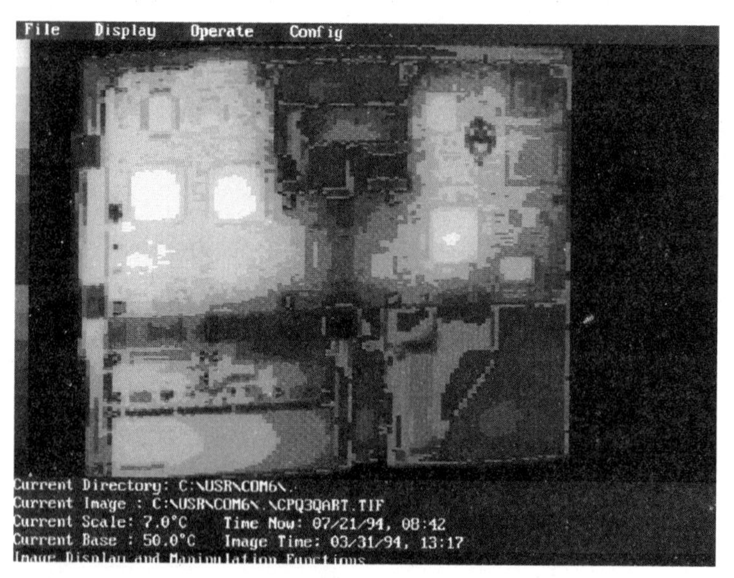

Figure 12-15. Temperature distribution over the board exiting from reflow oven—lighter area representing high temperature. (*Courtesy of Compix, Inc.*)

starts. The extent of the temperature gradient depends on the efficiency of the reflow oven, the operating parameters of the reflow oven, and the composition of the components and materials of the board. Figure 12-15 is shown as a black-and-white image of a circuit board coming out of the reflow oven, with the lighter area representing higher temperatures. The thermal image represents the temperature distribution which in turn is reflected by the color spectrum. Starting with 50°C as the baseline temperature, each color corresponds to a 7°C temperature increase from black, blue shades, purple shades, red and yellow groups, to white, with a total of 15 colors. The displayed temperature ranges from 50 to 165°C (122 to 329°F) with the microprocessor and other active devices showing the highest temperatures.[5]

Figure 12-16 presents the thermal image of four circuit boards with the lighter areas representing higher temperatures. The vertical line/cursor highlights a slice of board, and its corresponding thermal profile is shown at the right of the image. The range of the surface temperature shown is from 40 to 111°C (104 to 232°F).[5]

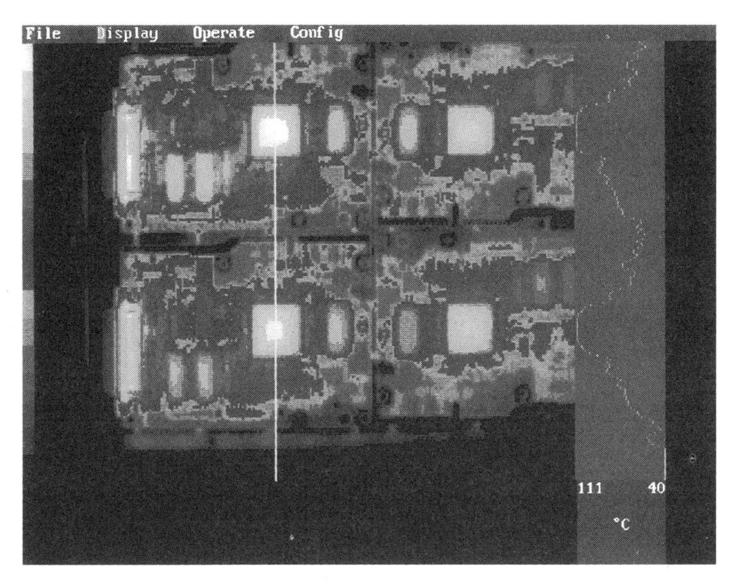

Figure 12-16. Thermal image of four circuit boards with the lighter area representing higher temperature. (*Courtesy of Compix, Inc.*)

12.8 Demanding Areas

With the anticipated emphasis on high-mix manufacturing and the availability of solder materials of varied chemical nature, such as water-soluble, no-clean, and solvent cleaning, functional flexibility of future soldering equipment is deemed to be the most important feature. Specific capabilities include providing a short or long preheating stage at different temperature levels, providing temperature profiles for fast or slow heating, and capability of maintaining precise atmospheric compositions with reduced cost. The capability of zone separation in the heating chamber of a forced-air convection oven is also in demand.

In addition, the compatibility of the heat transfer mechanism with the solder material is critical to soldering performance. One heat transfer method may benefit one solder material but not the other. Control and precision enabling the process to deliver consistent results are another two important features to be built in soldering methodologies. They are the prerequisite for a high-yield, quality production.

In addition to materials and process, successful surface mount soldering also depends upon the facility conditions, namely, temperature and humidity. Military standard MIL-STD-2000A currently states, "...the soldering facility shall be enclosed and be a slight positive pressure or be air conditioned. The temperature shall be maintained at 24°C ± 5 (79°F ± 8°) and the relative humidity shall not exceed 70% for more than 60 calendar days during a one-year period." Even by adhering to this specification, the temperature fluctuation between 21 to 29°C would be too big a variation for solder paste to endure without affecting its performance consistency. More controlled and consistent ambient facility conditions are in order.

12.9 References

1. Jennie S. Hwang, *Solder Paste in Electronics Packaging—Technology and Applications in Surface Mount, Hybrid Circuits, and Component Assembly,* Van Nostrand Reinhold, New York, 1989, pp. 186–187.
2. Ahmet N. Arslancan and David K. Flattery, "Infrared Reflow for SMT: Thermal and Yield Considerations," *Proceedings 1987 National Electronic Packaging and Production Convention-West,* p. 357.
3. Nile Plapp, "Condensation Inert Heating," *Electronic Packaging and Production,* December 1994, p. 17.
4. Jack Smith, "Vapor Phase System Reflows Solder Bumps Without Heat Damage," *Electronic Packaging and Production,* July 1993, p. 43.
5. James Wolcutt, Compix, Inc., Private Communication, 1995.

12.10 Suggested Readings

Ables, Billy, Wilton Herron, and Lance Scherer, "IR Controlled Induction Soldering of Microwave Cable Assemblies," *Connection Technology,* February 1990, p. 22.

Cox, Norman R., "Combining Radiation and Convection," *Circuits Assembly,* April 1991, p. 43.

Curran, Lawrence J., "Convection Dominates in Reflow Soldering," *Assembly,* February 1995, p. 24.

Dow, Stephen J., "The Use of Collimated Infrared Light," *Circuits Assembly,* November 1990, p. 28.

Dow, Stephen J., "Forced Convection for SMT Reflow Soldering," *Hybrid Circuit Technology,* October 1990, p. 47.

Dow, Stephen J., "Convection Full Force," *Circuits Manufacturing,* May 1990, p. 54.

Down, William H., "Achieving Consistency in IR Reflow," *Electronic Packaging and Production,* November 1991, p. 110.

Gross, P. M., B. F. Rothschild, and H. L. Garvin, "Evaluating Heater Bar Reflow Patterns with Solder Coated Test Panels," *Electronic Packaging and Production,* October 1989, p. 100.
Holloman, J., and D. Rall, "Advances in Reflow Soldering Process Control," *Proceedings, 1987 National Electronic Packaging and Production Convention West:* p. 393.
Huchins, Charles L., "Reflow Soldering," *Surface Mount Technology,* June 1994, p. 55.
Kazmierowicz, Philip C., "The Science Behind Conveyor Oven Thermal Profiling," KIC Oven Profiling Brochure, KIC, Wildomar, California, 1994.
Linman, Dale L., "High Temperature Reflow Soldering with Lead Free Solder," *Proceedings NEPCON-West,* 1994, p. 714.
Linman, Dale L., "Optimizing Vapor Phase Reflow Soldering," *Electronic Packaging and Production,* November 1990 (Suppl.), p. 43.
Lange, Roy, "Process Control in Surface-Mount Reflow," *Circuits Assembly,* October 1994, p. 64.
Marshall, Peter, "Establishing a Reflow Profile," *Circuits Assembly,* August 1994, p. 58.
Plapp, Nile, "A Vapor Phase Solution," *Surface Mount Technology,* August 1993, p. 48.
Prasad, Ray, and Raiyomand Aspandiar, "UPS and IR Soldering for SMAs," *Printed Circuit Assembly,* February 1988, p. 29.
Ruffing, John F., "Vapor Phase Reflow," *Circuits Assembly,* November 1993, p. 46.
Swanson, Donald, "Process Developments in Reflow Soldering," *Electronic Packaging and Production,* March 1995, p. 30.
Tuck, John, "Vertical Reflow on the Lauch Pad," *Circuits Assembly,* April 1995, p. 22.
Wurscher, Robert, "Conduction Soldering of SMT Assemblies," *Circuits Manufacturing,* November 1991, p. 123.
Zarrow, Phil, "IR Reflow Soldering Systems and Steps," *Circuits Manufacturing,* February 1990, p. 26.

13 Special Topics in Soldering-related Issues

13.1 Intermetallics Versus Solder Joint Formation

Intermetallics can be beneficial or detrimental to the formation of solder joints and their long-term reliability. Relating to electronics packaging and assembly, intermetallic compounds may come from one or more of the following processes and sources:

- Intermetallics are formed at the solder-substrate interface during soldering.
- Intermetallics are present in the interior of the solder joint as the inherent metallurgical phases of a given solder composition, such as 95 Sn/5 Sb and 96 Sn/4 Ag.
- Intermetallics are developed during a service life either along the interface and/or in the interior of the solder joint.

Metallurgically, an intermetallic compound is one type of intermediate phase that is a solid solution with intermediate ranges of composition. Intermetallic compounds may form when two metal elements have a limited mutual solubility. These compounds possess new compositions of a certain stoichiometric ratio of the two component elements for a binary system.

The new phases have different crystal structures from those of their elemental components. The properties of the resulting intermetallic compounds generally differ from those of the component metal, exhibiting less metallic characteristics, such as reduced ductility, reduced density, and reduced conductivity. Since the tin in a tin-lead solder reacts with most metals that are commonly used in electronics packaging and assembly, the presence and formation of intermetallic compounds are expected. In addition, the indium in indium-based solders is also readily reactive with copper, gold, and silver. The existence of intermetallic compounds has been observed at or near the interface of the solder and substrate as well as in the bulk of the solder joint.

By using phase diagrams, the potential formation of the intermetallics and their extent between known compositions of solder and substrates can be anticipated. Table 13-1 lists the compounds that have been identified and established under the equilibrium condition for common substrate metals in contact with tin-based and indium-based solders. It should be noted that the thermodynamically stable com-

Table 13-1. Thermodynamically Stable Intermetallic Compounds

Solder	Substrate	Intermetallic compounds
Sn-based	Au	AuSn, $AuSn_2$, $AuSn_4$
Sn-based	Ag	Ag_3Sn
Sn-based	Cu	Cu_4Sn, Cu_3Sn_8, Cu_3Sn, Cu_6Sn_6
Sn-based	Pd	Pd_3Sn, Pd_2Sn, Pd_3Sn_2, PdSn, $PdSn_2$, $PdSn_4$
Sn-based	Ni	Ni_3Sn, Ni_3Sn_2, Ni_3Sn_4
Sn-based	Pt	Pt_3Sn, PtSn, Pt_2Sn, $PtSn_2$, $PtSn_4$
Sn-based	Fe	FeSn, $FeSn_2$
In-based	Au	AuIn, $AuIn_2$
In-based	Ag	Ag_3In, $AgIn_2$
In-based	Cu	Cu_3In, Cu_9In_4
In-based	Pd	Pd_3In, Pd_2In, PdIn, Pd_2In_3
In-based	Ni	Ni_3In, NiIn, Ni_2In_3, Ni_3In_7
In-based	Pt	Pt_2In_3, $PtIn_2$, Pt_3In_7

pounds may not always be present and some intermetallic compounds that do not appear in the equilibrium phase diagram have been identified in soldered systems.

When solder is in contact with common substrates as listed in Table 7-1 for a long enough time at a high enough temperature, the potential for formation of intermetallic compounds may occur in a detectable amount within a practical time period.

At temperatures below the solder liquidus temperature, the formation is primarily a solid state diffusion process and is thus highly dependent on temperature and time. When the solder is in a molten state, the solubility of the element from substrate into molten solder is expected to accelerate the rate of intermetallic formation.

In addition to material compatibility, external factors also affect the rate of intermetallic compound formation, namely, temperature of exposure and time at elevated temperature. Thus, solder reflow conditions, such as peak temperature, time at peak temperature, and total dwell time at elevated temperatures, are expected to influence the rate and extent of intermetallic growth. In storage and service, the temperature exposure of the assembly is obviously another external factor related to intermetallic growth in the systems.

The growth of intermetallics between 60 Sn/40 Pb solder and copper is represented by Fig. 13-1 over a range of temperatures from 20 to

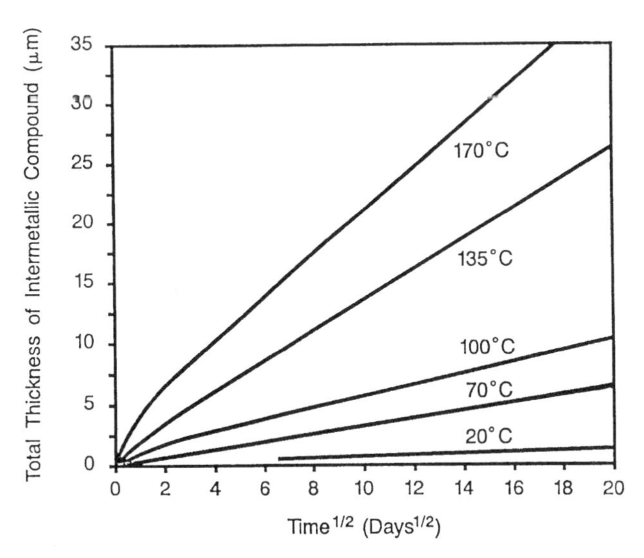

Figure 13-1. Intermetallic compound growth between 60 Sn/40 Pb and Cu.

170°C.[1] As expected, the thickness of growth is proportional to the square root of time, coinciding with the diffusion-controlled kinetics:

$$h \propto \sqrt{t}$$

where h is the thickness of the intermetallic layer and t is time.

As the temperature rises, the rate of formation increases, following the Arrhenius equation.

Within the series of tin-lead compositions, the growth rate of the intermetallic compound as measured by its thickness at a given temperature increases with increasing tin content due to the activity of tin with copper. Figure 13-2 illustrates the effect of tin content on the growth rate of intermetallics between 60 Sn/40 Pb and copper.[2]

A study was conducted on the copper dissolution rate in molten solder by dipping cleaned, oxygen-free copper wires with a nominal diameter of 0.5 mm (0.020 in) into a solder bath. The wire diameter versus time in the solder bath was measured with the aid of an optical microscope.[3] The dissolution rate of copper in 63 Sn/37 Pb is summarized in Fig. 13-3 over a temperature range from 275 to 425°C. Figure 13-4 gives the dissolution rate in 95 Sn/5 Sb over a temperature range of 260 to 429°C. The temperature dependency of the dissolution rate is evident.

It is found that the compositions of intermetallics at the interface dif-

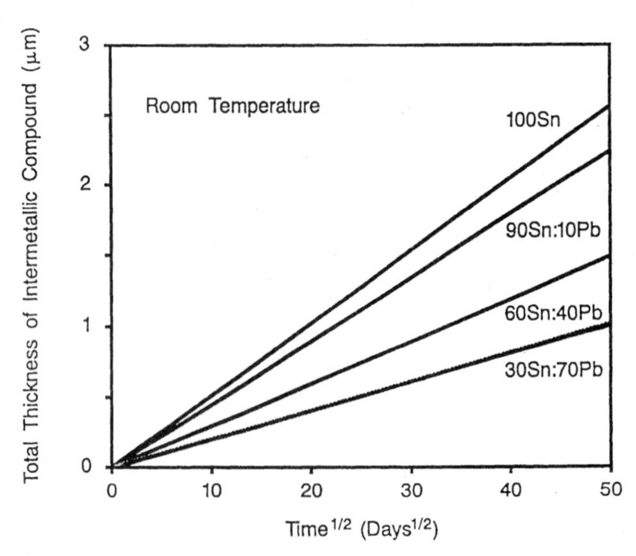

Figure 13-2. Tin content versus intermetallic compound growth between solder and Cu.

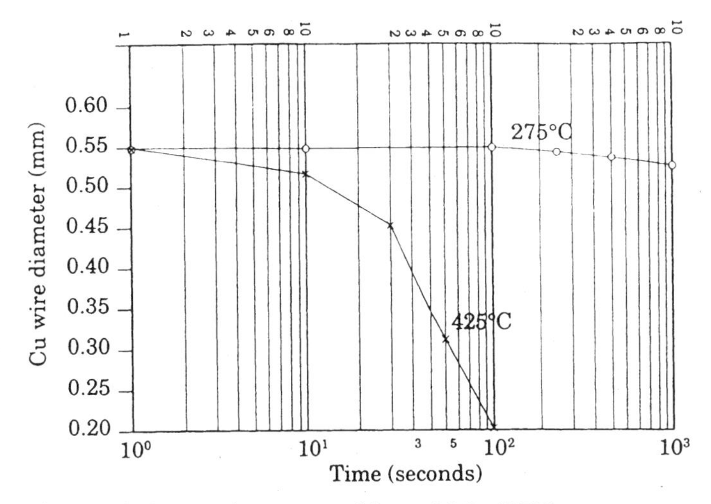

Figure 13-3. Dissolution rate of Cu in 63 Sn/37 Pb.

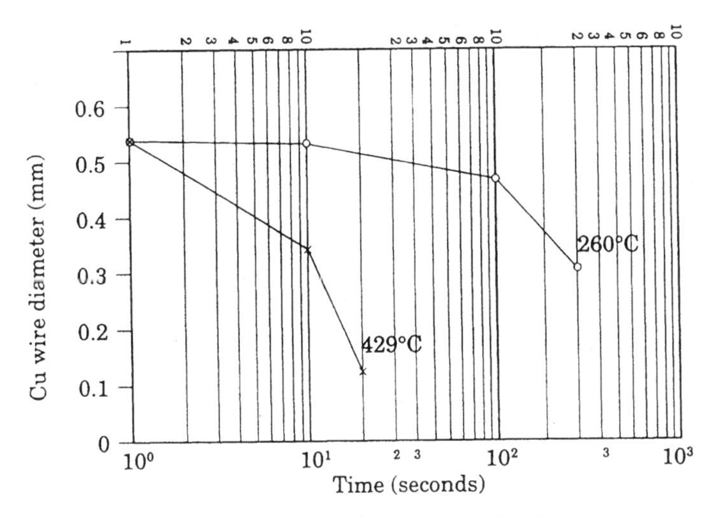

Figure 13-4. Dissolution rate of Cu in 95 Sn/5 Sb.

fer from those inside the solder joint.[4] Table 13-2 summarizes the compositions of the intermetallic layer at the interface and inside the solder joint for solder compositions of 5 Sn/95 Pb, 5 Sn/92.5 Pb/2.5 Ag, 92 Sn/8 Sb and 65 Sn/25 Ag/10 Sb on bare copper and nickel-plated substrates, respectively. It was also found that the oxidized surface showed a delayed development of the intermetallic phases, making a

Table 13-2. Formation of Intermetallic Compounds on Cu and Ni Substrates[4]

Solder composition	At interface		In the bulk of solder	
	Cu	Ni-plating	Cu	Ni-plating
5 Sn/95 Pb	Cu_3Sn	Ni_3Sn_4	0	0
5 Sn/92.5 Pb/2.5 Ag	Cu_6Sn_6	Ni_3Sn_2	Ag_3Sn	Ag_3Sn
92 Sn/8 Sb	Cu_3Sn, Cu_6Sn_6	Ni_3Sn_4	Cu_6Sn_5	$(NiCu)_3Sn$
65 Sn/25 Ag/10 Sb	Cu_3Sn, Cu_6Sn_5	Ni_3Sn_4	Ag_3Sn, CuSn	Ag_3Sn, $(NiCu)_3Sn$

thinner layer compared with the clean surfaces under otherwise identical conditions. Furthermore, it was found that the composition formed may vary with the surface condition and the availability of the tin content; for example, 5 Sn/95 Pb on a clean nickel surface, produced Ni_3Sn_4 intermetallics, while on an oxidized surface it resulted in the formation of Ni_3Sn_2. This is primarily attributed to the lower availability and less accessibility of tin to the oxidized surface. Comparing a high-lead solder (5 Sn/92.5 Pb/2.5 Ag) with a high-tin solder (65 Sn/25 Pb/10 Sb) on a nickel surface, different morphologies of intermetallics are detected; high-lead solder is associated with small, round intermetallic crystallites (Ag_3Sn) and high-tin solder with large, needlelike crystallites.

The effect of the reflow condition was studied by prolonging the dwell time at the peak temperature. Figure 6-14 is an SEM micrograph showing the microstructural development and intermetallic formation of 63 Sn/37 Pb solder on a Cu substrate under an extreme reflow condition that imposes a dwell time at peak temperature 3 times longer than that of a normal process, as shown in Fig. 6-13. The prolonged heating at peak temperature resulted in the development of a much coarser microstructure and a needle-shaped phase that is identified as Cu-Sn intermetallics by x-ray energy dispersion analysis as shown in Fig. 6-15. The examination of the cross section of the solder joint confirmed that intermetallics have moved away from the interface area and are distributed throughout the bulk of the solder joint.

In summary, the extent of intermetallic formation, the composition of intermetallic compounds and the morphology of intermetallics depend on internal as well as external factors. These include

- Metallurgical reactivity of solder composition with substrate
- Soldering (reflow) peak temperature
- Dwell time at peak temperature

- Surface condition of substrate—clean versus oxidized
- Solder composition
- Postsoldering storage conditions
- Service condition

As to the role of intermetallics at the interface or in the interior, beneficial or detrimental, it depends on some specific factors. Solder wetting on the substrate followed by the formation of a thin layer of intermetallic is a prevalent mechanism for solder to make a permanent bond with the substrate. Figure 13-5 shows the bonding layer between the 63 Sn/37 Pb solder and the substrate. However, as the intermetallic layer becomes thicker, adverse effects may result. The generally accepted thickness falls in the range of 1 to 5 μm.

The intermetallics in the bulk of the solder joint (away from the interface) may also play a positive or negative role. A proper morphology, size, and distribution of intermetallics in solder may act as a strengthening phase; in contrast, large needle-shaped compounds generally weaken the mechanical properties of the solder joint.

The formation of excessive intermetallic compounds has been demonstrated as one of the most likely sources of solder joint failure. When an appreciable amount of intermetallics is developed along the solder-substrate interface, cracks are often initiated around the interfacial area under stressful conditions. Figure 13-6 shows the development of Cu-Sn intermetallics after the solder joint of 96.5 Sn/3.5 Ag on Cu has been subjected to service life in an automobile under-the-hood application.[5] As can be seen, an extensive amount of needle-shaped intermetallics have formed. For the same assembly, solder joint cracking was detected near the solder joint interface, as shown in Fig. 13-7.

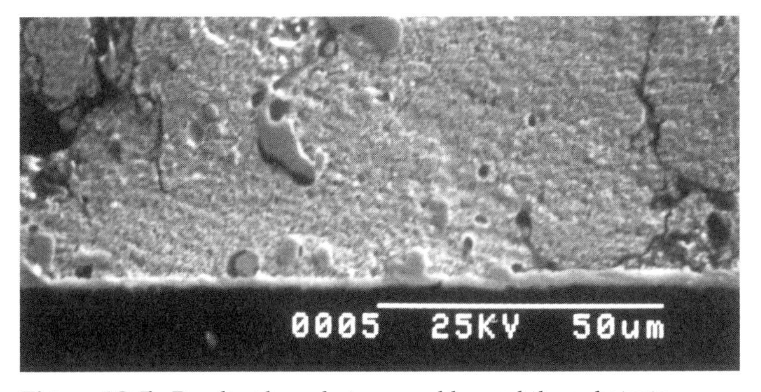

Figure 13-5. Bonding layer between solder and the substrate.

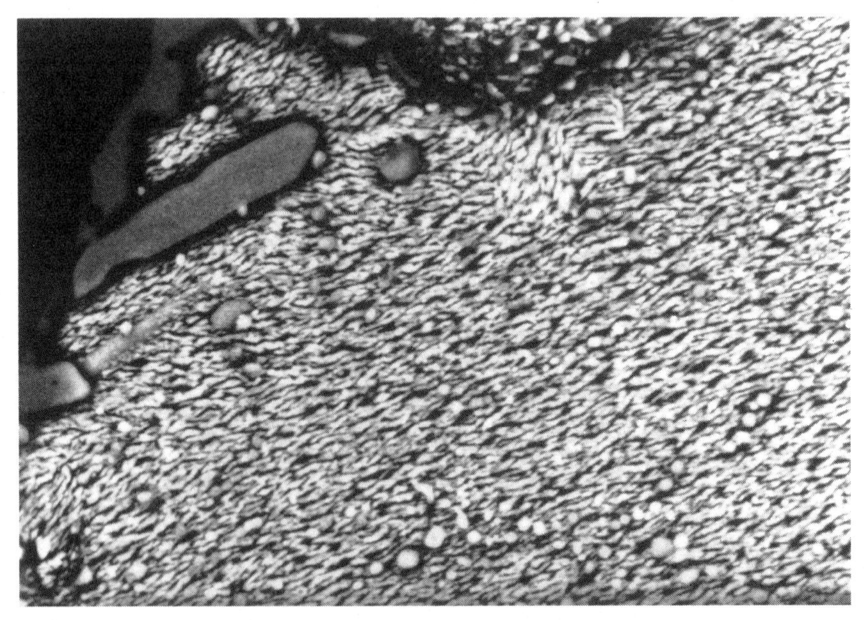

Figure 13-6. Intermetallic growth at 96.5 Sn/3.5 Ag and Cu joint after service life.

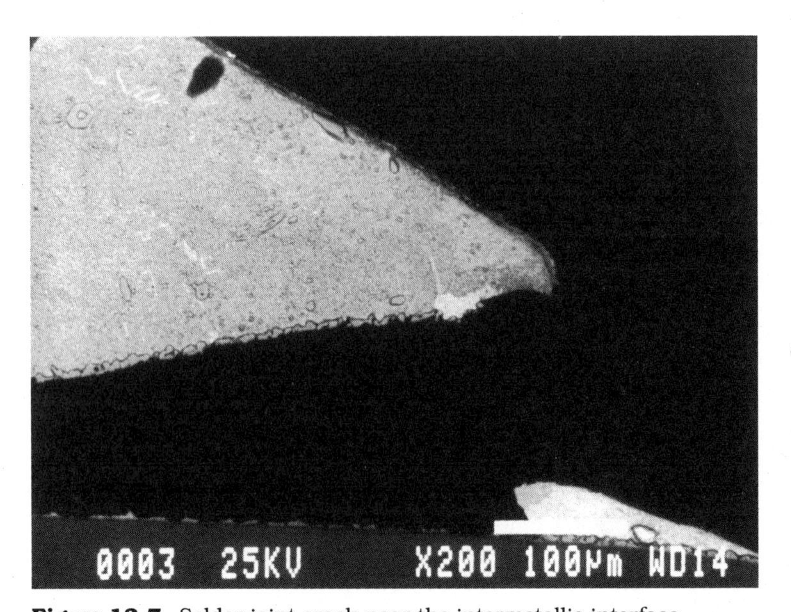

Figure 13-7. Solder joint crack near the intermetallic interface.

The adverse effect of intermetallic compounds on solder joint integrity is believed to be attributed to the brittle nature of such compounds and to the thermal expansion property, which normally differ from the bulk solder. The differential thermal expansion could contribute to internal stress development. In addition, excessive intermetallic compound which forms at the coating and substrate interface may impair the solderability of surfaces where the intermetallic compound depletes one element of the contact surface. One example is Sn depletion from the Sn-Pb coating on Cu leads, which causes the exposure of Sn-Cu intermetallic compound to oxidation, resulting in poor solderability of the component leads. In this case, the interface area is expected to have Cu_3Sn phase next to the copper substrate, followed by a Cu_6Sn_5 phase and a Pb-rich phase.

Excessive intermetallics may also alter the appearance of the solder joint, giving it a dull and rough surface. Figure 13-8 is the SEM micrograph of the free surface of a 63 Sn/37 Pb solder joint containing an excessive amount of intermetallics; the solder joint visually appears dull. In comparison, the shiny solder joint surface correlates with the SEM microstructure as shown in Fig. 13-9.

A newly developed process illustrates another positive role of intermetallics. A bonding process involves in situ formation of Cu-Sn intermetallics from a vapor-deposited Cu-Sn multilayer.[6] The unique features of the process are its controlled intermetallic bonding thickness and its fluxless nature. The process successfully bonds the die and the metallized substrate. In the process, the die is composed of a multilay-

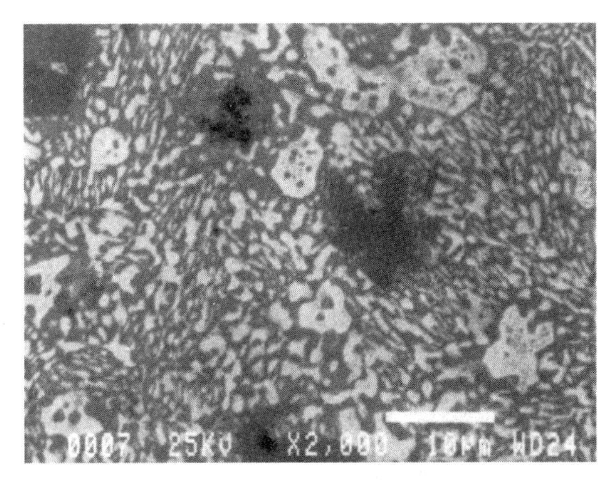

Figure 13-8. SEM micrograph of free surface of 63 Sn/37 Pb solder joint containing excessive intermetallics.

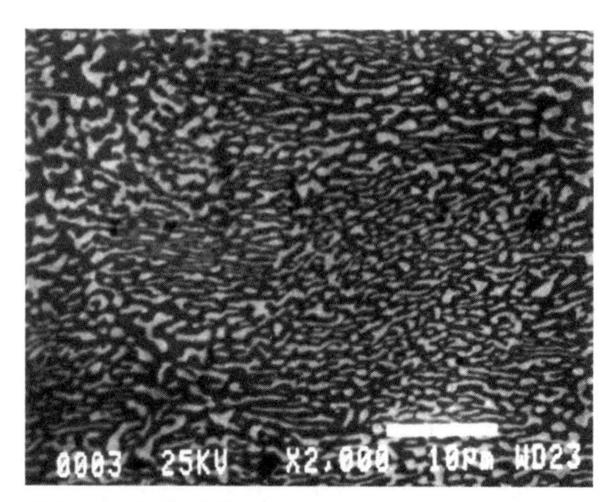

Figure 13-9. SEM micrograph of free surface of 63 Sn/37 Pb solder joint showing shiny surface.

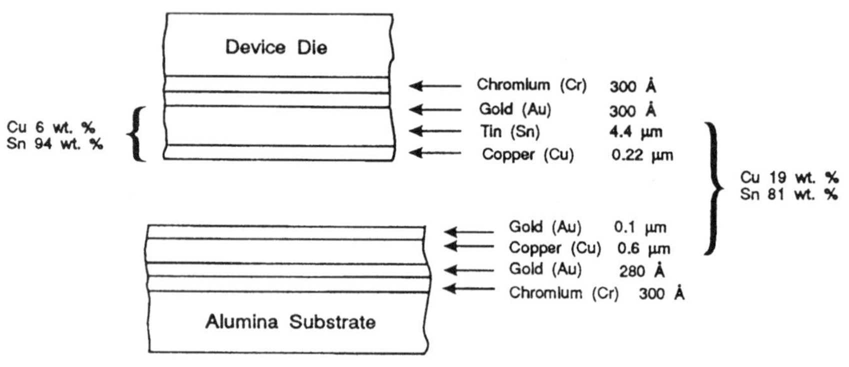

Figure 13-10. In situ bonding of die substrate by controlled intermetallic.

er Cr-Au-Sn-Cu structure, as shown in Fig. 13-10. Chromium and gold are first deposited directly on the polished side of a GaAs wafer in one vacuum cycle, each with a thickness of 300 Å. The purpose of the Cr layer is to enhance the adhesion, and the Au is to prevent Cr oxidation. Tin and copper are then deposited in the second vacuum cycle to a thickness of 4.4 and 0.22 μm, respectively. Upon deposition the intermetallic compound, Cu_6Sn_5, is immediately formed on the outer layer under a vacuum. The metallized substrate is coated with 300 Å Cr and 280 Å Au in one vacuum cycle. This is followed by the deposition of a 0.6-μm-thick layer of Cu and a 0.1-μm-thick layer of Au. The die is then placed on the substrate, and the bonding is formed by applying

40-lb/in^2 pressure at 250°C for 5 min in an N_2- and O_2-purged furnace. At 250°C, the diffusion bonding is facilitated by the presence of liquid Sn through the interdiffusion between liquid Sn and Cu with a faster rate than between solid Sn and Cu. The effect of an Au film on the solder joint is negligible because of its small amount, as concluded in studies on flip-chip soldering.[7,8] The resulting joint is composed of Sn and Cu_6Sn_5 which are uniformly distributed throughout the joint. The joint thickness is 4.5 μm.

Overall, the role of intermetallics is determined by the design of an assembly, the service conditions in relation to that design, and the control of the soldering (bonding) process. Understanding the relationship between them is the key to making reliable solder joints.

13.2 Gold-plated Substrate

The use of gold as a surface coating to resist oxidation of underlying metal has been a prevalent practice in semiconductor packages and electronics assemblies. Gold is also used as a diffusion-resistant barrier. Common applications include gold plating of component leads, solid gold (24 karat) wire bonding, and hard gold (cobalt- or nickel-gold) for edge fingers as connectors. Recently, printed wiring boards (PWBs) with copper-nickel and gold coating in lieu of hot-air solder leveling (HASL) on PWBs has drawn some attention in the industry. With all good use of gold, on the other hand, gold embrittlement in solder is a common concern in terms of the integrity of the solder joint and of the assembly. When a gold-coated substrate is in direct contact with tin-lead solders, the gold combines with the tin of the solder at a rapid rate due to the metallurgical affinity between tin and gold, forming gold-tin intermetallics. The presence of gold-tin intermetallics in solder affects the physical and mechanical properties of solder joints. It may also alter the appearance and microstructure of solder joints.

The following summarizes the effects of gold on tin-lead solders.

- *Ultimate tensile strength and shear strength.* It is found that the addition of gold to tin-lead solder may slightly increase the tensile strength initially and that the strength is sustained at above 8000 lb/in^2 until the gold content exceeds 10 wt%. Figure 13-11 summarizes the relationship of tensile strength versus gold content in solder.[9] The shear strength drops slowly with increasing content of gold, as shown in Fig. 13-12.
- *Hardness and ductility.* The hardness generally increases with the incorporation of gold, as shown in Fig. 13-13. As the gold content

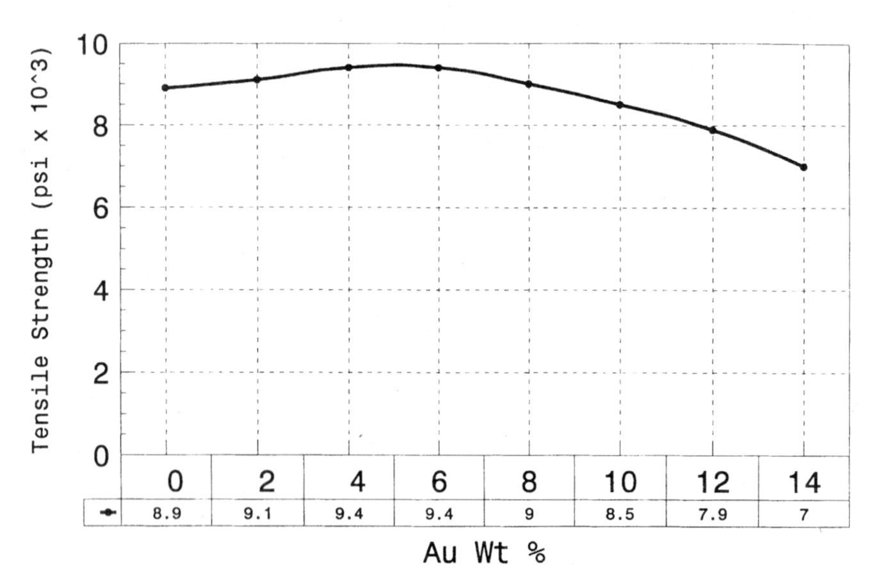

Figure 13-11. Gold content versus tensile strength of 63 Sn/37 Pb.

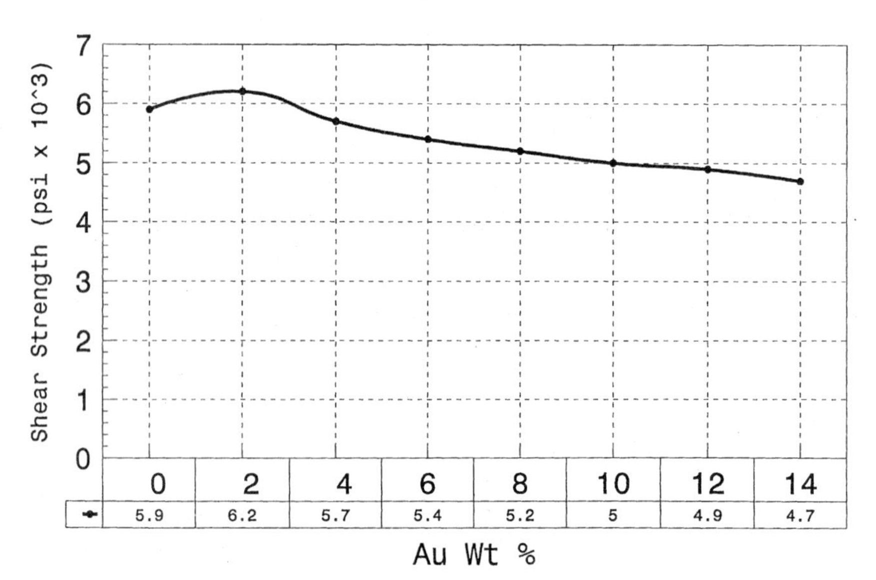

Figure 13-12. Gold content versus shear strength of 63 Sn/37 Pb.

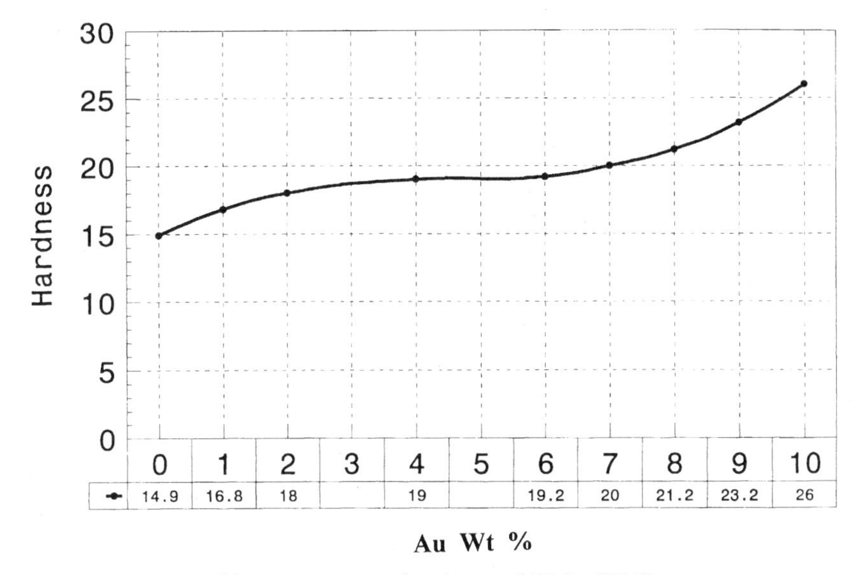

Figure 13-13. Gold content versus hardness of 63 Sn/37 Pb.

exceeds 7%, the effect is further enhanced. Figure 13-14 illustrates the slow reduction of elongation with the addition of gold up to 4 wt% and then the dramatic drop as the gold content exceeds 4 wt%.[10]

- *Microstructure.* The needle-shaped gold-tin intermetallic phase in eutectic solder can be readily detected in the microstructure as the gold content reaches 1 wt% weight. The formation of the hard needles increases with increasing gold content. The composition of the intermetallics is expected to be gold-tin and tin at room temperature.
- *Wettability and spread.* A study on the wettability of gold-plated copper by 63 Sn/37 Pb indicates that a pure gold coating provides better wetting and spread than alloy gold. Their relative spread factor is shown in Table 13-3.

It should be noted that in reflow processes where the heating time is long the quick dissolution of gold in the molten solder may cause the solder to wet directly on the base metal, not on the gold coating. The spreadability of 63 Sn/37 Pb is found to be unaffected by up to 2 wt% of gold content in the solder. Beyond 2 wt%, a reduction in fluidity and spread is expected. Although the inertness of gold is expected to provide full protection to the base metal underneath the gold coating, testing results on the aging of gold-electrodeposited copper indicate

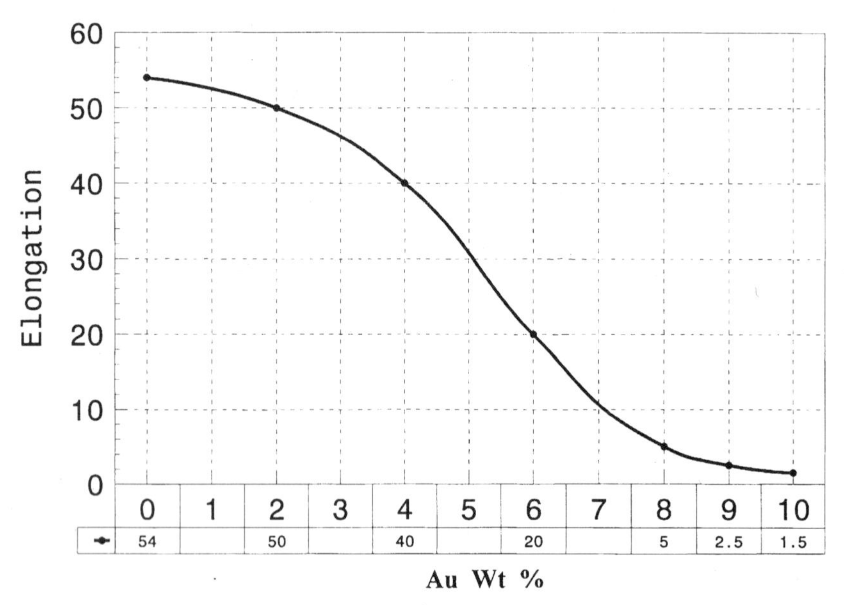

Figure 13-14. Gold content versus elongation of 63 Sn/37 Pb.

Table 13-3. Gold Plating Type versus Spread Factor[11]

Plating type	Spread factor
Pure Au	92–94
Alloy Au	81–90
Bare Cu	81

that the solderability as measured by wetting time degraded with aging at 170°F. Figure 13-15 shows the decayed solderability of copper with 50 μin gold. The observed degradation in solderability is attributed to two potential causes:

Diffusion of atmospheric contaminants through porous gold film, oxidizing the base material.

Diffusion of base metal through the coating reaching the surface. The diffusion rate may be associated with the grain size of the gold coating; it is suggested that the diffusion is favored by a smaller grain size.[11]

- *Solder composition versus dissolution rate.* The dissolution rate of gold in solder depends not only on temperature and time; the solder composition also plays a role. Figure 13-16 depicts the gold dissolu-

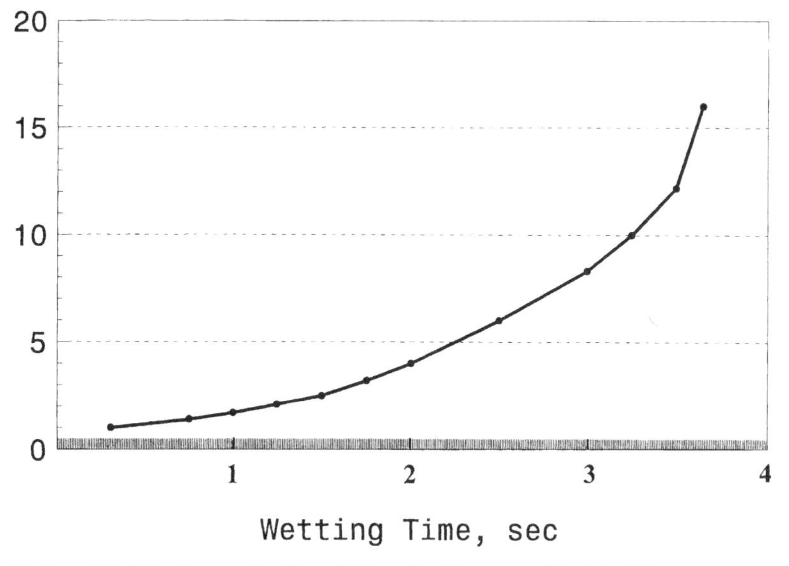

Figure 13-15. Decayed solderability of solder on gold-plated Cu.

tion rates in 60 Sn/40 Pb, 60 Sn/38 Pb/2 Zn, 55 Sn/35 Pb/10 Au, 37.5 Sn/37.5 Pb/25 In, and 53 Sn/29 Pb/17.5 In/0.5 Zn.[12] It is apparent that the presence of gold, indium, and zinc in tin-lead solder retards gold dissolution.

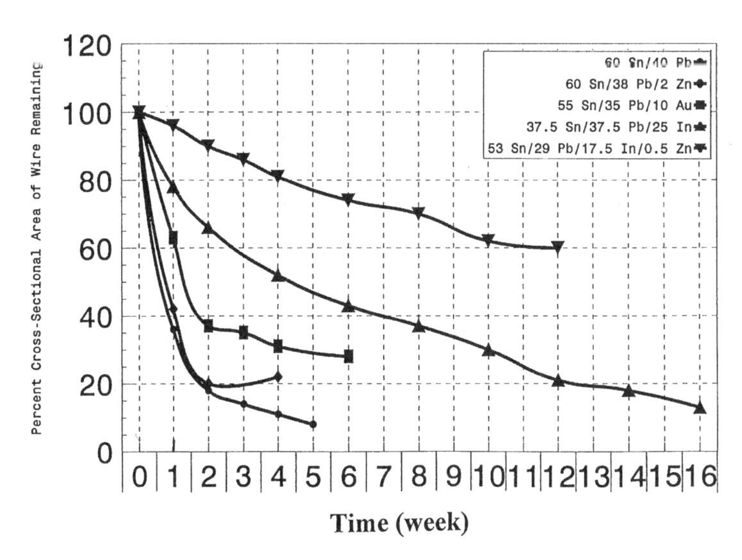

Figure 13-16. Gold dissolution rate in various solder compositions.

- *Electrical conductivity.* The electrical conductivity of solders is not sensitive to the gold concentration up to 10 wt% of gold.
- *Phase transition temperature.* The presence of gold tends to lower the solidus temperature and increase the liquidus temperature, thus widening the pasty range or creating a pasty range for eutectic solder. Table 13-4 lists the change in solidus temperature of 63 Sn/37 Pb with gold content.[9]

Overall, gold affects solder joints in the following areas:

- Fluidity
- Wettability and spread
- Mechanical properties
- Phase transition temperature
- Microstructure
- Appearance

In view of the extent of the effects, 3 wt% of gold is judged to be the upper limit of gold content when gold is uniformly and homogeneously distributed throughout the solder joint (excluding the situation that gold segregates at or near the interface). Above 3 wt%, deleterious effects may occur in one or more of the above areas.

As to the proper thickness of the gold coating, too thick a coating results in a higher concentration of gold in the solder as well as a higher material cost. Too thin a coating may prejudice the effectiveness of the surface protection. Further, the surface condition, particularly in

Table 13-4. Effect of Gold on the Solidus Temperature of 63 Sn/37 Pb

Au content, wt%	Solidus temperature, °C (°F)
0	183 (362)
0.327	182.2 (360)
1.0	180.6 (357)
2.0	178.9 (354)
3.0	177.8 (352)
5.0	177.8 (352)
7.0	176.7 (350)
10.0	177.8 (352)

terms of porosity, is equally important. The optimum lies in the balance among three areas: (1) surface intactness, (2) resulting concentration of gold in solder after dissolution, and (3) cost. Mechanically, gold may create

- Solder joint fracture due to embrittlement
- Void creation
- Microstructure coarsening

To ensure that the gold content does not exceed an alarming level, the industry standard calls for gold removal immediately prior to soldering. General guidelines are as follows:

- A double-dip tinning process or dynamic solder wave shall be used for proper gold removal.
- The gold removal procedure may be eliminated for through-hole components intended to be attached using dip or wave soldering processes provided that the gold on the leads is less than 100 μin (0.0025 mm) thick.
- For surface mount parts, the gold shall be removed from at least 95 percent of the surface to be soldered.

13.3 Solder Balling

When using solder paste, solder balling in the reflow process is a common phenomenon. It has been a continuous effort in soldering process control, in component and board quality, and solder paste design to minimize the occurrence of solder balling.

The *solder balling* phenomenon can be defined as the situation that occurs when small spherical particles with various diameters are formed away from the main solder pool during reflow and do not coalesce with the solder pool after solidification. Versatile manufacturing environments have revealed two distinct types of solder balling in terms of physical characteristics:

1. Solder balling around any components and over the board
2. Larger-size solder balls associated with small and low-clearance passive components (e.g., 0805, 1206); mostly larger than 0.005 in (0.13 mm) in diameter.

Type 1 solder balls normally can be removed during the cleaning process; type 2 solder balls, however, are difficult to remove using a normal cleaning process. With the implementation of the no-clean process, it is obviously desirable to avoid the occurrence of both types of solder balling. With the use of array packages (BGAs), solder balling also becomes more troublesome. In the presence of solder balls, the assembly may encounter the risk of electrical short when any solder balls become loose and mobile during service. Excessive solder balling may also deprive solder from making a good solder joint fillet. In general, type 1 solder balls can be formed for different reasons. The following are the likely sources to be considered:

- Solder paste with inefficient fluxing with respect to solder powder or substrate or reflow profile, resulting in discrete particles which do not coalesce, due either to paste design or to subsequent paste degradation
- Incompatible heating with respect to paste prior to solder melt (preheating or predry), which degrades the flux activity
- Paste spattering due to heating too fast, causing discrete solder particles or aggregates to form outside the main solder pool
- Solder paste contaminated with moisture or other high-energy chemicals that promote spattering
- Solder paste containing extra-fine solder particles that are carried away from the main solder pool by organic portion (flux/vehicle) during heating, resulting in small solder balls
- Interaction between solder paste and solder mask

The appearance of solder balls and their distribution often reveal the cause. Solder balls as a result of spattering are usually irregular and relatively large in size (larger than 20 μm is not uncommon) and are scattered over a large area of the board; solder balls caused by extra-fine powder in the paste often form a halo around the solder; ineffective or insufficient fluxing results in small solder balls scattered around the joint; and solder mask-related solder balls leave discoloration marks on the board.

Solder paste spattering during reflow can be caused by

- *Incompatibility between paste and reflow profile.* Examples are heating too fast and high volatile content in the paste.
- *Hygroscopicity of the paste.* When the open time during assembly exceeds the capability of the paste or the paste is exposed to a temperature and/or humidity beyond its tolerance level, the moisture absorbed by the paste can cause spattering.

To minimize solder balling during board assembly, several issues need to be addressed:

- Selection of a solder paste that is able to deliver performance under the specific production conditions.
- Understanding of the characteristics of the solder paste selected.
- Setup of the reflow process that best fits the solder paste selected.
- Assurance of consistency and quality of the solder substrate including boards and components.
- Control of ambient conditions (temperature and humidity).
- Control of open time that the paste can accommodate.
- Assurance of solder mask compatibility with the solder paste.
- Assurance of complete cure of the solder mask.

The formation of large solder balls associated with small passive components is largely attributed to the paste slump and flow under the component body between two terminations via capillary effect. The slump and flow dynamics can also be affected by the reflow temperature profile, the volume of paste, and component placement. In order to reduce the occurrence of these large solder balls, the following parameters are recommended for consideration:

- *Solder paste rheology.* Minimizing paste slump
- *Amount of solder paste deposit.* Avoiding excess paste
- *Component placement.* Avoiding paste spreading during placement
- *Reflow profile.* Reducing preheating temperature exposure

13.4 Voids in Solder Joint

Voids generated in solder connections can potentially create the following problems:

- Decreased electrical and thermal conductivity of the interconnecting path
- Possible initiative sites of mechanical failure

Generally, possible contributors to the formation of voids in solder joints are

1. *Wetting problem.* It is not well recognized that voids are closely linked to poor wetting, especially for closed (sandwiched) solder joints. Good wetting is a prerequisite for making solder joints with a minimal amount of voids (see Sec. 5.1.1 on wetting).

2. *Outgassing.* Incomplete outgassing leaves gases trapped in the solder joint. The efficiency of outgassing is related to the chemical makeup of flux and paste, alloy fluidity, and the geometry of the solder joint.

3. *Improper amount of solder volume.* To cover a fixed pad area of solder joint, the amount of voids generally decreases with increasing solder volume and reaches a minimum at the optimum volume. A further increase in volume may increase the amount of voids.

4. *Large solder pad and fillet.* For larger solder pad dimensions, voids are more aggravated.

5. *Excessive intermetallic compounds.* The formation of an excessive amount of intermetallics in solder may contribute to the formation of voids. The intermetallics are formed through the dissolution of foreign elements in the solder joint and metallurgical reactions between the solder and substrate which can be further promoted by an excessive temperature of the reflow process as discussed in Sec. 13.1.

6. *Grain boundary cavitation.* Under heat excursion and cyclic strain, solder joints may intrinsically undergo microscopic changes. When the conditions are met, cavitation will occur along grain boundaries.

7. *Stress-induced voids.* Stress fields created in solder are often associated with void formation.

8. *Gross volume shrinkage.* With a constrained structure, like most solder joints are, the volume shrinkage during liquid-solid transformation may create voids. It is estimated that there is a 3 to 3.5% volume shrinkage for 63 Sn/37 Pb going through the transformation. The cooling pattern affects the relationship of voids and shrinkage. A gradual outward cooling from interior to exterior minimizes the freeze-in shrinkage voids.

9. *Type of solder joint.* There is a significant difference in outgassing efficiency between closed (sandwiched) joints and open joints as shown in Fig. 13-17, and the outgassing efficiency is inversely related to the thickness of the joint. It is expected that the correlation of outgassing rate and thickness of the solder joint follows the gas diffusion theory.

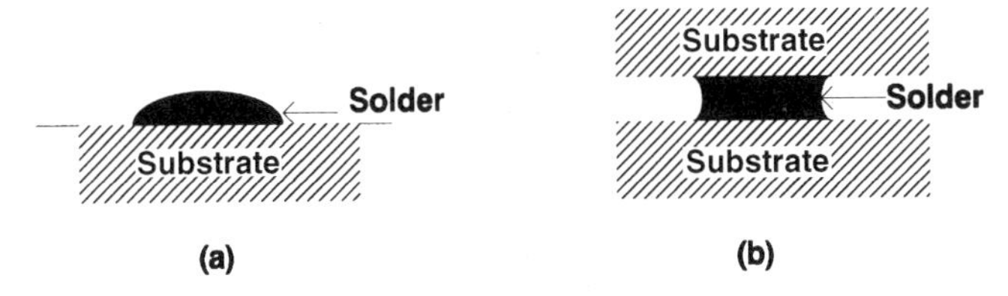

Figure 13-17. Closed (sandwich) and open solder joints.

Smaller solder joints in fine-pitch components make the effect of voids more pronounced. In order to minimize the voids in a lap/sandwich type of solder joint, as in those of surface mount/fine-pitch solder joints, four areas need to be scrutinized: fluxing, alloy composition, process, and design.

Fluxing. Always have the optimal wetting ability which is the prerequisite to achieve minimal voids.

Alloy. Select alloy composition with high fluidity and minimal potential for intermetallic formation between solder and substrates.

Process. Optimize the heating (reflow) profile. It should be compatible with the chemical makeup of solder paste, and sufficient dwell time in the molten state should be assured. Optimize the cooling rate to balance the gas release mechanism and microstructure development.

Design. Avoid excessively large areas or thicknesses of solder joints during each reflow.

13.5 Foreign Contamination

Two types of contaminations could potentially be of concern: chemical nature and metallurgical nature.

13.5.1 Chemical nature

The sources of the contamination can come from one or a combination of the following areas:

1. Flux and other contaminants introduced during component packaging and board fabrication

2. Flux used in board assembly
3. Bare board

The components must start clean. Flux used during board assembly should be thoroughly cleaned unless a proven no-clean chemistry is used.

Because of the process and chemicals involved in board fabrication, the presence of ionic contaminants on the bare board in excess of the allowable limit can occur. One test indicated that the major contaminants of a bare board, as summarized in Table 13-5, did exceed the allowable limit (10 μm of NaCl equivalent per square inch).

With an effective cleaning step, ionic contamination from board and component fabrication and from the board assembly process can be removed. If a no-clean system is elected, the use of a clean board and clean components are mandatory.

13.5.2 Metallurgical dissolution

Metallurgical dissolution has been well recognized as a problematic area in thick-film conductors of hybrid circuitry since the thick-film conductor was developed. It was found that noble metals, particularly Ag and Au, were deprived from the thick-film composition (as conductor pads on hybrid circuit board) or from the terminations of capacitors and resistors and dissolved into solder during soldering. This phenomenon is also referred to as a *leaching problem*. The dissolution could cause degradation in the thick-film composition and/or degradation of solder joints. Physically, the leached thick-film conductor may lose adhesion to the substrate and therefore lose the integrity of the circuitry. The solder with the elements leached in may form undesirable intermetallic compounds that may adversely affect the solder joint reliability.

The leaching process is metallurgical in nature and highly temperature dependent. Therefore, in addition to material factors, the soldering process in terms of peak temperature and dwell time at peak temperature is also an important factor to the extent of leaching.

The common metals encountered in electronics packaging and

Table 13-5. Ionic Contamination Level of Bare Board[13]

Cl^-	Na^+	NH^{4+}
4.9 μg/in^2	6.6 μg/in^2	2.6 μg/in^2

assembly include Sn, Cu, Au, Ag, Pd, Pt, Ni. These metals are normally used for one or two of the following functions:

1. To make the base substrate solderable without appreciable solubility in the molten solder.
2. To serve as a thick film protecting the underlying substrate. Among these metals, their dissolution rate in Sn-Pb solder is in the following descending order:

$$Sn > Au > Ag > Cu > Pd > Ni > Pt$$

The measured dissolution rate of each of the elements[14–18] is summarized in Figs. 13-18 through 13-23. The data are useful to aid in selecting a proper metal layer either as a surface protecting film or as the solderable surface of the substrate. They are also helpful in designing a correct layer thickness suitable for a specific set of conditions. The activation energies of the dissolution kinetics are listed in Table 13-6.[14–18]

These elements after being dissolved in Sn-Pb solder were found to form the intermetallic compounds as outlined in Table 13-7. At the eutectic reflow temperature, 210 to 225°C (435 to 455°F), the dissolution rates in solder are shown in Table 13-8. Both Ni and Pt have negligible dissolution rates until the temperature reaches 350°C (700°F) (Ni: 1.7 μin/s; Pt: <1 μin/s).

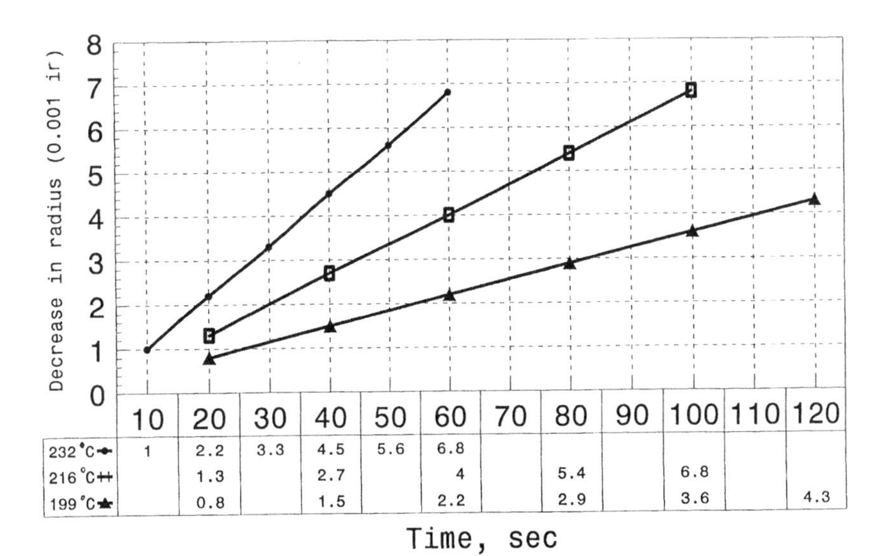

Figure 13-18. Measured dissolution rate of Au in Sn-Pb solder.

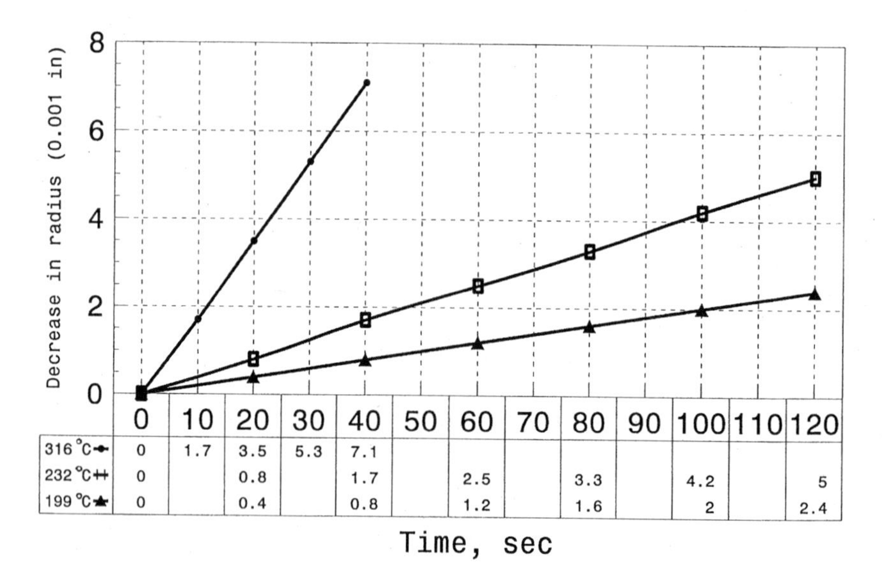

Figure 13-19. Measured dissolution rate of Ag in Sn-Pb solder.

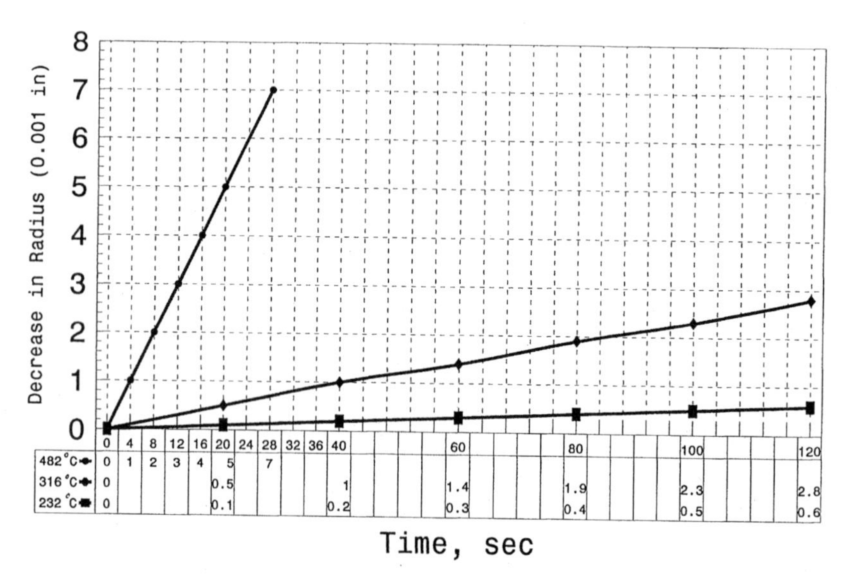

Figure 13-20. Measured dissolution rate of Cu in Sn-Pb solder.

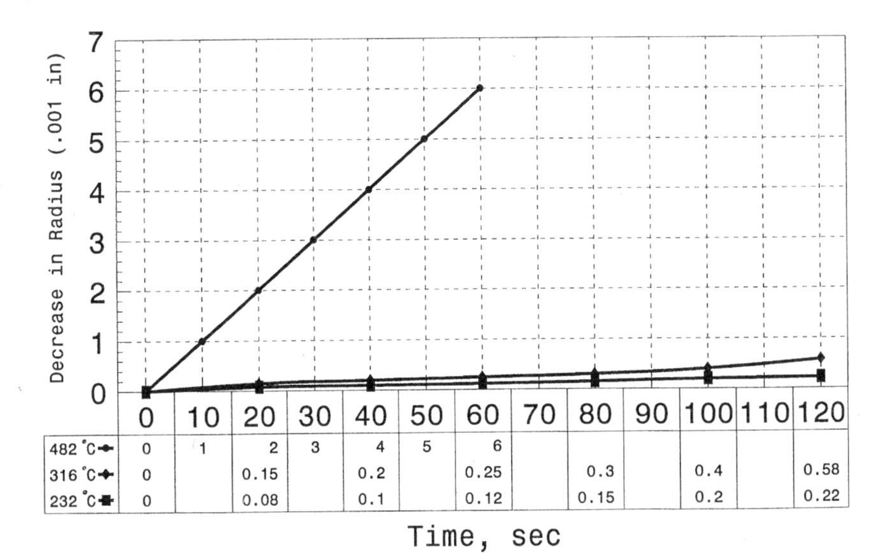

	0	10	20	30	40	50	60	70	80	90	100	110	120
482 °C	0	1	2	3	4	5	6						
316 °C	0		0.15		0.2		0.25		0.3		0.4		0.58
232 °C	0		0.08		0.1		0.12		0.15		0.2		0.22

Figure 13-21. Measured dissolution rate of Pd in Sn-Pb solder.

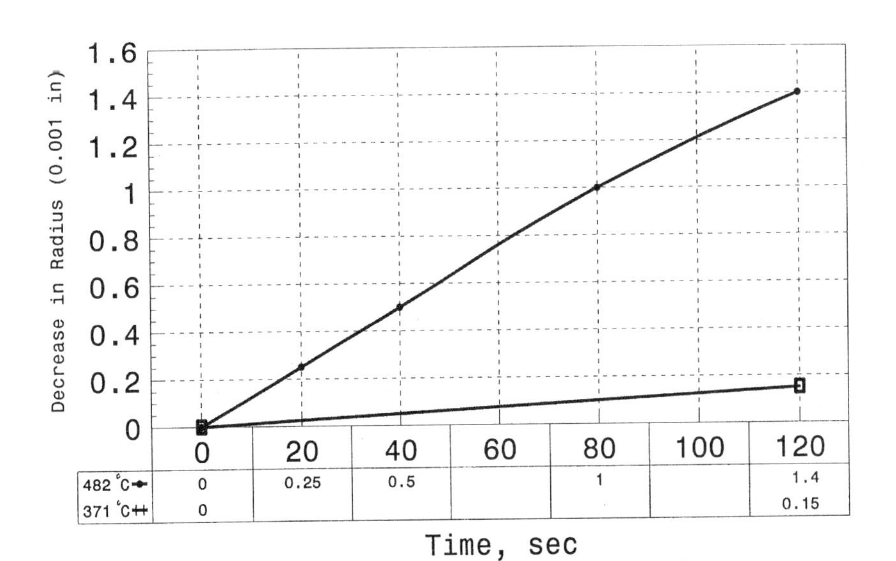

	0	20	40	60	80	100	120
482 °C	0	0.25	0.5		1		1.4
371 °C	0						0.15

Figure 13-22. Measured dissolution rate of Ni in Sn-Pb solder.

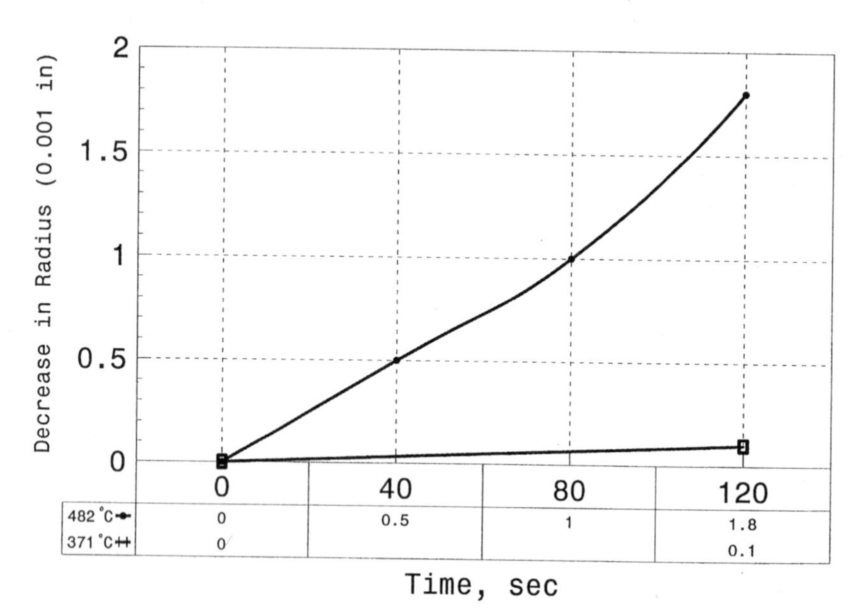

Figure 13-23. Measured dissolution rate of Pt in Sn-Pb solder.

Table 13-6. Activation Energies of Dissolution Kinetics of Elements in 60 Sn/40 Pb Solder

Element	Activation energy, kcal/mol
Au	13.6
Ag	9.5
Cu	11.3
Pd (<525°F)	18.0
Pd (>600°F)	16.0
Ni	15.9
Pt	22.8

13.6 Plastic Package Cracking

The plastic molded IC packages are prone to moisture absorption. The absorbed moisture induces package cracking during the reflow process. The problem had been recognized and remedies have been and are still being developed since the late 1980s. Figure 13-24 displays typical moisture absorption rates under ambient temperature and

Table 13-7. Intermetallics Formed at the Sn-Pb Solder Interface

Elements	Intermetallics
Au	$AuSn_2$, $AuSn_4$
Ag	Ag_3Sn
Cu	Cu_6,Sn_5
Pd	$PdSn_2$, $PdSn_3$, $PdSn_4$
Ni	Ni_3Sn_4
Pt	$PtSn_4$

Table 13-8. Relative Dissolution Rate of Elements in 60 Sn/40 Pb Solder at 450°F

Elements	Radial dissolution rate, 10^{-6} in/s
Au	117.9
Ag	43.9
Cu	4.1
Pd	1.4
Ni	<1.0
Pt	Negligible

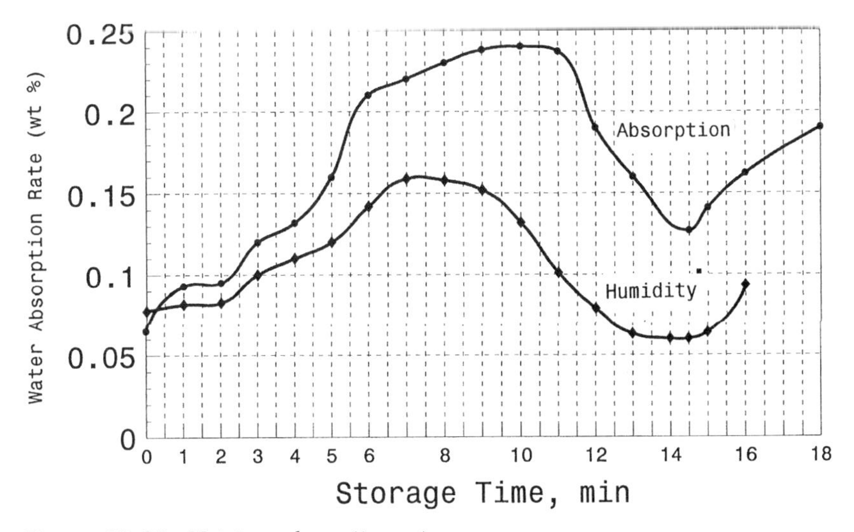

Figure 13-24. Moisture absorption rate.

humidity.[19] The moisture level can reach 0.24 wt% within a short time period. As expected, the moisture absorption increases as the ambient humidity increases and/or the temperature rises.

During reflow, the absorbed moisture vaporizes and generates pressure at metal-to-plastic interfaces in the package potentially causing the loss of encapsulant adhesion to die and lead frame, and cracks. The cracks propagate either through the body of the plastic or along the lead frame and open the package to ionic contaminants at the die surface. The most insidious danger is that cracks rarely cause immediate electrical failure and, therefore, the cracked packages remain undetected and are used. The proposed package cracking mechanism[20] is illustrated in Fig. 13-25; it consists of the diffusion of moisture from the ambient environment into the plastic molding, condensation of moisture at the die paddle and plastic interface, and the buildup of vapor pressure during reflow, resulting in cracks at the corner of the interface propagating through the bottom part of the plastic package. The moisture absorption rate changes with the ambient environment and the

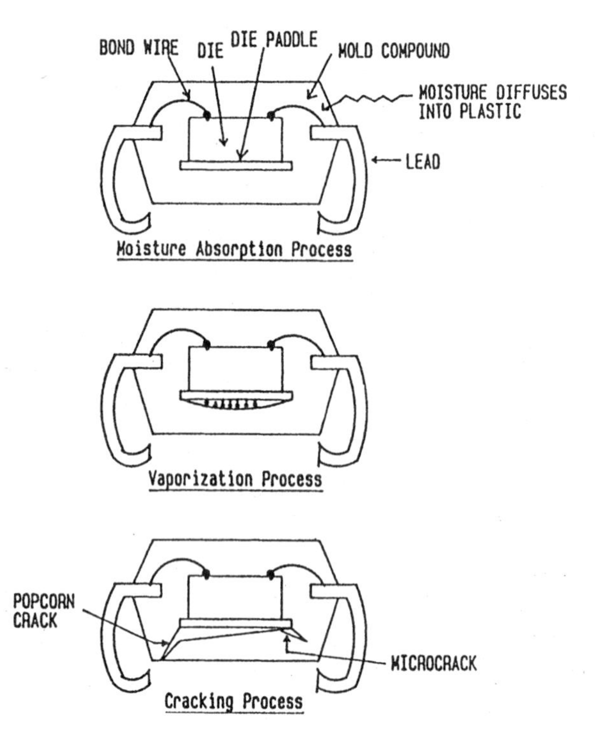

Figure 13-25. Proposed package cracking mechanism.

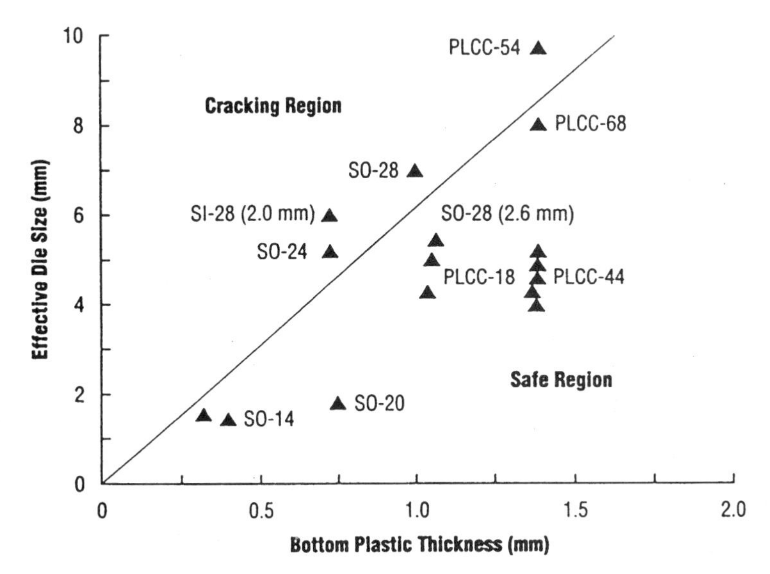

Figure 13-26. Relative crack resistance of various packages.

structure and dimension of plastic packages with a given polymer system. It is found that the thickness of the package bottom is a major factor to the crack resistance. Figure 13-26 compares the relative crack resistance of various packages. The thicker small outline (SO) packages such as SO-14 and SO-20 have better crack resistance than thinner packages with comparable chip size.[19] It should be noted that P-BGA packages are about one-third as thick as a standard QFP; thus a P-BGAs moisture absorption is expected to be accelerated. A study using a 225-pin OMPAC exposed to 30°C and 70% relative humidity (RH) had a weight gain of about 0.15 percent after 24 h and 0.21 percent after 72 h. Crack failure was found to occur at 0.24% water absorption and die bond delamination occurred at 21%. It is recommended that parts exposed to more than 48 h of floor life be baked in an N_2 purged oven at 125°C for 8 h.

A study has shown that the reflow heating rate has a direct bearing on the formation of the plastic package cracks.[21] Preconditioned PQFP packages under different environmental treatments were subject to reflow at various heating rates. The packages were then examined by means of code-scanning acoustic microscopy (C-SAM). C-SAM nondestructively analyzes the package integrity for any delamination or cracks before preconditioning and after reflow. Three preconditionings are: (1) 85°C/85% RH, 168 hours; (2) 30°C/60% RH, 168 hours; and (3) 30°C/60% RH, 72 hours, which result in 0.24%, 0.0175%, and 0.005%

Table 13-9. Plastic Package Crack versus Reflow Heating Rate

Ramp rate (°C/s)	Preconditioning		
	85°C/85% RH, 168 h	30°C/60% RH, 168 h	30°C/60% RH, 72 h
10.0	—	No crack	No crack
5.0	Crack	—	—
2.0	Crack	—	—
0.75	Crack	—	—
0.50	No crack	—	—
0.25	No crack	—	—

moisture content, respectively. Table 13-9 summarizes the results of three series of tests, indicating that there is a threshold heating rate below which no crack exists, e.g., the threshold heating rate for a moisture content of 0.24% is 0.5°C/s, and that such a threshold heating rate increases as the total moisture absorbed in the package decreases.

13.7 Thermal Management

To meet the end-use performance of electronics products, active IC packages face increasing challenges in managing heat dissipation. The integrity and service life of the system may very much depend on the efficiency of heat dissipation and indirectly on the external temperature. The heat generated in the package during functioning must be effectively carried away from the die to the package surface and then to the ambient environment. There are three basic heat transfer mechanisms: conduction, radiation, and convection.

Through conduction, the heat flow per unit area Q is related to the thermal conductivity of the material κ and the temperature T gradient in a given distance x, as experienced by Fourier's Law:

$$Q = -\kappa \frac{\delta T}{\delta x}$$

The energy of thermal radiation e is proportional to the fourth power of its absolute temperature T in K.

$$e = bT^4$$

where b is the Stefan-Boltzman constant. The heat transfer by convection is normally expressed as a heat transfer coefficient h between two temperatures T_1 and T_2 and is related to the heat transfer rate R:

$$h = \frac{R}{T_2 - T_1}$$

Although all three heat transfer mechanisms may contribute to the overall heat transfer of a system, conduction is primarily the main mechanism for transferring heat from the die to the package.

Thus, at a given period of time, the efficiency of heat dissipation is largely dependent on the conductivity of the materials involved and the length of the heat transfer path. The shortest heat path and highest conductivity of materials are obviously the best combination. Table 13-10 lists the thermal conductivity of common materials incorporated in BGAs. Comparing C-BGA with P-BGA, C-BGA is expected, because of its ceramic substrate, to deliver more efficient heat dissipation in an otherwise similar system. Solder interconnections are expected to conduct heat better than adhesive joints.

Overall, the factors affecting heat dissipation include

- Chip
- Air flow
- Molding compound
- Package substrate material
- Board material
- Power plane
- Ground plane

Table 13-10. Thermal Conductivity of Common BGA Materials

Material	Thermal conductivity, W/(m · K)
Copper	400
Gold	320
Lead frame	160
Silicon	80
63 Sn/37 Pb solder	50
Die-attach adhesive	5
Aluminum oxide (96%)	35
FR-4 or BT laminate	0.2
Molding compound	0.6
Air	0.03

- Heat sink
- Thermal vias
- External temperature

For a given IC chip, the increased air flow increases the heat dissipation, consequently lowering the temperature rise of the chip. It is found that the thermal conductivity of the board onto which the IC devices are attached and the conductivity of the molding compound affect the maximum chip temperature rise above ambient.[22–24] The heat dissipation increases as the thermal conductivity of the board or the molding compound increases. However, the effect of the thermal conductivity of the board is much more significant than that of the molding material. Both power and ground planes improve the thermal performance. It is reported that a substrate without power and ground planes has a much higher temperature gradient across its surface and a larger increase in chip temperature than one incorporating a power plane and a ground plane.

An array of thermal vias is another means of improving the heat dissipation. These are constructed as holes through the substrate from the back of the die to the back of the package and are plated in copper. The effectiveness of thermal vias depends on the area of copper in the array and is inversely proportional to the thickness of the substrate.[25]

High-power chips and a package or module housing more than one chip (MCM) dissipate more power. Increased efficiency of heat removal from these packages is obviously needed. The heat resistance, depending on the design of the package and the mother board, can range from 2 to 100°C/W.[26–28]

Since the path of heat transfer is shorter in BGA packages from the chip substrate to the board through solder ball interconnections than in QFPs, the thermal performance of BGAs is expected to be superior to that of QFPs. Figure 13-27 illustrates the relative junction-to-PCB thermal resistance between an OMPAC and a QFP.[29] Thermal vias have offered lower thermal resistance for 225-OMPACs. With thermal vias, a 225-OMPAC had 20 percent lower thermal resistance than a 208-lead QFP. Since OMPACs have a lower junction-to-PCB thermal resistance than QFP packages, a higher thermal conductivity in the PCB makes a more pronounced improvement in the thermal performance of the OMPAC.

13.8 Electromigration

The electromigration of metals in electronics has been recognized for over 30 years. The phenomenon causes electrical shorts as two adja-

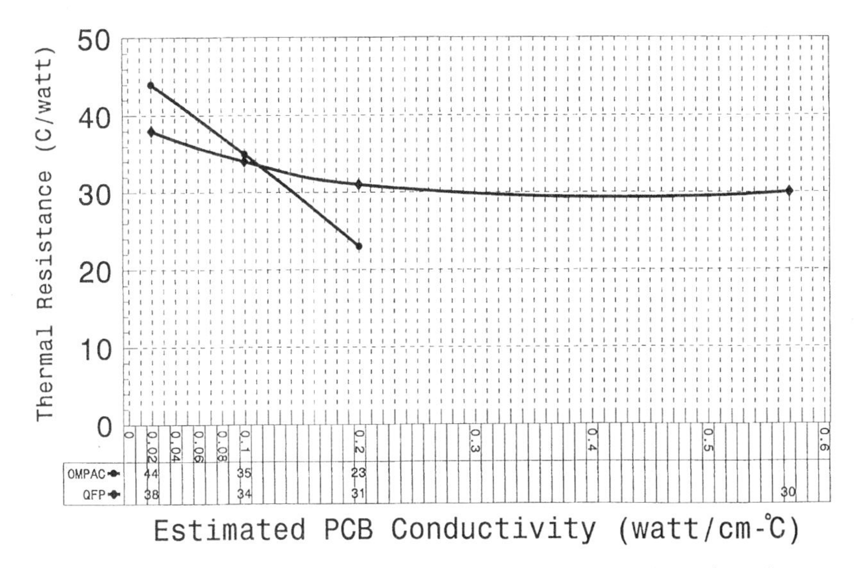

Figure 13-27. Junction-to-PCB thermal resistance versus thermal conductivity of PCB.

cent conductor paths are bridged together through the growth of metal dendrites. In principle, a metal which can readily form ions in the presence of moisture and have reasonable mobility in a medium under an electrical voltage potential would exhibit electromigration. Thus the requirements of electromigration of metals are electrolyte, electrical voltage, time, and temperature. To monitor the level of electric current between two conductors consisting of the metal to be tested and separated at a specified distance in an electrolytic environment under applied voltage is normally measured. The electrolytic environment is typically moisture with or without ionic species.

Under favorable conditions, the metal ions can migrate from cathode (−) to anode (+) and form a dendritelike structure between two electrodes. Table 13-11 lists the results of electromigration among different metals. The metals are grouped into three classes in terms of susceptibility to electromigration. Electromigration of the first group can occur with the presence of only moisture and voltage. The second group has the additional requirement of needing halogen-containing contaminants. The third group is quite passive in a normal environment. Among the metals, silver is most prone to electromigration, due to its high activity in the presence of moisture. Except for silver, the metals related to soldering such as lead, tin, copper, and bismuth all have electromigration potential when conditions are met. For thick-film conductors, the silver-based systems are normally built with plat-

Table 13-11. Electromigration Conditions for Different Metals

Metals that migrate with distilled water	Metals that migrate with distilled water and halogen contaminant	Metals that may need other conditions to migrate
Bismuth	Gold	Aluminum
Cadmium	Indium	Antimony
Copper	Palladium	Chromium
Lead	Platinum	Iron
Silver		Nickel
Tin		Rhodium
Zinc		Tantalum
		Titanium
		Vanadium

inum and/or palladium to improve electromigration resistance (also leaching resistance).

The resistance can be further enhanced by using coprecipitated silver-palladium or silver-platinum systems. The copper-based thick film is considered to deliver better migration resistance. The study indicates that soldered copper thick-film systems are not totally immune from electromigration, and lead migration is found in the soldered copper thick film. The same study on dendritic growth on bulk solder and soldered thick film indicates that dendrites originate from the solder/thick film interface, and therefore the migration can be reduced by completely covering the thick film by solder. Among the bulk solder alloys, the study shows that migration is retarded by high-tin-containing solders due to the formation of a nonconductive tin oxide film. Table 13-12 contains the results of migration tests among solder alloys.[30,31] It is indicated that the key element which is necessary for the occurrence of electromigration in circuits is moisture. Figure 13-28 illustrates the effect of encapsulant on the high-voltage circuits. The observed difference in leakage currents under accelerated aging and bias (85°C, 85 percent relative humidity, 1000 V) is attributed to the thickness of room temperature vulcanized (RTV) encapsulant which serves as a sealing material for circuits.[32] The thinner encapsulant provided less protection from moisture permeation, and therefore higher electric leakage occurs in the thinner encapsulant package.

Table 13-12. Electromigration Time of Some Solder Alloys

Solder alloy	Typical migration time, s	Dendrites identity
96 Sn/4 Ag	>720	Tin oxide filaments
30 Sn/70 Pb	200	Lead
10 Sn/88 Pb/2 Ag	150	Lead
10 Sn/90 Pb	125	Lead
93 Pb/5 In/2 Ag	40	Lead

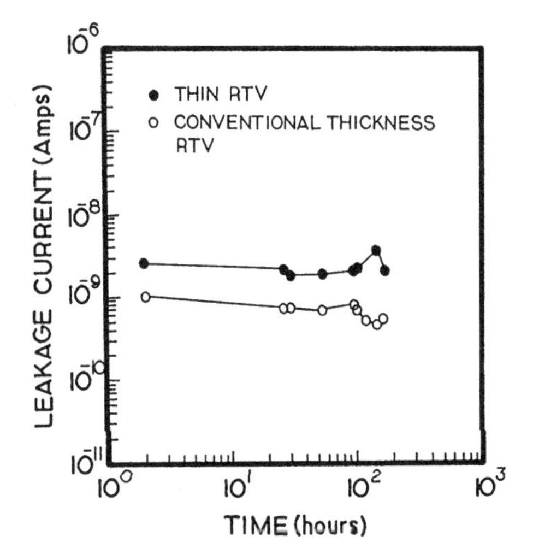

Figure 13-28. Effect of encapsulant on high-voltage circuits.

Electrical failures due to excessive current leakage were observed in hybrids after board-level testing.[33] Failure analysis indicated that silver migrated from the die-attach adhesive. These failures occurred in hermetic package devices in which the moisture level was quite low (<2000 ppm), contrasting to typical metal migration failure. A possible explanation for this phenomenon is that outgassing from the adhesive occurred during burn-in which generated significant levels of ammonia, methyl alcohol, and carbon dioxide along with the low level of water. When the hybrid device was cooled during subsequent board-level testing, these items condensed on the surfaces of the device, affecting surface conductivity.[33]

13.9 References

1. J. P. Kay, *Transactions of the Institute of Metals,* vol. 54, 1976, p. 68.
2. D. Unsworth, *Transactions of the Institute of Metal Finishing,* vol. 51, 1973, p. 85.
3. D. Blair et al., "Copper Dissolution in Molten Solders," *Proceedings NEP-CON-West,* 1994, p. 543.
4. C. Luchinger et al., "Inside Solder Connections," *Advanced Packaging,* January/February 1994, p. 14.
5. J. S. Hwang, *Solder Paste for Electronics Packaging,* Van Nostrand Reinhold, New York, 1989, p. 282.
6. G. Matijasevic, Y. Chen, and C. C. Lee, "Copper-Tin Multilayer Composite Solder," *Proceedings IEPS Conference,* 1993, p. 264.
7. Y. Tagunia et al., "Application of DC Magnetron Sputtered Cr-Cu-Au Thin Films for Flip-Chip Solder Terminal Contact," *Proceedings IEPS Conference,* 1991, p. 619.
8. G. Matijasevic and C. C. Lee, "Void-Free Au-Sn Eutectic Bonding of GaAs Dice and its Characterization Using Scanning Acoustic Microscopy," *Journal of Electronic Materials,* vol. 18, 1989, p. 327.
9. L. Cangemi and G. F. Pittinato, "The Effect of Gold Plate on Solder Joint Integrity," MPR 6-117-001, Rocketdyne Division of North American Aviation, January 1966.
10. F. G. Foster, "Embrittlement of Solder by Gold from Plated Surfaces," *65th Annual Meeting of ASTM,* Special Technical Publication, no. 319, June 1992, p. 13.
11. W. B. Harding, "Soldering to Gold Plating," *Proceedings American Electroplaters Society,* June 1963, p. 90.
12. J. Braun, "An Improved Soft Solder for Use with Gold Wire," *Transactions of the American Metals Society,* vol. 57, no. 2, June 1964, p. 568.
13. Ruth Anderson, "Manufacturing Experience with Water-soluble Solder Paste," *Proceedings SMI,* 1991, p. 483.
14. Thwaites, *Transactions of the Institute of Metal Finishing,* vol. 43, 1965, pp. 143–152.
15. Bailey and H. C. Watkins, *International Institute of Metals,* vol. 80, 1951, p. 1325.
16. W. R. Lewis, "Notes on Soldering," *Tin Research Institute,* October 1959, p. 73.
17. Elliott, *Constitution of Binary Alloys, First Supplement,* McGraw-Hill, New York, 1956.
18. Cooper, "The Use of Phase Diagrams in Dissolution Studies," NASA Report N 68-28917, 1987.

19. Nicolaou, "Plastic Package Failure Mode," *Surface Mount Technology*, November 1993, p. 45.
20. David Cheng et al., "Moisture Sensitivity of Surface Mount Plastic Packages," *Proceedings SMI*, 1992, p. 421.
21. P. Yalamanchili et al., "Optimum Processing Prevents PQFP Popcorning," *Surface Mount Technology*, May 1995, p. 39.
22. D. B. Walshak and H. Hashemi, "Thermal Modeling of a Multichip BGA Package," *Proceedings NEPCON-West*, 1994, p. 1266.
23. P. E. Rogren, "MCM-L Built on Ball Grid Array Formats," *Proceedings NEPCON-West*, 1994, p. 1277.
24. S. Mulgaonher et al., "Thermal Performance Limits of the QFP Family," *Proceedings 9th IEEE Semi-Therm Symposium*, 1993, p. 166.
25. P. E. Rogren, "Thermal and Electrical Performance Enhancements in Plastic Ball Grid Array Packages," *Proceedings NEPCON-East*, 1994, p. 289.
26. M. Hu, "Choosing Laminates," *Advanced Packaging*, Fall 1993, p. 16.
27. P. E. Rogren, "The Current State of Laminate Based, Molded MCM Technology," *IEEE International Electronics Manufacturing Technology Symposium*, 1993, p. 485.
28. G. Stout, "Identifying the Building Blocks in Leading-Edge PCBs," *Electronic Packaging and Production*, May 1993, p. 17.
29. P. Lin, F. D. J. Martin, and H. P. Wilson, "Manufacturability of Plastic Ball Grid Array (PBGA) vs. QFP Packages," *3rd International Microelectronics and Systems Conference*, Kuala Lumpur, Malaysia, 1993, p. 1215.
30. A. Dermarderosian, "The Electrochemical Migration of Metals," *Proceedings Annual ISHM Symposium*, 1987, p. 134.
31. R. G. Gehman, "Dendritic Growth Evaluation of Soldered Thick Film," *Proceedings Annual ISHM Symposium*, 1983, p. 239.
32. S. D. Schlough and E. F. O'Connell, "Cleaning Processes for High Voltage Circuits with Soldered Chip Carriers," *Proceedings Annual ISHM Symposium*, 1981, p. 28.
33. Richard C. Benson, et al., "Electromigration of Silver in Low-Moisture Hybrids," *Proceedings Annual ISHM Symposium*, 1993, p. 530.

13.10 Suggested Readings

Banks, Sherman, "Reflow Soldering to Gold," *EP&P*, June 1995, p. 69.

Jozefowicz, Miklolaj, and Ning-Cheng Lee, "Electromigration vs. SIR," *Proceedings Annual ISHM Symposium*, 1993, p. 537.

LaCap, Efren M., and Junaid I. Khan, "Baking Eliminates Plastic Package Cracks," *Semiconductor International*, March 1990, p. 92.

Maxwell, John, "Cracks: The Hidden Defect," *Circuits Manufacturing*, November 1988, p. 26.

McKenna, Robert, "Surface Mount Device Cracking: An Overview," *Journal of SMT*, October 1989, p. 20.

Moore, T. M., S. J. Kelsall, and R. G. McKenna, "The Importance of Delamination in Plastic Package Moisture Sensitivity Evaluation," *Proceedings SMI*, 1991, p. 1231.

14 Strengthened Solders

14.1 Scientific Approaches

From a material point of view, the crystalline alloys can deform via one or a combination of mechanisms: (1) slip, (2) dislocation climb, (3) shear on grain boundary, (4) in-grain vacancy or atomic diffusion. It is generally understood that fatigue failure or cracking of metals is often caused by dislocation slip and the localization of plastic deformation. It is also generally understood that the plastic deformation kinetics follows the power-law dislocation climb-controlling mechanism at high-stress/low-temperature conditions. At a low-stress region and high temperature, the grain boundary sliding becomes a rate-controlling process. Therefore, in order to strengthen the performance of conventional solders that are subject to stressful conditions as a result of external temperature fluctuation and/or in-circuit power dissipation and power on-off of electronic circuit boards, several approaches as listed below can be considered. Alloy 63 Sn/37 Pb is selected as the reference because of its wide usage and acceptance in modern electronic packaging and assemblies.

1. Microscopic incorporation of nonalloying dopant
2. Microstructural strengthening
3. Alloy strengthening
4. Macroscopic blend of selected fillers

Approach 2 is essentially process-based when the elemental composition is given, although other approaches may involve the alteration in microstructural evolution; approaches 1, 3, and 4 are largely material-based. In reality, strengthening mechanisms often include more than one basic approach. Material properties to be addressed are

- Phase transition temperature
- Thermal conductivity
- Electrical conductivity
- Surface tension
- Wetting ability
- Thermal expansion coefficient
- Intrinsic strength
- Creep resistance
- Isothermal low-cycle fatigue resistance
- Thermomechanical fatigue resistance
- Metallurgical stability (microstructural stability)
- Environmental stability

14.2 Microscopic Incorporation of Nonalloying Dopant

The solder matrix is doped with foreign compounds (*dopants*) which are homogeneously distributed and embedded in the matrix. The *foreign compounds* are defined as the compounds that do not form a significant solid solution with the matrix composition. In the material system, the matrix is comparatively soft and the dopants are fine and hard particulates. The typical size of the dopant particles is smaller than 30 μm. The dosage of doping addition is usually in the range of 0.01 to 5 wt% based on solder matrix weight (weight percentage is used for convenience, volume percentage is a more accurate representation). The process to make the mixture is vitally important; it must be capable of constituting a uniform and physically stable system. Under this approach, the physical properties of the resulting compositions are expected to be unchanged from those without dopants and the mechanical properties are targeted to be improved. Some empirical data are presented as follows. The mechanical properties are monitored by monotonic shearing, creep in shear, and isothermal low-cycle fatigue for 200-μm (0.008-in) thick solder joints on tin-plated copper substrate.[1]

The monotonic shear flow resistance of the composition containing 1% doping addition (62.5 Sn/36.5 Pb/1.0 additive) in comparison with that of 63 Sn/37 Pb composition is shown in Fig. 14-1. A significant improvement in shear strength of the doped composition over the 63 Sn/37 Pb composition was observed. The dependency of shear strength on the concentration of the doping additive over a range of concentration from 0.1 to 5% is summarized in Fig. 14-2. The magnitude of improvement in shear strength reached the maximum between 0.1 to 1% and started to drop at above 1%, as shown in Fig. 14-3. The doped solder joints also exhibited a higher creep resistance than the 63 Sn/37 Pb solder joints. Figures 14-4 through 14-7 illustrate the creep behavior of the doped 63 Sn/37 Pb composition under various stresses at ambient temperature (300 K). As discussed in Sec. 3.3.2, the plastic deformation resistance under creep can be monitored by parameters such as creep rupture time, total creep rupture ductility, and steady-state plastic deformation rate. For the high-temperature deformation

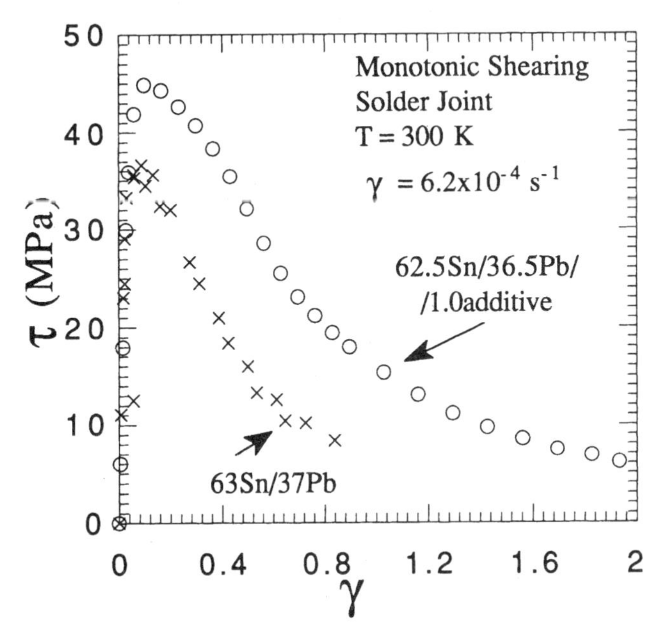

Figure 14-1. Monotonic shear flow resistance of 62.5 Sn/36.5 Pb/1.0 additive.

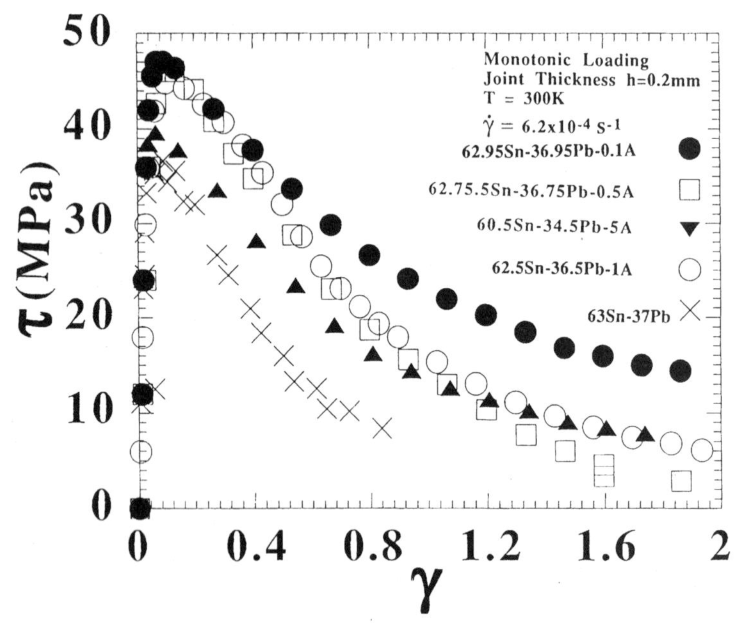

Figure 14-2. Shear flow resistance versus concentration of additive.

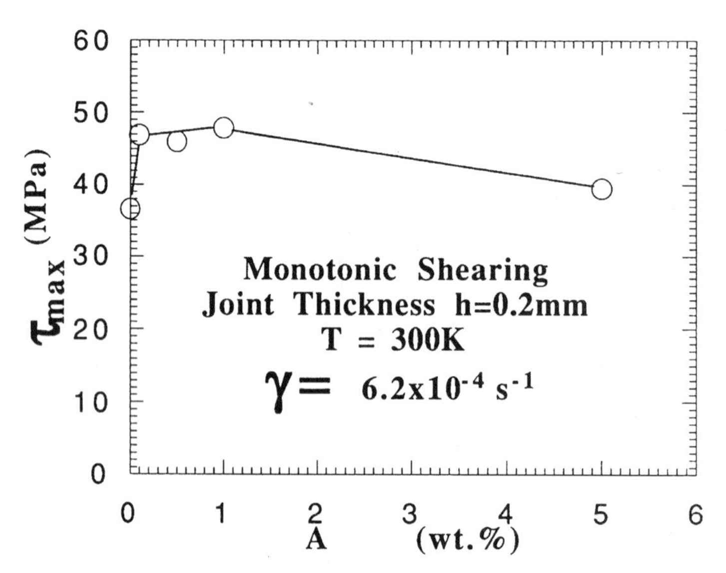

Figure 14-3. Shear strength maximum versus concentration of additive.

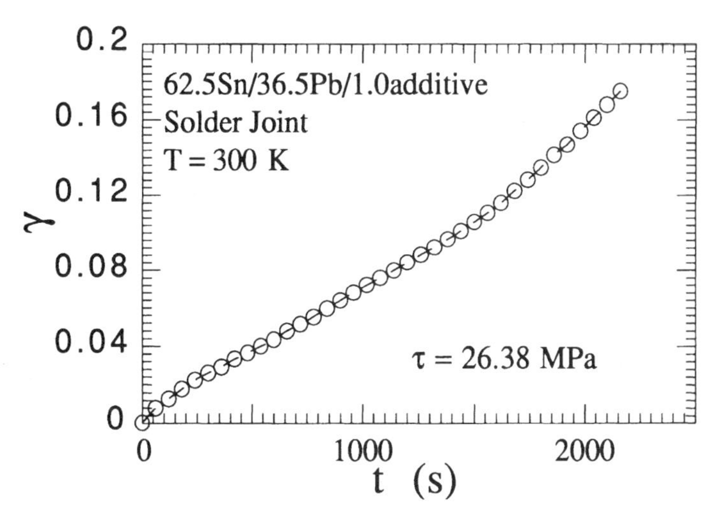

Figure 14-4. Creep curve of 62.5 Sn/36.5 Pb/1.0 additive at stress of 26.38 MPa.

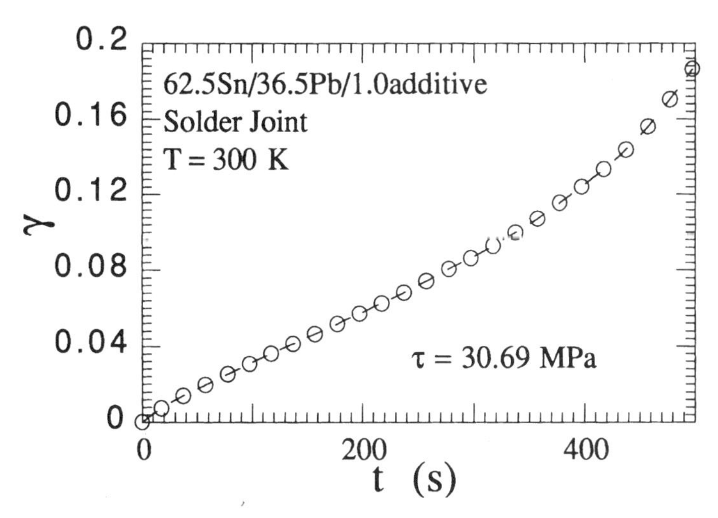

Figure 14-5. Creep curve of 62.5 Sn/36.5 Pb/1.0 additive at stress of 30.69 MPa.

mechanism that is expected to dominate in solder failure involving fatigue-creep interaction, the steady-state plastic deformation rate implicitly governs the other creep parameters and is considered to be an important kinetic parameter for phenomenological analysis of the

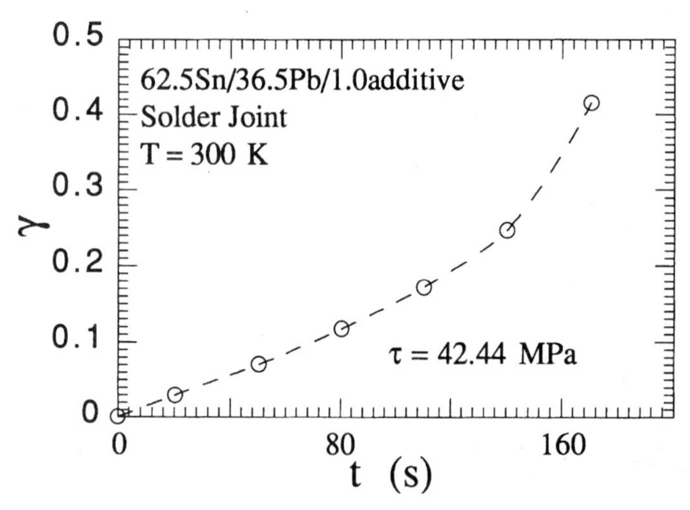

Figure 14-6. Creep curve of 62.5 Sn/36.5 Pb/1.0 additive at stress of 42.44 MPa.

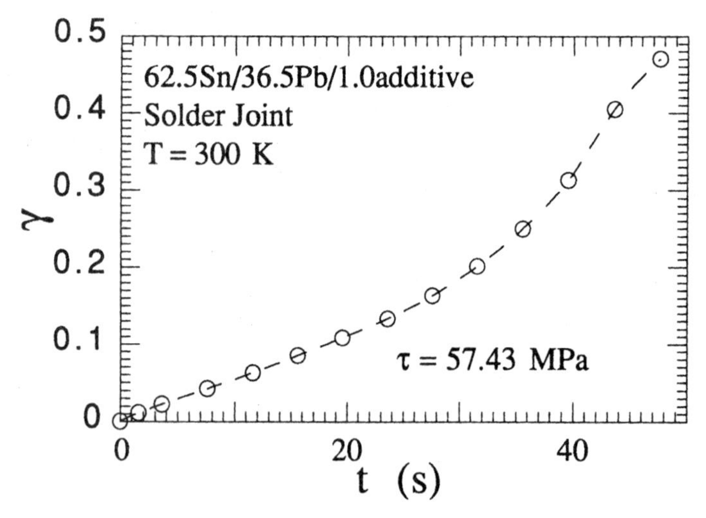

Figure 14-7. Creep curve of 62.5 Sn/36.5 Pb/1.0 additive at stress of 57.43 MPa.

high-temperature deformation mechanism. Accordingly, plastic deformation rates ($\dot{\gamma}_p$) derived from the creep curves of the doped compositions, as shown in Figs. 14-4 through 14-7, are related to the corresponding applied stress τ based on the Dorn equation as follows:

$$\dot{\gamma}_\rho = \gamma_{\rho_0} \tau^n \exp\left(-\frac{\Delta H}{kT}\right)$$

where n = stress exponent
γ_{ρ_0} = kinetics factor related to material intrinsic parameters such as lattice constant, diffusion coefficient, and shear modulus
ΔH = activation energy of deformation
k = Boltzmann's constant
T = absolute temperature

As shown in Fig. 14-8, the plastic deformation rates of the doped solder joints are about an order of magnitude lower than those of the 63 Sn/37 Pb solder joints at higher stress regions. At lower stress regions, the plastic deformation for the doped solder exhibits a stress exponent factor of 2, indicating the possible superplastic deformation mechanism.[2] This is in contrast to the stress exponent of 4 or higher for stan-

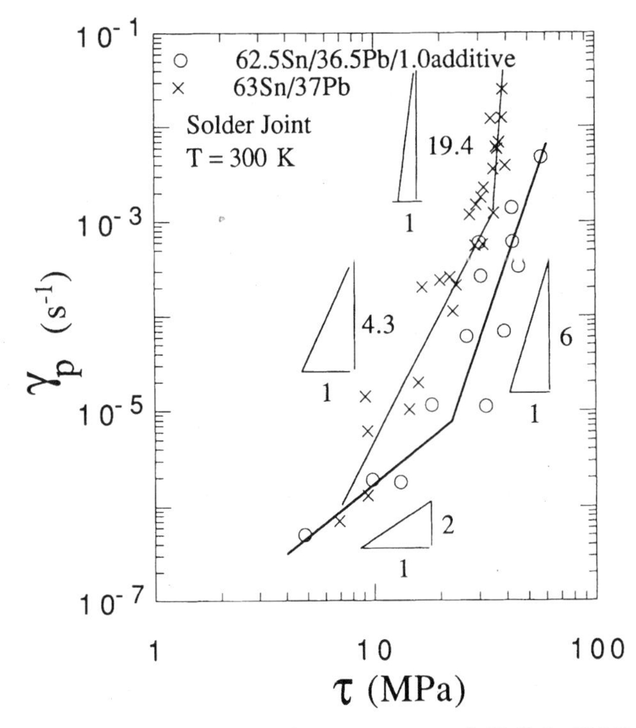

Figure 14-8. Plastic deformation rate of 62.5 Sn/36.5 Sn/1.0 additive in comparison with 63 Sn/37 Pb.

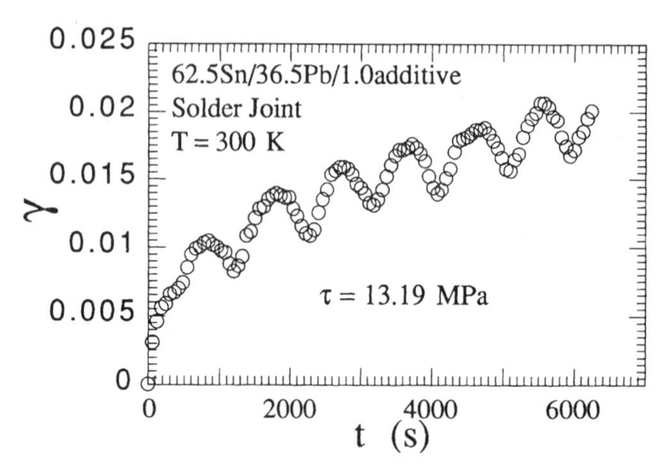

Figure 14-9. Oscillating creep behavior of 62.5 Sn/36.5 Pb/1.0 additive at very low stress.

dard 63 Sn/37 Pb; its relationship between the plastic deformation rates and the stress shows two regions: power-law creep ($3 < n < 10$) at lower stresses and power-law breakdown creep ($n > 10$) at higher stresses. It is interesting to note that the creep behavior of doped solder exhibits an oscillating pattern in the secondary stage when subjected to lower constant stresses (< 15 MPa) as shown in Fig. 14-9. This occurrence is not fully understood at this time. However, a similar phenomenon has occurred for other materials and has been attributed to the process of dynamic recovery and recrystallization.[2]

The response of the cyclic load ΔP (displacement or strain $\Delta\gamma$) hysteresis loop with cycling under total strain control $\Delta\gamma_t$ for 63 Sn/37 Pb eutectic solder joints is shown in Fig. 14-10. For a particular total strain range $\Delta\gamma_t$, the load amplitude (ΔP = the height of the loops) gradually decreases and the plastic strain range $\Delta\gamma_t$, steadily develops. It was established that this decrease of cyclic load-bearing capacity with cycling was not a phenomenon of cyclic softening but an indication of subcritical fatigue crack growth.[3,4] Thus, fatigue lives (N_f) can be determined through a failure criterion defined as a load drop parameter ϕ:

$$\phi = \frac{\Delta P_I - \Delta P_N}{P_I}$$

where ΔP_I is the load amplitude of the first cycle and ΔP_N the load amplitude of the Nth cycle.

As shown in Fig. 14-11*a*, *b*, and *c* for the doped solder at three failure

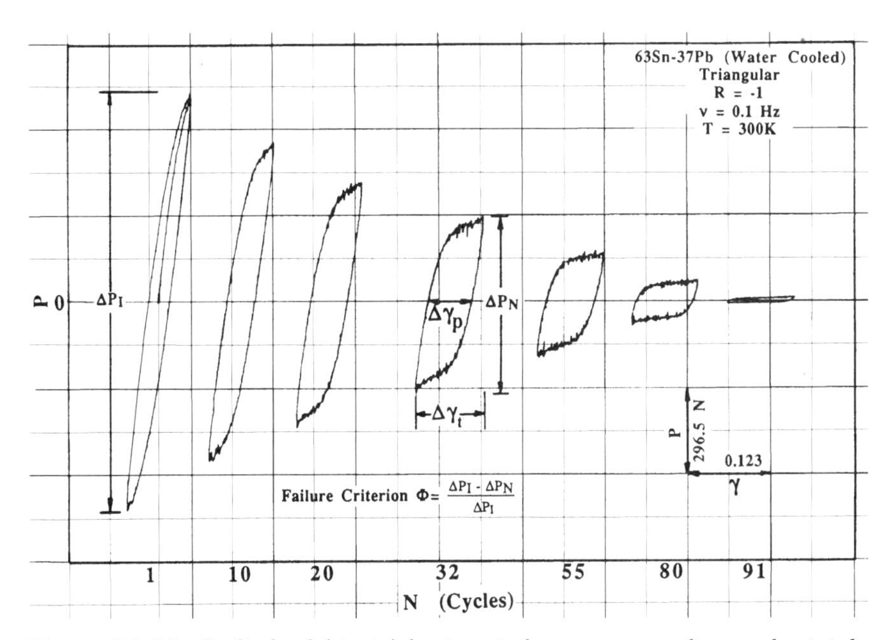

Figure 14-10. Cyclic load (strain) hysteresis loop versus cycling under total strain control.

criteria $\phi = 25$, 50, and 95 percent, respectively, the fatigue life N_f is related to the total strain range $\Delta\gamma_t$ in the form of the Coffin-Manson (C-M) equation as follows:

$$N_f^q \, \Delta\gamma_t = C$$

where q and C are material constants. These different failure criteria reflect the different stages of fatigue crack growth or failure, which in turn provide the indication for engineering design.

The fatigue lives of doped solder joints as evaluated on the C-M curves for the three failure criteria at 0.1 Hz and 300 K do not show significant improvement over those of the standard 63 Sn/37 Pb solder joints. Figure 14-12 shows the reduction of fatigue life of the doped solder at a given total strain range with decreasing cyclic frequency from 0.1 to 10^{-4} Hz. As the cyclic frequency decreases to 10^{-4} Hz, the doped solder experiences less reduction in fatigue life than Pb-Sn eutectic solder as indicated in Fig. 14-13.[4] This may suggest that the low-cycle fatigue resistance of the doped solder would be more enhanced at low-stress or high-temperature conditions.

This study reveals that the fatigue life of the doped solder joints undergoes a transition in the frequency range of 10^{-2} to 10^{-3} Hz. In

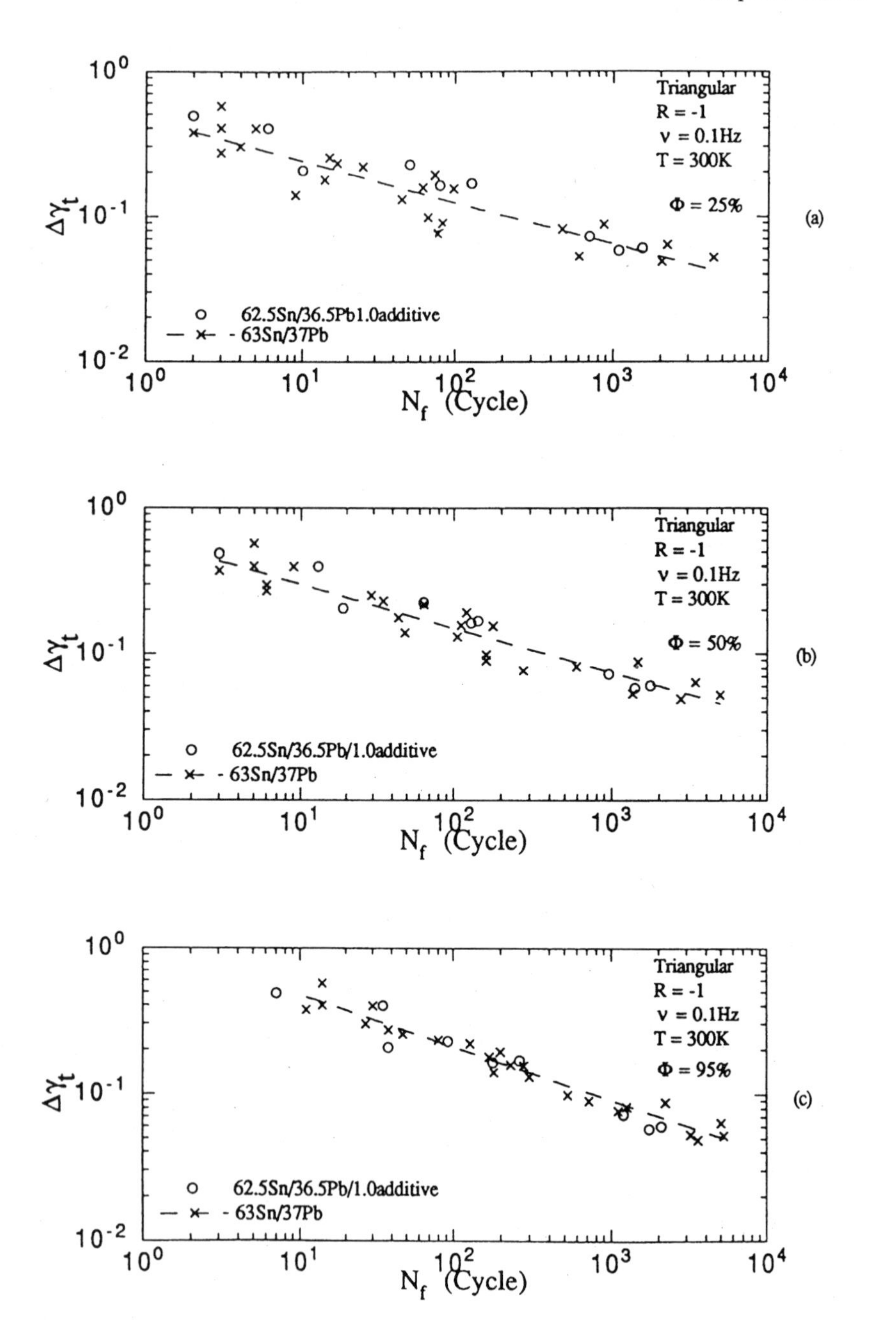

Figure 14-11. Coffin-Manson total shear strain range versus number of cycles at (*a*) $\phi = 25\%$, (*b*) $\phi = 50\%$, (*c*) $\phi = 95\%$.

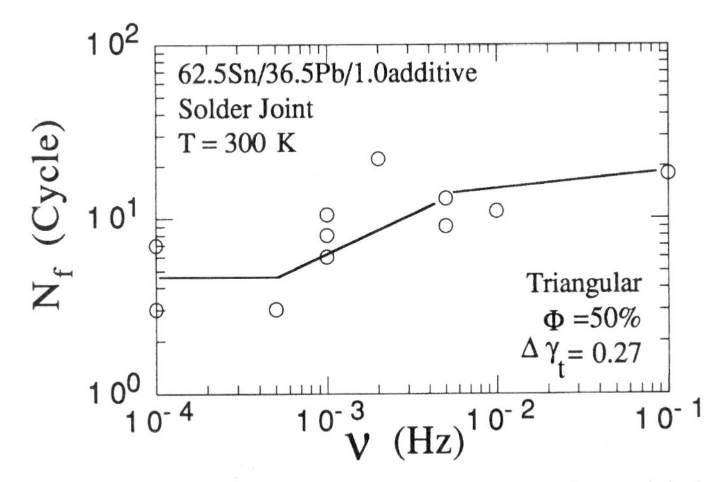

Figure 14-12. Reduction of fatigue life of 62.5 Sn/36.5 Pb/1.0 additive solder joint with decreasing cyclic frequency.

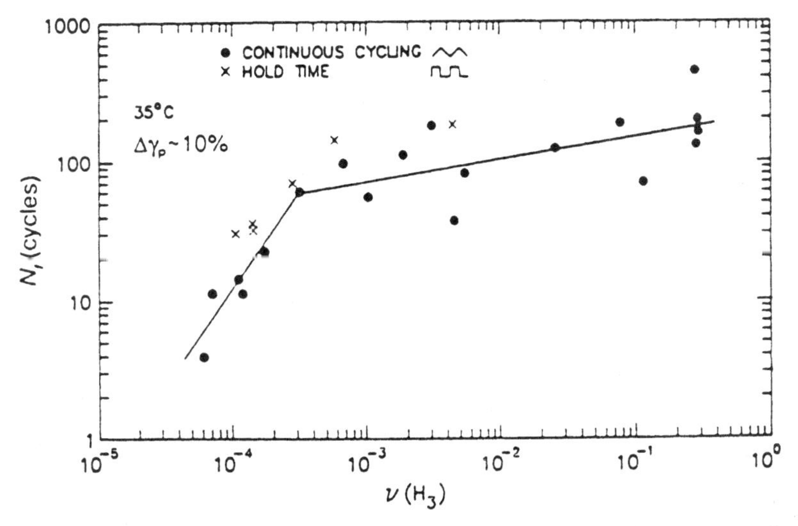

Figure 14-13. Reduction of fatigue life of 63 Sn/37 Pb solder joint with decreasing cyclic frequency.

general, the cyclic frequency at above ~1 Hz and below $\sim 10^{-4}$ Hz has no effect on the number of cycles to failure or fatigue life for most metallic materials. As the frequency decreases from 1 to 10^{-4} Hz, fatigue life generally drops abruptly,[5] although the specific transition point depends on the nature of the materials. It is suggested that the fatigue life saturation at the high level of frequencies is attributed to a thermal failure in nature and that at the low level of frequencies is a thermally activated kinetic process or subcritical crack growth.[6] The fatigue life endurance of the doped solder under this study is improved about 2 times over the 63 Sn/37 Pb solder as the frequency approaches 10^{-4} Hz. This echoes the findings that the doped solder imparts superior plastic deformation resistance to 63 Sn/37 Pb solder. This further suggests that the superplastic grain boundary sliding retards the fatigue crack growth at low frequencies by homogenizing the dislocation distribution caused by cyclic fatigue and by minimizing the deleterious localization of strain into slip bands.

The as-reflowed microstructure of the doped eutectic solder joint is essentially unchanged as compared with the one which has no dopant. The strengthening effect is attributed to the microstructural evolution in response to stress in the scale of submicrostructure. It can be explained by the Orowan dislocation theory as expressed by

$$\tau = \tau_o + \alpha \frac{\mu b}{L}$$

where τ = flow (yield) stress of the material
τ_o = flow stress of unstrengthened matrix
μ = shear modulus
b = Burgers' vector
L = average interparticle spacing
α = geometrical constant

Orowan dislocation bowing and looping at the doping sites and dislocation cross slipping and prismatic looping around the doping sites result in work hardening in the early stage of deformation and, therefore, a higher creep resistance. When stresses are not great enough for Orowan dislocation looping between the sites, the dislocation will bypass the sites through climb, as indicated by the power-law creep for the strengthened solder. The doping site, however, elevates the activation energy for the dislocation movement in the localized field, offering a higher resistance to dislocation climb.

At a lower stress region or correspondingly higher temperature as indicated by the stress exponent of 2, the grain boundary sliding operates as the creep-rate–controlling process. Inasmuch, as the stress decreases, the doping sites in the matrix can be rendered ineffective and the creep resistance of the doped solder thus approaches that of eutectic 63 Sn/37 Pb solder.

The doping additive is believed to affect the distribution of β-Sn precipitates from the Pb-rich oversaturated solid solution. It is further believed that the doping additive promotes the homogeneous nucleation of β-Sn precipitates in the matrix in lieu of the heterogeneous nucleation on grain boundaries which often occurs in nondoped systems. This coincides with the findings in plastic deformation kinetics that superplastic grain boundary sliding is the rate-controlling process for the doped solder.

There is no significant difference in the fracture surface of the early-stage low-cycle fatigue between doped (Fig. 14-14*a* and *b*) and un-

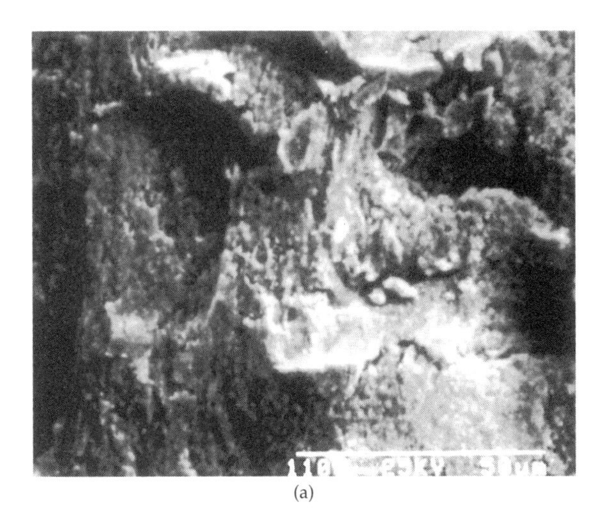

(a)

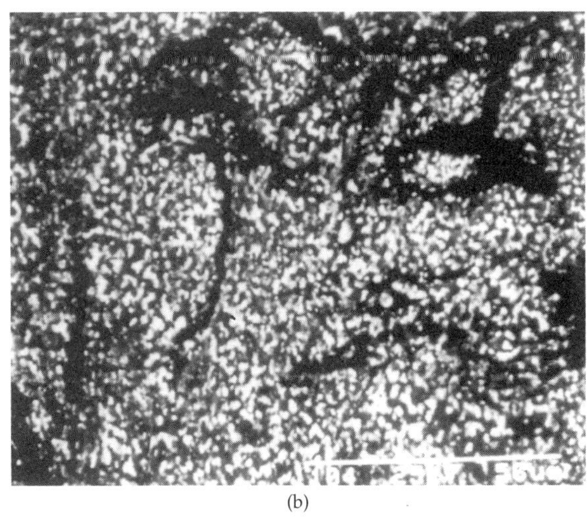

(b)

Figure 14-14. SEM fractograph of early-stage, low-cycle fatigue of 62.5 Sn/36.5 Pb/1.0 additive: (*a*) secondary electron image, (*b*) backscattered electron image.

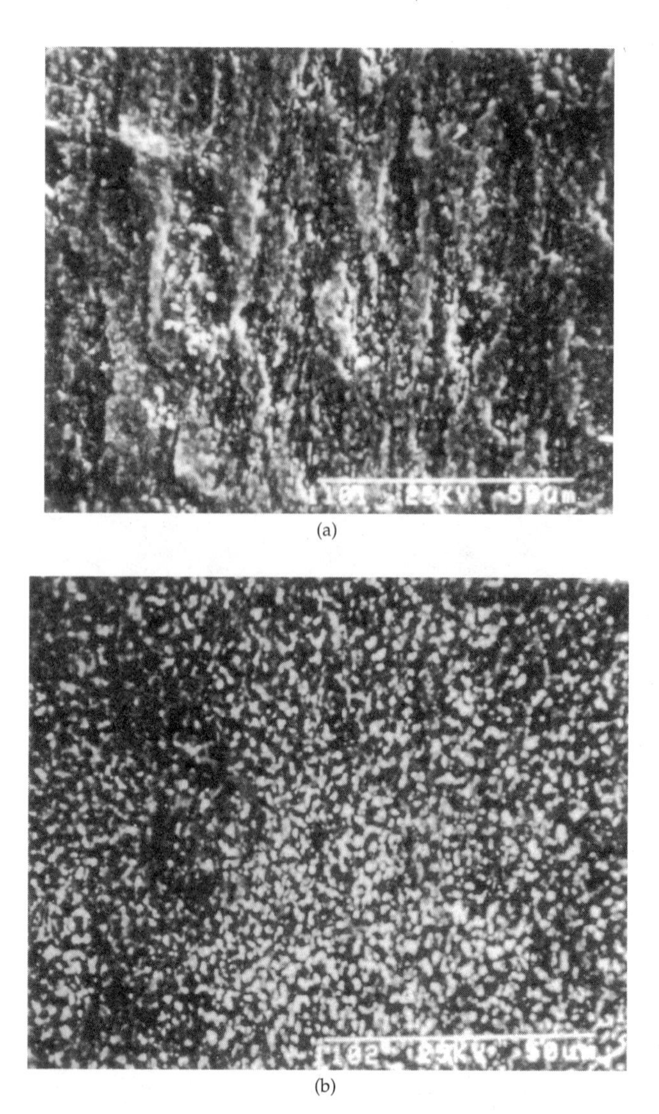

(a)

(b)

Figure 14-15. SEM fractograph of later-stage low-cycle fatigue of 62.5 Sn/36.5 Pb/1.0 additive: (*a*) secondary electron image, (*b*) backscattered electron image.

doped 63 Sn/37 Pb (Fig. 3-43*a* and *b*). However, in the later-stage low-cycle fatigue, no obvious presence of large primary Pb-rich dendrites on the fracture surface of the doped 63 Sn/37 Pb was observed as shown in Fig. 14-15*a* and *b*. This is in direct contrast with conventional 63 Sn/37 Pb as shown in Fig. 3-43*a* and *b*.

14.3 Microstructure Strengthening

To demonstrate the relationship between microstructure and mechanical performance and to identify the strengthening effect in relation to microstructure, the following study was conducted for solder joints (63 Sn/37 Pb on Cu) that were made under identical conditions except for the cooling rate used for solder joint formation during solidification. Solder joints were solidified in five different manners, 0.1°C/s, 1.0°C/s, 50°C/s, 230°C/s, and a two-step (uneven) cooling condition resulting in an average cooling rate of 12°C/s.[7] The microstructures at these cooling rates are illustrated in Sec. 6.4.

In the range of cooling rates performed in this study, the stress-versus-strain curves of 63 Sn/37 Pb solder joints under monotonic shearing exhibited a general pattern that consisted of a negligible elastic region, a short strain-hardening region, and a rapid drop in stress necessary for additional deformation after reaching maximum stress, as shown in Fig. 14-16. In comparison, the highest shear strength was obtained at the slowest cooling rate, 0.1°C/s. As the cooling rate increased from 0.1°C/s to 1.0°C/s, the maximum strength dropped from 45 to 37 MPa. At cooling rates faster than 1°C/s, the shear strength was generally insensitive to the variation of cooling rate. However, as the cooling rate increased to 230°C/s, the maximum strength was regained to 43 MPa. Figure 14-17 depicts the relationship between maximum strength and cooling rate. Creep curves of 63

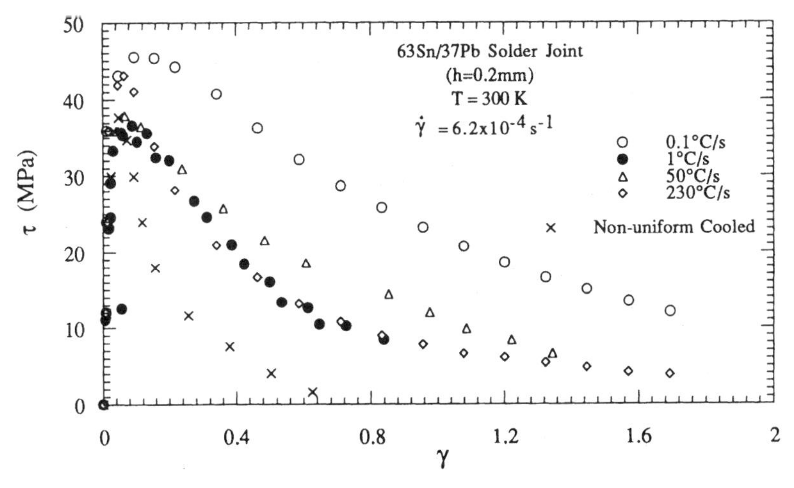

Figure 14-16. Stress versus strain curve of 63 Sn/37 Pb solder joints made with different cooling rates.

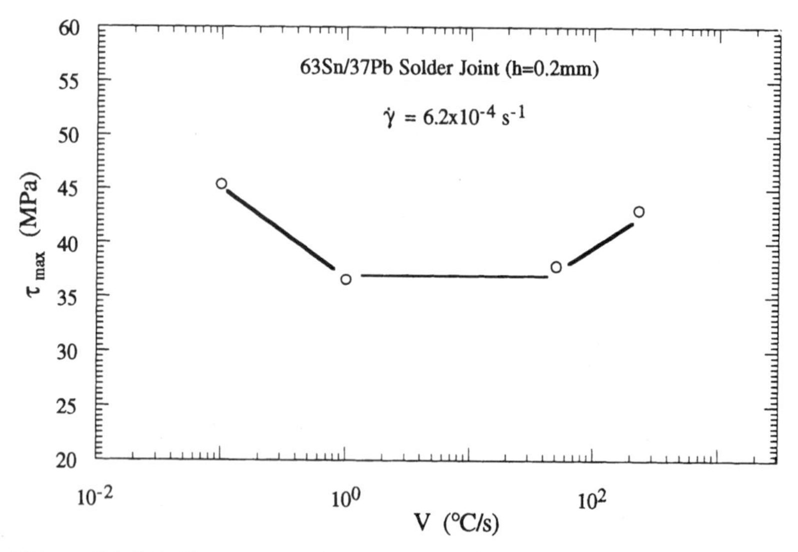

Figure 14-17. Shear strength maximum of 63 Sn/37 Pb versus cooling rate.

Sn/37 Pb solder joints formed at different cooling rates over a range of stresses are shown in Figs. 14-18 to 14-22.

The relationship between the plastic strain rate as derived from creep curves and the stress at five different cooling rates is summarized in Fig. 14-23. It is indicative that the plastic strain rate increased with increasing cooling rate, which, in turn, illustrates that the creep resistance decreased with increasing cooling rate. Based on Fig. 14-23, the stress exponent n falls in the range of 3 to 10 for all cooling rates 0.1 to 230°C/s. Figure 14-24 shows the relationship of the stress exponent with the cooling rate.

The lowest value of n is around 4, suggesting that the plastic deformation mechanism for the fast cooling rate of 230°C/s still follows the power-law creep and that the dislocation climb/glide with lattice or vacancy diffusion is a rate-controlling step. As the cooling rate increases, the vacancy concentration increases, and thus creep resistance decreases. This is in contrast to the previous findings[8] suggesting that there is a superplastic deformation mechanism involved for the solder joint that is cooled very quickly.

As shown in Fig. 14-25, the low-cycle fatigue life decreases as the cooling rate increases from 1.0 to 230°C/s and as the cooling rate decreases from 1.0 to 0.1°C/s.

Fracture surfaces of the early-stage fatigue crack growth of solder joints formed at a slow cooling rate (0.1°C/s) exhibit larger rough

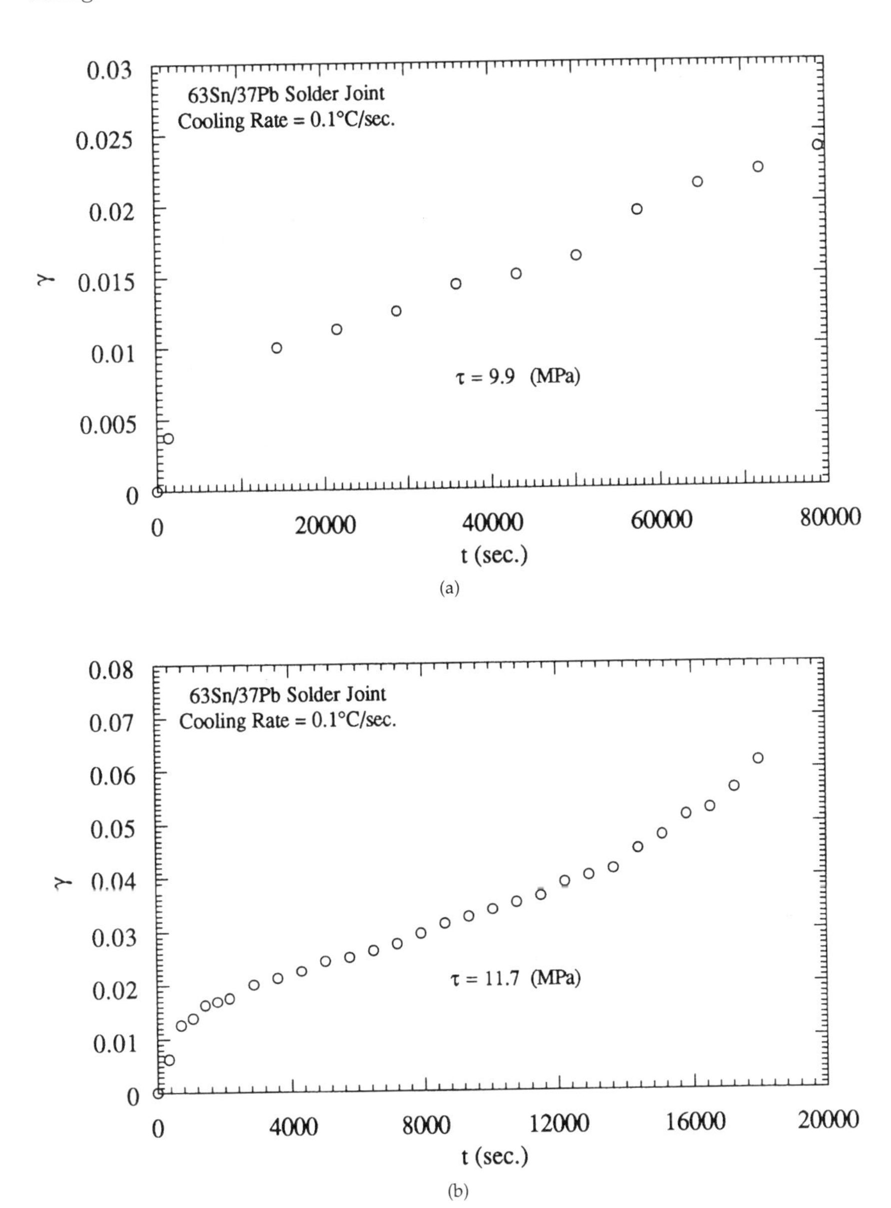

Figure 14-18. Creep curve of 63 Sn/37 Pb solder joint formed at 0.1°C/s under (*a*) 9.9 MPa, (*b*) 11.7 MPa, (*c*) 22.9 MPa, (*d*) 36.5 MPa, (*e*) 50.4 MPa.

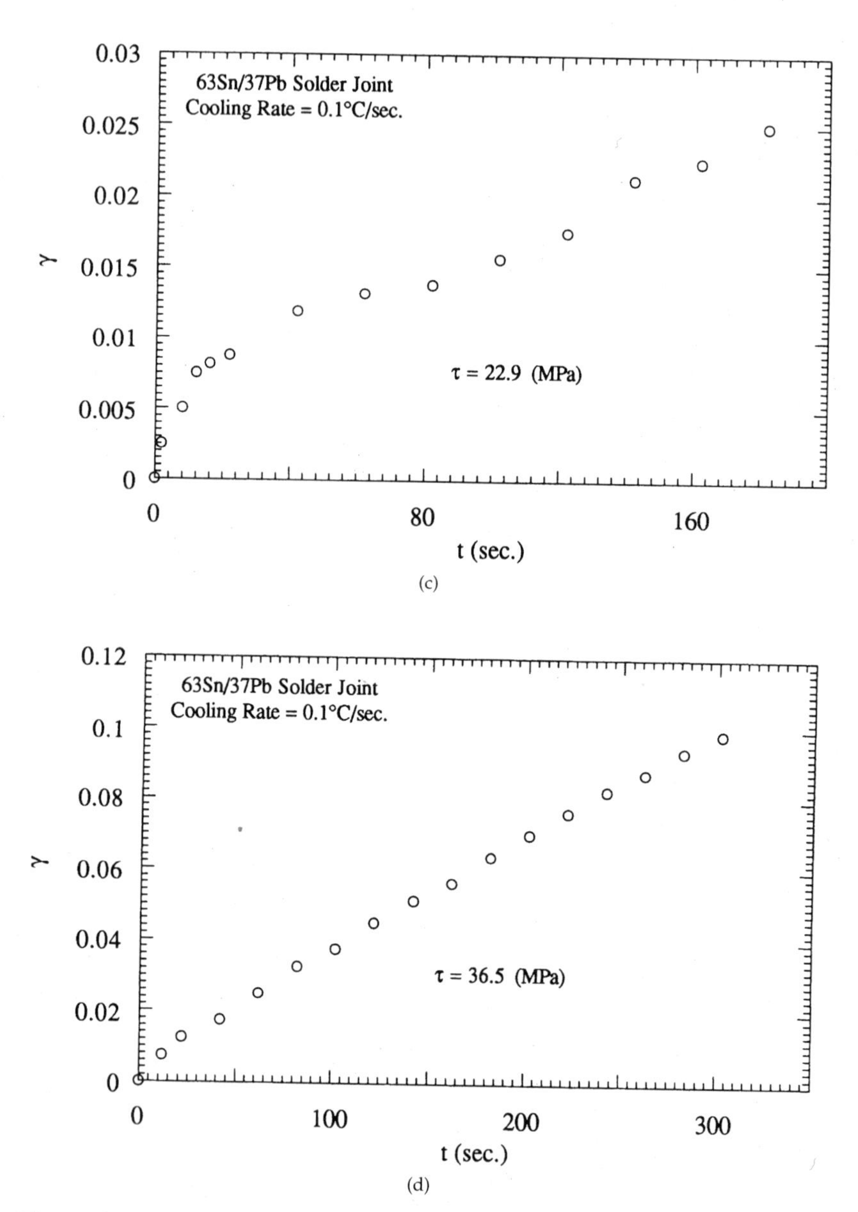

Figure 14-18. (*Continued*) Creep curve of 63 Sn/37 Pb solder joint formed at 0.1°C/s under (*a*) 9.9 MPa, (*b*) 11.7 MPa, (*c*) 22.9 MPa, (*d*) 36.5 MPa, (*e*) 50.4 MPa.

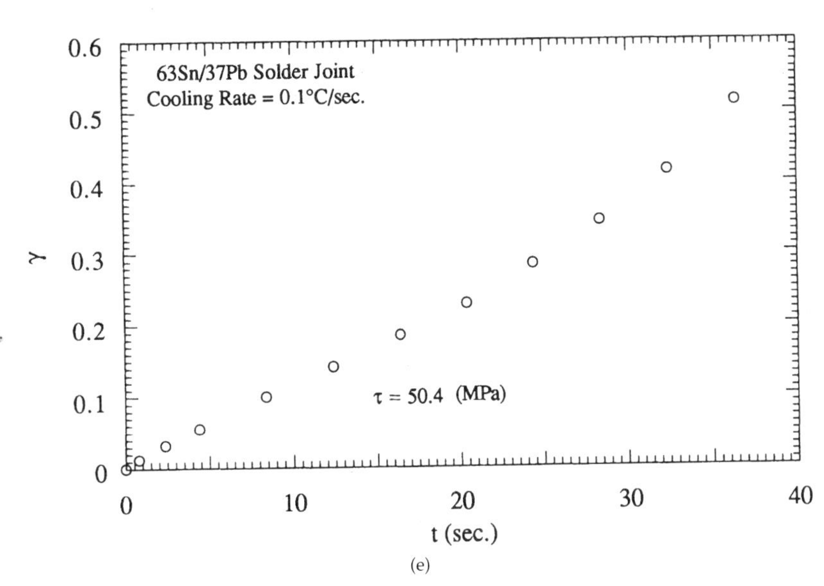

Figure 14-18. (*Continued*) Creep curve of 63 Sn/37 Pb solder joint formed at 0.1°C/s under (*a*) 9.9 MPa, (*b*) 11.7 MPa, (*c*) 22.9 MPa, (*d*) 36.5 MPa, (*e*) 50.4 MPa.

facets (Fig. 14-26*a* and *b*) than those formed at faster cooling rates. This is due to the presence of large eutectic colony size which leads to a faster overall crack rate at lower strain range. At later-stage crack growth or a higher strain range, the fatigue crack growth become transgranular without the preference of microstructure (Fig. 14-27*a* and *b*). The fracture surface is irregularly rough and the secondary cracks are observed.

It is further observed that the rough facets reconciled with the dimension of eutectic colonies and the secondary cracks were associated with the colony boundaries, suggesting that fatigue fracture units or plastic zones at the tip of fatigue cracks were limited by the scale of the eutectic colony at the early stage of crack growth.

Fatigue fracture surfaces at the later stage of fatigue crack growth appeared to be flat and with a shear-deformed morphology as shown in Fig. 14-27*a*. This indicated that fatigue cracks propagated transgranularly as confirmed by electron backscattering (EB) micrograph in Fig. 14-27*b*.

As a fatigue crack grows longer, it is expected that the plastic zone at the tip of the fatigue crack is greater and on a larger scale than the dimension of eutectic colonies. Thus, the fatigue crack propagates without the preference of microstructure.

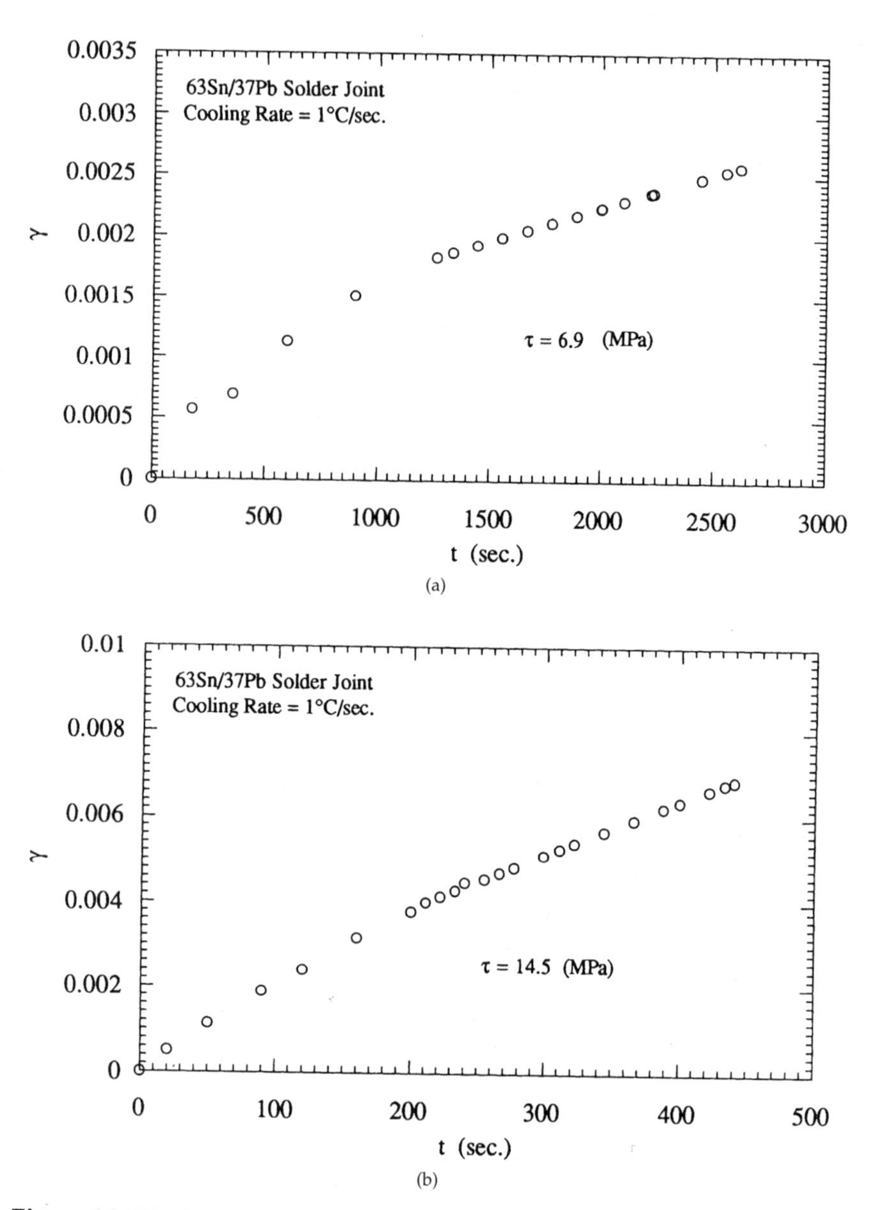

Figure 14-19. Creep curve of 63 Sn/37 Pb solder joint formed at 1.0°C/s under (*a*) 6.9 MPa, (*b*) 14.5 MPa, (*c*) 22.9 MPa, (*d*) 31.4 MPa, (*e*) 39.7 MPa.

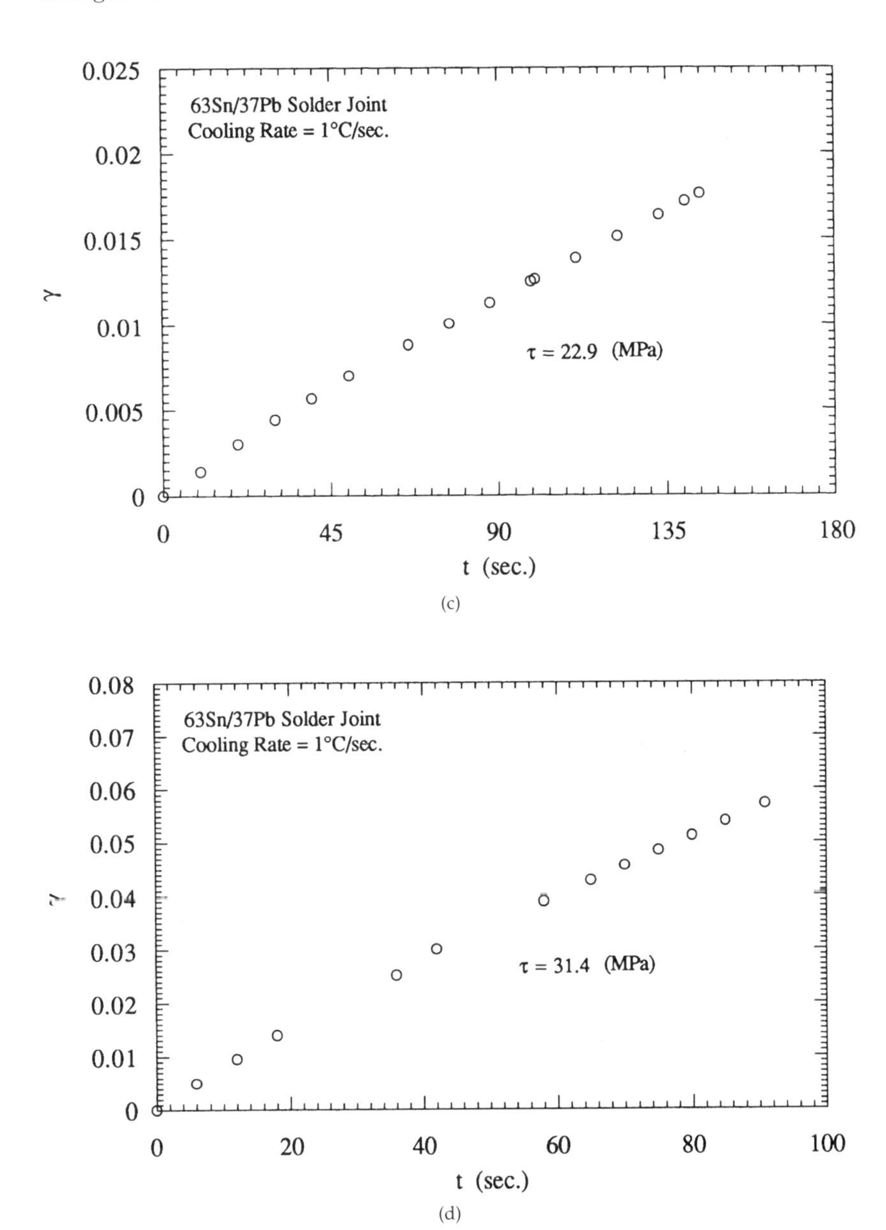

Figure 14-19. (*Continued*) Creep curve of 63 Sn/37 Pb solder joint formed at 1.0°C/s under (*a*) 6.9 MPa, (*b*) 14.5 MPa, (*c*) 22.9 MPa, (*d*) 31.4 MPa, (*e*) 39.7 MPa.

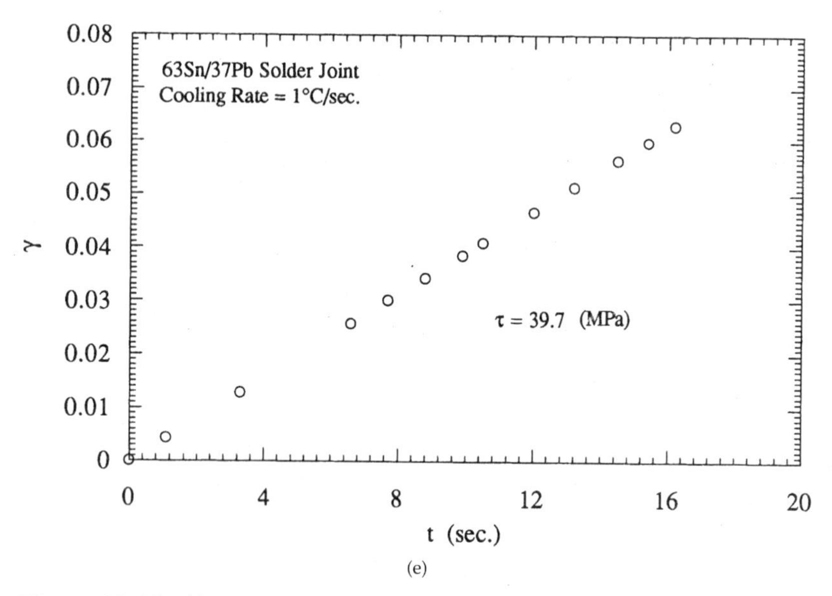

Figure 14-19. (*Continued*) Creep curve of 63 Sn/37 Pb solder joint formed at 1.0°C/s under (*a*) 6.9 MPa, (*b*) 14.5 MPa, (*c*) 22.9 MPa, (*d*) 31.4 MPa, (*e*) 39.7 MPa.

It should be noted that the large primary Pb-rich phase which constitutes the coarsened bands of the as-reflowed microstructure also appeared on the fracture surface as shown by the light phase of Fig. 14-27*b*. This indicates that the fatigue crack paths were indeed through the coarsened bands near Cu-solder interfaces.

As the cooling rate increased from 0.1 to 1.0°C/s, the solder joint fracture surface at the early stage of the fatigue crack possessed smaller rough facets when comparing Fig. 14-28*a* with Fig. 14-26*a*, due to the smaller colony size. Figure 14-29*a* and *b* shows the fracture surfaces of solder joints at a cooling rate of 1.0°C/s at later crack growth. The result is that the fatigue life of a solder joint at this cooling rate of 0.1°C/s approaches that at a cooling rate of 1.0°C/s. This indicates that the fatigue crack growth stage was transgranular without the preference of microstructure at a later crack growth stage as shown in Fig. 14-27*b* and 4-29*b*.

As the cooling rate increased to 230°C/s, the fracture surface as shown in Fig. 14-30*a* was very smooth with no rough morphology, which is consistent with what was observed from the microstructure of the solder joint. The fracture surface again exhibited Pb-rich dendrites as shown in Fig. 14-30*b*, indicating that the fatigue crack paths were also through the coarsened bands.

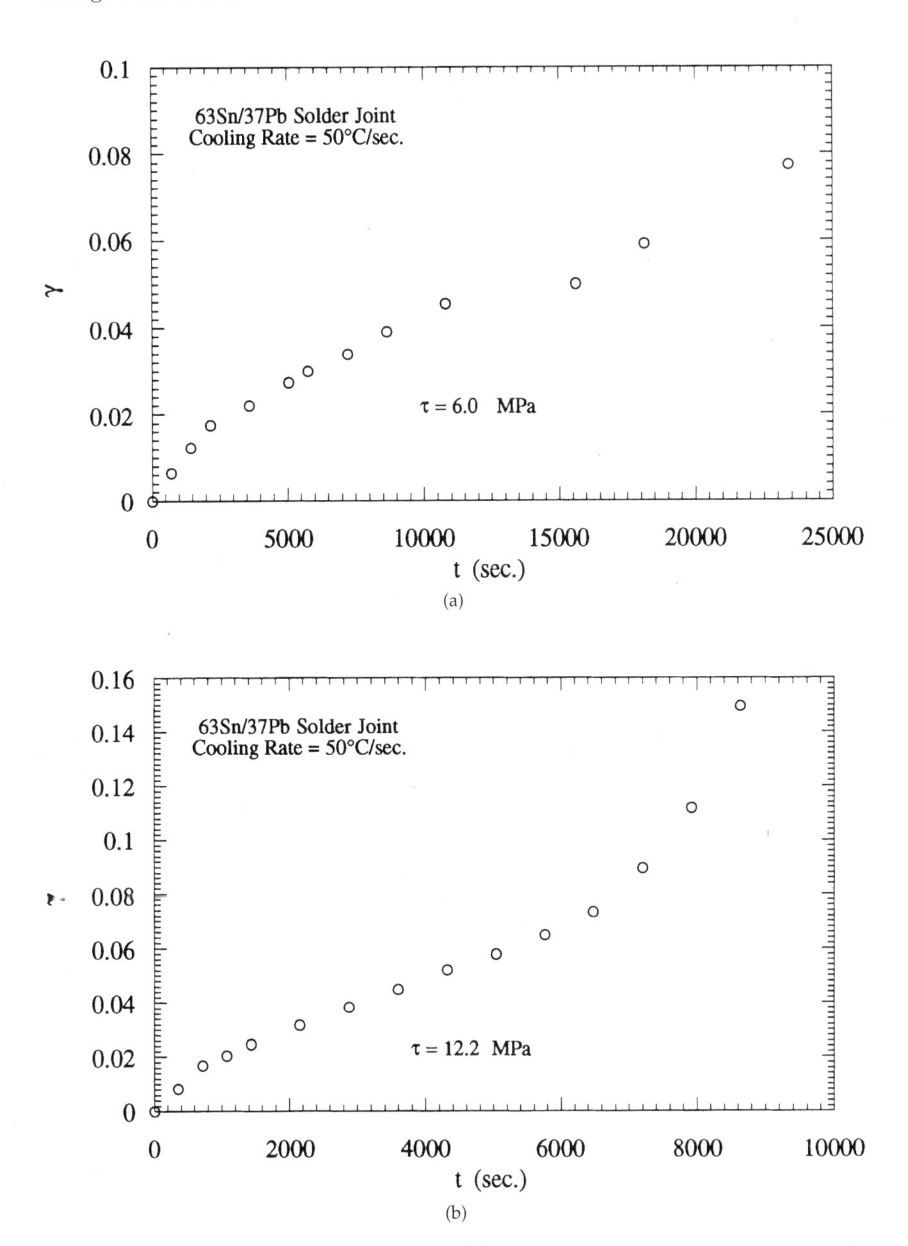

Figure 14-20. Creep curve of 63 Sn/37 Pb solder joint formed at 50°C/s under (*a*) 6.0 MPa, (*b*) 12.2 MPa, (*c*) 23.9 MPa, (*d*) 38.4 MPa, (*e*) 44.4 MPa.

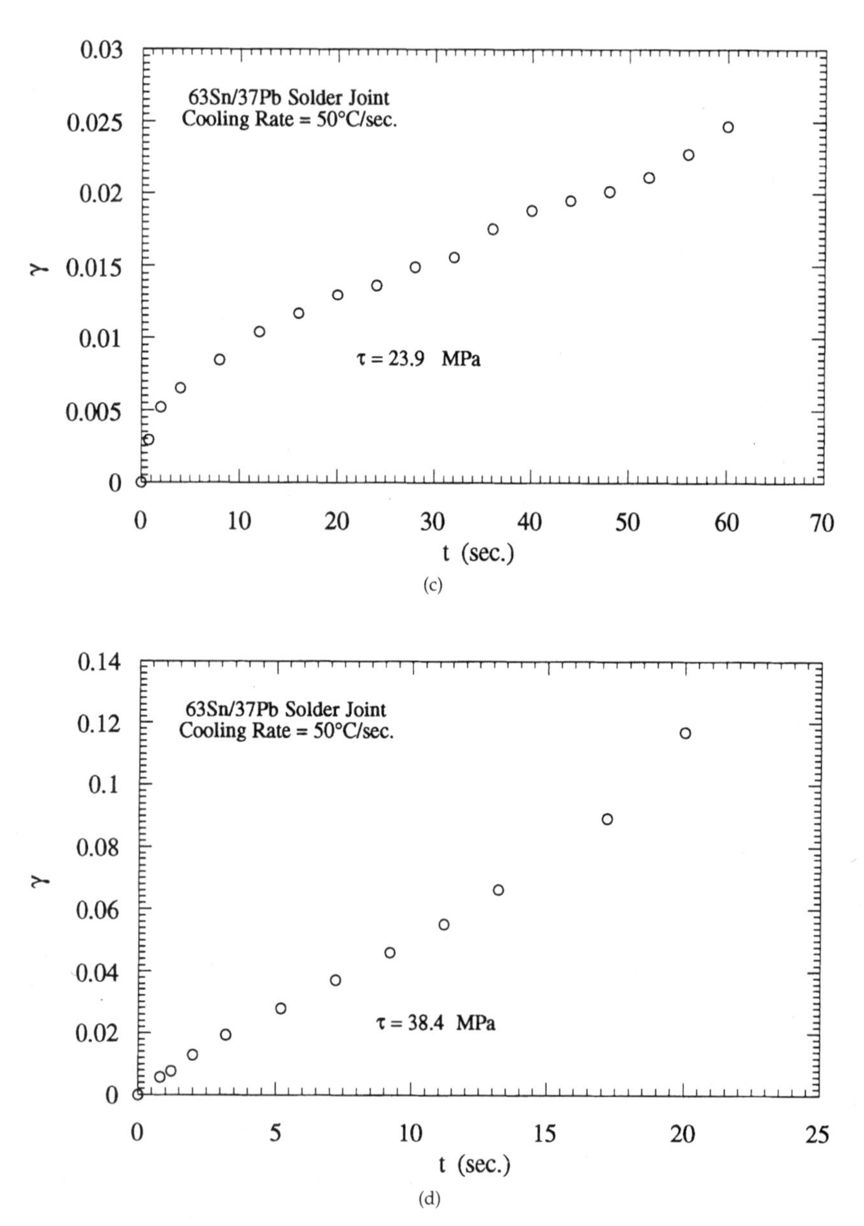

Figure 14-20. (*Continued*) Creep curve of 63 Sn/37 Pb solder joint formed at 50°C/s under (*a*) 6.0 MPa, (*b*) 12.2 MPa, (*c*) 23.9 MPa, (*d*) 38.4 MPa, (*e*) 44.4 MPa.

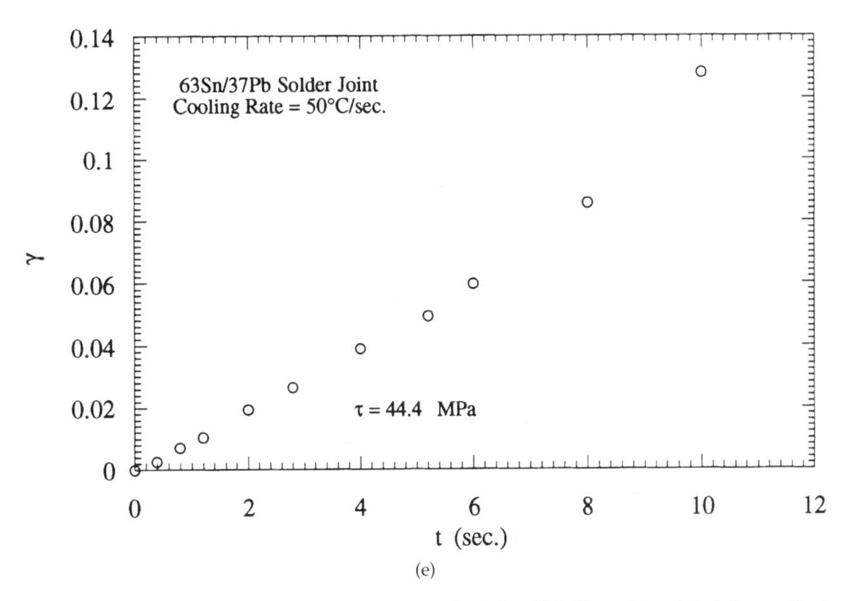

Figure 14-20. (*Continued*) Creep curve of 63 Sn/37 Pb solder joint formed at 50°C/s under (*a*) 6.0 MPa, (*b*) 12.2 MPa, (*c*) 23.9 MPa, (*d*) 38.4 MPa, (*e*) 44.4 MPa.

The solidification of the solder joint through two-step (nonuniform) cooling resulted in a highly inhomogeneous microstructure in the solder joint as shown in Fig. 14-31*a* and *b* for the center of the solder joints and in Fig. 14-32*a* and *b* for the area near the free surface of the solder joint. The gradient in the microstructure across the solder joint was noted by comparing Fig. 14-31*a* with Fig. 14-32*a*, and Fig. 14-31*b* with Figure 14-32*b*. The microstructure near the center of the solder joint was closer to that at a slower cooling rate, and the microstructure near the surface was closer to that at faster cooling rate.

The stress versus strain of the two-step cooled solder joint is shown in Fig. 14-16. Note that the flow stress after τ_{max} decreased much faster than those solder joints made under other cooling conditions. This suggests that the uneven cooling may have promoted both cracking and dynamic recovery.

Low-cycle fatigue fractography reveals that the fatigue failure mode is associated with eutectic colony boundaries. Therefore, the very slow cooling rate (0.1°C/s) which facilitates the development of eutectic colonies produced solder joints with inferior fatigue resistance. At an extremely fast cooling rate (230°C/s), the formation of an inhomogeneous mixture of irregularly branched Pb-rich dendrites and the two-

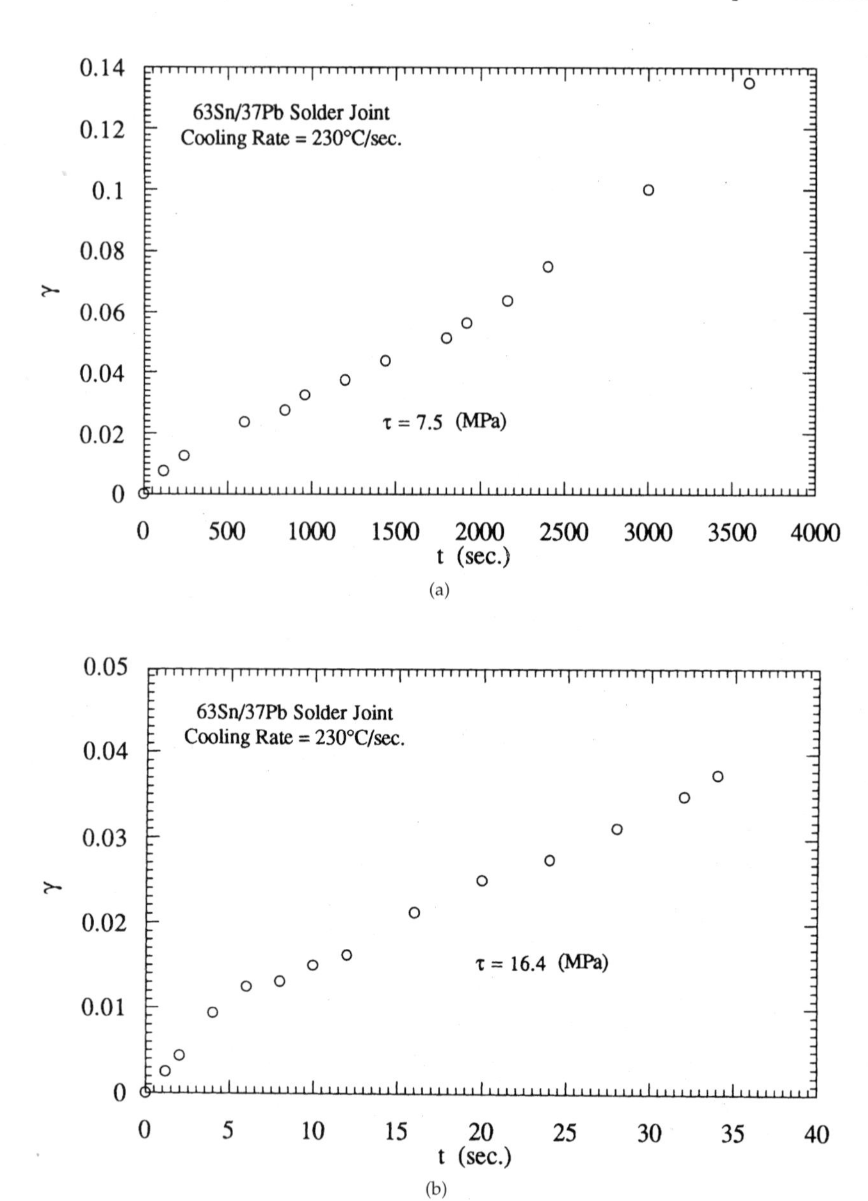

Figure 14-21. Creep curve of 63 Sn/37 Pb solder joint formed at 230°C/s under (*a*) 7.5 MPa, (*b*) 16.4 MPa, (*c*) 22.7 MPa, (*d*) 31.0 MPa, (*e*) 43.9 MPa.

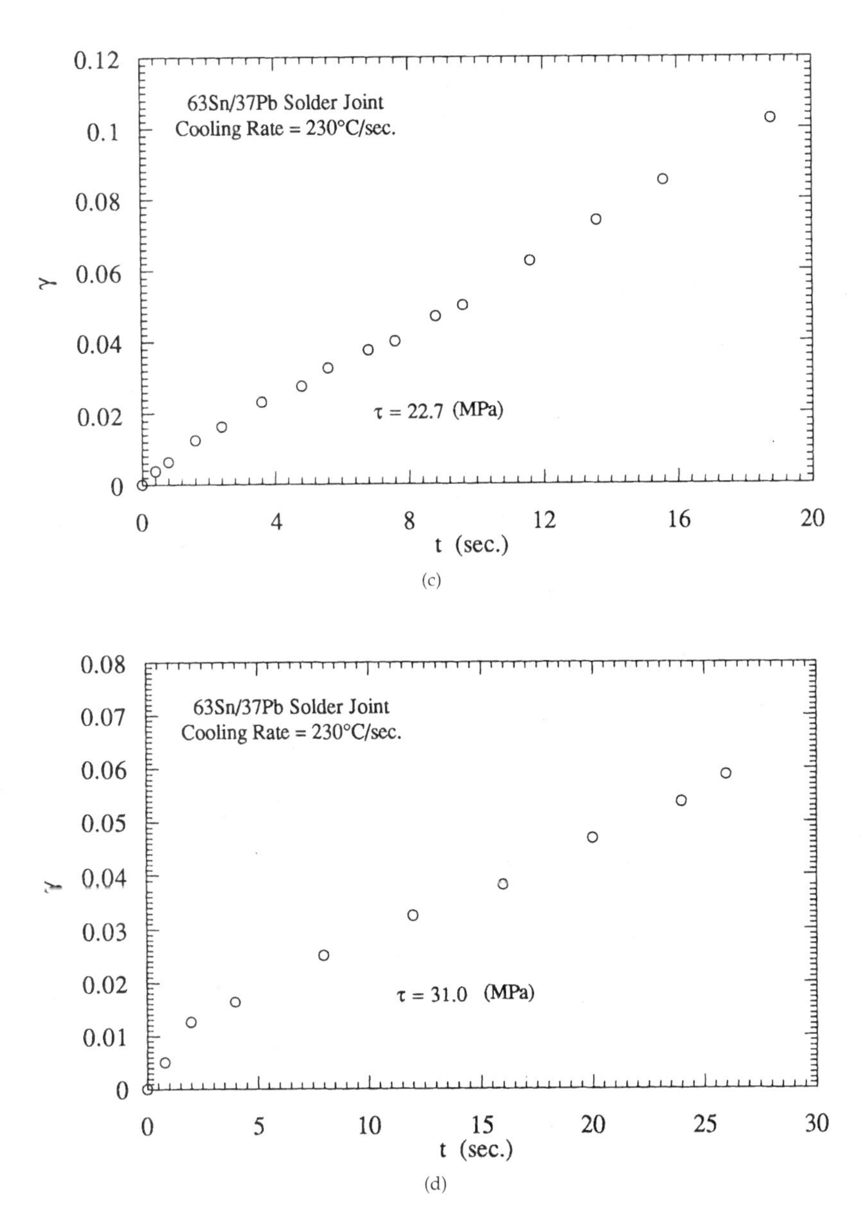

Figure 14-21. (*Continued*) Creep curve of 63 Sn/37 Pb solder joint formed at 230°C/s under (*a*) 7.5 MPa, (*b*) 16.4 MPa, (*c*) 22.7 MPa, (*d*) 31.0 MPa, (*e*) 43.9 MPa.

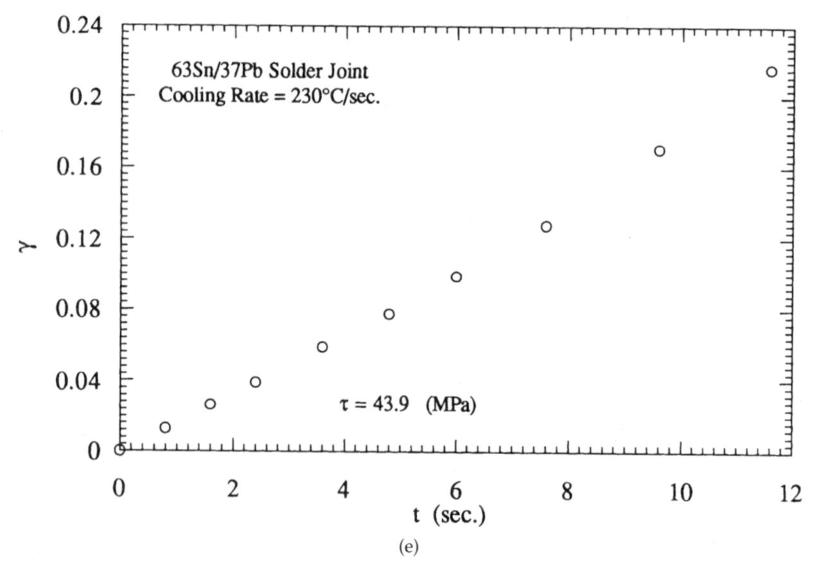

Figure 14-21. (*Continued*) Creep curve of 63 Sn/37 Pb solder joint formed at 230°C/s under (*a*) 7.5 MPa, (*b*) 16.4 MPa, (*c*) 22.7 MPa, (*d*) 31.0 MPa, (*e*) 43.9 MPa.

phase mixture is considered to be the culprit for the inferior low-cycle fatigue resistance. In addition, the likely buildup of elastic strain field and vacancy clusters could also be the contributor to the reduction of low-cycle fatigue resistance. Consequently, the balance of smaller eutectic colonies and the moderate two-phase mixture as can be obtained by a moderate cooling rate (1.0°C/s), is the basis to optimize the low-cycle fatigue resistance of solder joints at ambient temperature.

Under the thermomechanical fatigue mode, the crack path to the thermally activated microstructures of solder joints is found to go without preference through Sn-rich matrix grains, the Pb-rich second phase, as well as the interface boundaries as shown in Fig. 14-33 for the cooling rate of 0.1°C/s after 269 cycles (−30 to 130°C, 12 min/12 min dwell time). Figure 14-34 exhibits the thermally fatigued solder joint (353 cycles, −30 to 130°C, 12 min/12 min) which was formed at a cooling rate of 230°C/s. It is also found that cracks are initiated predominantly at the free surface of the solder joint. By comparing Fig. 14-33 with Fig. 14-34, it is indicative that the kinetic process in microstructure damaging is slower for finer microstructures. It is postulated that the thermomechanical fatigue-induced recrystallization and grain growth is a diffusion-controlled phase transformation. A

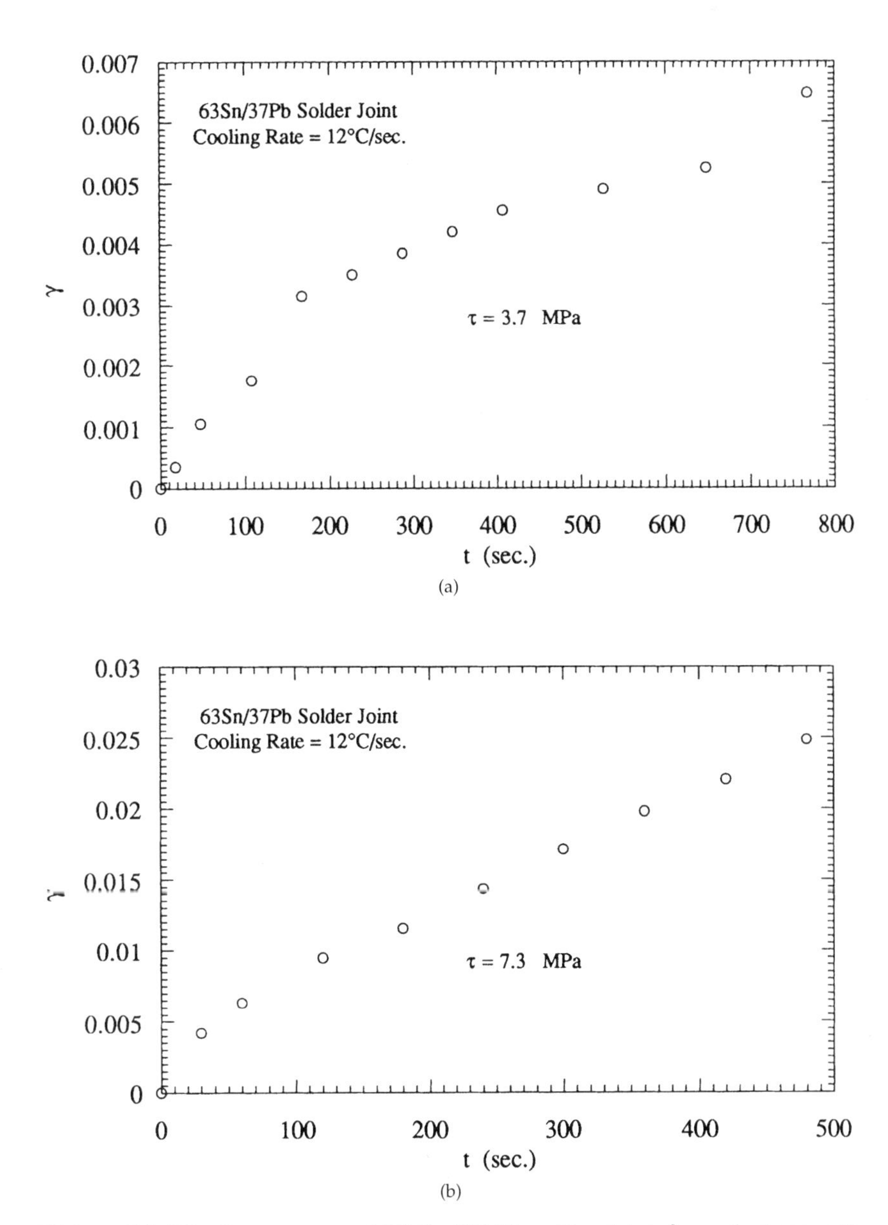

Figure 14-22. Creep curve of 63 Sn/37 Pb solder joint formed at average 12°C/s under (*a*) 3.7 MPa, (*b*) 7.3 MPa, (*c*) 12.2 MPa, (*d*) 25.1 MPa, (*e*) 33.3 MPa.

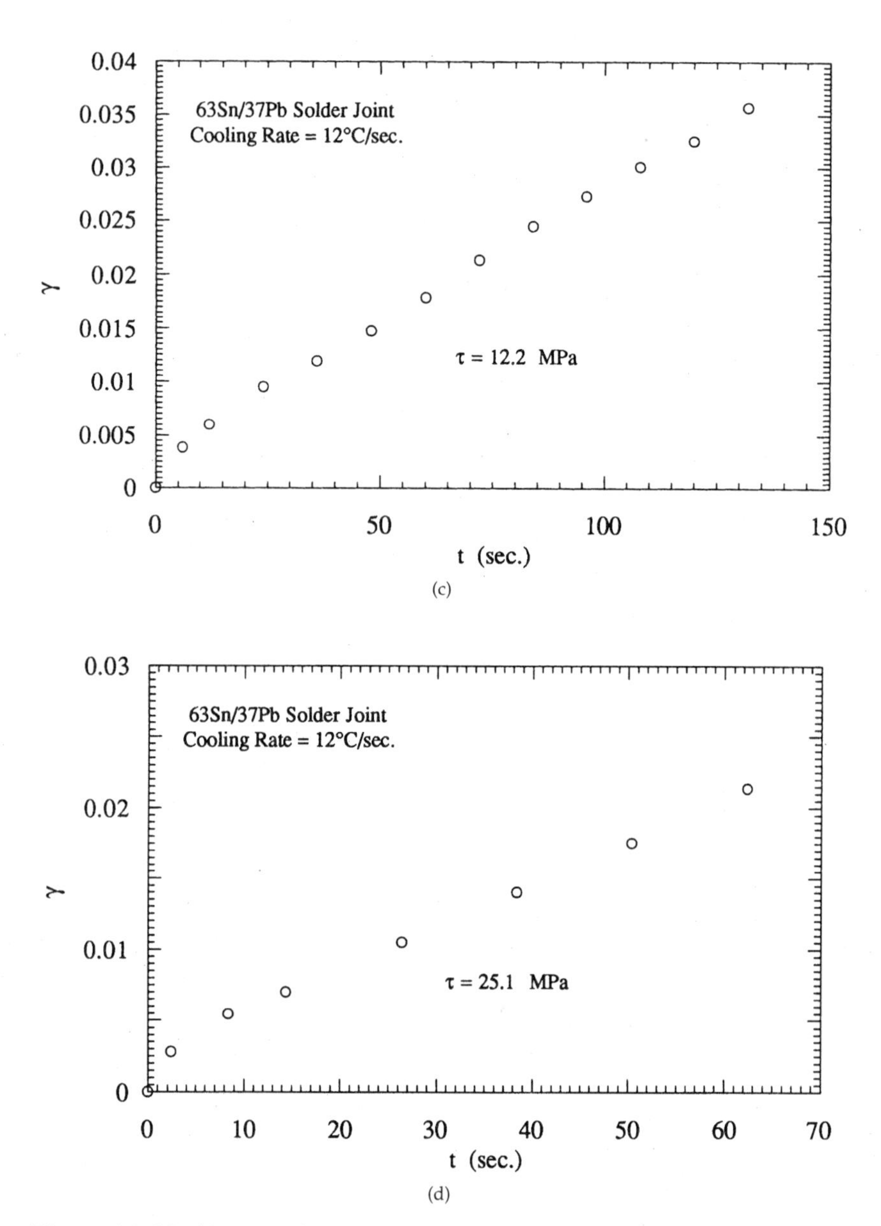

Figure 14-22. (*Continued*) Creep curve of 63 Sn/37 Pb solder joint formed at average 12°C/s under (*a*) 3.7 MPa, (*b*) 7.3 MPa, (*c*) 12.2 MPa, (*d*) 25.1 MPa, (*e*) 33.3 MPa.

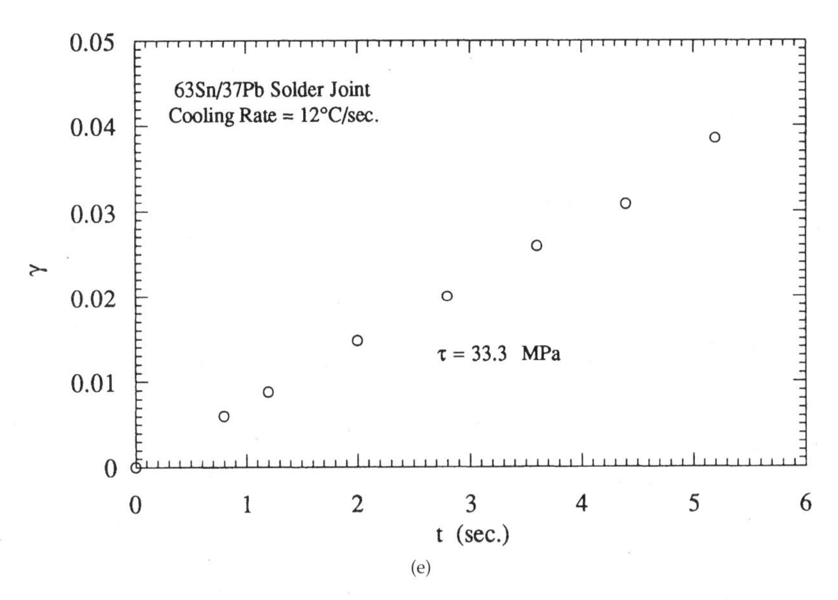

Figure 14-22. (*Continued*) Creep curve of 63 Sn/37 Pb solder joint formed at average 12°C/s under (*a*) 3.7 MPa, (*b*) 7.3 MPa, (*c*) 12.2 MPa, (*d*) 25.1 MPa, (*e*) 33.3 MPa.

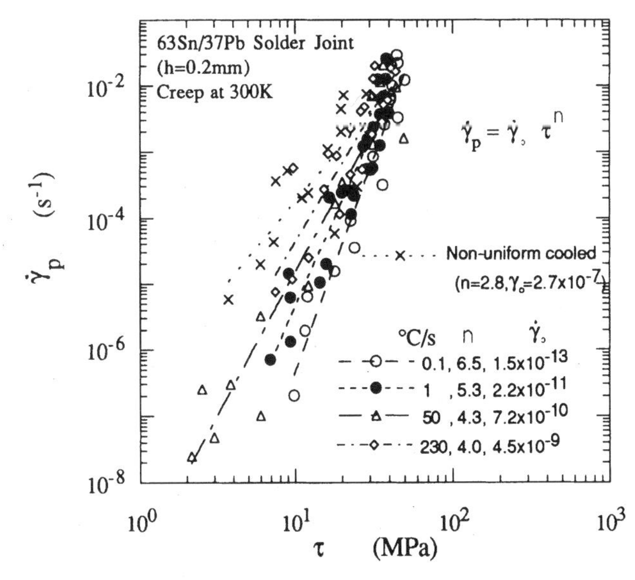

Figure 14-23. Plastic deformation rate versus stress for 63 Sn/37 Pb solder joints formed at various cooling rates.

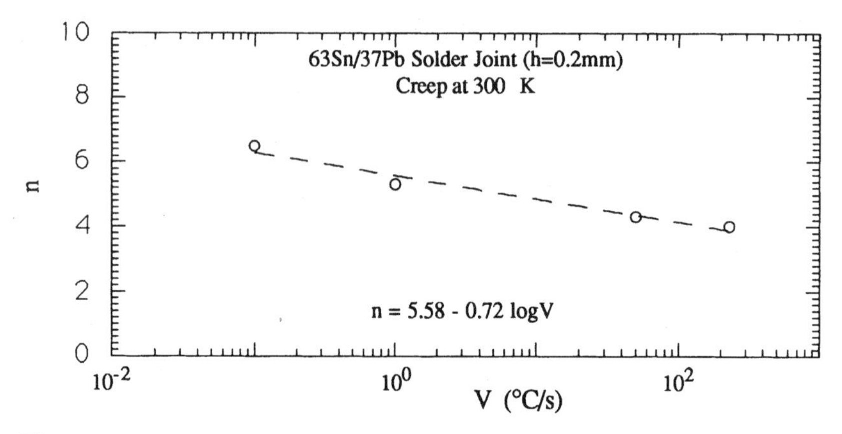

Figure 14-24. Stress exponent versus cooling rate used to form 63 Sn/37 Pb solder joints.

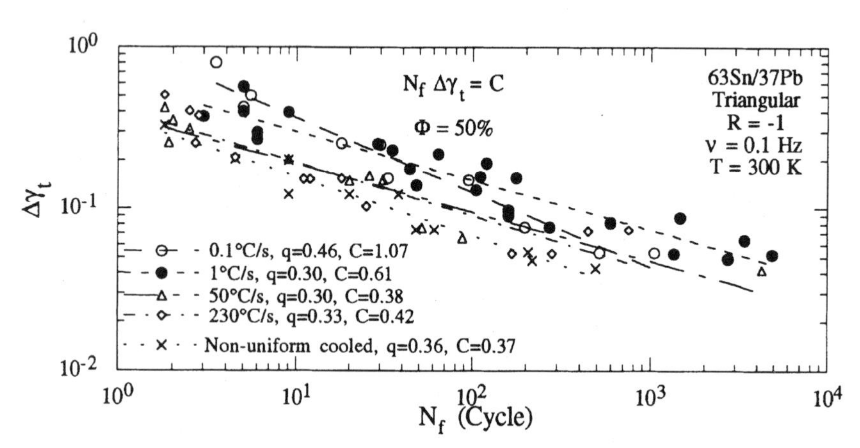

Figure 14-25. Fatigue life of solder joints formed at various cooling rate of ϕ = 50 percent.

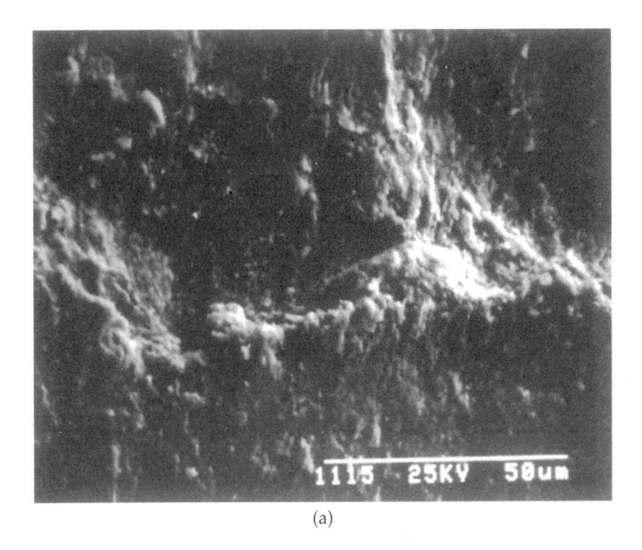

(a)

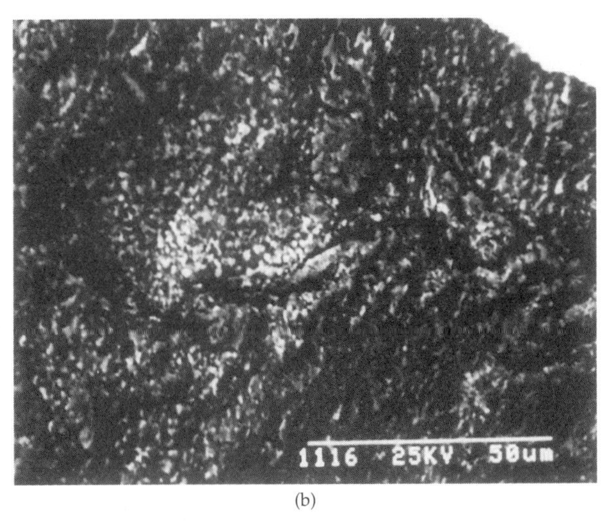

(b)

Figure 14-26. SEM fractograph of 63 Sn/37 Pb solder joint formed at 0.1°C/s at early-stage, low-cycle fatigue: (*a*) secondary electron image, and (*b*) backscattered electron image.

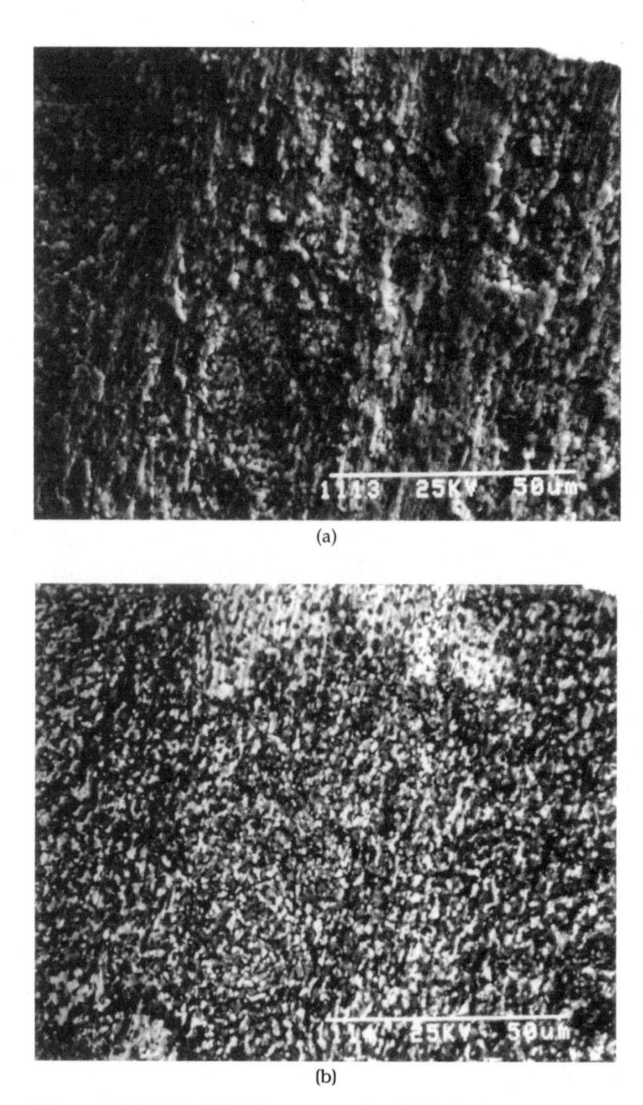

Figure 14-27. SEM fractograph of 63 Sn/37 Pb solder joint formed at 0.1°C/s at later-stage, low-cycle fatigue: (*a*) secondary electron image, and (*b*) backscattered electron image.

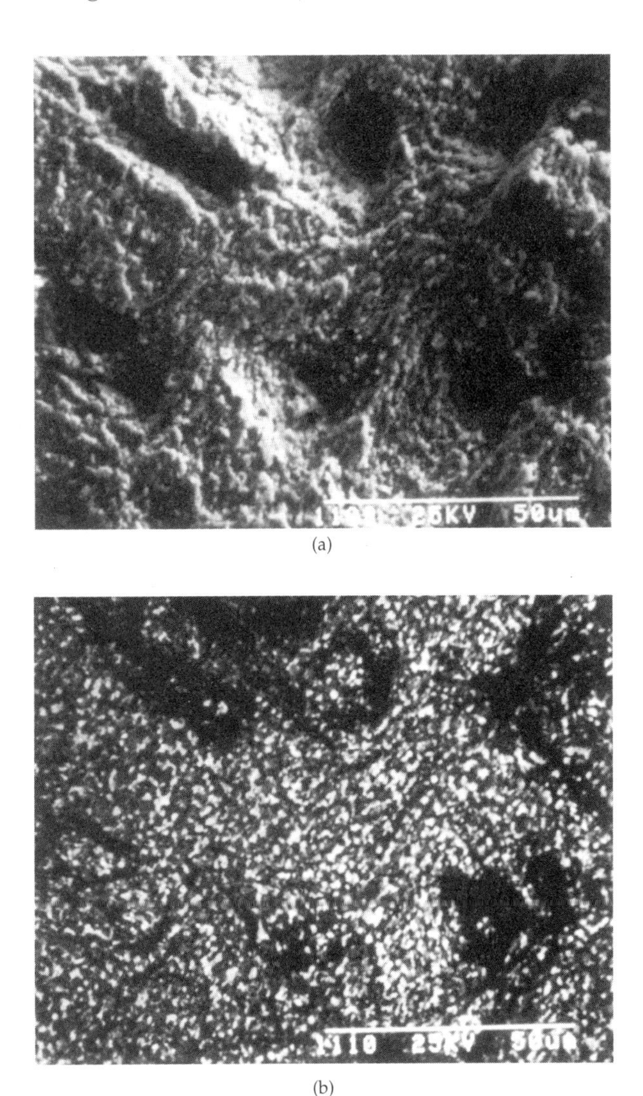

(a)

(b)

Figure 14-28. SEM fractograph of 63 Sn/37 Pb solder joint formed at 1.0°C/s at early-stage, low-cycle fatigue: (*a*) secondary electron image, and (*b*) backscattered electron image.

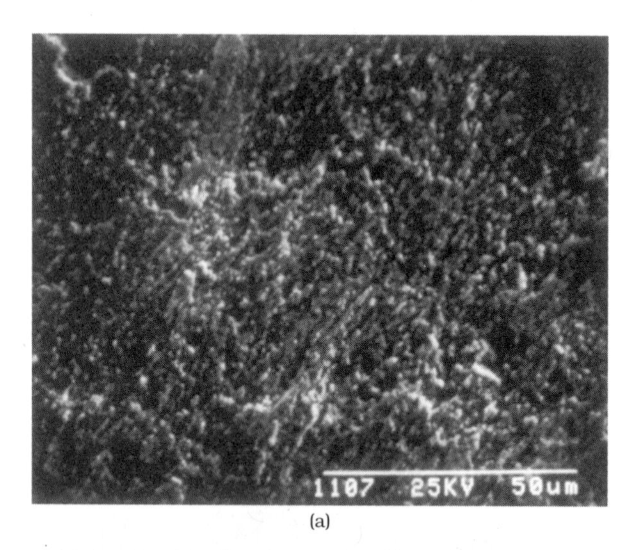

(a)

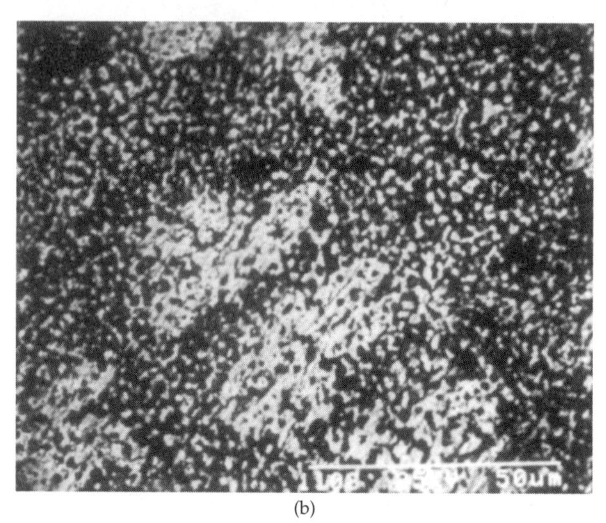

(b)

Figure 14-29. SEM fractograph of 63 Sn/37 Pb solder joint formed at 1.0°C/s at later-stage, low-cycle fatigue: (*a*) secondary electron image, and (*b*) backscattered electron image.

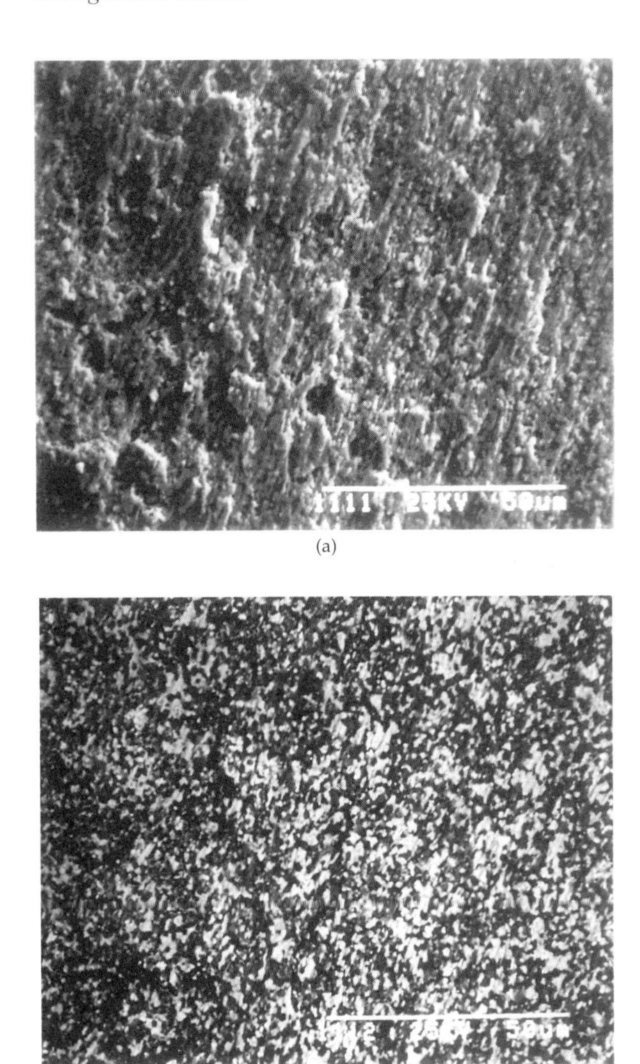

(a)

(b)

Figure 14-30. SEM fractograph of 63 Sn/37 Pb solder joint formed at 230°C/s at early-stage, low-cycle fatigue: (*a*) secondary electron image, and (*b*) backscattered electron image.

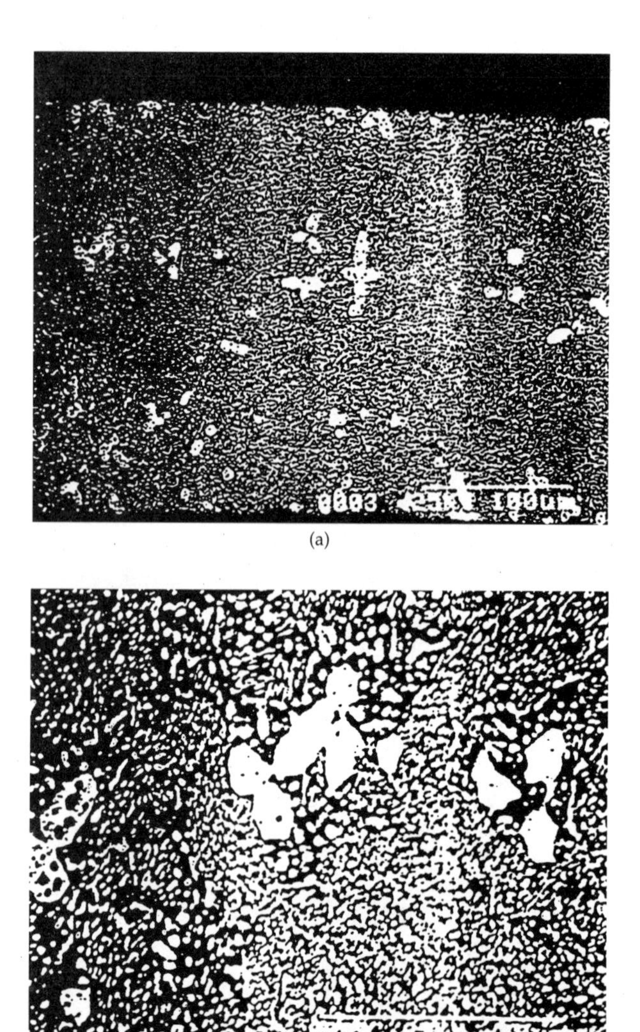

(a)

(b)

Figure 14-31. SEM microstructure near the center of the solder joint with two-step cooling: (*a*) secondary electron image, and (*b*) backscattered electron image.

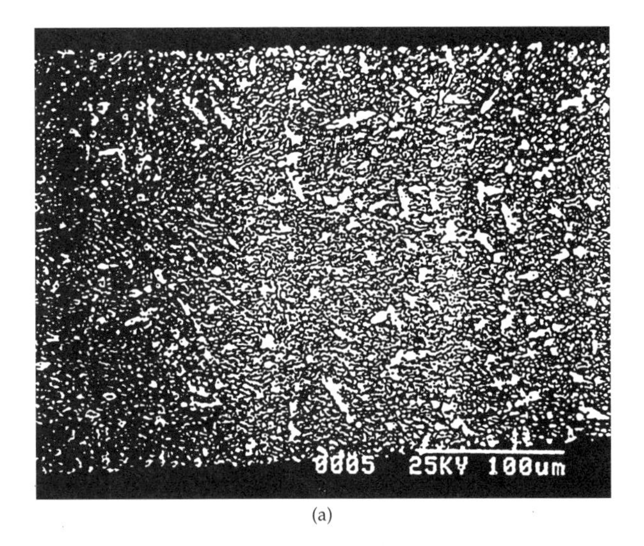

(a)

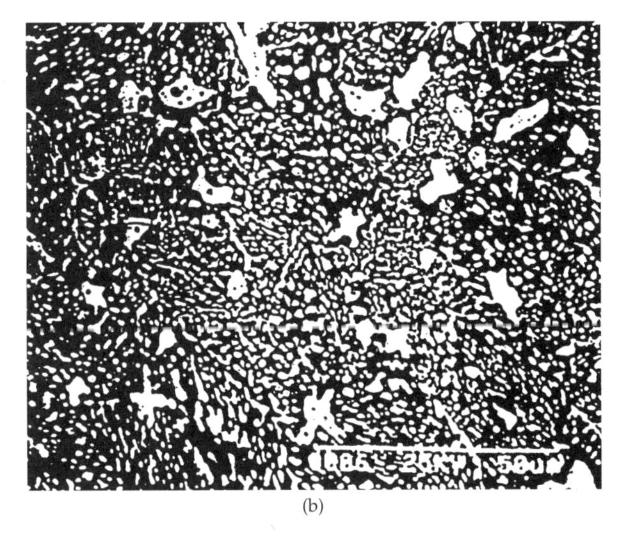

(b)

Figure 14-32. SEM microstructure near the free surface of the solder joint with two-step cooling: (*a*) secondary electron image, and (*b*) backscattered electron image.

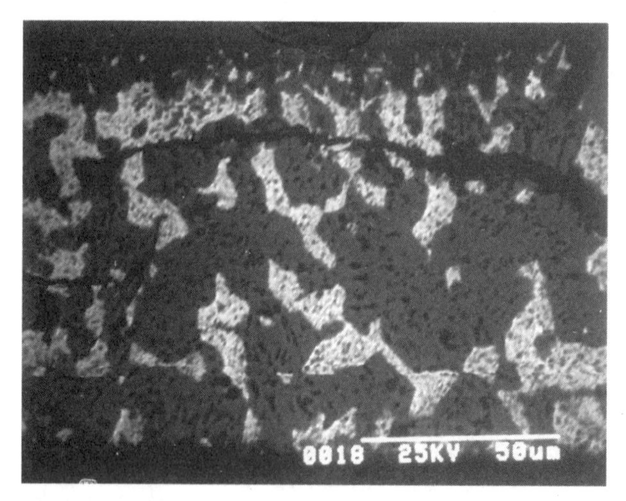

Figure 14-33. SEM micrograph of thermomechanical fatigued microstructure ($N = 269$ cycles) for 0.1°C/s cooled solder joint.

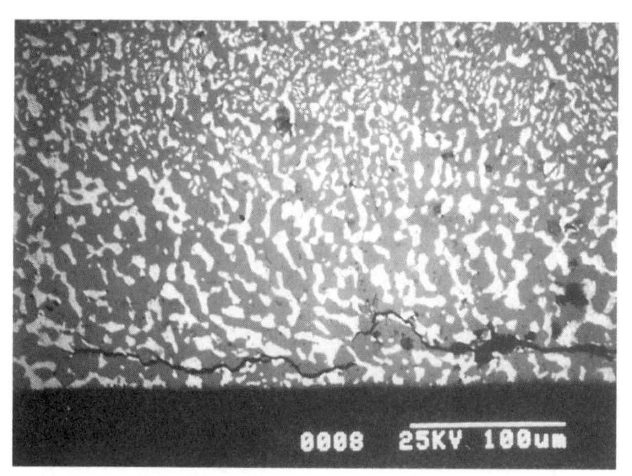

Figure 14-34. SEM micrograph of thermomechanically fatigued microstructure ($N = 353$ cycles) for 230°C/s cooled solder joint.

coarser microstructure is expected to be more extensively deformed, leading to a more thermodynamically favorable nucleation condition, and thus a faster nucleation rate. The large second phase also leads to a shorter diffusion rate.

Fractographs further reveal that fracture surfaces near the free surfaces at the early stage of thermomechanical crack growth appear rough (Fig. 14-35*a* and *b*) as compared with the later crack propagation further away from the free surface (Figs. 14-36*a* and *b* and Fig. 14-37*a* and *b*). This indicates a closer interaction of crack process with microstructure at early stages. At later stages, the fracture surface is relatively flat with distinctive shear-deforming marks which is more

Figure 14-35. SEM fractograph of solder joint formed at 1.0°C/s after early-stage thermomechanical fatigue (N = 360 cycles): (*a*) secondary electron image, and (*b*) backscattered electron image.

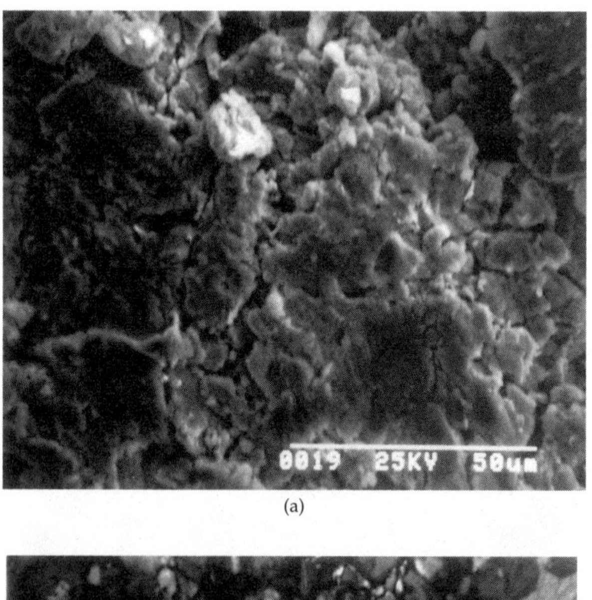

(a)

(b)

Figure 14-36. SEM fractograph of solder joint formed at 1.0°C/s after early-later-stage thermomechanical fatigue ($N = 360$ cycles): (*a*) secondary electron image, and (*b*) backscattered electron image.

obvious toward the interior of the solder joint. Microstructurally, the second Pb-rich phase (light color) appears equiaxed, typical of the recrystallized and coarsened second phase. This is in contrast with the fracture surface of low-cycle fatigue at room temperature where eutectic colony boundaries, Pb-rich proeutectic dendrites, and fine Pb-rich

(a)

(b)

Figure 14-37. SEM fractograph of solder joint formed at 1.0°C/s after later-stage thermomechanical fatigue (N = 360 cycles): (a) secondary electron image, and (b) backscattered electron image.

particles are still distinctive (as shown in the as-reflowed microstructure). This illustrates that the microstructure of 63 Sn/37 Pb eutectic solder is relatively stable under mechanical fatigue at room temperature. Presumably the same would occur at the low temperature end of thermomechanical cycling. The microstructural damage thus largely occurs during the high-temperature end of thermal cycling. The observation of more secondary cracks in Sn-rich matrix boundaries than at the interphase boundary suggests that interphase boundaries are stronger than grain boundaries. Since the failure mode is transgranular rather than the interlinking of grain boundary cavitation, grain boundaries are stronger than the intrinsic matrix.

Under thermomechanical fatigue, the finer microstructure in the solder joint reduces the kinetics of its overall degradation (Pb-rich particle coarsening and Sn-rich matrix grain growth). As the microstructure becomes finer through a faster cooling rate, the increased area of external grain boundary aids the grain matrix to resist deformation through a dislocation pileup mechanism at lower-temperature regions. It can also release grain matrix damage through grain boundary sliding accommodation at high-temperature regions. If the grain boundary sliding does not cause the early thermomechanical crack at the boundary, the event will enhance fatigue resistance. It is noted that the high thermomechanical resistance obtained by a faster cooling rate may not coincide with the results obtained under low-cycle fatigue at room temperature. This may be attributed to the higher internal residual stress and the microstructural inhomogeneity induced by a faster cooling rate. These inherent characteristics from the fast cooling rate can be erased at the high-temperature end of thermal cycling under thermomechanical conditions. When the favorable microstructure evolves, fatigue resistance increases.

14.4 Alloy Strengthening

The strengthening of alloys may be approached (1) from a macroscopic scale or (2) by incorporating a small amount of dopant. Approach 1 is primarily based on the conventional metallurgical alloying principle: to develop a composition and microstructure that is stronger and/or more stable under stressful conditions. The effect of adding Sb to Sn-Pb solders as shown in Fig. 3-18 is one example. Approach 2 involves an insignificant compositional change to the solder alloy to be modified. The dopant which is usually a small percentage of the total composition promotes a more homogeneous microstructure and/or drives the microstructure evolution under imposed stress in a more favorable direction. The dopant is capable of alloying with one or more compositional elements. A small addition of dopants, such as Cd and In to Sn-Pb solder were studied.[9]

14.5 Macroscopic Blend of Selected Fillers

The formation of a composite by adding fillers to an Sn-Pb solder matrix can be achieved through various techniques using different types of filler materials. Common techniques include physical blend, in situ melting, and solidification. The filler can range from surface-activated inert material such as copper-plated carbon fibers to alloyable compounds such as intermetallics which are reactive to Sn-Pb solder. The pivotal property required is the system stability of the filler-matrix system. Tin-lead solder containing 8 to 54 vol% continuous undimensional copper-plated carbon fibers demonstrated improved mechanical properties in tensile strength, tensile module, and thermal fatigue life over Sn-Pb solder without fibers.[10] Compositions made of a physical blend of 63 Sn/37 Pb and Cu_6Sn_5 powder particulates in various concentrations demonstrated an improvement in creep resistance, as shown in Fig. 14-38, and yield strength while maintaining high ductility.[11,12]

Sn-Pb-Cu and Sn-Pb-Ni compositions containing 2.5 to 7.5 vol% of

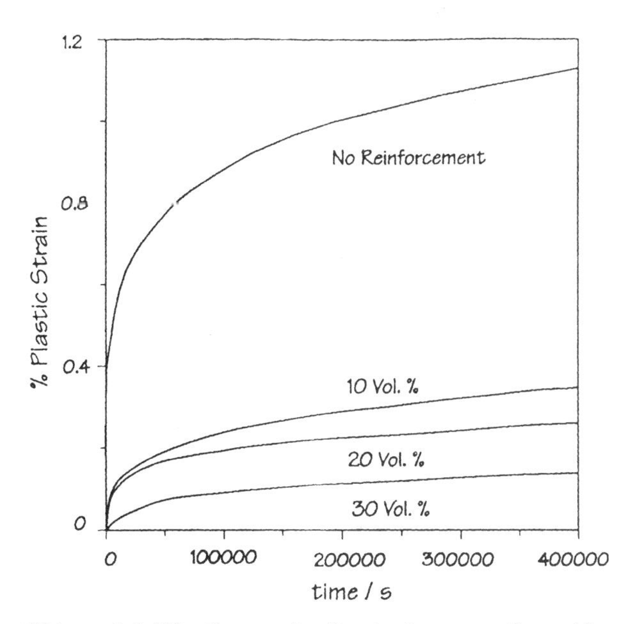

Figure 14-38. Creep plastic strain curve for solder matrix composites at room temperature and 5 MPa.

Cu and Ni in place of Pb in 63 Sn/37 Pb form in situ composites. These composites demonstrated superior strength in comparison with 63 Sn/37 Pb which is processed in the same manner.[13] The process involves melting, atomization, and rapid solidification. A separate study on the composition 60 Sn/40 Pb with 5 and 10 vol% Cu_9NiSn also showed improved properties over 60 Sn/40 Pb without additives.[14]

14.6 References

1. Jennie S. Hwang and Z. Guo, "Strengthened Solder Materials for Electronic Packaging," *Proceedings SMI,* 1993, p. 662.
2. H. J. Frost and M. F. Ashy, *Deformation Mechanism Maps,* Pergamon Press, New York, 1982.
3. A. F. Guo, Sprecher, and H. Conrad, "Crack Initiation and Growth During Low Cycle Fatigue of Pb-Sn Solder Joints," *Proceedings Electronics Components Conference,* 1991, p. 658.
4. H. D. Solomon, "Low Cycle Fatigue of 60/40 Solder-Plastic Strain Limited vs. Displacement Limited Testing," *Electronic Packaging and Processes,* ASM, 1986, p. 29.
5. S. S. Manson, "Separation of the Strain Components for Use in Strain Range," *Advances in Design for Elevated Temperature Environment,* ASME, New York, 1975, p. 17.
6. A. S. Kransz and K. Kransz, *Fracture Kinetics of Crack Growth,* Kluwer Academic Publishers, The Netherlands, 1988.
7. Jennie S. Hwang and Z. Guo, "Reflow Cooling Rate vs. Solder Joint Integrity," *Proceedings NEPCON-West,* 1994, p. 1095.
8. D. Mei and J. W. Morris, "Fatigue Lives on 60 Sn/40 Pb Solder Joints Made with Different Cooling Rate," *Journal of Electronic Packaging,* vol. 114, June 1992, p. 104.
9. D. Tribula, Ph.D. Thesis, University of California-Berkeley, 1990.
10. C. T. Ho and D. D. L. Chung, "Carbon Fiber Reinforced Tin-Lead Alloy as a Low Thermal Expansion Solder Preform," *Journal of Materials Research,* vol. 5, no. 6, 1990, p. 1266.
11. R. B. Clough et al., "Preparation and Property of Reflowed Paste and Bulk Composite Solders," *Proceedings NEPCON-West,* 1992, p. 1256.
12. T. L. Marshall et al., "Composite Solders," *Proceedings of the 41st Electronic Component and Technical Conference,* Atlanta, GA, May 1991, p. 59.
13. S. Sastry et al., "Microstructures and Mechanical Properties of In-Situ Composite Solders," *Proceedings NEPCON-West,* 1992, p. 1266.

14. Hemant S. Betrabet and Susan McGee, "Towards Increased Fatigue Resistance in Sn-Pb Solders by Dispersion Strengthening," *Proceedings NEPCON-West*, 1992, p. 1276.

14.7 Suggested Readings

Engelmaier, W., "Fatigue Life of Leadless Chip Carrier Solder Joints During Power Cycling," *IEEE Transactions on Components, Hybrids and Manufacturing Technology*, vol. CHMT-6, no. 3, September 1983, p. 232.

Marshall, James J., Jennifer Sees, and Jose Caleron, "Microcharacterization of Composite Solder," *Proceedings NEPCON-West*, 1993, p. 1278.

Pnizotto, Russell F., et al., "Microstructural Development in Composite Solders Caused by Long Time, High Temperature Annealing," *Proceedings NEPCON-West*, 1992, p. 1284.

Wasnczuk, J. A., and George K. Lucey, "Shear Creep of Cu_6Sn_5/Sn-Pb Eutectic Composites," *Proceedings NEPCON-West*, 1993, p. 1245.

Wong, B., et al., "Fatigue-Resistant Solder," *Proceedings SMI*, 1993, p. 485.

15

Lead-free Solders

15.1 Driving Forces

Lead-containing solders have been used as an interconnecting and surface coating material in various applications for decades. Two driving forces that grossly affect today's and future requirements of solders for electronic and microelectronic applications are (1) the heightened demands on the level of performance due to the increased density and complexity of circuitry, and (2) the concern of the toxicity and health hazard of lead. This latter concern has led to government legislation and regulations that have continued to impact the future of lead usage.

Research activities for alternative technologies for nonsolder materials, such as metallic-particle-filled adhesives, anisotropic conductive adhesives, and conductive polymers, have been undertaken, in order to develop materials which are capable of providing the interconnecting function equal to or better than solders. Some progress has been made in this area over the past decade, yet the demonstration of practical performance is much limited when all processing, economic, and functional requirements for surface mount/fine-pitch manufacturing are considered. Consequently, the development of lead-free solders becomes a necessary industry effort.[1–13]

As discussed in Chap. 14, most lead-containing soft solders, even under ambient temperatures (298 ± 5 K), reach a homologous temperature T/T_m well beyond 0.5. It is therefore expected that the properties and behavior of solders could alter significantly when they are exposed to temperature changes and/or stresses.[14–24]

Thus, while developing lead-free compositions, it would be a timely opportunity to revisit the solder technology in an effort to further strengthen the performance level of solder materials. In other words, enhancing the performance of solders can be a complementary objective to finding a replacement for lead solders.

15.2 Lead Use and Resources

Lead has been used for applications in glass, glazes, and enamels since ancient times. The oldest artifacts that were made of lead reportedly date back to 1600 to 2000 BC.[25] Today the largest use is in the manufacture of storage batteries, which accounts for 60 percent of all lead products. Other major uses of lead include ammunition, paints, cable sheathing, sheet lead, and solders.

The United States is the world's leading producer of primary lead. Australia, Canada, Peru, Mexico, and Yugoslavia follow, in descending order. The U.S. primary lead production,[15] which constitutes about 20 percent of world production, declined during the period of 1980–1991, as shown in Fig. 15-1. The sources of lead supplies of primary, secondary, and concentrate form in the United States in 1991 are summarized in Table 15-1.

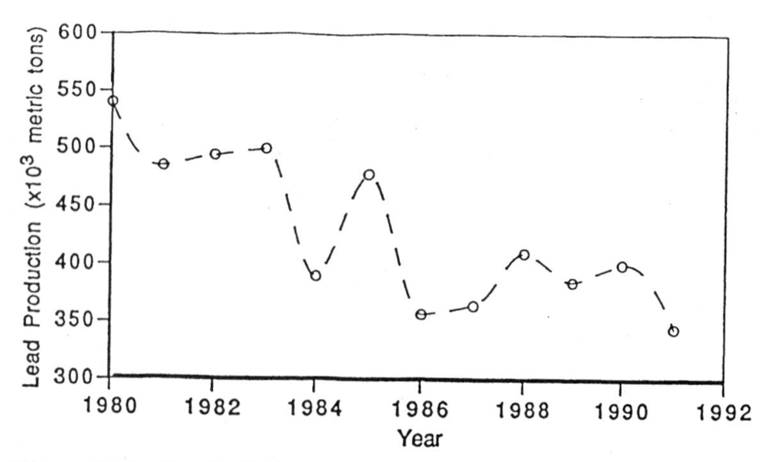

Figure 15-1. Trend of U.S. primary lead production during the period of 1980–1991.

Table 15-1. Lead Supplies in the United States in 1991[26]

Source	Metric tons
Mine production (lead in concentrates)	477,000 (estimated)
Refined primary lead	345,714 (actual)
Refined secondary lead	865,600 (estimated)
Imported refined lead	116,473 (actual)
Imported concentrates	12,437 (actual)
U.S. consumption	1,200,000 (estimated)

According to the American Industrial Hygiene Association (AIHA), the potential health hazards of lead include disorders of nervous and reproductive systems, delayed neurological and physical development, reduced production of hemoglobin, anemia, and hypertension.

The main concerns about the use of lead in the industry stem from possible landfill contamination and the effluent discharge from production processes. In addition, worker exposure to lead through fumes, dust inhalation, and direct ingestion is also of concern.

15.3 Legislation Status

The use of lead in plumbing, gasoline, and the paint industry has been heavily regulated. For example, the use of lead in consumer paints has been banned since 1978. The present campaign against lead poisoning from housings with lead-based paints could put a high price tag on the paint industry. A bill at the congressional level (HR 2478, Lead-Based Paint Hazard Abatement Trust Fund Act, Representative Ben Cardin) was introduced in 1993 to remove the lead from housings with a raised fund of $10 billion over a period of 10 years. The Department of Housing and Urban Development estimated that as many as 57 million private-sector homes still contain lead-based paint, and the cost to remove those paints would exceed $500 billion.[27] The paint industry has responded to this finding with the statement: "There is a great chance that the source of lead poisoning comes from diverse sources as plumbing pipe, solder, smelting, industrial emissions and lead-based gasoline."[27]

Lead and its compounds have been rated as one of the top 17 chemicals that impose the greatest threat to human health along with the following chemicals.[28]

Benzene

Cadmium and cadmium compounds

Carbon tetrachloride

Chloroform

Chromium and chromium compounds

Cyanides

Dichloromethane

Mercury and mercury compounds

Methyl ethyl ketone

Nickel and nickel compounds

Tetrachloroethylene

Toluene

1,1,1-trichloroethane

Trichloroethylene

Methyl isobutyl ketone

Xylene

Recently, public-health officials have classified lead as the number one environmental threat to children.[29]

It is reported that the U.S. Food and Drug Administration is about ready to recommend that the United States join Europe in banning the use of tin-coated lead foil capsules for wine bottles.[30]

Los Angeles County released a proposal to change the California OSHA lead standard. This proposal recommends changes in several requirements for employers in California with respect to lead exposure:

- Decrease the air lead action level (the lead level when action needs to be taken) from 0.03 to 0.025 mg/m^3
- Decrease permissible exposure limits (PEL) from 0.05 to 0.04 mg/m^3
- Decrease the employee blood lead action level of 0.025 mg/dL

On another front, the assertion that lead poisoning is responsible for damaging the intelligence of children, which prompted U.S. environmental and health officials to tighten the guidelines on acceptable levels of lead, is being challenged by a different group of scientists.[31] The

Lead Industry Association, Inc., reported that, "all current uses of lead combined have an insignificant health effect on the most sensitive population—young children," as concluded in a recent study by Environ Corporation.

It is evident that recent findings have provided the basis for the formation of two opposite positions regarding the hazards and regulations of lead. Nevertheless, the Center for Disease Control has lowered the threshold definition of dangerous levels of lead in the bloodstream by 60%, from 25 to 10 μm/dL.

The public mood and political climate appear to exert increasing pressure on the use of lead regardless of industry and end-use products. As to current U.S. legislation/regulations regarding the use of lead, the Lead Exposure Reduction Act (S. 729) was introduced in April, 1993. Under this Act, in addition to banning lead solders for plumbing, inventory and concern lists, new-use notification requests, and product labeling are included. The EPA must inventory all lead-containing products and then develop a *concern list* of all products that may reasonably be anticipated to present an unreasonable risk of injury to human health or the environment. Any person can petition the EPA at any time to add a product to the concern list. Anyone who manufacturers or imports a lead-containing product that is not on the inventory list must submit a notification to the EPA. The products on the concern list must be labeled.

On May 25, 1994, the Lead Exposure Reduction Act (S. 729) passed the Senate floor. Furthermore, the Lead Tax Act (Representative Ben Cardin, HR 2479; Senator Bill Bradley, S. 1347) was also introduced in June 1993 and would place a 45 cent per pound tax on all lead smelted in the United States and on the lead content of all imported products.

The Resource Conservation and Recovery Act (RCRA) classifies solder skimmings and solder pot dumpings as scrap metal not as hazardous waste, if it can be shown that they are recycled. However, if they are disposed of as a waste, they must pass the Toxic Characteristic Leaching Procedure (TCCP) test with the established maximum concentration of lead by TCCP testing being 5 ppm; solder skimming and pot dumpings are expected to fail. Therefore, they are not exempted from RCRA hazardous waste regulations.

Waste solder materials are not considered hazardous if they are returned to a reclaimer or the supplier. A container with less than 3% by weight of its total content is considered empty and is not subject to hazardous waste regulations.[32,33]

15.4 Targeted Properties

Within the scope of electronics applications, solder has been primarily serving two functions: (1) interconnection for mechanical linkages as well as for electrical and thermal conduction, and (2) surface coating for surface protection and/or solderability. Solder interconnections can be formed in various fillet configurations with a wide range of solder volumes, including through-hole solder joints; surface mount J-lead, gull-wing-lead, passive chip termination solder joints; and BGA solder bumps by using various physical forms of solder. They range from molten solder wave, wire, solder paste, preforms, and solder balls. Solder coating works as a surface finish for component leads, terminations, and the circuit traces and solder pads of circuit boards. The coating is normally applied by electroplating or hot solder dipping techniques. Therefore, to replace lead in solders, new solder alloys must possess characteristics that are compatible with the practical techniques and be able to remain stable and intact in the designated physical form under the common application conditions.

The fundamental material properties to be gauged include

- Phase transition temperatures (liquidus and solidus temperature) to be equivalent to Pb-bearing counterparts
- Suitable physical properties, specifically electrical and thermal conductivity, and thermal expansion coefficient
- Compatible metallurgical properties with the interfacial substrates of components and boards
- Adequate mechanical properties including shear strength, creep resistance, isothermal fatigue resistance, thermomechanical fatigue resistance, and microstructural stability
- Intrinsic wetting ability
- Environmental shelf stability

The phase transition temperature is not only related to the overall performance of a solder alloy but is also an important application parameter. For example, the cost-effective PCB material, FR-4, can withstand an application temperature up to approximately 240°C. Considering the temperature variation over the different locations of the board and components under a normal reflow process, the liquidus temperature of the solder alloy that is used for board assembly should not exceed 200°C. Its solidus temperature and its relation to the liquidus temperature (paste range) also plays a role to the long-range performance of the solder joint, as well as to the flow property of the molten solder at a given temperature. In terms of flow property, a eutectic composition which is featured with a single melting point imparts the highest fluidity in the molten state of an alloy system. This is exemplified by an Sn-Pb system where the eutectic composition 63 Sn/37 Pb has the highest fluidity. Eutectic composition is also associated with other unique properties in mechanical behavior.

High thermal and electrical conductivities are always the desired physical characteristics of solder interconnections. However, a material with a lower coefficient of thermal expansion has a better suitability for functioning in a solder joint in most assemblies. Good melting ability and metallurgical compatibility with the commonly used metal substrates of electronic packages and assemblies are important factors to form sound solder joints. Resistance to environmental conditions is also a necessary property to be a useful material. Seeking maximum strength and maximum resistance to creep and fatigue under isothermal and temperature cycling conditions are other targets.

15.5 Scientific Approaches

From a materials point of view, alloy metallurgy and microstructure evolution of solder compositional systems are two important scientific building blocks. Therefore, technical approaches resort to the following areas:

1. Alloy strengthening
2. Microstructure strengthening

Approach 1 is essentially material design and approach 2 is related to both the process and the material.

In designing new alloy systems, the selection of a base metal element and alloying elements is the first step. The basic characteristics of metals for use in soldering are

- Relatively low melting temperature for a base metal element
- Relatively low surface tension and good wetting ability of the alloy system
- Reasonably stable under ambient environment
- Low toxicity
- Easy to use

Considering the above requirements, elements with high melting points, alkali metals, Cd, and Tl are automatically excluded. The candidate elements are then limited. Table 15-2 tabulates the candidate

Table 15-2. Candidate Elements and Key Properties

Element	Melting point, °C	Crystal structure*	CTE at 293 K, °C × 10^{-6}	Thermal conductivity at 300 K, W/(m·K)
In	156.4	FC tetra	32.1	82.0
Se	217.0	Cubic	37.0	0.76
Sn	232.0	BC tetra	22.0	73.0
Bi	271.3	Rhombic	13.4	8.0
Zn	419.5	HCP	30.2	116.0
Sb	630.5	Rhombic	11.0	24.0
Ag	960.6	FCC	18.9	429.0
Cu	1083.0	FCC	16.5	401.0
Ni	1455.0	FCC	13.4	91.0

*BC = body-centered; FC = face-centered; FCC = face-centered cubic; HCP = hexagonal-cubic pack.

elements and their key physical properties. Based on these selected elements, alloys can be derived.

15.6 Alloy Systems

The potential alloy systems are grouped in binary, ternary, quaternary, and pentanary alloys. These systems have been designed and disclosed primarily for electronics or plumbing uses; some of the systems may be in the development state. The list is by no means exclusive, and the proprietary compositions are obviously omitted.

15.6.1 Binary alloys

System	Known composition*	Melting temperature, °C
Sn-Sb	99 Sn/1 Sb (E)	235
	95 Sn/5 Sb	232–240
Sn-Ag	96.5 Sn/3.5 Ag (E)	221
	95 Sn/5 Ag	221–240
Sn-Bi	58 Bi/42 Sn (E)	138
Sn-In	48 Sn/52 In (E)	118
	Other Sn/In ratios	——
Sn-Cu	99.3 Sn/0.7 Cu (E)	227
Sn-Zn	91 Sn/9 Zn (E)	199

*E = eutectic.

The above are relatively more established binary alloy compositions. Each has its characteristics. In general, the Sn-Zn system is prone to oxidation and often imparts poor wetting ability. Compositions of Sn-Cu containing Cu beyond the eutectic content have too high of a melting point to be useful for printed circuit assembly. In Sn-In and Sn-Bi systems, alloy compositions with a significant amount of In or Bi fall in the group of low-temperature alloys. In comparison with Sn-Pb alloys, Sn-In generally possesses inferior strength; however, Sn-Bi has a high tensile strength at ambient room temperature. Eutectic 58 Bi/42 Sn may impart a higher tensile strength than 63 Sn/37 Pb under certain load conditions. However, at elevated temperatures (approaching 100°C), both Sn-In and Sn-Bi are drastically weakened. The melting temperature of the Sn-Ag system falls in the range of 221 to 250°C, which is significantly higher than that of Sn-Pb eutectic solder.

15.6.2 Ternary alloys

System	Known compositions	Melting temperature, °C
Sn-Bi-Ag	91.8 Sn/4.8 Bi/3.4 Ag[34]	211
	94 Sn/4 Bi/2 Ag[35]	203.6–231.1
	92.5 Sn/6 Bi/1.5 Ag[35]	187.6–228.9
	90.5 Sn/7.5 Bi/2 Ag	207–212
Sn-In-Ag	77.2 Sn/20 In/2.8 Ag[34]	178
	>90 Sn/<81 In/Ag[36]	195
Sn-Cu-Ag[37]	96.75 Sn/2 Cu/1.25 Ag	224–260
	96.5 Sn/3 Cu/0.5 Ag	225–296
	95.65 Sn/4 Cu/0.35 Ag	227–332
	95.5 Sn/4 Cu/0.5 Ag	227–330
	94.75 Sn/4 Cu/1.25 Ag	223–329
Sn-Cu-Se (or Te)[38]	95 Sn/4.75 Cu/0.25 Se	210–217
Sn-Sb-Ag	——	——

The Sn-Cu-Se (or Te) system covers the compositional range as listed in Table 15-3. More than approximately 1.0 wt% of Se or Te may cause the solder to become excessively fluid, and more than approximately 6.0 wt% of copper causes the pasty range to be too large to use the solder as a convenient application material. However, if the Cu content is less than 3.0 wt%, the flowability of the solders is drastically reduced to an unacceptable level. The Sn-Cu-Ag system is disclosed with the general compositional range as shown in Table 15-4.

Compositions of Sn-Cu-Ag tend to possess a relatively high melting

Table 15-3. Compositional Range of the Sn-Cu-Se (Te) System[38]

Element	% by weight
Sn	93.0–97.0
Cu	3.0–6.0
Se (or Te)	0.1–1.0

Table 15-4. Compositional Range of the Sn-Cu-Ag System[37]

Element	% by weight
Sn	92–99
Cu	0.7–6
Ag	0.05–3

temperature compared with 63 Sn/37 Pb and a large pasty range. Table 15-5 provides a comparison between the Sn-Cu-Ag and Sn-Cu-Se (Te) systems in percent of elongation, hardness, and shear strength.[38]

The comparison in physical and mechanical properties between Sn-Bi-Ag and Sn-In-Ag remains to be obtained. A relatively lower melting temperature can be derived from the Sn-Bi-Ag and Sn-In-Ag systems than from the Sn-Cu-Ag or Sn-Cu-Se (Te) systems. It should be cautioned that a significant amount of Bi or In is needed in order to achieve a low melting point; however, a too large amount of Bi may adversely affect the mechanical property of Sn-Bi-Ag compositions and the high content of In renders the solder more expensive.

Table 15-5. Comparison of Sn-Cu-Ag and Sn-Cu-Se (Te)

Property	Sn-Cu-Ag	Sn-Cu-Se (Te)
Percent elongation, 1″	22.1	19.3
Hardness (ball Brinell)	14.8	15.1
Shear strength (bulk), lb/in^2	6020	5970

15.6.3 Quaternary alloys

System	Known composition	Melting temperature, °C
Sn-Bi-Ag-Cu	95.9 Sn/1 Bi/0.1 Ag/3 Cu[39]	206–223
	90 Sn/7.5 Bi/2 Ag/0.5 Cu[40]	207–212
Sn-Ag-Sb-Cu	92.2 Sn/3.5 Ag/0.5 Sb/0.8 Cu[34]	217
Sn-Ag-Bi-Zn[35]	93.5 Sn/1.5 Ag/4 Bi/1 Zn	201.4–228.5
	91.5 Sn/1.5 Ag/6 Bi/1 Zn	192.2–225.2
	93 Sn/2 Ag/4 Bi/1 Zn	189.6–226.7
	91 Sn/2 Ag/6 Bi/1 Zn	185–225.2
	89.5 Sn/1.5 Ag/7 Bi/2 Zn	190.4–223.4
Sn-Ag-Bi-Sb[41]	90 Sn/0.5 Ag/4.5 Bi/5 Sb	228–234
Sn-Ag-Cu-Ni[42]	95.5 Sn/0.2 Ag/4.0 Cu/0.3 Ni	238–377
Sn-Ag-Sb-Zn[43]	96.5 Sn/1.0 Ag/0.5 Sb/2.0 Zn	197–220
	94.5 Sn/1.0 Ag/3.0 Sb/1.5 Zn	199–227

Quaternary alloys contain four elements and are quite complex metallurgically. Sn-Ag-Bi-Zn and Sn-Ag-Bi-Cu systems were primarily designed for plumbing uses. Tables 15-6 and 15-7 summarize the general compositions of these two systems, respectively. The Sn-Ag-Bi-Zn system generally has a moderate melting temperature, in the range of 220 to 230°C, and a wide pasty range, in the neighborhood of 30°C.

Table 15-6. Compositional Range of Sn-Ag-Bi-Zn System[43]

Element	% by weight
Sn	87–97
Ag	0–3
Bi	3–7
Zn	0–3

Table 15-7. Compositional Range of Sn-Ag-Bi-Cu System[39]

Element	% by weight
Sn	88–99.35
Ag	0.05–3
Bi	0.1–3
Cu	0.5–6

Table 15-8. Compositional Range of Sn-Ag-Sb-Zn System[43]

Element	% by weight
Sn	90–98.5
Ag	0.5–2
Sb	0.5–4
Zn	0.5–4

Table 15-9. Compositional range of Sn-Ag-Bi-Sb System[40]

Element	% by weight
Sn	90–95
Ag	0.1–0.5
Bi	1–4.5
Sb	3–5

The Sn-Ag-Sb-Zn system was designed to be a replacement for 50 Sn/50 Pb.[43] Its general compositional range is shown in Table 15-8.

Another lead-free quaternary system is shown in Table 15-9.

Copper and/or nickel are the elements that widen the pasty range of Sn-Ag solder, thus facilitating solder flow into the poorly fitted wide gaps. Using a solder with a wide pasty range in conjunction with the soldering temperature just above the solidus temperature is a viable technique to fill poorly fitted large gaps. Some of the designed compositions of the Sn-Ag-Cu-Ni system are listed in Table 15-10.

Table 15-10. Compositional Range of Sn-Ag-Cu-Ni System[42]

Element	% by weight
Sn	92.5–96.9
Ag	0–5.0
Cu	3.0–5.0
Ni	0.1–2.0

15.6.4 Pentanary alloys

System	Known composition	Melting temperature, °C
Sn-Ag-Sb-Cu-Zn[44]	90.8 Sn/1.0 Ag/4.0 Sb/0.2 Cu/4.0 Zn	199–238
	97.2 Sn/0.5 Ag/1.0 Sb/0.8 Cu/0.5 Zn	211–226

Pentanary alloys contain five elements. Their phase diagrams are not available. The Sn-Ag-Sb-Cu-Zn system was designed as the lead-free replacement for the 50 Sn/50 Pb alloy. The compositional range is covered in Table 15-11.

15.7 Wetting Properties

Wetting is an important surface phenomenon in making solder joints; it is primarily driven by the relative magnitudes of thermodynamic surface energy and interfacial energy before and after a wetting process.[45] As discussed in Chap. 5, in order for a system with liquid to wet the solid substrate, the spreading occurs only if the surface energy of the substrate to be wetted is higher than that of the liquid to be spread.

The metallurgical reaction is considered the primary bonding mechanism between solder and the metal substrate. Aside from surface tension and metallurgical reactions, flux chemistry which affects the surface energy and cleanliness of the substrate and operating temperature which can alter the fluxing activity and surface tension are also important factors to the wettability of a given solder alloy.

Wettability measurement of lead-free solders has been carried out under several studies. Tests were conducted on 58 Bi/42 Sn and 96.5 Sn/3.5 Ag in comparison with 63 Sn/37 Pb. The results indicated that

Table 15-11. Compositional Range of Sn-Ag-Sb-Cu-Zn System[44]

Element	% by weight
Sn	86.8–98.8
Ag	0.1–3.0
Sb	0.5–4.0
Cu	0.1–2.0
Zn	0.5–4.0

these two lead-free compositions exhibited less wetting (spreading) on 100% Sn-plated leads than 63 Sn/37 Pb with the rosin fluxing chemistry.[46]

Another study on the wettability of 96.5 Sn/3.5 Ag and 58 Bi/42 Sn solders to the gold-plated substrate as compared with 60 Sn/40 Pb solder demonstrated that 60 Sn/40 Pb produced the greatest extent of spreading.[47]

Comparing the wetting ability of 95 Sn/5 Sb, 58 Bi/42 Sn, and 50 Sn/50 In with 60 Sn/40 Pb on Cu substrate using OA and RMA flux chemistries revealed that none of the alloys tested wet as well as 60 Sn/40 Pb and that the OA flux produced better wetting than the RMA flux.[48]

Quantitative results using wetting balance were obtained for a series of alloys, including 60 Sn/40 Pb, 58 Bi/42 Sn, 91 Sn/9 Zn, 96.5 Sn/3.5 Ag, 99.3 Sn/0.7 Cu, 95 Sn/5 Sb, 90 Sn/7.5 Bi/2 Ag/0.5 Cu, and a novel Sn-rich composition.[49] It was found that 60 Sn/40 Pb performed the best among all on Cu substrates, that all alloys showed good solderability on Sn-plated substrates, and that 99.3 Sn/0.7 Cu was benefitted the most by inert atmosphere soldering.

Compared to 63 Sn/37 Pb, the lead-free solder compositions, such as 91.8 Sn/4.8 Bi/3.4 Ag, 77.2 Sn/20 In/2.8 Ag, and 96.2 Sn/2.5 Ag/0.8 Cu/0.5 Sb exhibited reduced spreading on imidazole-protected or tin-plated Cu substrate.[34]

15.8 Mechanical Properties

An example of the shear versus strain relationship for 63 Sn/37 Pb, 58 Bi/42 Sn, 96 Sn/4 Ag and a proprietary lead-free alloy is shown in Fig. 15-2.[1]

Mechanical properties were also compared among 63 Sn/37 Pb, 96.5 Sn/3.5 Ag, and 58 Bi/42 Sn solders used for J-lead and chip capacitor solder joints.[46] Under thermal cycling, the J-lead pull strength of all three solders decreased by 25 percent from the as-reflowed after being imposed with 5000 thermal cycles (0 to 100°C, 30 min). However, the shear strength of 96.5 Sn/3.5 Ag chip capacitor solder joints were not sensitive to the thermal cycling condition. In this study, the fatigue damage was attributed to grain (phase) boundary sliding in 96.5 Sn/3.5 Ag and 58 Bi/42 Sn solder joints, and the damage was essentially found in the solder between the lead (or termination) and the pad. Furthermore, 58 Bi/42 Sn displaced a greater extent of microstructural fatigue damage as the result of thermal cycling.

A study on the elongation development with gold dissolution in 96.5 Sn/3.5 Ag revealed interesting results. A 3 wt% gold dissolved in 63 Sn/37 Pb reduced its elongation from 40 to 15 percent whereas 96.5

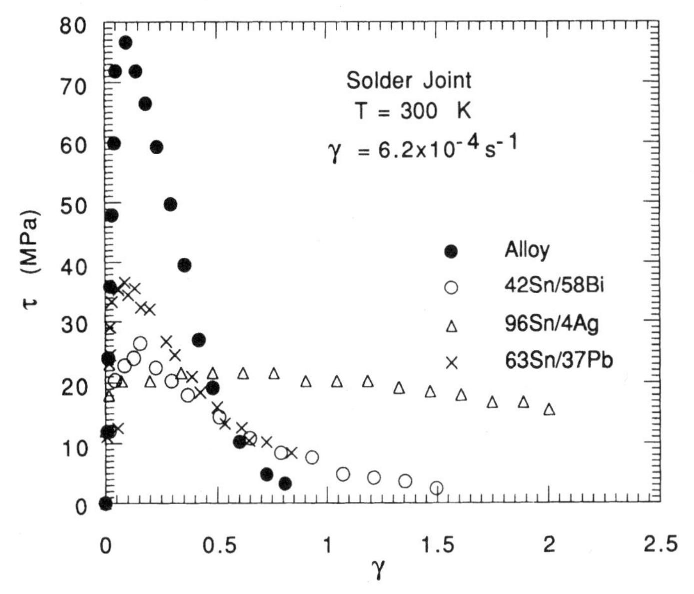

Figure 15-2. Shear stress versus strain for 63 Sn/37 Pb, 58 Bi/42 Sn, 96 Sn/4 Ag, and a proprietary lead-free alloy.

Sn/3.5 Ag with 3 wt% gold had considerably higher elongation at fracture of approximately 50 percent. At 5 wt% gold, 96.5 Sn/3.5 Ag sustained a 40 percent elongation at fracture and 63 Sn/37 Pb a 10 percent elongation.[50] This is congruent with the appearance of the fracture surface, showing that 63 Sn/37 Pb corresponds to a brittle surface and 96.5 Sn/3.5 Ag a ductile one.

The performance of the 58 Bi/42 Sn solder joint between the Cu strap and copper-plated duroid was compared with an In-based solder (80 In/15 Pb/5 Ag) under power cycling (circuit on/off, 0.8 amp, temperature rise 82°C) and temperature cycling (−55 to 85°C, 10°C/min transition, 15-min dwell). The test results indicated the following:[51]

1. A power-cycled 58 Bi/42 Sn solder joint showed no grain size change, no increase in intermetallic layer, and minimal void after 500 cycles.
2. A temperature-cycled 58 Bi/42 Sn solder joint had tears in the copper strap. No other changes were observed.
3. A power-cycled 80 In/15 Pb/5 Ag solder joint exhibited grain coarsening, an increased intermetallic layer, and fractures through the intermetallic layer at 100 cycles; the severity increased with increasing number of cycles.

4. A temperature-cycled 80 In/15 Pb/5 Ag solder joint showed Cu-In intermetallics dispersed through the whole solder. Grain coarsening, however, was not observed. Under the testing conditions of this specific study, In-based solder performed poorly and was recommended not to be used with copper substrate.

15.9 Surface Finish

Lead-free solder is also being sought after for use as a surface finish for component leads and pads on the board. Comparison in solderability of ceramic dual-in-line packages constructed with Alloy 42 lead frames which were hot-dipped with 63 Sn/37 Pb, 95 Sn/5 Sb, 96.5 Sn/3.5 Ag, 77.2 Sn/20 In/2.8 Ag, respectively, was conducted.[52] The results indicated that lead-free compositions in this study are suitable for use as a lead finish, albeit the components coated with 63 Sn/37 Pb produced the best solder joint fillet on the board.

As an alternative to Sb-Pb solder coating, Pd-plated leads or lead frames have been developed.[53] The coating is generally thinner in the range of 1 to 5 μin. In comparison with the Sn-Pb dipping process, the performance and characteristics are satisfactory.[53]

15.10 Cost and Availability

Since lead is one of the more abundant natural resources and is also one of the elements in the lowest-cost category, it is prudent to consider the world capacity and availability of potential lead substitutes and their relative cost. Table 15-12 summarizes the world production of the potential elements, and Table 15-13 provides the unit cost of the elements of interest and their respective cost relative to lead.

As shown in Table 15-13, a substitute which is selected to replace lead will inevitably increase the raw materials cost. By taking into consideration the cost factors involved in the hierarchy of actual manufacture from the element raw materials to the useful product, below is one scenario which illustrates the role of raw material cost.

- Assume that the cost of raw materials for a new lead-free solder composition is 400 percent that of 63 Sn/37 Pb.
- Assume that the cost of raw materials is commercially represented in unit weight.
- Assume that the use product is a solder paste.

Then, the calculation, based on the prevailing cost of raw materials,

Table 15-12. World Production and Capacity Data for Raw Matcrial Elements (1993)

Element	World production, ton	World capacity, ton
Ag	13,500	15,000
Bi	4,000	8,000
Cu	8,000,000	10,200,000
In	80–100	200
Sb	78,200	122,300
Sn	160,000	281,000
Zn	6,900,000	7,600,000

NOTE: Present world solder consumption = 60,000 ton or 6,600,000 L. (*Data from the U.S. Bureau of Mines.*)

Table 15-13. Relative Cost of Selected Elements (1994)

Element	Cost per pound, $	Cost ratio
Ag	73.0	183.0
Bi	3.9	9.8
Cu	0.9	2.3
In	70.0	175.0
Sb	0.8	2.0
Sn	3.6	9.0
Zn	0.5	1.2
Pb	0.4	1

processes involved, and the actual cost of production for the key steps of the manufacturing cycle as shown in the flowchart below, leads to the following cost-estimate summary.[1]

The 400 percent cost of elemental raw materials is substantially diluted to a range of 140 to 180 percent at step IV and is further diluted to 100 to 130 percent at step V. Since each solder joint is gauged by volume rather than weight, the correction factor from solder weight to solder volume, which is estimated to fall in the range of 10 to 15 percent, will bring down the cost of lead-free solder by another 10 to 15 percent for making solder joints with equivalent volume. In addition, the volume required for each solder joint can likely be reduced when the solder material is intrinsically strengthened. Therefore, the cost of lead-free solder per solder joint could be equal to or less than that of 63 Sn/37 Pb, even if the cost of starting materials is 4 times that of 63 Sn/37 Pb. Overall, the cost of raw material stock plays a minor role, especially if it is used for an alloying element in a small percentage of the total composition.

The above assessment, focusing on the segments of the manufacturing cycle from raw materials to solder joints, does not include other cost factors such as taxes, fees, administration cost, and waste-disposal cost which might likely be incurred in the use of lead in the future. The inclusion of these factors will render the cost of lead-free solders even more favorable.

Based on the estimated usage of solders and the availability of material resources as listed in Table 15-12, the elements of Ag, Cu, Sb, and Zn are expected to be in adequate supply when using them as alloying elements. The use of In, however, is limited to below 0.5 wt% and Bi to below 15 wt%.

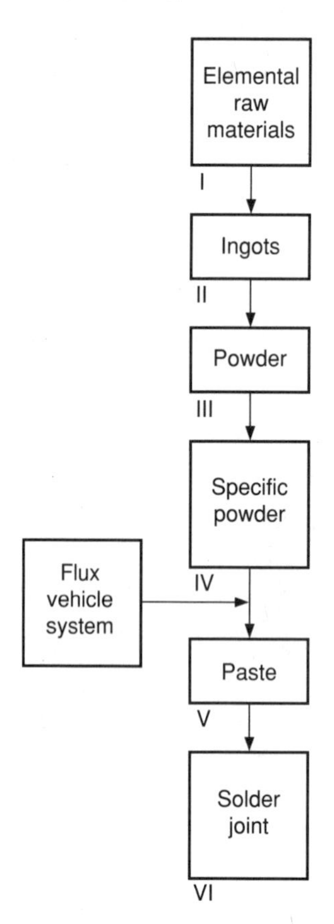

15.11 Manufacturing Issues

Manufacturing-related areas that may be affected by using lead-free solder alloys are

- Flux chemistry
- Metallurgical reactions with substrate
- Solder alloy liquidus temperature versus reflow peak temperature
- Reflow temperature in relation to the thermal stability of board material and other materials

The selected elements that constitute the making of the lead-free alloys may demand a different flux chemistry, which in turn may require different temperature conditions for the proper flux activation and activity. During reflow, the kinetics of the thermodynamic-driven metallurgical reactions between the lead-free solder and substrate may proceed at a different extent from that between 63 Sn/37 Pb and substrate. In addition to anticipating the potential intermetallic compounds that may form between the solder and substrate, the understanding of the temperature effect on the formation of intermetallic compounds is important.

As the liquidus temperature of the solder alloy increases from that of 63 Sn/37 Pb (higher than 183°C), the reflow profile may need to be established to assure a good wetting process. If the required reflow profile calls for a higher peak temperature, the thermal stability of the board material, other board fabrication material, and components has to be considered. Attention should also be paid to the lowest phase transition temperature and the possible formation of low-temperature binary or ternary eutectic composition when the solidus is in direct contact with the substrate. When any new phase is formed and if it melts at a much lower temperature than the liquidus temperature of the solder, adverse effects on the solder joint may result. Depending on the distribution and location of the new phase in the solder joint, it may contribute to early softening or weakening of the solder joint.

Another practical issue is the relative toxicity of the selected elements. Table 15-14 lists the toxicity in terms of the OSHA permissible exposure limit (PEL). The rank of toxicity follows the order of

$$\text{Bi} < \text{Zn} < \text{Cu} < \text{Sb} < \text{In, Ag}$$

Table 15-14. OSHA PEL

Element	PEL, mg/m^3
Bi	None
Zn	5
Sn	2
Cu	1
Sb	0.5
In	0.1
Ag	0.1

15.11.1 Case study: Use of low-temperature 58 Bi/42 Sn solder[54]

Description

The assembled high-density PCB measured about 24×24×0.33 in and weighed 40 lb. The board consisted of 50 layers. Each assembly contained 90,000 solder joints, 52,000 of which were hidden. There were 210 pin grid arrays (PGAs) plus power plugs on the top side of each board, and 420 fine-pitch surface mount components on the bottom side.

Because of the high density and large mass, the board required a higher operating temperature (250°C) than a normal process using the conventional solder (63 Sn/37 Pb). However, the maximum allowable solder temperature for the assembly is 210°C.

Approach

In order to lower the required operating temperature, thus minimizing thermal stresses, a solder composition with a lower melting temperature was needed.

Process

Initially, a prototype process where components were pretinned with 58 Bi/42 Sn solder was developed as shown in the following flowchart. For production, the modified assembly process was used, as represented below.

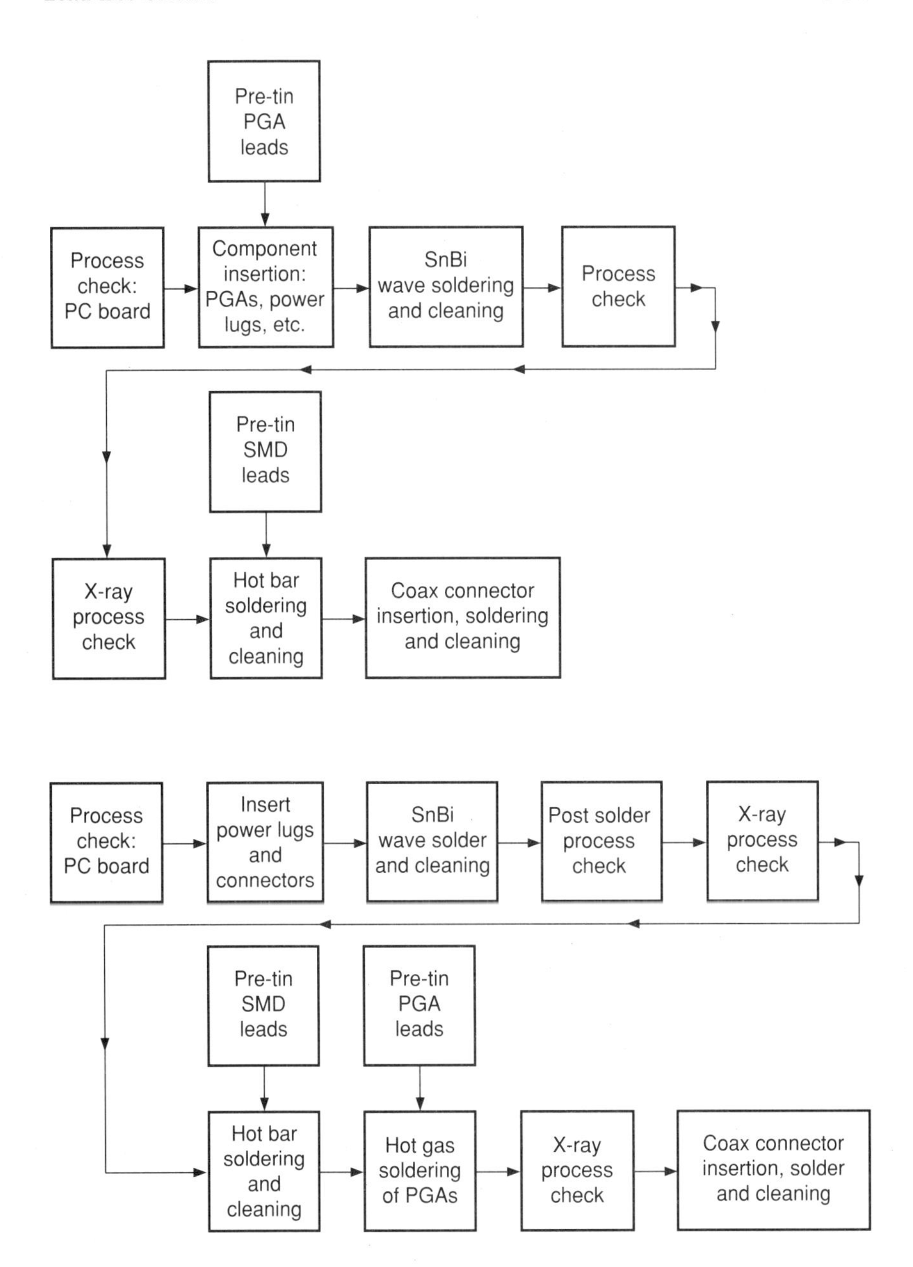
Pre-tin
PGA
leads
Process
check:
PC board
Component
insertion:
PGAs, power
lugs, etc.
SnBi
wave soldering
and cleaning
Process
check
Pre-tin
SMD
leads
X-ray
process
check
Hot bar
soldering
and
cleaning
Coax connector
insertion, soldering
and cleaning
Process
check:
PC board
Insert
power lugs
and
connectors
SnBi
wave solder
and cleaning
Post solder
process
check
X-ray
process
check
Pre-tin
SMD
leads
Pre-tin
PGA
leads
Hot bar
soldering
and
cleaning
Hot gas
soldering
of PGAs
X-ray
process
check
Coax connector
insertion, solder
and cleaning

Test results

Among the potential solder compositions (Sn-Pb-In, Sn-Pb-Bi, Sn-Bi), the eutectic 58 Bi/42 Sn yielded the best results in the combined performance. Low-temperature soldering and lead-free solder with a composition of 58 Bi/42 Sn were demonstrated to be the choices of process and material system for this application.

15.12 Present and Future Perspective

Development work on lead-free solders has been launched, thus far, by a number of organizations and individuals through either formal partnership or professional alliance in the United States and Europe. Metallurgical alloying and microstructure evolution and their relationship to the soldering process and service conditions are critical scientific bases from which to develop new lead-free solders.

A major effort has been focused on developing a direct replacement for Sn-Pb or Sn-Pb-Ag eutectic compositions to fit in the mainstream surface mount manufacturing. It is indicative that the introduction of lead-free solder compositions having melting points of 200 to 240°C is proliferating. However, compositions delivering a liquidus temperature in the range of 175 to 190°C are scarce; a particular challenge goes to the development of a eutectic composition in this temperature range. A technical challenge lies upon the intricate design of a composition with a melting temperature of 175 to 190° without incurring a cost increase. A cursory cost analysis reveals that the cost of raw material can absorb up to 400 percent of the cost of 63 Sn/37 Pb solder without consequently resulting in a significant increase in the cost of the end-use product such as solder paste.

The elements of Ag, Cu, Sb, and Zn are expected to be in adequate supply for use as alloying elements. The use of In, however, is limited to below 0.5 wt% and Bi to below 15 wt%, assuming that 45 percent of all lead-free solders contain In and/or Bi elements.

In terms of physical and mechanical properties, certain parameters of the known lead-free compositions reach the level of performance on a par with that of the lead-containing counterparts, yet other aspects of performance fall behind. To achieve the level of wetting ability of Sn-Pb eutectic alloy on common solderable substrates which are being used in the industry is another technical challenge. In this regard tailored flux chemistry and, in some cases, moderately modified soldering temperature profile play complementary roles to the overall wetting performance.

Technical data as well as practical application information continue to be generated at a reasonably rapid pace. It is anticipated that the implementation of the use of lead-free solders in the electronics assembly is imminent. The selection of the lead-free solder composition should be determined by the need of the performance level of a specific application.

With the increased demand on the performance and reliability of interconnections in electronics and microelectronics assemblies, the task of developing lead-free solders should be taken as a timely opportunity; the opportunity to further enhance the intrinsic properties of solder materials in the meantime alleviate the toxicity issue of lead. It is a mission of the science and technology community.

With respect to lead as a health hazard, the *New York Times* of July 27, 1994, reported that a U.S. government study has found that the amount of lead circulating in the bloodstreams of Americans has dropped by 78 percent. In people ages 1 to 74, the level has dropped from 12.8 μg of lead per deciliter of blood to 2.8 μg. The decline of the concentration of lead in the bloodstream is attributed to the removal of lead from gasoline since 1976 and from the solder of the canning industry. Experts in this field continue to assert that young children are at a greater risk for lead poisoning because they tend to put everything into their mouths; and young children quickly absorb lead at a time when their brains are developing rapidly and can easily become poisoned.

15.13 References

1. Jennie S. Hwang and Z. Guo, "Lead-Free Solders for Electronic Packaging and Assembly," *Proceedings SMI Conference,* 1993, p. 732.
2. B. R. Allenby et al., "An Assessment of the Use of Lead in Electronic Assembly," *Proceedings SMI Conference,* 1992, p. 1.
3. Jennie S. Hwang et al., "Futuristic Solders—Utopia or Ultimate Performance," *Surface Mount Technology,* September 1991, p. 40.
4. William B. Hampshire, "The Search for Lead-Free Solders," *Proceedings SMI Conference,* 1992, p. 729.
5. John B. Greaves, "Evaluation of Solder Alternatives for Surface Mount Technology," *Proceedings NEPCON-West,* 1993, p. 1479.
6. Debra K. Kopp and Matthew G. Bevan, "Conductive Adhesive Substitutes for Tin-Lead Solder," *Proceedings NEPCON-West,* 1993, p. 1501.
7. K. Gilleo et al., "High Stability Solderless Junctions Using Advanced Conductive Adhesives," *Proceedings NEPCON-West,* 1991, p. 193.
8. K. Gilleo, "The Polymer Electronics Revolution," *Proceedings NEPCON-West,* 1992, p. 1390.

9. Chang et al., "Design Considerations for the Implementation of Anisotropic Conductive Adhesive Interconnections," *Proceedings NEPCON-West,* 1992, p. 1381.
10. David H. Benton, "Lead Removal from a New Aqueous Cleaning Agent Before Discharge," *Proceedings NEPCON-West,* 1994, p. 1346.
11. Craig Dawson and Connie J. Schultz, "An Assessment of Lead Exposure to Wave Solder Machine and Solder Pot Preventive Maintenance Workers in an Electronics Assembly Operation," *Proceedings NEPCON-West,* 1994, p. 619.
12. Warwick et al., "Screening Studies on Lead-Free Solder Alloys," *Proceedings NEPCON-West,* 1994, p. 874.
13. Duane Napp, "NCMS Lead-Free Electronic Interconnect Program," *Proceedings SMI,* 1994, p. 425.
14. Jennie S. Hwang, "Innovation, Leadership and Competitives," *Surface Mount Technology,* May 1993.
15. Grivas, K. L. Murty, and J. W. Morris, "Deformation of Sb-Sn Eutectic Alloys at Relatively High Strain Rates," *Acta Metall,* vol. 27, 1979, p. 731.
16. Tribula, D. Grivas, and R. D. Frear, "Microstructure Observation of Thermomechanically Deformed Solder Joints," *Welding Research Supplement,* October 1978, p. 404s.
17. Jennie S. Hwang and R. M. Vargas, "Solder Joint Reliability—Can Solder Creep?" *International Symposium on Microelectronics,* ISHM, 1989, p. 513.
18. T. S. E. Summece and J. W. Morris, "Isothermal Fatigue Behavior of Sn-Pb Solder Joints," *Transactions of ASME,* vol. 112, June 1990, p. 94.
19. Z. Guo, A. F. Sprecher, and H. Conrad, "Plastic Deformation Kinetics of Eutectic Pb-Sb Solder Joints in Monotonic Loading and Low-Cycle Fatigue," *Journal of Electronic Packaging,* Transactions of ASME, vol. 114, June 1992, p. 35.
20. S. Vaynman, M. F. Fine, and D. A. Jeannott, "Low-Cycle Isothermal Fatigue Life of Solder Materials," Chap. 4 in D. R. Frear et al., *Solder Mechanics,* The Minerals, Metals and Materials Society, Warrendale, Pa., 1991, p. 155.
21. D. R. Frear, "Thermomechanical Fatigue in Solder Material," Chap. 5 in D. R. Frear et al., *Solder Mechanics,* The Minerals, Metals and Materials Society, Warrendale, Pa., 1991, p. 192.
22. James L. Marshall et al., "Fatigue of Solders," *International Journal for Hybrid Microelectronics,* 1987, p. 261.
23. Jennie S. Hwang, Z. Guo, and G. Lucy, "Strengthened Solder Materials for Electronic Packaging," *Proceedings SMI,* 1993, p. 662.
24. Jennie S. Hwang, *Solder Paste In Electronic Packaging—Technology and Applications in Surface Mount, Hybrid Circuits, and Component Assembly,* Von Nostrand Reinhold, New York, 1989, Chap. 9, p. 282.

25. Nordyne, "Lead Products," *Ceramic Bulletin*, vol. 70, no. 5, 1991, p. 872.
26. U.S. Bureau of Mines.
27. Kenneth Silverstein, "Lead Issues Gain Steam at State and Federal Levels," *Modern Paint and Coatings*, April, 1993, p. 30.
28. Environmental Protection Agency, Washington, D.C.
29. Steven Waldman, "Lead and Your Kids," *Newsweek*, July 15, 1991, p. 42.
30. *Packaging Digest*, February 1993, p. 2.
31. Gary Putka, "Research in Lead Poisoning is Questioned," *The Wall Street Journal*, March 6, 1992, p. 8.
32. CFR 40, Federal Regulations, Edition-July 1992.
33. N. Irving Sax et. al, *Dangerous Properties of Industrial Materials*, 6th ed., Van Nostrand Reinhold, New York, 1989.
34. Artaki, A. M. Jackson, and P. T. Viance, "Fine Pitch Surface Mount Assembly with Lead-Free, Low Residue Solder Paste," *Proceedings SMI*, 1994, p. 449.
35. European patent publication number 0499452 A1.
36. Patent #5,229,070.
37. Patent #4,778,733.
38. Patent #5,102,748.
39. Patent #4,879,096.
40. P. G. Harris and M. A. Whitmore, "Alternative Solders for Electronic Assemblies," *Circuit World*, vol. 19, no. 2, 1993, p. 25.
41. Patent #4,806,309.
42. Patent #4,758,407.
43. Patent #4,670,217.
44. Patent #4,695,428.
45. Jennie S. Hwang, *Solder Paste in Electronic Packaging Technology—Applications in Surface Mount, Hybrid Circuits, and Component Assembly*, Van Nostrand Reinhold, New York, 1989, Chap. 4.
46. Paul T. Viano, Iris Artaki, and Anna M. Jones, "Assembly Feasibility and Reliability Studies of Surface Mount Boards Manufactured with Lead-Free Solders," *Proceedings SMI Conference*, 1994.
47. Cindy Melton, Andrew Skipor, and John Thome, "Material and Assembly Issues of Non-Lead Bearing Solder Alloys," *Proceedings NEPCON-West*, 1993, p. 1489.
48. Micael Hosking et al, "Wetting Behavior of Alternative Solder Alloys," *Proceedings SMI Conference*, 1993, p. 476.
49. Vincent et al., "Alternative Solder for Electronics Assemblies," *Circuit World*, vol. 19, no. 3m, 1993, p. 32.

50. Harada and R. Satoh, "Mechanical Characteristics of 96.5 Sn/3.5 Ag Solder in Microbonding," *IEEE Transactions on Components, Hybrids and Manufacturing Technology*, vol, CHMT-13, no. 4, 1990, p. 736.
51. H. Getty, "The Effect of Power and Temperature Cycling on Tin/Bismuth and Indium Solders," *IPC Technical Review*, December 1991, p. 14.
52. Mark A. Kwoka and Dawn M. Foster, "Lead Finish Comparison of Lead-Free Solders Versus Eutectic Solder," *Proceedings SMI Conference*, 1994.
53. Donald C. Abbott et al., "Palladium as a Lead Finish for Surface Mount Integrated Circuit Packages," *IEEE Transactions on Components, Hybrids and Manufacturing Technology*, vol. 14, no. 3, September 1991, p. 567.
54. Hilal Erdogan Grudul and Richard R. Carlson, "Low Temperature Lead-Free Wave Soldering for Complex PCBs," *Electronic Packaging and Production*, September 1992, p. 31.

16

Solder Joint Reliability and Failure Mode

This chapter is designed to highlight the factors that are important to the reliability of solder interconnections and the prevalent failure modes which are associated with solder fillets of common IC components, discretes, and other joint structures. The information and data presented are largely based on Sn-Pb solder, mostly eutectic 63 Sn/37 Pb composition unless otherwise specified. Manufacturing perspectives will be emphasized.

16.1 Factors of Solder Joint Integrity

Solder joint integrity relies on not only the intrinsic material properties but also on the design, the component, the solder-substrate interface, the process, and service conditions. The assurance of solder joint integrity warrants the step-by-step evaluation of the following items:

- Suitability of solder alloy in its melting point for required service temperature
- Suitability of solder alloy for required mechanical properties
- Anticipated metallurgical compounds formed between solder and substrates

- Adequacy of solder wetting on substrates
- Design of joint configuration in shape, thickness, and fillet area
- Optimum soldering technique including reflow method, reflow process, and parameters
- Conditions of storage in relation to aging effect on solder joint
- Conditions of actual service in terms of upper temperature, lower temperature, temperature cycling frequency, vibration, or other mechanical loading
- Effects from component characteristics including heat dissipation, stress distribution, and coplanarity
- Effects of characteristics of leads including hardness and dimensions for leaded components
- Performance requirements under conditions of actual service
- Design of viable accelerated testing conditions that correlate to actual service conditions

16.1.1 Failure process

In general, the basic processes or initiators that are frequently linked to the failure of solder interconnections include

- Inadequate material strength at given service conditions
- Excessive disparity in the coefficients of thermal expansion among components and materials involved
- High-temperature creep due to global plastic deformation
- Mechanical and/or thermal fatigue due to cyclic strain accumulation
- Corrosion-accelerated fatigue
- Excessive intermetallic compound formed at interface
- Undesirable intermetallic phases formed in the bulk of the solder joint
- Detrimental microstructure development
- Presence of excessive or large-size voids

Intermetallics and voids are covered in Secs. 13.1 and 13.4, respectively, and the role of microstructure is discussed in Chap. 6. When metallurgical reactions are not predominant, solder intrinsically may often undergo both creep and fatigue processes in an interactive manner under most service conditions of electronic circuit boards.

Although creep fatigue is considered to be a prevalent mechanism leading to eventual solder joint failure, separate test schemes in creep and fatigue are often conducted in order to facilitate the data interpretation and the understanding of the material behavior.

16.1.2 General phenomena of fatigue and creep material behavior

Macroscopically, the fracture surface as a result of fatigue shows combined regions of smooth appearance and rough appearance due to initial slow crack spreading and final rapid crack propagation.[1] The fatigue failures most commonly start at the free surface and are very sensitive to stress raisers, such as surface defects and inclusions. In contrast to creep, fatigue failures do not involve macroscopic plastic flow, resembling brittle fractures in this regard. The movement of a crack involves very little plastic deformation of the metal adjacent to the crack, and fatigue fractures can occur at very low temperatures. At higher temperatures, more factors are involved. One of the most important considerations is the global plastic deformation or creep effect.

It is important to recognize that a small plastic strain, applied only once, does not cause any substantial changes in the substructure of materials, particularly ductile materials. However, repetition of this very small plastic deformation may lead to fracture failure.

Thermal fatigue is directly related to the total strain encountered. The strain level is determined by

- Extent of thermal expansion coefficient mismatch
- Temperature gradient in an assembly
- Temperature range
- Upper temperature
- Dwell temperature at upper temperature
- Geometry of the solder joint
- Degree of compliancy of the system

Thermal fatigue occurs under repeated applications of thermal stress, as a result of temperature changes. Imposed cyclic strains $\Delta\gamma$ in solder joints can be estimated according to

$$\Delta\gamma = \frac{d\Delta\alpha\Delta T}{h}$$

where d = distance of a solder joint from mechanical neutral point
$\Delta\alpha$ = differential thermal expansion coefficient
ΔT = temperature differential
h = solder joint height

As the temperature fluctuates, the creation of temperature gradients may cause thermal strains and associated thermal stresses across the section of the material or assembly, resulting in the initiation of local plastic deformation. When there is no temperature gradient, the thermal strain can also be introduced to the assembly if external constraints exist and the assembly consists of component materials with different expansion coefficients.

In addition to the thermal expansion coefficient, the magnitude of thermal stress development depends on the amplitude of the temperature change; on the heating and cooling rate; and on material properties such as thermal conductivity, heat capacity, density, mechanical properties such as yield point and elastic modulus, and geometrical factors. Therefore, it is apparent that a precise thermal analysis for an assembly that is composed of different materials is immensely difficult. In practice, the complexity is further enhanced by the temperature dependency of the fatigue process and creep process activated by the same temperature effect, in addition to thermal stresses induced by temperature cycles.

Phenomenologically, the fatigue undergoes progressive events involving microcrack initiation, crack propagation, and the growth of microcracks into a macroscopic crack that advances under every cycle until failure.

16.2 Bulk Solder Versus Solder Joint

It should be noted that the interfacial boundary of solder joints can play a major role in the solder failure mechanism, separating its mechanical behavior from that of bulk solder. It should also be noted that the nature of the substrate and its metallurgical reactivity with the solder composition could alter the failure mechanism and the fatigue life, even if the solder composition and other conditions are kept identical. For example, solder joints made of 63 Sn/37 Pb on Cu may behave differently from those in Ni under fatigue conditions.

This is presumably due to the following causes:

- The presence of a high ratio of substrate surface to solder volume, resulting in a large number of heterogeneous nucleation sites during solidification
- The presence of concentration gradient of elemental or metallurgical composition when the solder joint is formed

Either one of the above conditions leads to an inhomogeneous structure. As the thickness of the solder joint decreases, the interfacial effect is more pronounced. Accordingly, the properties and failure mechanism of the solder joint may be incongruent with those derived for bulk solders.

16.3 Fatigue Life Measurement

The fatigue life of solder materials is generally measured by a low-cycle isothermal fatigue mode and/or by a thermomechanical mode. Both methods impose strain cycling on the materials. However, low-cycle isothermal fatigue is performed at a constant temperature, and thermomechanical fatigue is conducted under a continuous spectrum of temperature cycling between two selected temperature extremes. It is considered that low-cycle fatigue is a more convenient method, since it is more straightforward to relate the fatigue phenomenon to fundamental material behavior at a fixed temperature. In comparison, the thermomechanical test mode is closer to the actual service conditions of electronic packages and assemblies. From the material point of view, thermomechanical testing involves a process integrating the deformation mechanisms generated at all temperatures within the lower and upper limits. In addition to temperature, common testing variables include waveform, loading amplitude, frequency, and strain amplitude. All these variables can affect the test results and the interpretation.

Fatigue failure criteria may vary. Categorically, it can be assessed as follows:

- *Catastrophic failure.* The specimen breaking into two pieces
- *Mechanical cracks.* Detected with or without visual aids
- *Load drop.* Monitored by the number of cycles needed to reduce the maximum stress to 25, 50, or 95 percent of its initial value
- *Electrical failure.* Determined by the specific increase of electrical

resistance

Commonly the mechanical cracks are to be inspected with a visual aid at intervals of a small number of cycles, e.g., every 50 cycles (20 to 60 times). The failure of a solder joint may be defined as the point when total solder joint opening occurs between the lead and the pad: the number of cycles-to-failure before the first solder joint failure. When adopting electrical criteria, the electrical resistance of each solder joint is to be monitored and compared with the selected threshold resistance. The criteria for threshold resistance need to be predetermined.

According to IPC-SM-785, "Guidelines for Accelerated Reliability Testing of Surface Mount Attachment," a failure is defined as the first interruption of electrical continuity that is confirmed by nine additional interruptions within an additional 10 percent of cycle life. Usually, solder joint cracks can be detected before an electrical failure. Different failure criteria lead to different conclusions. Thus, it would be prudent to compare data from separate studies performed under different sets of conditions and failure criteria.

Fatigue lives under low-cycle isothermal and thermomechanical testing modes are expected to be comparable in most cases.[2] Yet, disparity in solder joint fatigue obtained by these two methods is often observed. A real comparison between low-cycle and thermal cycling needs to be performed in a continuous integration manner, i.e., integrating the isothermal fatigue lives at a fixed strain range over the entire temperature limits of the thermal cycling.

Nonetheless, with a similar operative mechanism, the low-cycle fatigue life or fatigue resistance is expected to be comparable to that under thermal cycling when the testing temperature of low-cycle fatigue is equivalent to the maximum temperature of the thermal cycling. As the operative mechanism is different between these two test schemes, such comparable performance may not exist. This is primarily the result of thermodynamic and kinetic development under the effect of operating temperature, leading to different microevents or submicroevents as reflected in the evolution of microstructure.

Generally, the conditions that facilitate the formation of fine precipitates, particle dispersion, homogenization, and granulation tend to favor fatigue resistance;[3] those conditions that generate needle morphology and the formation of brittle phases, excessive voids, and stress concentration sites reduce the fatigue life.

The combined results under these two fatigue test modes in conjunction with the results under the creep test and the relationship of monotonic stress versus strain will lead to an understanding of the operative mechanism. For example, a system may demonstrate a high thermomechanical fatigue life, but if it has an inferior low-cycle fatigue life, then it may imply that a thermally activated microevent is predominant, directing a favorable microstructural evolution at a high temperature; or a system having high thermal fatigue resistance as well as high creep resistance may suggest the prevalence of a submicroevent involving the development of fine precipitates and dispersion. The result of high low-cycle fatigue resistance at ambient temperature but low thermal fatigue resistance may suggest a deleterious development of microstructure at elevated temperatures. Under thermomechanical cycling, the solder degradation mechanism via microstructural evolution is illustrated in Sec. 14.3.

16.4 Fatigue Life Prediction

Solder material generally falls into a low-cycle fatigue regimen where elastic strain is negligible and plasticity is encountered during the range of cycling. The number of cycles is relatively small, less than 10,000. This is in contrast with the high-cycle fatigue where the number of cycles is larger than 10,000 and strains in small amplitude are elastic in nature. Fatigue life depends on temperature, strain range, strain rate, loading wave, and other metallurgical factors.

The ultimate goal in studying fatigue phenomena is to accomplish the following:

- Understand the material behavior under cyclic strains
- Develop or improve the material in its resistance to degradation under cyclic strains
- Predict the fatigue life of a material so that its performance reliability at a given set of service conditions can be designed and assured

Numerous fatigue life prediction methods have been proposed.[4–10] Some examples are

- *Frequency-modified Coffin-Manson equation.*[11] The dependence of fatigue life on the waveform and temperature is implied in the frequency component

$$N_f \, \nu^{a-1} = \frac{C}{\Delta r_p^q}$$

where N_f = fatigue life
ν = frequency

$\Delta\gamma_p$ = cyclic inelastic strain
a = frequency exponent
q and c = material constants

- *Strain range partition methodology.*[12] It is assumed that the inelastic strain in a hysteresis loop during fatigue can be partitioned into four strain components, which can be obtained either experimentally or analytically. The four components are loading plastic strain balanced by unloading plastic strain Δr_{pp}, loading creep strain balanced by unloading plastic strain $\Delta\gamma_{cp}$, loading plastic strain balanced by unloading creep strain Δr_{pc}, and loading creep strain balanced by unloading creep strain Δr_{cc}:

$$N_{pp}{}^{q} = \frac{C}{\Delta r_{pp}} \qquad N_{cp}{}^{q} = \frac{C}{\Delta r_{cp}}$$

$$N_{pc}{}^{q} = \frac{C}{\Delta r_{pc}} \qquad N_{cc}{}^{q} = \frac{C}{\Delta r_{cc}}$$

Fatigue life can be predicted through the interactive rule:

$$\frac{1}{N_f} = \frac{F_{pp}}{N_{pp}} + \frac{F_{cp}}{N_{cp}} + \frac{F_{pc}}{N_{pc}} + \frac{F_{cc}}{N_{cc}}$$

where F_{ij} is the fraction of inelastic strain components.

- *Norris and Landzberg temperature factor into frequency-modified Coffin-Manson equation.*[13]

$$N_f\, v^{a-1} \exp\left(\frac{-\Delta H}{kT}\right) = \frac{C}{\Delta\gamma_p}$$

where ΔH is the activation energy and k is the Boltzman constant.

- *Two-dimensional fatigue cracked area growth rate.*[14]

$$\frac{\Delta A_c}{\Delta N} = B\, v^{-k} \exp\left(\frac{-\Delta H_a}{RT}\right) \Delta W_p^b$$

where $\Delta A_c/\Delta N$ = fatigue cracked area growth rate
v = cyclic frequency
ΔH_a = apparent activation energy
ΔW_p = plastic work per cycle
B and b = material constants
k = frequency exponent
R = gas constant
T = temperature

Predicting the fatigue life of solder joints under actual service conditions is a daunting task due to the complex nature of solder joints as well as the presence of various parameters and their interactive effects on the solder behavior. Listed below are some important areas that have either not been included or not adequately covered in existing models. This contributes to the limitation of using the models in real-world applications.

- Effect of initial microstructure
- Effect of inhomogeneity in microstructure
- Change in microstructure versus external parameters
- Nonlinear fracture mechanics
- Assumption of crack-free material as a starting point
- Assumption of the size of existing cracks
- Effect of large strain involved
- High homologous temperature of solders
- Effect of fillet geometry
- Multiaxial stress imposed
- Effect of interfacial metallurgical interaction
- Lack of sufficient field data
- Variation in test data versus test conditions
- Thinner solder joint imposing increased interfacial effect and decreased conventional fatigue-creep phenomenon

A working fatigue-life prediction model that is capable of predicting service life of solder joints under a specific set of conditions must consider and cover all the above areas.

16.5 Component Lead Effect[15]

Consistency among components in their intended lead dimensions and consistency among the leads of a component are crucial to the quality of solder joints and to the overall yield of manufacturing surface mount assemblies. Although specifications for lead dimensions exist, deviations from these specifications or variations within the specifications in commercial component supplies often contribute to manufacturing problems in terms of quality and yield. This is because physical characteristics affect the long-term performance of solder joints. The effect can come from various sources:

- Lead material
- Lead length
- Lead width
- Lead thickness
- Lead height
- Lead coplanarity
- Lead material

As discussed in Sec. 5.3, common lead materials include copper, Alloy 42, and Kovar. Lead stiffness varies with the design of the component package; however, the intrinsic stiffness follows the general order:

$$\text{Copper} < \text{Kovar} < \text{Alloy 42}$$

It is believed that less stiff or more compliant lead materials are more favorable to the fatigue life of solder joints, all other conditions being equal.

- *Lead length.* For QFPs, the lead length is measured from the toe to the contact point with the package body in a horizontal direction. The fatigue life of solder joints was found to increase by 67 percent as the lead length was increased from 0.085 to 0.1125 in (2.13 to 2.82 mm).[16]
- *Lead width.* Figure 16-1 shows the effects of lead width on solder joint fatigue life;[16] fatigue life decreases as lead width increases. It was found that fatigue life is more sensitive to lead thickness than to lead width. The fatigue life drops rapidly as lead thickness increases, as shown in Fig. 16-2.
- *Lead height.* Figure 16-3 shows the effect of lead height as measured from the contact point of the lead and package body to the solder pad in a vertical direction. As can be seen, solder fatigue life increases with increasing lead height.
- *Coplanarity.* Production defects are often related to the coplanarity of component leads, which includes starved solder joints and open joints. To avoid problems caused by poor coplanarity, it is advisable to maintain lead coplanarity in the range of 0.002 in (0.05 mm) for fine-pitch components, although 0.004 in (0.1 mm) [0.003 in (0.075 mm) for fine-pitch devices] seems to be an industry-accepted value. JEDEC 95 specifies coplanarity of individual components.

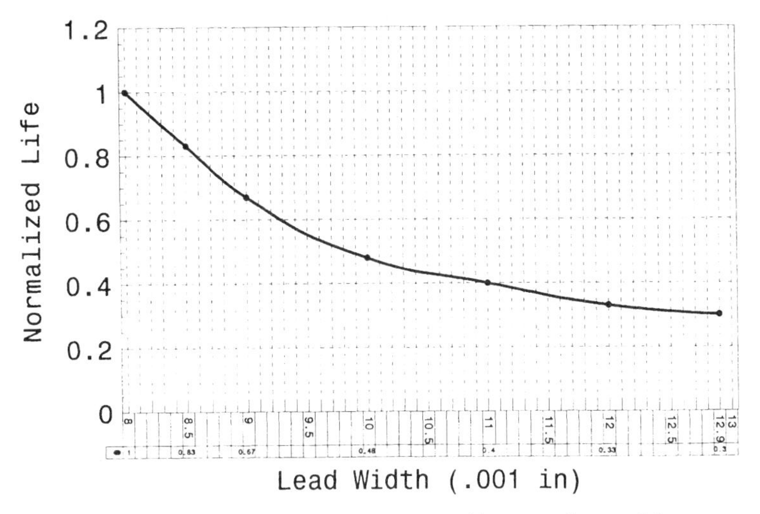

Figure 16-1. Effect of QFP lead width on solder joint fatigue life.

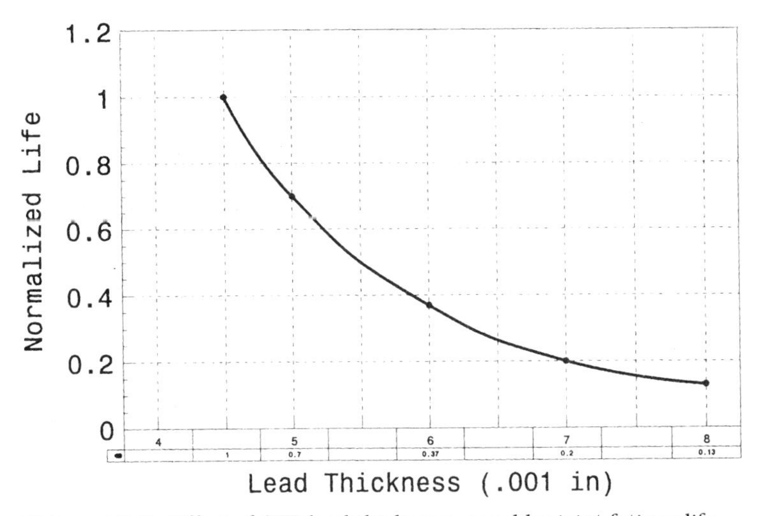

Figure 16-2. Effect of QFP lead thickness on solder joint fatigue life.

To meet the PCMCIA requirements regarding thickness and size of components, a standard thin small outline package (TSOP) is used. It is found that TSOP solder joints have a shorter fatigue life than QFPs on a similar assembly due to the facts that[17,18]

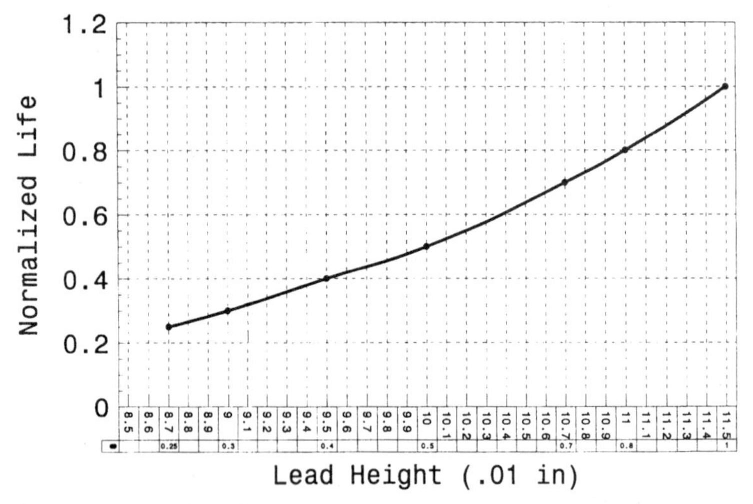

Figure 16-3. Effect of QFP lead height on solder joint fatigue life.

- The TSOP has shorter leads that are less compliant than those of a QFP.
- The TSOP has a lower ratio of plastics to silicon; thus silicon dominates the overall CTE of the package. This creates a larger difference in CTE between the component and the board.

Furthermore, PCMCIA cards are very thin (Sec. 2.9) and become warped more easily. The warpage may adversely affect the fatigue life of solder joints.

16.6 Solder Joint Volume and Configuration

The shape of configuration of solder joints can change the stress distribution and consequently affect failure mode development. Solder joint volume contributes to the kinetics of solder joint crack propagation. For BGAs, uniformity and consistency in volume and configuration among array solder joints within a package are important.

For leadless chip carriers, it is well established that higher solder joint volume provides better fatigue resistance.[19,20] The ability of large-volume solder joints to resist fatigue failure is viewed from the following two aspects. Assuming that the solder joint failure mechanism is primarily the result of crack initiation and propagation, then

- The larger solder volume is able to distribute the stress imposed, thus prolonging the fatigue life.
- The larger solder volume provides a greater distance for crack propagation, thus slowing down the crack fracture, resulting in a prolonged fatigue life.

However, for fine-pitch QFP assemblies with gull-wing solder joints, the geometrical characteristics of the solder fillet may apply. Solder joints of QFPs [256 pins, 28-mm body size, 0.4-mm (0.016-in) pitch] were studied by monitoring solder joint failure versus the number of thermal cycles (−40 to 125°C). Solder joint failure in this study was defined as total solder joint opening, and the number of cycles as the first solder joint failure.[21] The effects of standoff height and the height of the solder heel on solder fatigue life were obtained. The fatigue life increases with increasing standoff height from 0.05 mm to 0.10 mm (0.0002 to 0.004 in). At a low standoff, 0.045 mm (0.018 in) and less, fatigue life appears to be insensitive to the height of the solder heel. At a standoff of 0.10 mm (0.004 in) fatigue life increases as the height of the solder heel increases from 0.10 to 0.25 mm (0.004 to 0.010 in). The thermal fatigue life was predicted in this study using plastic shear strain as a failure index. The maximum plastic strain was located at the solder standoff. This was attributed to the short foot length used in this study, 0.5 mm (0.020 in), where deformation could be accommodated by the entire deformation. As the foot length increases, the effect of the standoff height is expected to decrease since most deformation will be concentrated near the heel.

American National Standards Institute ANSI/J-STD-001, Sec. 9.2.6.2, provides guidelines for the solder fillets of gull-wing leads. Parameters recommended are maximum side overhang, maximum toe overhang, minimum end joint width, minimum side joint length, maximum heel height, minimum heel height, and minimum thickness (standoff height), as illustrated in Fig. 18-4.

Dimensional criteria for these parameters are summarized in Table 18-3. The criteria are set for three end-product classes of assemblies based on producibility, complexity, functional performance requirements, and verification (inspection/test) frequency.

IPC-SM-782, "Surface Mount Design and Land Pattern Standard," Subsection 11.2 SQFP/QFP (square) and Subsection 11.3 SQFP/QFP (rectangular), specifies component dimensions, land pattern dimensions, and solder joint analysis. As examples, component dimensions for three SQFP/QFP (square) components in the range of 0.50 mm to 0.30 mm (0.020 to 0.012 in) are listed in Table 16-1 and their corresponding land patterns in Table 16-2.

Table 16-1. Nominal Component Dimensions for Three Square QFPs

Components	L (mm)	S (mm)	W (mm)	T (mm)	P (mm)	H (mm)
SQFP 28×28-208	30.00±0.20	28.55±0.35	0.20±0.10	0.60±0.30	0.50	3.75
SQFP 20×20-256	22.00±0.20	20.55±0.35	0.10±0.05	0.60±0.20	0.30	2.70
SQFP 24×24-304	26.00±0.20	24.55±0.35	0.10±0.05	0.60±0.20	0.30	3.20

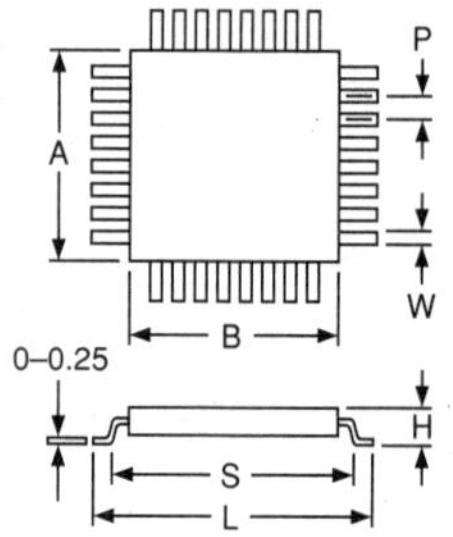

Table 16-2. Land Pattern for Three Square QFPs

Components	Z (mm)	G (mm)	E (mm)
SQFP 28×28-208	30.90±010	27.90±0.10	0.50
SQFP 20×20-256	22.90±0.10	19.90±0.10	0.30
SQFP 24×24-204	26.90±0.10	23.90±0.10	0.30

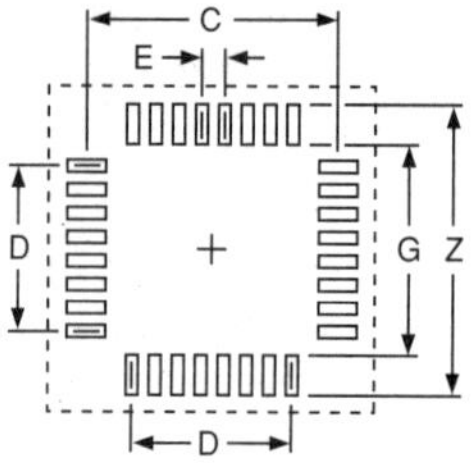

Table 16-3. Solder Joint Fillet Criteria of IPC-SM-782

Component pitch, mm	Solder joint toe, mm	Heel, mm	Side, mm
0.50	0.36 to 0.60	0.17 to 0.55	−0.02 to 0.15
0.40	0.36 to 0.60	1.17 to 0.55	−0.04 to 0.13
0.30	0.36 to 0.60	0.17 to 0.55	−0.08 to 0.08

Based on component dimension and land pattern, solder joint fillets in toe, heel, and side are provided in Table 16-3.

Uniformity and consistency in volume and configuration among array solder joints within a package are also important as discussed in Sec. 11.4.4.

The thickness (height) of solder joints between BGAs and the board is much greater than that of fine-pitch QFPs. The actual solder joint height depends on the diameter of the bumped balls and dimensions of the solder pads; for example, the 0.022-in (0.55-mm) BGA solder height compares with a 0.003-in (0.08-mm) height for the QFP. Since the solder height for BGAs is larger, the effect of the intrinsic properties of solder material for BGAs is expected to be more pronounced than for QFPs.

16.7 Conformal Coating

The beneficial effect of conformal coating on the reliability of solder joints that are exposed to large thermal stresses is attributed to the role the coating plays as

- A protective layer, preventing direct exposure to atmospheric oxygen and contaminants
- A stress absorber, distributing stress over a larger area

In order to benefit from the stress distribution, the coating must be in good contact with the solder joint. The performance of a protective barrier depends on adhesion, film integrity, permeability, moisture resistance, and chemical characteristics.

Much work has been conducted in relation to the effect of conformal coating on the solder joint reliability of LCCCs mounted on PCBs. It has been found that parylene conformal coating significantly improves the fatigue life of solder joints by 1.8. Table 16-4 compares the mean cycles-to-failure of both coated and uncoated solder joints for 32-pin LCCCs under thermal cycling from −55 to + 100°C with 24-min dwell time to the extremes and total cycle time of 2 h.[22] Another study compares the relative effect of two polymer coatings on the reliability of LCCC solder joints on FR-4 board.[23] The results indicate that acrylic conformal coating increased solder joint reliability by a factor of about 1.5, while parylene conformal coating improved solder joint reliability

Table 16-4. Mean Cycles-to-Failure of Coated and Uncoated Solder Joints for 32-Pin LCCCs

Solder joint	±cycles to-failure	Mean number of cycles
Uncoated	363, 374, 410, 410 506, 506, 506	439
Coated	626, 696, 697, 802 802, 824, 828, 834 850, 886, 909, 922 946	817

by a factor of about 2.5, as measured by the average number of cycles to electrical failure. However, when the failure is measured by mechanical cracks rather than the increase in electrical resistance, the effect of parylene is much more pronounced. Table 16-5 summarizes the percentage of visual cracks prior to the first electrical failure of solder joints coated with parylene.

The effect of coating on metal fatigue is well established. One example is illustrated in Fig. 16-4 representing the effect of humidity on the fatigue behavior of an aluminum alloy in *S*/*N* curves. It is understood that the fatigue strength of metals is often reduced by the presence of a corrosive environment, so-called corrosion-enhanced fatigue. Oxygen and water are considered corrosive elements to metals. As can be seen, the fatigue life in the presence of a corrosive atmosphere has dropped dramatically. The crack rate in an aggressive environment is always higher than that in an inert environment under otherwise identical conditions. The lowest crack rate is normally observed in a completely inert environment such as in a vacuum or dry inert gas. Liu and

Table 16-5. Percentage of Visual Cracks Prior to First Electrical Failure

Component	FR-4 (no coating)	FR-4 coated with parylene
LCC-44	51	5
LCC-28	31	5
LCC-20	80	3

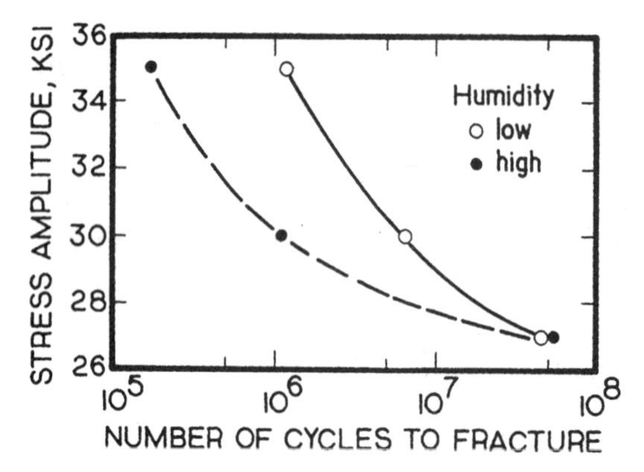

Figure 16-4. Effect of humidity on fatigue life of an aluminum alloy.

Corten indicated that the mean fatigue life of an aluminum alloy wire could be changed by a factor of 2 to 3 by the moisture content of air.[24]

A corrosive environment was found to affect both the nucleation and the propagation stages of crack development by shortening the nucleation period and accelerating the propagation rate. Generally, corrosion media can be in either a gaseous or a liquid state. A gaseous environment is less aggressive than a liquid environment. In a liquid environment local etching could possibly produce pits which act as stress raisers either at selective places of higher slip activity or at nonselective places on the surface. The corrosion environment enhances the formation of slip stress raisers. It is believed that, in higher slip activity areas, slight differences in electrochemical potential inside or outside the slip band promote the process of stress-raiser formation.

16.8 Solder Joint Failure Mode

Solder joint failure has been observed through one or more of three basic modes related to creep/fatigue processes or intermetallic-induced cracks:

- Cracks along or near the interface
- Cracks (microcracks) appearing in the bulk of the solder joint (away from the interface), which proliferate and propagate until failure
- Cracks initiated on the free surface and propagating through the solder joint along transgranular or intergranular paths or by a combined transgranular and intergranular path without preference

16.8.1 Cracks along or near the interface

In general terms, solder joints with high levels of stress due to a high strain rate being imposed or solder joints made of an intrinsically highly creep resistant alloy are prone to fractures near and along the interface. Two types of crack developments along or near the interface, leading to eventual solder joint failure, have been observed. In the first case, the crack path for 63 Sn/37 Pb solder on a Cu substrate is through the coarsened band near the solder/Cu interface. The coarsened band, as shown in Fig. 16-5, is composed of a large primary Pb-rich phase, which is considered a weak zone developed under cyclic strains. The examination of fatigue fractographs reveals that fracture surfaces of 63 Sn/37 Pb solder joints corresponding to early low-cycle

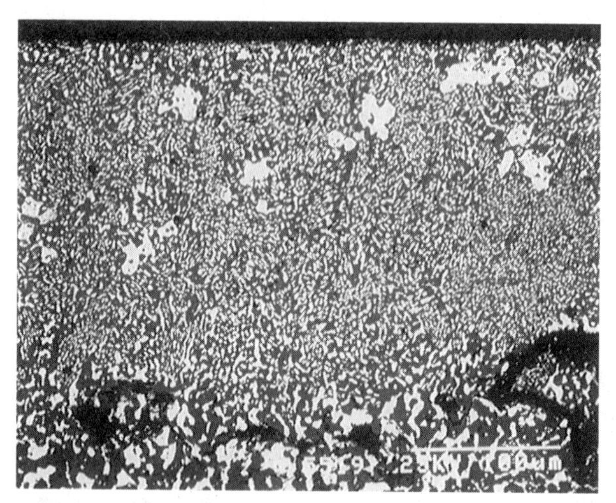

Figure 16-5. Coarsened band developed near 63 Sn/37 Pb solder/Cu interface.

fatigue crack growth were rough and irregular with many secondary cracks observed through secondary electron (SE) signals under an SEM. Analysis of backscattered electron (BE) signals on fracture surfaces revealed that these rough facets were related to the dimension of the eutectic colonies, and the secondary cracks were associated with the colony boundaries. Figure 14-28*a* and *b*, shows SE topography and BE morphology at the same site of the fracture surface for eutectic 63 Sn/37 Pb solder joints. This suggests that fracture units or plastic zones at the tip of fatigue cracks were limited within the scale of the eutectic colony at the earlier stage of crack growth.

Fatigue fracture surfaces at the later stages of fatigue crack growth were rather flat and had a shear-deformed morphology, as shown in Fig. 14-29*a*. This indicated that fatigue cracks propagated transgranularly as confirmed by BE micrograph (Fig. 14-29*b*). When the fatigue crack grows longer, it is expected that the plastic zone at the tip of the fatigue crack will become more intensive and extensive than the rest of the structure. Consequently, the fatigue crack propagates dynamically without the preference of microstructure. It should be noted that large primary Pb-rich dendrites present in the coarsened bands of the as-reflowed microstructure also appear on the fracture surface, indicating that the fatigue crack paths are indeed through the coarsened bands near the Cu/solder interfaces.[25]

In the second case, the crack path is at or along the intermetallic interface where an excess thickness of intermetallics is formed. Figure 13-6 shows the formation of excessive intermetallic compounds during service life, and Fig. 13-7 shows the eventual solder joint crack open.

16.8.2 Cracks appearing in the bulk of the solder joint

It is believed that cracks occur in the bulk of the solder joint when the interface area is relatively strong without the formation of a coarsened band or the accumulation of excess intermetallics. It is often the result of a large differential in the coefficients of thermal expansion between the component and the board. Figure 16-6 shows the cracks occurring in the bulk of the solder.

Cracks can propagate through a transgranular path or an intergranular path. Conditions that promote intergranular cracks include

- High operating temperature
- Lower strain rate
- Lower strain range
- Longer dwell time at high temperature
- Corrosive environment

Solder alloy composition can also affect the crack propagation path, generally depending on the microstructure, particularly the relative strength of the phase matrix and phase boundaries.

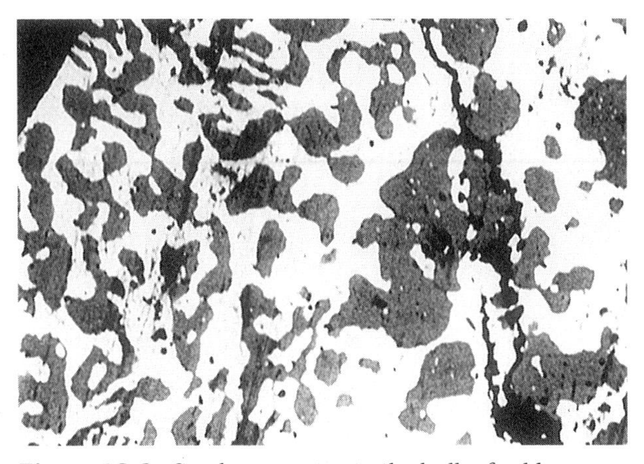

Figure 16-6. Cracks occurring in the bulk of solder.

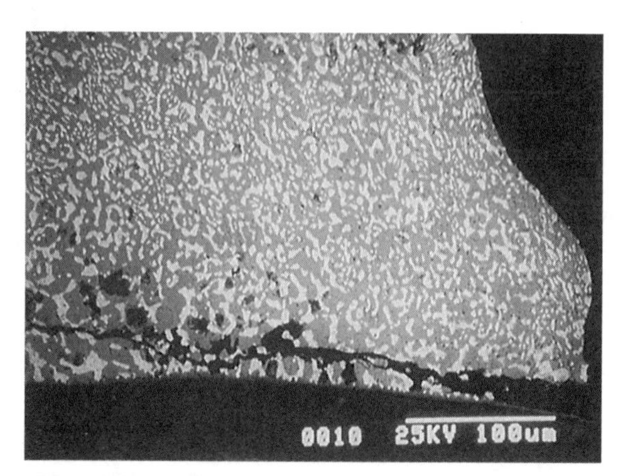

Figure 16-7. Crack initiated at the free surface of a solder joint.

16.8.3 Cracks initiated on the free surface

Under the fatigue conditions of cyclic strains, the free surface is often the crack initiation site. Figure 16-7 illustrates the phenomenon of cracking initiated at the free surface of the solder joint and then propagating through the solder joint with both primary and secondary cracks occurring.

16.8.4 Failure mode versus solder joint configuration

16.8.4.1 Gull-wing leaded component on a PCB. It is often observed that solder joint cracks of tin-lead eutectic or equivalent alloy compositions start near the heel of the solder fillet with secondary cracks occurring at the toe area.[26,27] The initial cracks then propagate inward and meet in the center of the joint, causing failure. The relationship of the equivalent plastic strain (defined as proportional to the maximum normal strain consisting of in-plane thermally induced displacement and out-of-plane displacement between the component and the PCB) and the distance from the heel of a 128-pin QFP corner solder joint is plotted in Fig. 16-8, based on finite-element analysis.[21] The analysis indicates that the equivalent strain decreases as the distance from the heel increases.

16.8.4.2 Leadless ceramic chip carrier on a PCB. For LCCCs, solder joint cracks are often detected at the solder standoff closer to the interface of the chip and the solder. Cracks then propagate through the solder standoff and fillet.[28–31]

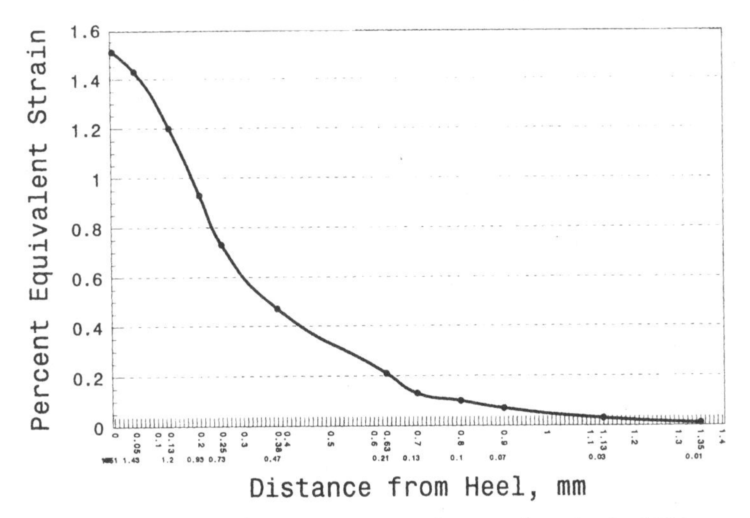

Figure 16-8. Equivalent strain versus distance from heel of 128-pin QFP corner solder joint.

16.8.4.3 Leadless discrete component on a copper heat sink. Solder joint failure for a leadless discrete component attached on a copper heat sink started at the free surface and propagated through the solder joint essentially with a primary crack, as shown in Fig. 16-9. The solder composition used in this assembly was 96.5 Sn/3.5 Ag.

16.8.4.4 Chip capacitor and resistors on a ceramic substrate. Under temperature cycling (−65 to 150°C), solder joint cracking initiated at the free surface near the middle of the solder fillet slope and then

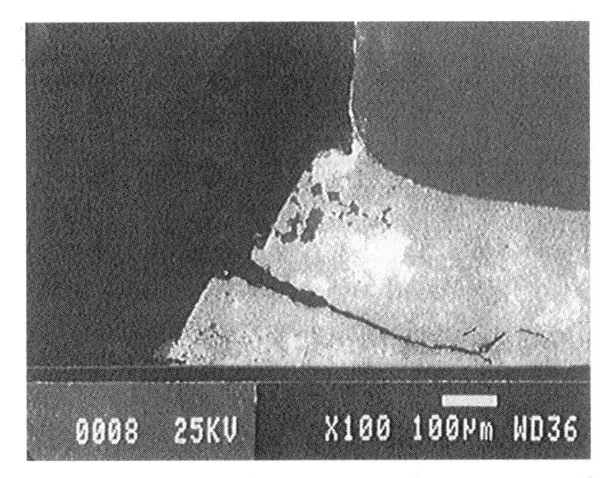

Figure 16-9. Crack initiation and propagation of a solder joint between leadless discrete on copper heat sink.

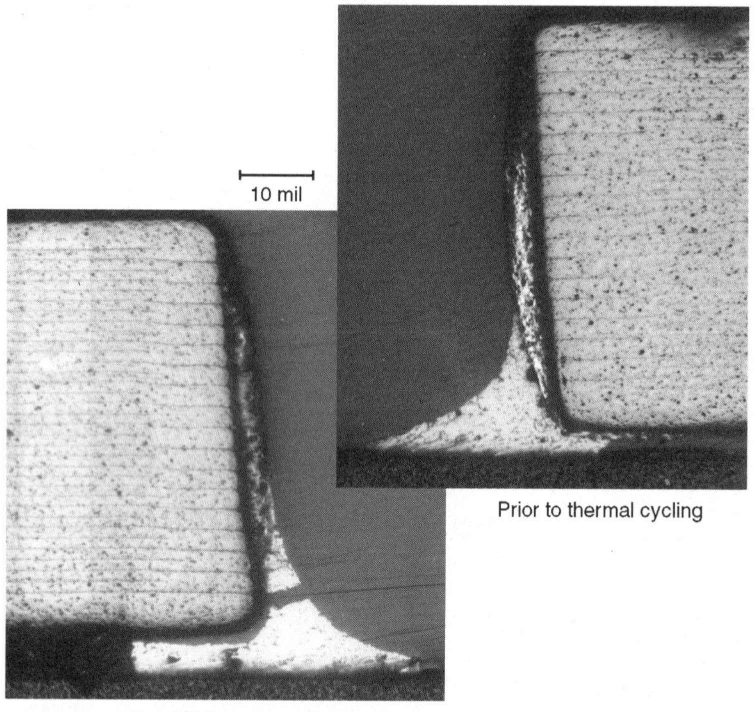

Figure 16-10. Solder joint crack of ceramic capacitor on ceramic substrate.

propagated through the solder fillet and standoff near solder-chip interface as shown in Fig. 16-10.

16.8.4.5 Sandwich versus free surface solder joint. A structure comprises unconstrained 63 Sn/37 Pb solder on top of an Alloy-42 lead that is connected to copper-clad PWB with a 63 Sn/37 Pb solder joint in-between as illustrated in Fig. 16-11. Under temperature cycling, the top solder which is unconstrained remains intact, while a large number of solder joints between Alloy 42 and Cu cracked near the solder–Alloy-42 interface as shown in Fig. 16-12. Further examination, as exhibited in Fig. 16-13, reveals that a coarsened band near the solder–Alloy-42 interface proceeds the crack and that the crack path preferentially goes through the coarsened region.

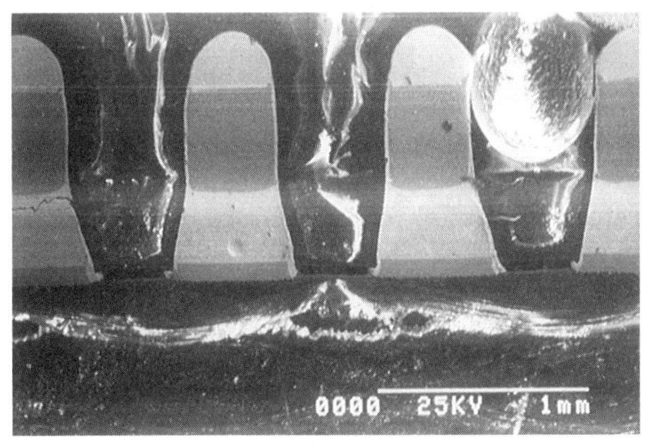

Figure 16-11. Solder structure of unconstrained solder/Alloy 42/solder/PWB.

Figure 16-12. Preferential solder crack between unconstrained top solder and solder joint.

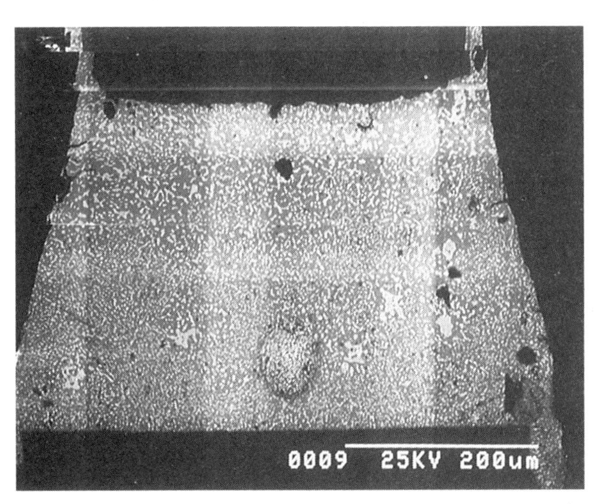

Figure 16-13. Coarsened band appearing near solder/Alloy 42 interface.

16.8.4.6 Array C5 on a PCB. For array solder joints of a ceramic controlled collapse chip corner connector (C5) component on PCB, the test results indicated that the initiation of cracks occurred near the interface of the component and solder at the outside edge. Secondary cracks then occurred at the inside edge of the solder joint.[32]

16.8.4.7 Array C-4 flip chip and C-BGA/C-CGA on a ceramic substrate. The small C-4 solder joints were reported to have cracks starting at the opposite corners near the two interfaces.[13,26,33] It was also found that cracks occurred near interfaces of C-BGA solder joints when tested under 0 to 100°C temperature cycling after 3000 cycles, as shown in Fig. 16-14. Figure 16-15 shows the fatigue fracture of a C-CGA solder joint, demonstrating that failure occurred through the 10 Sn/90 Pb solder with necking.

16.8.4.8 Array P-BGA on a PCB. Solder joint failure occurred at the solder sphere-to-package interface as discussed in Sec. 11.4. It was also found that solder cracks near the solder sphere–package interface (not solder sphere–board interface) depended on the package design, specifically the solder mask used to define the geometry of the solder pad on the package[34,35] The solder joint with solder mask–defined geometry showed earlier failure than that without solder mask restriction. However, the failure rate of the solder mask–defined solder joint is slower than that of the assembly using the package without a solder mask–defined pad.

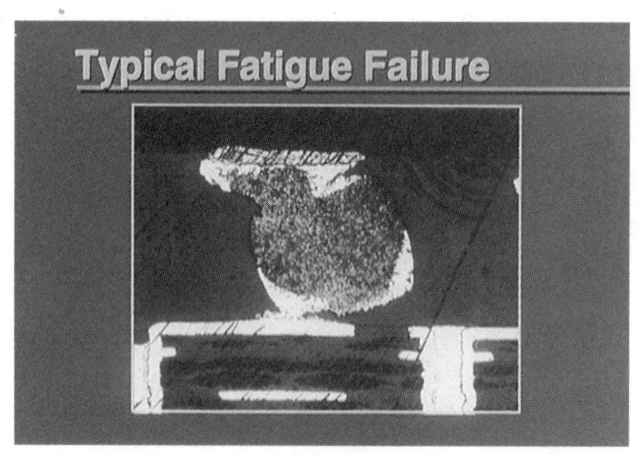

Figure 16-14. Fatigue crack of C-BGA under 0 to 100°C temperature cycling.

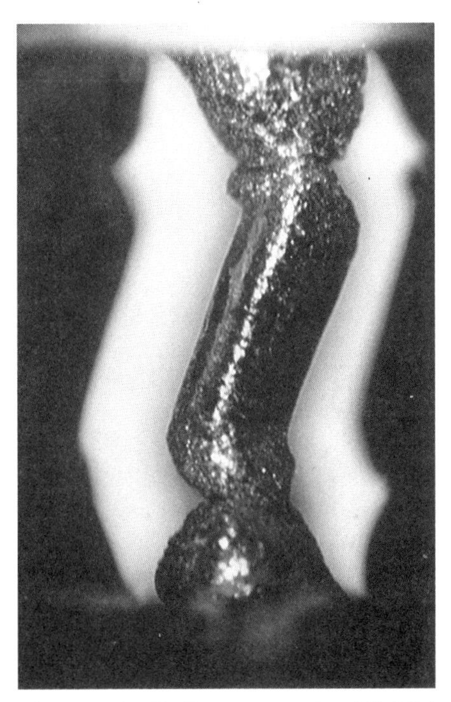

Figure 16-15. Fatigue crack of C-CGA under 0 to 100°C temperature cycling.

In an effort to minimize fatigue life, the relative displacement of the PAC component and PCB was constrained by using larger corner solder pads. Test results, however, indicated that larger corner pads did not significantly improve the fatigue life of 62 Sn/36 Pb/2 Ag solder joints.[36] Comparing BGAs and QFPs under a fatigue environment, Fig. 16-16 illustrates the relative fatigue rate among C-BGAs, C-CGAs, and QFPs.[37]

It should be noted that the above illustrations are some examples that are considered to be prevalent failure modes for various solder configurations. As failure modes largely depend on the test (service) conditions and on the intrinsic properties of the solder compositions, deviations from the above illustrations are not uncommon.

16.8.4.9 Other solder joint failure modes. Solder joint failure can occur without the involvement of a typical creep and/or fatigue process. The following case study illustrates solder joint failure caused at the manufacturing stage.

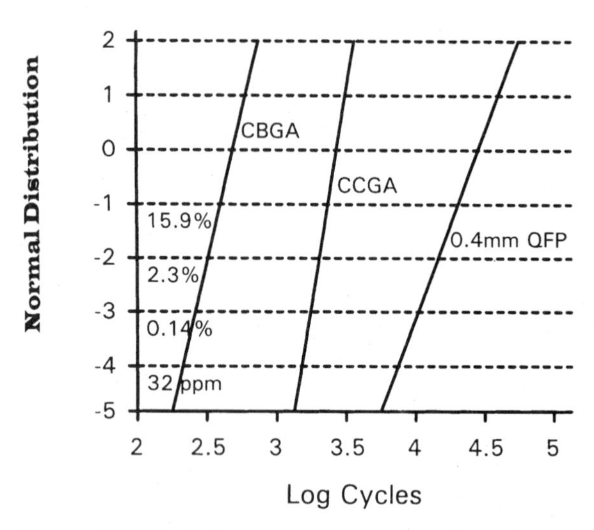

Figure 16-16. Failure rate comparison of C-BGA, C-CGA, and 0.4-mm QFP.

Case study: Solder joint failure (open joint) on production floor

Description

An OEM manufacturing 12-layer PCBs for supermicrocomputers (four coprocessors) encountered functional failure (rejects of circuit boards during electrical functional tests). The board was double-sided, epoxy-fiber based containing a mix of surface mount components and through-hole components. It had many via holes of different sizes. The finest-pitch surface mount component was QFP, 20 mil (0.50 mm). The solder pads on the board were typically Sn-Pb electroplated Cu.

The manufacturing system consisted of four primary processes:

1. Dry the board overnight at around 110°C.
2. Assemble the bottom side of the board by the steps of printing solder paste, applying adhesive, placing chip components, and curing/reflow. A preheat stage at about 150°C during the reflow process also cures the adhesive.
3. Assemble the top side of the board by the typical surface mount steps, including printing solder paste, placing components, and infrared reflow with a peak temperature around 200°C as measured by thermocouple.
4. Insert through-hole components, and drag solder by floating the board with the bottom side in contact with the molten solder at a peak temperature of about 230°C for 10 to 15 s. On completion of

the drag soldering, all via holes are plugged so that the probe needle will not be caught, and all through-hole components are soldered.

Observations

After process 3, all solder joints were examined and found to be shiny, smooth and to have good wetting and solder fillets. After process 4, the board underwent an electrical function test and failed.

The failure was detected as open solder joints which were found to be associated with 20-mil- (0.50-mm-) pitch surface mount components. It was further detected that the open solder joints occurred only with those solder joints that were parallel to the direction of board travel.

The examination also showed that the open joints with the solder fillet primarily remaining on the leads of the component appeared to be shiny and intact. The portion of the solder fillet sandwiched between the lead and pad looked dull, rough, and soft.

Analysis

It was first thought that the repeated heat excursion may have promoted excessive intermetallics, therefore causing degraded wettability of the pads. It was also thought that the electroplated solder pads may have a poor solder surface. Attempts were made to shorten the heating cycles and to replace electroplated boards with solder-dipped/air knife boards. The problem persisted, indicating that the solderability versus intermetallics would not be the main source of the open solder joints.

Excess heat exposure during process 4 was suspected. High-temperature tape was used to cover the bottom side of the 20-mil-(0.50-mm-) pitch component before going through the drag solder pot. The problem was alleviated. It was found that the relatively large size (0.40-in diameter) and large number of via holes (240) near the four sides of the component allowed sufficient heat to be transferred to the solder joints which approached melting temperature. The heat transfer was facilitated by the travel direction and the relative location of via holes. This accounted for the occurrence of open joints aligned longitudinally with, rather than perpendicular to, the travel direction.

Moral

Any step in the process could contribute to the overall manufacturing quality and long-term reliability. If the electrical function test did not detect the failure and the assembly was not rejected, it would have caused early failure in the field. The initial circuitry design may not always be compatible with the manufacturing process selected.

16.9 References

1. H. J. Grover, S. A. Gordon, and L. R. Jackson, "The Fatigue of Metals and Structures," U.S. Government Printing Office, 1954.
2. A. Miller and R. H. Priest, "Materials Response to Thermal-Mechanical Strain Cycling," in R. P. Skelton, ed., *High Temperature Fatigue—Properties and Prediction,* Elsevier Applied Science, New York, 1987, pp. 113–175.
3. J. S. Hwang and Z. Guo, "Lead Free Solder for Electronics Packaging," *Proceedings Surface Mount International,* 1993, p. 733.
4. Solomon, "Review of Critical Variables Determining Solder Fatigue Lives," *Material Research Society Symposium Proceedings,* vol. 154, 1989, p. 4411.
5. H. D. Solomon, V. Brozowski, and D. G. Thompson, "Prediction of Solder Joint Fatigue Life," GE Report #88CRD101, 40th ECTC, May 1990, p. 351.
6. H. D. Solomon, "High and Low Temperature Strain Life Behavior of a Pb-Rich Solder," *ASME Journal of Electronic Packaging,* vol. 112, June 1990, p. 123.
7. J. P. Clech et al., "Surface Mount Solder Attachment Reliability Figures of Merit," *1989 SMART V Conference,* New Orleans, LA.
8. J. Lau and D. W. Rice, "Solder Joint Fatigue in Surface Mount Technology: State of the Art," *Solid State Technology,* October 1985, p. 99.
9. W. Engelmaier, "Fatigue Life of Leadless Chip Carriers Solder Joints During Power Cycling," *IEEE Transactions on Components, Hybrids and Manufacturing Technology,* CHMT-6, September 1983, p. 232.
10. F. Lovasco, "Solder Joint Modeling Using Finite Element Analysis," *Proceedings 1988 International Electronic Packaging Conference,* Dallas, TX, November, p. 410.
11. L. F. Coffin, *Metal Transactions,* vol. 2, 1971, pp. 3105–3113.
12. S. S. Manson, G. R. Halford, and H. M. Hirshberg, "Design for Elevated Temperature Environment," *Transactions of ASME,* 1971, p. 250.
13. K. C. Norris and A. H. Landzberg, "Reliability of Controlled Collapse Interconnections," *IBM Journal of Research and Development,* 13, 1969, p. 266.
14. Z. Guo, Ph.D. Dissertation, North Carolina State University, 1993.
15. Jennie S. Hwang, *Ball Grid Array and Fine Pitch Peripheral Interconnections,* Electrochemical Publications, LTD, 1995, p. 185.
16. D. B. Barker et al., "Effect of SMC Lead Dimensional Variabilities on Lead Compliance and Solder Joint Fatigue Life," *Journal of Electronic Packaging,* vol. 114, June 1992, p. 177.
17. J. Lau et al., "Experimental and Analytical Studies of 28-pin Thin Small Outline Package (TSOP) Solder Joint Reliability," *Journal of Electronic Packaging,* vol. 114, no. 2, 1992, p. 169.

18. P. Viswanadham, A, Emerick, and R. Haggett, "Reliability Assessments of Thin Small Outline Packages for Memory Card Applications—A Comparative Study," IBM Technical Report TR-01. C4919, 1992.
19. F. J. Liotine, "Screen/Stencil Printing of Large Solder Volumes for Extending LCC Solder Joint Reliability," *Journal of Surface Mount Technology*, July 1990, p. 19.
20. F. J. Liotine, "Lead-Less Chip Carrier Solder Joint Reliability: A Comparison Study for High Strain Ranges," *Journal of Surface Mount Technology*, July 1991, p. 5.
21. J. Lau et al., "Reliability of 0.4 mm Pitch 256-Pin Plastic Quad Flat Pack No-Clean and Water-Clean Solder Joints," *Proceedings IEEE/EIA Electronic Components and Technology Conference*, June 1993, p. 39.
22. M. Bevan, "Effect of Conformal Coating on the Reliability of LCCC Solder Joints Exposed to Thermal Cycling," *Proceedings Surface Mount International*, 1991, p. 737.
23. D. Ostendorf, "Solder Joint Reliability of Leaded and Leadless Components Mounted on Multiple Substrate Materials: A Case Study," *Proceedings Surface Mount International*, 1992, p. 453.
24. W. H. Liu and H. T. Corten, Report no. 566, Theoretical and Applied Mechanics Department, University of Illinois, December 1958.
25. J. S. Hwang, Z. Guo, and G. Lucy, "Strengthened Solder Materials for Electronic Packaging," *Proceedings Surface Mount International*, 1993, p. 662.
26. R. Satoh et al., "Thermal Fatigue Life of Pb-Sn Alloy Interconnections," *IEEE Transactions on Components, Hybrids, and Manufacturing Technology*, vol. CHMT-14, no. 1, March 1991, p. 224.
27. M. McShane et al., "Lead Configurations and Performance for Fine Pitch ST Reliability," *Proceedings NEPCON-West*, 1990, p. 238.
28. B. Wong and D. E. Helling, "A Mechanic Model for Solder Joint Failure Prediction Under Thermal Cycling," *Journal of Electronic Packaging*, vol. 112, June 1990, p. 104.
29. R. N. Wild, "Some Factors Affecting Leadless Chip Carrier Solder Joint Fatigue Life II," *Circuit World*, vol. 14, no. 4, 1988, p. 29.
30. W. M. Wolverton, "The Mechanism and Kinetics of Solder Joint Degradation," *Brazing and Soldering*, no. 13, Autumn 1987, p. 33.
31. C. J. Brierley and D. J. Pedder, "Surface Mounted IC Packages—Their Attachment and Reliability on PWB's," *Circuit World*, vol. 10, no. 2, 1984, p. 28.
32. R. Darveaux, "Crack Initiation and Growth in Surface Mount Solder Joints," *Proceedings ISHM*, 1993, p. 86.
33. B. N. Agarwala, "Thermal Fatigue Damage in Pb-In Solder Interconnections," *International Reliability Physics Symposium*, 1985, p. 198.

34. Abbas I. Attarwala, "Failure Mode Analysis of a 540 Pin Plastic Ball Grid Array," *Proceedings SMI*, 1994, p. 252.
35. C. Ramirez and S. Fauser, "Fatigue Life Comparison of the Perimeter and Full Plastic Ball Grid Array," *Proceedings SMI*, 1994, p. 258.
36. K. Moore et al., "Solder Joint Reliability of Fine Pitch Solder Bumped Pad Array Carriers," *Proceedings NEPCON-West*, 1990, p. 264.
37. W. E. Bernier et al., "BGA vs. QFP," *Circuits Assembly*, March 1995, p. 38.

16.10 Suggested Readings

Adams, P. J., "Thermal Fatigue of Solder Joints in Micro-Electronic Devices," M.S. thesis, MIT, 1986.

Baik, S., and R. Raj, "Mechanism of Creep-Fatigue Interaction," *Metallurgy Transactions*, vol. 13A, 1982, p. 1215.

Baker, E., "Stress Relaxation in Tin-Lead Solders," *Material Science and Engineering*, vol. 38, 1979, p. 241.

Burkhart, A., "Recent Developments in Flip Chip Technology," *Surface Mount Technology*, July 1991, pp. 41–44.

Clatterbaugh, G. V., and H. K. Charles, "Thermomechanical Behavior for Soldered Interconnects for Surface Mounting, A Comparison of Theory and Experiment," *Proceedings 35th IEEE Electronic Components Conference*, 1985, p. 60.

Clech, J. P., and J. A. Augis, "Temperature Cycling, Structural Response and Attachment Reliability of Surface-Mounted Leaded Package," *Proceedings IEPS Conference*, 1988, p. 305.

Coffin, L. F., "A Note on Low Cycle Fatigue Laws," *Journal of Materials*, vol. 6, no. 2, 1971, p. 388.

Engelmaier, W., "Solder Interconnects Component to Board," *Interconnection Technology*, March 1994, p. 16.

Engelmaier, W., "Surface Mount Attachment Reliability of Clip-Leaded Ceramic Chip Carrier on FR-4 Circuit Board," *IEPS Journal*, vol. 9, no. 4, January 1988, p. 3.

Engelmaier, W., "Test Method Considerations for MST Solder Joint Reliability," *Proceedings IEPS*, 1984, p. 360.

Enke, N. F., et. al., "Mechanical Behavior of 60/40 Tin-Lead Solder," *Proceedings 39th Electronic Components Conference*, 1989, p. 264.

Evans, J. W., "Grain Growth in Eutectic Solder: Impact on Accelerated Testing," *Proceedings IPC Meetings*, 1990, p. 58.

Frear, D., "Thermomechanical Fatigue in Solder Material," Chap. 5, in *Solder Mechanics*, The Minerals, Metals and Materials Society, Warrendale, Pa., 1992, p. 192.

Frost, H. J., R. T. Howard, and G. J. Stone, "Effects of Thermal History on Microstructure and Mechanical Properties of Solder Alloys," *Proceedings ASM 2nd International Electronic Materials and Processing Congress*, 1989, p. 121.

Greenfield, L. T., and P. G. Forrester, "The Properties of Tin Alloys," International Tin Research Institute Publication No. 155, 1947.
Grivas, D., K. L. Murty, and J. W. Morris, "Deformation of Pb-Sn Eutectic Alloys at Relatively High Strain Rate," *Acta Metallurgica*, vol. 27, 1989, p. 731.
Guo, Z., A. F. Sprecher, and H. Conrad, "Plastic Deformation Kinetics of Eutectic Pb-Sn Solder Joints in Monotonic Loading and Low-Cycle Fatigue," *Journal of Electronic Packaging*, vol. 114, June 1992.
Hall, P. M., "Forces, Moments and Displacements During Thermal Chamber Cycling of Leadless Ceramic Chip Carriers Soldered to Printed Boards," *IEEE Transactions on Components, Hybrids and Manufacturing Technology*, vol. 7, no. 4, 1984, p. 314.
Harper, C. A., *Handbook of Materials and Processes for Electronics*, McGraw-Hill, New York, 1970.
Howard, R. T., S. W. Sobeck, and C. Sanatra, "A New Package Related Failure Mechanism for Leadless Ceramic Chip Carriers Solder Attached to Alumina Substrates," *Solid State Technology*, February 1983, p. 115.
Hwang, J. S., *Solder Paste in Electronic Packaging—Technology and Applications in Surface Mount, Hybrid Circuits, and Component Assembly*, Van Nostrand Reinhold, New York, Chap. 9, 1989, p. 261.
Hwang, J. S., and R. M. Vargas, "Solder Joint Reliability—Can Solder Creep?" *International Symposium of Microelectronics*, 1989, p. 513.
Inone, H., Y. Kurihara, and H. Hachino, "Pb-Sn Solder for Die Bonding of Silicon Chips," *IEEE Transactions on Components, Hybrids and Manufacturing Technology*, vol. 9, no. 2, 1986, p. 190.
Johnson, R., et. al., " Feasibility Study of Ball Grid Array Packaging," *Proceedings NEPCON-East*, 1993, p. 413.
Karjalainen, L. P., "Relation Between Microstructure and Fracture Behavior in Surface Mounted Joints," *Brazing and Soldering*, no. 15, 1988, p. 37.
Kashyap, B. P., and G. S. Murty, "Experimental Constitutive Relations for High Temperature Deformation of a Pb/Sn Eutectic Alloy," *Materials Science and Engineering*, vol. 50, 1981, p. 205.
Kitano, M., T. Shimizu, and T. Kumazava, "Statistical Fatigue Life Estimation: The Influence of Temperature and Composition on Low-Cycle Fatigue of Tin-Lead Solders," *Current Japanese Materials Research*, vol. 2, 1987, p. 235.
Knott, J. F., *Fundamentals of Fracture Mechanics*, John Wiley, New York, 1973.
Kotlowwitz, R. W., "Comparative Compliance of Generic Lead Designs for Surface Mounted Components," *Proceedings International Electronic Packaging Conference*, 1987, p. 965.
Lam, S. T., A. Arieli, and A. K. Mukherijee, "Superplastic Behavior of Pb-Sn Eutectic Alloy," *Materials Science and Engineering*, vol. 40, 1979, p. 241.
Liejstrand, L. G., and L. O. Anderson, "Accelerated Thermal Fatigue Cycling of Surface Mounted DWB Assemblies in Telecom Equipment," *Circuit World*, vol. 14, no. 3, 1988, p. 69.
Manson, S. S., *Thermal Stress and Low-Cycle Fatigue*, McGraw-Hill, New York, 1966.
Manson, S. S., "Interfaces Between Fatigue Creep and Fracture," *International Journal of Fracture Mechanics*, vol. 2, no. 1, 1966, p. 327.

Marshall, J. L., and S. R. Walter, "Fatigue of Solder," *International Journal for Hybrid Microelectronics,* 1987, p. 11.

McCreary, J., et al., "Flip Chip Solder Bump Fatigue Life Enhanced by Polymer Encapsulation," *Proceedings 40th Electronic Components and Technology Conference,* 1990, p. 158.

Meleka, A. H., "Combined Creep and Fatigue Properties," *Metallurgical Review,* vol. 7, 1962, p. 43.

Miner, M. A., "Cumulative Damage in Fatigue," *Journal of Applied Mechanics,* vol. 12, 1945, p. 452.

Ng, K. G., "Assessing Surface Mount Attachment Reliability of Ceramic Cylindrical Resistor on FR-4 Circuit Boards," *Proceedings Expo SMT,* 1988, p. 1.

Oyan, C., et al., "Role of Strain-Partitioning Analysis in Solder Life Prediction," *International Journal for Hybrid Microelectronics,* ISHM, vol. 14, no. 2, June 1991, p. 37.

Pao, Y. H., E. Jin, and V. Reddy, V., "Development of a Computer Aided System for Electronic Packaging Interconnect Reliability Prediction," *Proceedings NEPCON-East,* 1994, p. 298.

Raman, V., and T. C. Reileg, "Cavitation and Cracking in As-Cast and Superplastic Pb-Sn Eutectic During High-Temperature Fatigue," *Journal of Materials Science Letters,* vol. 6, no. 15, 1987, p. 549.

Seraphim, D., R. Lasky, and C. Ki, *Principles of Electronic Packaging,* McGraw-Hill, New York, 1989.

Shah, H. J., and J. H. Kelly, "Effect of Dwell Time on Thermal Cycling of the Flip-Chip Joint," *Proceedings ISHM,* 1970, p. 341.

Shine, M. C., and L. R. Fox, "Fatigue of Solder Joint in Surface Mount Devices," American Society of Testing and Materials, ASTM STP 942, 1987, p. 588.

Solomon, H. D., "Strain Life Behavior in 60/40 Solder," *Journal of Electronic Packaging,* ASME, vol. 111, no. 2, 1989, p. 75.

Solomon, H. D., "Effect of Hold Time and Fatigue Cycle Wave Shape on the Low Cycle Fatigue of 60/40 Solder," *Proceedings 38th Electronic Components Conference,* 1988, p. 7.

Solomon, H. D., "Creep, Strain Rate Sensitivity and Low Cycle Fatigue of 60/40 Solder," *Brazing and Soldering,* 1986, p. 68.

Stone, D., et. al., "An Investigation of the Creep Fatigue Interaction in Solder Joints," *Proceedings NEPCON,* 1986, p. 175.

Stone, D., et al., "The Mechanism of Damage Accumulation in Solders during Thermal Fatigue," *Proceedings,* 36th Electronics Components Conference, 1986, p. 630.

Subrahmanyan, R., J. R. Wilcox, and Cy Li, "A Damage Integral Approach to Thermal Fatigue of Solder Joints," *Proceedings Electronic Component Conference,* 1989, p. 240.

Summece, T. S. E., and J. W. Morris, "Isothermal Fatigue Behavior of Sn-Pb Solder Joints," *Transactions of ASME,* vol. 112, June 1990, p. 94.

Tien, J. K., et al., "Creep-Fatigue Interactions in Solder," *Proceedings 39th Electronic Components Conference,* 1989, p. 259.

Tribula, D., D. Grivas, and R. D. Frear, "Microstructure Observation of

Thermomechanically Deformed Solder Joints," *Welding Research,* (Suppl.) October 1989, p. 404s.

Tummala, R., and E. J. Rymaszewski, eds., *Microelectronics Packaging Handbook,* Van Nostrand Reinhold, New York, 1989.

Vaynman, S., M. F. Fine, and D. A. Jeannott, "Low-Cycle Isothermal Fatigue Life of Solder Materials," in Chap. 4 of *Solder Mechanics,* The Minerals, Metals and Materials Society, p. 155.

Vaynman, S., M. E. Fine, and D. A. Jeannotte, "Isothermal Fatigue of Low Tin Lead-Based Solder," *Metallurgy Transactions,* vol. 19A, 1988, p. 1051.

Wild, R. N., "Some Factors Affecting Leadless Chip Carrier Solder Joint Fatigue Life II," *Circuit World,* vol. 14, no. 4, 1988, p. 29.

Wolverton, W. M., "The Mechanisms and Kinetics of Solder Joint Degradation, *Brazing and Soldering,* no. 13, 1987, p. 33.

Wong, B., D. Helling, and R. W. Clark, "A Creep-Rupture Model for Two Phase Eutectic Solder," *IEEE Transactions on Components, Hybrids and Manufacturing Technology,* vol. 11, 1988, p. 305.

Yamada, T., R. Doyle, and J. Barrett, "A Fracture Surface Examination of Fine Pitch Surface Mount Solder Joints," *International Journal of Microcircuits and Electronic Packaging,* vol. 17, no. 2, 1994, p. 184.

17

New and Revised Standards and Test Methods

As discussed in Chaps. 1 and 2, the driving forces of the electronics industry continue to change the requirements and/or demands of products and their performance in electronics packaging and circuit board assemblies. This chapter is intended to provide an overview of specifications and standards that are related to the subject of the book as well as to highlight key aspects of developments of standards and test methods.

17.1 Key Issues

In developing new specifications or standards or revising the existing standards, several challenges have emerged; these include

- Necessary and sufficient data
- Obtaining consensus
- Verifying viability
- Achieving practical adoption
- Industry standards versus U.S. Military standards
- U.S. standards versus International standards

The level of performance and reliability of product systems varies with the manufacturers of different segments of the industry and with the end-use applications. This, coupled with individuals' diverse experiences and perceptions makes the consensus among the parties involved hard to come by. This is particularly true for areas where sufficient hard data are lacking. Surface mount technology was implemented for PCB assembly in the early 1980s; long-term real-world performance results are still limited. Extrapolation, deduction, and assessment may often be needed to form a standard. As the initial draft of standards is composed, extensive discussion, debate, comments gathering, and sometimes further data collection constitute the review process. After the completion of the review process, which can be very lengthy sometimes, a standard will be put in the format by an accredited standards-developing organization for its issuance.

An issued standard then faces its adoption by users and suppliers. Only those standards that fulfill the goal of maximizing and controlling the quality and performance of products without imposing unrealistic cost to the yield and productivity of the manufacturing operation would be widely adopted.

United States Military standards are still apart from industry standards. In the past, military standards have been the driver and have served as the basis for setting industry standards due largely to the research and technology leadership provided by the military sector. With the rapid changes in technologies, products, and marketplace, the industry must and has been responding to the changing demands. Recently, the industry has taken the lead to revise existing standards to fit the marketplace and to develop new standards to meet new product needs. As of the time of this writing military programs have recently started using commercial standards and specifications (except a small number of applications from which no commercial products will meet the requirements). Accordingly, most of joint industry standards are equally applicable to military uses, in lieu of previous military standards.

Another industry trend in market globalization prompts the development of international standards. Efforts required to make standards internationally acceptable and usable are obviously immense. On the bright side, diligent work has been ongoing and much progress has been made.

17.2 Specifications and Standards Organizations

The institutions and organizations that have made a significant effort in setting standards that are related to solder interconnections of electronic packaging and assemblies are listed below, with their addresses and their respective standards designations included under each institution or organization name.

- Electronic Industries Associations (EIA)

  2001 Eye Street, N.W., Washington, DC 20006
  JEDEC- (Joint Electronic Device Engineering Council)

- Institute for Interconnecting and Packaging Electronic Circuits (IPC)

  7380 N. Lincoln Avenue, Lincolnwood, IL 50546
  IPC-

- Joint coordination by EIA and IPC

  J-STD- (Joint Industry Standards)
  ANSI/J-STD- (American National Standards Institute)

- Department of Defense and Federal Government

  Standardization Documents Order Desk, Building 4D, 700 Robbins Avenue, Philadelphia, PA 19111-5094
  MIL-

- American Society for Testing and Materials (ASTM)

  1916 Race Street, Philadelphia, PA 19103-1187
  ASTM-

- International Electrochemical Commission (IEC)

  3 Rue de Varembe, 1211 Geneva, Switzerland
  IEC-

- International Organization for Standards (ISO)

  1, Rue de Varembe, Case Postale 56, CH-1211 Geneve 20, Switzerland
  ISO-

17.3 Applicable U.S. Specifications and Standards

The U.S. specifications and standards applicable to the scope of this work include

- ANSI/J-STD-001 Requirements for Soldered Electrical and Electronic Assemblies.
- ANSI/J-STD-002 Solderability Tests for Components, Leads, Terminations, Lugs, Terminals and Wires
- ANSI/J-STD-003 Solderability Tests for Printed Boards
- ANSI/J-STD-004 Requirements for Soldering Fluxes
- ANSI/J-STD-005 Requirements and Test Methods for Solder Paste
- ANSI/J-STD-006 General Requirements and Test Methods for Electronic Grade Solder Alloys and Fluxed and Non-fluxed Solid Solders for Electronic Soldering Applications
- MIL-F-14256F Flux, Soldering, Liquid, Paste Flux, Solder Paste and Solder-Paste Flux (for Electronic/Electrical Use), General Specifications for
- IPC-SM-782 Surface Mount Design and Land Pattern Standard
- MIL-STD-2000A Standard Requirements for Soldered Electrical and Electric Assemblies
- MIL-P-28809A Printed Wiring Assemblies
- MIL-P-55110 Printed Wiring Boards, General Specification for
- MIL-STD-202 Test Methods for Electronic and Electrical Component Parts
- Belcore TR-TSY-00078 General Physical Design Requirements for Telecommunications Products and Equipment
- QQ-S-571 Solder, Tin Alloy, Lead Tin Alloy, and Lead Alloy
- JEDEC-22 Test Method B-108 Coplanarity Test Method for Surface Mount Devices
- JEDEC-95 Component Standards
- IPC-A-600 Acceptability of Printed Boards
- IPC-A-610 Acceptability of Printed Board Assemblies
- IPC-SM-785 Guidelines for Accelerated Reliability Testing or Surface Mount Solder Attachments
- IPC-SM-786 Recommended Procedures for Handling of Moisture Sensitive Plastic IC Packages

- IPC-SM-817 General Requirements for Dielectric Surface Mounting Adhesives
- IPC-SM-840 Qualification and Performance of Permanent Polymer Coating (Solder Mask) for Printed Boards
- IPC-TM-650 Test Methods Manual

  2.2.14 Solder Powder Particle Size Distribution—Screen Methods
  2.2.14.1 Solder Powder Particle Size Distribution—Measuring Microscope Method
  2.2.14.2 Solder Powder Particle Size Distribution—Optical Image Analyzer Method
  2.2.14.3 Determination of Maximum Solder Powder Particle Size
  2.2.20 Solder Paste Metal Content by Weight
  2.3.13 Determination of Acid Value of Liquid Solder Flux—Potentiometric and Visual Titration Methods
  2.3.25 Resistivity of Solvent Extract
  2.3.26 Ionizable Detection of Surface Contaminants (Dynamic Method)
  2.3.26.1 Ionizable Detection of Surface Contaminants (Static Method)
  2.3.27 Cleanliness Test—Residual Rosin
  2.3.32 Flux Induced Corrosion (Copper Mirror Method)
  2.3.33 Presence of Halides in Flux, Silver Chromate Method
  2.3.34 Solids Content, Flux
  2.3.35 Halide Content, Quantitative (Chloride and Bromide)
  2.3.35.1 Fluorides by Spot Test, Fluxes—Qualitative
  2.3.35.2 Fluoride Concentration, Fluxes—Quantitative
  2.3.38 Surface Organic Contaminant Detection Test (In-House Method)
  2.4.14.2 Liquid Flux Activity, Wetting Balance Method
  2.4.34 Solder Paste Viscosity—T-Bar Spin Spindle Method
  2.4.34.1 Solder Paste Viscosity—T-Bar Spindle Method
  2.4.34.2 Solder Paste Viscosity—Spiral Pump Method
  2.4.34.3 Solder Paste Viscosity—Spiral Pump Method (Applicable at Less than 300,000 cps)
  2.4.34.4 Paste Flux Viscosity
  2.4.35 Solder Paste—Slump Test
  2.4.43 Solder Paste—Solder Ball Test
  2.4.44 Solder Paste—Tack Test
  2.4.45 Solder Paste—Wetting Test
  2.4.46 Spread Test, Liquid Solder Flux
  2.4.47 Flux Residue Dryness
  2.4.48 Spitting of Flux-Cored Wire Solder
  2.4.49 Solder Pool Test

2.6.3.3 Surface Insulation Resistance, Fluxes
3.6.1 Fungus Resistance Printed Wiring Materials

- ASTM-B-32 Standard Specification for Solders
- ASTM-B-212 Test for Apparent Density of Metal Powders
- ASTM-B-213 Test for Flow Rate of Metal Powders
- ASTM-B-214 Sieve Analysis of Granular Metal Powders
- ASTM-D-465 Acid Number of Rosin, Test Methods for
- ASTM-D-1210 Fineness of Dispersion of Pigment Vehicle Systems

17.4 Applicable International Standards and Test Methods

Any participating country can submit its national standards for use as international standards, subject to a thorough review process by all participating countries for comments, changes, verification, and agreement. Some examples of standards and test methods are

IEC-68-2-54 Environmental Testing. Part 2: Tests. Test Ta: Solder-ability Testing by the Wetting Balance Method.

IEC-68-2-58 Environmental Testing. Part 2: Tests. Test Td: Solder-ability Resistance to Dissolution of Metallization and to Soldering Heat of Surface Mounting Devices (SMD).

IEC-1191-1 Generic Standard, Requirements for Soldered Electrical and Electronic Assemblies Using Surface Mount and Related Assembly Technology.

IEC-8-2-20 Basic Environmental Testing Procedures

ISO-9453-1990 Soft Solder Alloys—Chemical Compositions and Forms

ISO-9000 Quality Management and Quality Assurance Standards Guidelines Selection and Use

ISO-9002 Quality Systems—Model for Quality Assurance in Production and Installation

17.5 Examples of Revisions of U.S. Specifications and Standards

As to the general requirements for soldered electronic assemblies, it is intended to integrate the pertinent information and data from MIL-

STD-2000A, Belcore TR-TSY-000078, relevant IPC standards, and the most-recent data and information on the subject into one document, i.e., ANSI/J-STD-001. Using a similar technique, the part of QQ-S-571 that is related to fluxes, IPC-SP-818 (General Requirements for Electronic Soldering Fluxes), and the part of MIL-F-14256F that is related to fluxes are merged into one document, ANSI/J-STD-004. The portion of QQ-S-571 that is related to solder paste, IPC-SP-819 (General Requirements for Electronic Grade Solder Paste), and applicable parts of MIL-F-14256 are combined into ANSI/J-STD-005. The portion related to solder alloys of QQ-S-571 becomes ANSI/J-STD-006.

Some examples of new or significant changes which are incorporated into the above industry standards are summarized here.

17.5.1 Solder alloys

In addition to the widely increased listing of solder compositions, the level of impurity elements has been altered. Table 17-1 lists the impurity elements and their permissible levels in solder (per ANSI/J-STD-006).

Three classes of alloys designated as A, B, C are included; they contain three allowable levels of Sb content. Table 17-1 applies for alloys identified as E category except that the Pb level is 0.1 max and the Sb level is 0.2 max. For alloys identified as D alloys (ultrapure, intended

Table 17-1. Level of Impurity Elements in Solders

Element	Percentage by mass
Ag	0.05
Al	0.005
As	0.03
Au	0.05
Bi	0.10
Cd	0.002
Cu	0.08
Fe	0.02
In	0.10
Ni	0.01
Pb	0.02
Sb (A)	0.5 max
Sb (B)	0.2 max
Sb (C)	0.05 max
Sn	0.25
Zn	0.03

for use in barrier-free die attachment applications), specifications are

Total percentage by mass of all impurity elements is 0.05 max.
Total percentage by mass of Be, Hg, Mg, and Zn is 0.0005 max.
Total percentage by mass of As, Bi, P, and Sb is 0.0005 max.

17.5.2 Fluxes

A new nomenclature for fluxes is proposed in J-STD-004. Flux types are listed as

RO Rosin
RE Resin
OR Organic
IR Inorganic

The activity levels are classified as

L Low
M Medium
H High
O Halide absent
I Halide present

Fluxes are further grouped into six ranges of concentration:

1 1.1 to 2.0
2 2.1 to 3.0
3 3.1 to 4.0
4 4.1 to 5.0
5 5.1 to 6.0
6 6.1 to 7.0

These new flux types can be grossly related to the conventional fluxes as follows:

LO All R, some RMA, some low solids, no-clean
L1 Most RMA, some RA
MO Some RA, some low solids, no-clean
M1 Most RA
HO Some water-soluble
H1 Most water-soluble and synthetic activated

Table 17-2. Flux Type Versus Test Requirements

Flux type	Copper mirror	Silver chromate (Cl, Br)	Spot test (F)	Quantitative (halide)	Corrosion test
L0	No evidence of mirror break-through	Pass	Pass	0	No evidence of corrosion
L1		Pass	Pass	<0.5%	
M0	Breakthrough in less than 50% of test area	Pass	Pass	0	Minor corrosion acceptable
M1		Fail	Fail	0.5 to 2.0%	
H0	Breakthrough in more than 50% of test area	Pass	Pass	0	Major corrosion acceptable
H1		Fail	Fail	>2.0%	

Further changes that incorporate the new flux types to be defined by requirements of established halide and corrosion tests are summarized in Table 17-2 (per ANSI-STD-004).

Test procedures for the above requirements specified in ANSI/J-STD-004 are the copper mirror test: IPC-TM-650 Test Method 2.3.32; the silver chromate test for chlorides and bromides: IPC-TM-650 Test Method 2.3.33; the Spot Test for fluorides: IPC-TM-650 Test Method 2.3.35.1; quantitative halide content: IPC-TM-650-Test Method 2.3.35 for chlorides and bromides and IPC-TM-650 Test Method 2.3.35.2 for fluorides; corrosion test: IPC-TM-650 Test Method 2.6.15.

17.5.3 Solder powder particle size distribution

Six classes of solder powder particle size distribution are offered in ANSI/J-STD-006 as listed in Table 17-3.

Table 17-3. Six Classes of Solder Powder Particle Size Distribution

Powder size symbol	None larger than, μm	Less than 1% larger than, μm	At least 80% between, μm	At least 90% between, μm	Less than 10% smaller than, μm
1	160	150	150–75	150–20	20
2	80	75	75–45	75–20	20
3	50	45	45–25	45–20	20
4	40	38		38–20	20
5	28	25		25–15	15
6	18	15		15–5	5

17.5.4 Solderability test and criteria for components and terminations

Four alternative tests are included in ANSI/J-STD-002 for leaded and leadless components. These are

- Dip and look for both leaded and leadless components
- Wetting balance for leaded components
- Wetting balance for leadless components
- Globule test for leaded components

For the dip-and-look test, accept/reject criteria are that all leads or terminations shall exhibit a continuous solder coating free from defects for a minimum of 95 percent of the surface area of any individual lead or termination. The standard does not provide the accept/reject criteria for wetting balance and the globule test. However, the suggested criteria for the wetting balance test are offered as

For leadless components

- Coefficient of Wetting > 150 μN/s where the coefficient of wetting is the equilibrium wetting force divided by the time to reach the equilibrium force.

For leaded components

- The wetting curve (force versus time) should cross the buoyancy-corrected zero axis at, or before, 1 s from the start of the test.
- The wetting force should reach a positive value of 200 μN/mm at, or before, 2.5 s from the start of the test.
- The wetting force should remain above the positive value of 200 μN/mm at 4.5 s from the start of the test.

17.5.5 Solderability test and criteria for printed boards

ANSI/J-STD-003 provides the following five alternate tests:

- Edge dip
- Rotary dip
- Solder float

- Wave solder
- Wetting balance

The accept/reject criteria when using the first four test techniques are generally that a minimum of 95 percent of the surface being tested shall exhibit good wetting for surface evaluation (does not apply to related through-holes).

17.5.6 Surface Insulation Resistance (SIR)

Current test parameters as outlined in IPC-TM-650 Test Method 2.6.3.3 specify a bias potential of 40 to 50 V dc at 65 ± 2°C, 85 percent relative humidity for 168 h, and a test voltage of 100 V dc for all three classes of products. This is in contrast with the previous test conditions with classes 1 and 2 at 50 ± 2°C, 90 percent relative humidity, and class 3 at 85 ± 2°C, 85 percent relative humidity. Comparing IPC's SIR test with the electromigration test as specified by Bellcore TR-O-TSY-000078, the major differences are a much lower applied bias potential of 10 V and the test temperature at 85°C as called for in the latter specification. Either applied bias voltage or testing temperature could affect the data of resistance measurements. It would be prudent to interpret and compare the test results with consideration of test conditions in the SIR test or electromigration test. The prevalent acceptance of the SIR value is 500 MΩ for military applications, and 100 MΩ for commercial applications. Because of the expected reduction of testing temperature from 85 to 65°C in the IPC test method, it has been proposed that the acceptance value of SIR be increased to 500 MΩ for commercial applications as well.

17.5.7 Polyglycols

Based on some test data, the presence of one or more chemical ingredients that are in the category of polyglycols in flux or solder paste systems has generated a great deal of concern. The data shows that some products containing polyglycols lower the SIR value, and thus are considered detrimental to electronic circuitry. This has led to restrictions and exclusion of the use of polyglycols in certain grades of flux and solder in MIL-F-14256F and other specifications (see Sec. 8.1.6).

17.6 Examples of the Development of International Standards and Test Methods

Many IEC standards and test methods are concurrently under review including Requirements for Surface Mount Connections, Requirements for Soldered Electrical and Electronic Assemblies Using Surface Mount and Related Assembly Technologies, Requirements for Terminal Solder Connections, Test Method for Evaluating the Ability of Discrete SMDs to withstand Mechanical Forces, Test Method for Determining Pull-off Strength of Discrete SMDs, Test Method for Shear (Adhesion of Discrete SMDs), Test Method of Discrete SMD Body Strength, Test Method of Resistance to Ultrasonic Cleaning.

The set of conditions for testing the component's resistance to ultrasonic cleaning has been a concern for a long time. Current consensus is summarized as follows:

Frequency, kHz	Output of transducer, W/L	Duration, s	Temperature, °C
21–29	20	55–65	40 or below
36–48	20	110–130	40 or below

The distance between the specimen and the bottom of the tank shall be a half wavelength of an ultrasonic wave.

The IEC test method is intended to provide a concise procedure by which the properties or constituents of a material, a printed board, or an electronic assembly can be examined. It is not intended to contain acceptability levels for performance. Documents under IEC test methods are generally laid out in the format of

Slope

Normative references and applicable documents

Test conditions

Apparatus or material

Procedures

Notes

17.7 Basis of Setting Standards

Standards are primarily based on

Hard data

Knowledge

Judgment

Setting a working standard takes a great deal of effort to integrate and assess the hard data, to apply knowledge, and to exercise judgment. The latest trends and philosophy in setting standards have shifted and become directed to more flexibility in obtaining the desired results and to incorporate less specific mandates in materials, processes, or means of assuring the quality as well as in manufacturing competitiveness. It is also intended to do away with rejects based on appearance; however, cosmetic appearance is not to be ignored and shall be used as a process indicator. Process, rather than inspection/rework, has gained increased emphasis.

Among the standards formed by different organizations, disparity and discrepancy may exist. It is the user's decision to elect the most suitable standard for the specific requirements. To unify the standards is certainly the ultimate goal.

In line with market globalization, fast technology change, competitive pricing, and increased quality demand, a standard needs to be periodically reviewed and updated to maintain its viability and intended functions. A good standard is the document that will maximize the quality and performance of a product and minimize the cost of manufacturing operation when there is effective communication and total conformance with the well-established requirements of the document.

18 Solder Joint Quality

Studies on mechanical properties of bulk solders and conventional structural solder joints, including through-hole solder joints, have been carried out over several decades. Information and data derived from such studies are useful references for surface mount solder joint design and construction; yet in most cases data cannot be directly transferred or extrapolated to the behavior of surface mount solder joints. The differences are primarily a result of new and different conditions under which solder joints must perform, new materials, and new processes, namely,

- Small volume of solder fillet
- Restricted boundaries of solder joint
- Potential metallurgical interactions between substrate surface and solder alloy
- Effect of augmented temperature fluctuation and heat dissipation due to intrinsic high-density circuitry
- Effect of diverse materials that make up the circuit assembly
- New solder alloy compositions rather than tin-lead

Understanding the impact of the above-mentioned factors is the key to effectively extrapolating the existing information for its use for new applications in conjunction with the use of new data obtained.

18.1 Factors

The integrity of surface mount solder joints can be affected by the intrinsic properties of the solder alloy, the substrates in contact with the solder, the joint design or structure, and the joint-making process, as well as the external environment to which the solder joint will be exposed. Regardless of array solder interconnection or peripheral solder joints, the assurance of solder joint integrity relies on the step-by-step evaluation of the following items:

- Suitability of solder alloy for required mechanical properties
- Suitability of solder alloy for substrate compatibility
- Anticipated metallurgical compounds formed between solder and substrates,
- Adequacy of solder wetting on substrates
- Design of joint configuration in shape, thickness, and fillet area
- Optimum soldering technique including reflow method, reflow process, and parameters
- Conditions of storage in relation to aging effect on solder joint
- Conditions of actual service in terms of upper temperature, lower temperature, temperature cycling frequency, vibration, or other mechanical loading
- Effects from component characteristics including heat dissipation, stress distribution, and coplanarity
- Performance requirements under conditions of actual service
- Design of viable accelerated testing conditions that correlate to actual service conditions

18.2 Solder Joint Appearance

Despite the advances in chemical spectroscopy and electron microscopy, verification of the quality of solder joints requires either destructive techniques or time-consuming and costly examination. Visual appearance of solder joints continues to be a useful indicator about the quality and state of solder joints. Thus, this section is dedicated to the solder joint appearance. The factors which affect solder joint appearance in terms of luster, texture, and intactness are

- Inherent alloy luster
- Inherent alloy texture

- Residue characteristics after soldering from flux or paste
- Degree of surface oxidation
- Completeness of solder powder coalescence (if solder paste is in use)
- Microstructure
- Mechanical disturbance during solidification
- Foreign impurities in solder
- Level and distribution of intermetallic compounds
- Cooling rate during solidification
- Heat excursion after the solder joint is formed, including aging, temperature cycling, power cycling, higher-temperature storage

Solder alloy compositions determine the inherent luster and texture of solder joints. For example, Sn-Pb eutectic solder appears shiny, bright, and smooth, whereas a deviation from the eutectic composition such as 25 Sn/75 Pb or 10 Sn/90 Pb makes the solder joint look dull. Although Sn-Pb eutectic solder is normally shiny and smooth, other eutectic compositions in Sn-Ag, Sn-Bi, Sn-In systems have a matte surface finish. In addition to material aspects, the process of soldering also plays a role in the resulting appearance of solder joints. As discussed in Chap. 6, the process of solidification from melt is crucial to the microstructure development of an alloy composition. Microstructure in turn affects the mechanical behavior of solder joints as well as their surface appearance. Figure 18-1 shows a 63 Sn/37 Pb microstruc-

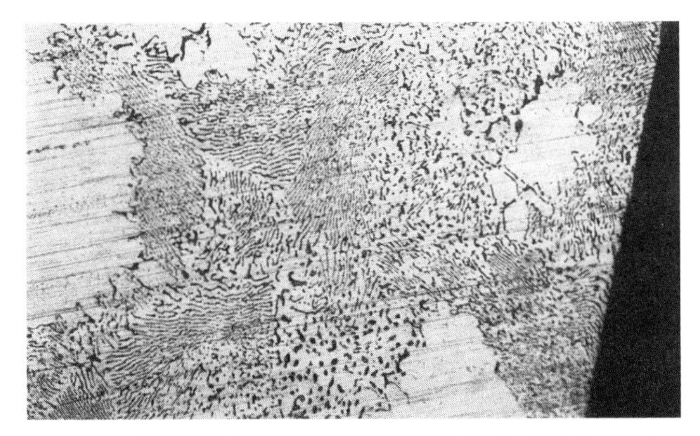

Figure 18-1. SEM microstructure of 63 Sn/37 Pb solder joint appearing dull.

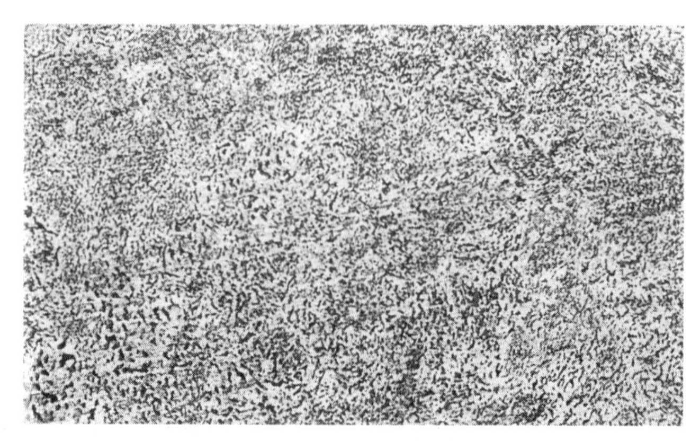

Figure 18-2. SEM microstructure of 63 Sn/37 Pb solder joint appearing shiny.

ture of a solder joint made of a solder paste containing a two-metal powder mix which was reflowed under the same conditions as that in Fig. 18-2 which exhibits the microstructure of a solder joint made of the same solder paste chemistry with the prealloyed powder. The solder joint of Fig. 18-2 appears shiny and smooth, yet the solder joint of Fig. 18-1 looks dull. Relating to Figs. 18-1 and 18-2, the following case study is illustrated.

Description

An SMT production line elected to continue using vapor phase reflow as the reflow method to manufacture SMT printed circuit boards.

In order to prevent the known J-lead wicking while assembling the large PCBs, a solder paste with a special alloy blend was used as a remedy for the wicking phenomenon. The special alloy consisted of a two-part blend of pure Sn and a high-lead Sn-Pb alloy in a specific ratio, so that the resulting composition of the blend is exactly 63% Sn and 37% Pb.

The purpose of using a two-part blend instead of prealloyed 63 Sn/37 Pb is to retard the alloy from reaching the melting stage, and thus providing the extra time for the large board to be heated up. By doing so, the leads of the components, which normally heat up much faster than the large board, would not draw the molten solder up to the top portion of the leads, causing starved or open joints.

The reflow equipment was normally used to produce a high mix of products, large and small boards, using different grades of 63 Sn/37 Pb paste. The production floor reported that the solder joints appeared dull and rough, and therefore did not pass the inspection. The production was not able to continue.

Analysis and remedy

The production personnel first suspected that the solder paste did not provide good fluxing. The examination of the cross section of the dull solder joint revealed a microstructure which significantly deviated from that of the eutectic Sn-Pb solder. The anomaly of the microstructure, composed of large segregations of high-lead phase and a small amount of eutectic phase, was attributed to the lack of heat and/or time at peak temperature during reflow. Consequently, the two-part blend was not able to alloy into a 63 Sn/37 Pb composition.

A change in the reflow process by slowing down the belt speed brought the solder joints appearance back to normal.

Moral

The appearance of solder joints can serve as a process indicator. Reflow parameters, such as peak temperature and residence time at peak temperature affect the appearance of solder joints which in turn reflects the microstructure of solder joints. Microstructure is closely linked with the mechanical properties of solder joints as well.

The characteristics of after-soldering residue are basically controlled by the chemical and physical properties of the flux/vehicle system and by the soldering parameters. As to its relation to solder joint appearance, residue may remain on the top of the solder pool or break away from the solder and settle on the outside of the solder pool. When the residue sits on the molten solder surface during solidification, it could contribute to the rough texture on the solder surface before or after its removal.

For the use of solder paste, inadequate heating time and/or temperature will cause incomplete coalescence of solder particles, which also contributes to an unsmooth surface and possibly to an inferior solder joint. During soldering, any mechanical agitation which disrupts the solidification process may lead to an uneven solder surface. Appreciable amounts of impurities and intermetallic compound formation due to foreign elements may result in a dull appearance. After the completion of reflow and solidification, heat excursion is expected to have a significant impact on solder joint appearance, whether it is only a surface reaction or an internal structural change. Heat excursion can come from different sources, such as high-temperature storage, aging, temperature fluctuation during service, and power cycling during functioning. The effects of heat excursion have been discussed in other sections.

In many cases, a duller and/or unsmooth joint may not be necessarily a defective joint in a functional sense. Tests may also confirm that the duller and unsmooth joints have equivalent mechanical strength. However, one must be cautious in drawing the conclusion regarding the joint reliability in relation to joint surface appearance. It should be noted that the surface condition of metal is considered to be one of the variables that affect failure mechanisms of the alloy. Thoroughly designed tests, considering all variables relevant to the service condition of solder joints, will provide significant information to predict the solder joint reliability, although a test to precisely simulate the conditions is quite difficult. In the author's view, the smoothness and brightness to the level that the alloy should exhibit, reflect a proper joint having been produced from a proper process.

With respect to the specification of solder joint appearance, ANSI/J-STD-001A offers the following guidelines:

The acceptable solder connection must indicate evidence of wetting and adherence where the solder blends to the soldered surface, forming a contact angle of 90° or less. The quantity of solder should not result in a contour which extends over the edge of the land. Occurrences of a contact angle greater than 75° shall be recorded as a process indicator. The solder joints should have a generally smooth appearance. A satin luster is permissible.

There are solder alloy compositions, lead or printed board platings, and special soldering processes, e.g., slow cooling with large-mass printed wiring boards (PWBs) that may produce dull, matte, gray, or grainy appearing solders that are normal for the material or process involved. These solder joints are acceptable.

A smooth transition from land to the connection surface or component lead must be evident. A line of demarcation or a transition zone where applied solder blends with solder coating, solder plate, or other surface material is acceptable provided that wetting is evident. In the case of fused solder coatings, the presence of the applied solder above the rim of the hole is not required if the hole wall and component lead exhibit good wetting. Marks or scratches in the solder joint shall not degrade the integrity of the connection. The following types of solder joints are considered defective:

- Fractured solder connections
- Disturbed solder connections
- Cold solder connections
- Excess solder which contacts the component body except otherwise noted
- Insufficient gold removal

ANSI/J-STD-001A, however, specifies the following conditions as acceptable, but they shall be considered as process indicators:

> Voids and blow holes where wetting is evident and which do not reduce solder volume below the allowable minimum.
>
> Voids detected by automated inspection techniques. An outline of lead not visible in the solder joint because of excess solder.

18.3 Solder Joint Fillet

Today, the requirements for solder joint fillets are not unified. They vary with specific standards or specifications. There are discrepancies among the standards originating from different countries and organizations. United States industry standards still stand different from those of military standards. Chapter 19 will provide an overview of U.S. industry standards and international standards in solder or solder-related areas.

The U.S. industry standard, ANSI/J-STD-001A, provides guidelines for solder joint fillets in three classes of assemblies which are based on producibility, complexity, functional performance requirements, and verification (inspection/test) frequency. In order to refer to the guidelines, users need to determine for which class the assemblies are intended. The three classes are defined as

> *Class 1: General electronic products.* Includes consumer products, some computers and computer peripherals, and hardware suitable for applications where the major requirement is function of the completed assembly.
>
> *Class 2: Dedicated service electronic products.* Includes communications equipment, sophisticated business machines, and instruments where high performance and extended life are required and for which uninterrupted service is desired but not critical. Typically, the end-use environment would not cause failures.
>
> *Class 3: High-performance electronic products.* Includes equipment for commercial and military products where continued performance or performance-on-demand is critical. Equipment downtime cannot be tolerated, end-use environment may be uncommonly harsh, and the equipment must function when required, such as life-support systems and critical weapons systems.

The recommended parameters for solder joint fillet are maximum side overhang, maximum toe overhang, minimum end joint width, minimum side joint length, maximum heel height, minimum heel height, and minimum thickness (standoff height).

18.3.1 J-lead

Table 18-1 summarizes the solder fillet parameters for the J-lead according to ANSI/J-STD-001A; Fig. 18-3 describes the parameters.

The document that was submitted by the Netherlands for visual inspection of solder joints proposed three conditions: preferred, acceptable, and unacceptable (rework) as follows:

Preferred	Solder fillet at both sides of bend,thickness of lead (equivalent to F of Fig. 18-3)
Acceptable	Solder fillet at both sides of bend $\geq \frac{1}{2}$ thickness of lead
Unacceptable	Solder fillet at both sides of bend $< \frac{1}{2}$ thickness of lead

United States Military standard MIL-STD-2000A specifies:

- The solder fillet shall cover at least 75 percent of the lead width. (This is equivalent to maximum side overhang *A* of ANSI/J-STD-001A class 3, as shown in Fig. 18-3, wherein maximum side overhang = $\frac{1}{4}$ lead width.)

Table 18-1. Solder Fillet Criteria for J-Leads (ANSI/J-STD-001A)

Feature	Dimension parameter	Class 1	Class 2	Class 3
Maximum side overhang	A	$\frac{1}{2}W$	$\frac{1}{2}W$	$\frac{1}{4}W$
Maximum toe overhang	B	*	*	*
Minimum end joint width	C	†	$W-A$	$W-A$
Minimum side joint length	D	†	$1\frac{1}{2}W$	$1\frac{1}{2}W$
Maximum fillet height‡	E	‡	‡	‡
Minimum heel fillet height	F	†	$G + \frac{1}{2}T$	$G + T$
Minimum thickness	G	†	†	†

*Unspecified parameter.
†Properly wetted fillet evident.
‡Maximum fillet—solder fillet *shall not* touch package body.
NOTE: W = lead width, T = lead thickness. All dimensions are in millimeters.

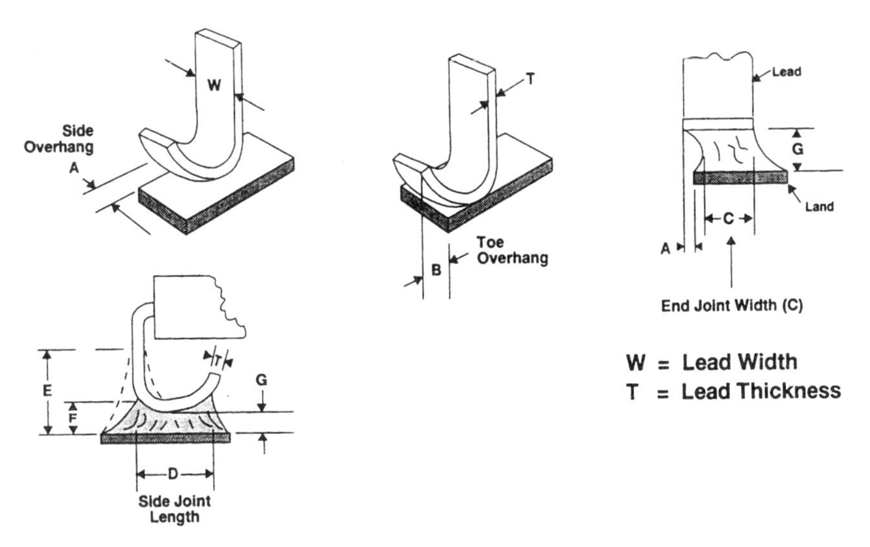

Figure 18-3. Schematic of J-lead parameters.

- The solder fillet shall be at least two lead widths along the J curve. (The criterion is equivalent to minimum side joint length *D* of ANSI/J-STD-001A, wherein the minimum side joint length of class 3 is 1½ lead width.)
- The solder shall extend up the side of the lead to the height of the inner surface of the lead, preferably the solder extending onto the inner surface of the lead but less than one lead thickness above the inner surface of the lead. (ANSI/J-STD-001A does not specify a requirement.)

18.3.2 Flat and gull-wing leads

Solder fillet requirements for flat and gull-wing leads as illustrated in Fig. 18-4 are summarized in Table 18-2 according to ANSI/J-STD-001A.

Under International Electrotechnical Commission (IEC) review, the recommended requirements deviated somewhat from the U.S. standard as shown in Table 18-3. It is to be noted that three classes of assemblies are designated as levels A, B, and C.

The Netherlands submitted a document proposing the following visual criteria:

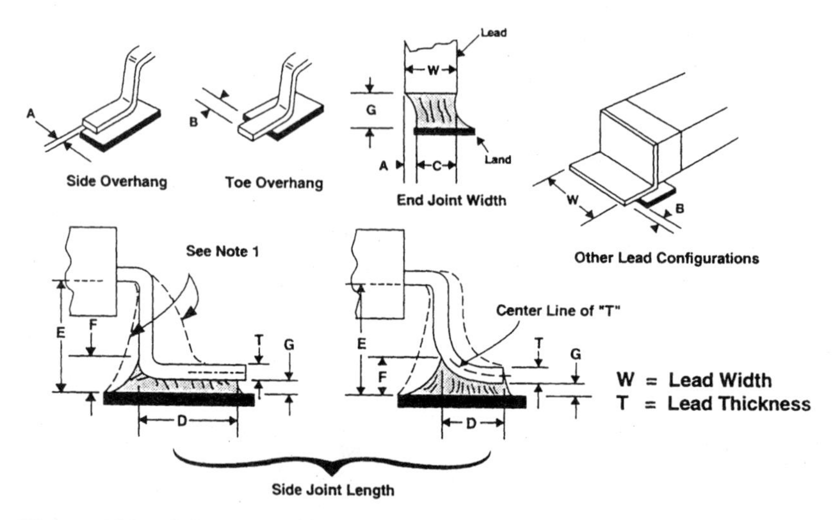

Figure 18-4. Schematic of flat and gull-wing lead parameters.

Table 18-2. Solder Fillet Criteria for Flat and Gull-Wing Leads (ANSI/J-STD-001A)

Feature	Dimension	Class 1	Class 2	Class 3
Maximum side overhang	*A*	½*W* or 0.5, whichever is less	½*W* or 0.5, whichever is less	¼*W* or 0.5, whichever is less
Maximum toe overhang	*B*	*	*	*
Minimum end joint width	*C*	½*W*	*W*—*A*	*W*—*A*
Minimum side joint length	*D*	*W* or 0.5, whichever is less	*W*	*W*
Maximum heel fillet height‡	*E*	†	‡	‡
Minimum heel fillet height	*F*	†	*G* + ½*T*	*G* + *T*
Minimum thickness	*G*	†	†	†

**Shall not* violate design conductor spacing.

†Properly wetted fillet evident.

‡Solder fillet may extend through the top bend. Solder *must not* touch package body or end seal, except for low profile SMD devices, e.g., SOICs and SOTs. Solder should not extend under the body of low-profile surface mount components whose leads are made of Alloy 42 or similar metals.

NOTE: *W* = lead width, *T* = lead thickness. All dimensions are in millimeters.

Table 18-3. Solder Fillet Criteria for Flat and Gull-Wing Leads (IEC)

Feature	Dimension	Level A	Level B	Level C
Maximum side overhang	A	½W or 0.5,* whichever is less, ⅓W below 0.5-mm pitch devices	½W or 0.5,* whichever is less, ⅓W below 0.5-mm pitch devices	¼W or 0.5, whichever is less, ⅓W below 0.5-mm pitch devices
Maximum toe overhang	B	½W*	½W*	½W*
Minimum end joint width	C	½W or 0.5, whichever is less	$W - A$, ⅔W below 0.5-mm pitch devices	$W - A$, ⅔W below 0.5-mm pitch devices
Minimum side joint width	D	W or 0.5, whichever is less	W, 2 W below 0.5-mm pitch devices	W, 2W below 0.5-mm pitch devices
Maximum heel fillet height	E	†	‡	‡
Minimum heel fillet height	F	†	G + ½T	$G + T$
Minimum thickness	G	†	†	†

*Shall not violate minimum design conductor spacing.

†Properly wetted fillet evident.

‡Solder fillet may not extend through the top bend.

NOTE: W = lead width, T = lead thickness. All dimensions are in millimeters.

For components with long flexible leads

Preferred

- Solder heel fillet height ≥ thickness of lead (equivalent to F of Fig. 18-4)
- Solder over total length of lead (equivalent to D of Fig. 18-4)
- Solder fillet at toe is not required

Acceptable

- Solder heel fillet height ≥ ½ thickness of lead
- Soldered over 75 percent of total length of lead
- Solder fillet at toe is not required

Unacceptable

- Solder heel fillet height < ½ thickness of lead
- Not soldered over 75 percent of total length of lead

For components with short stiff leads

The leads of these components are curved up at the end, forming a wedge-shaped space between the underside of the lead and the solder land.

Preferred

- Solder heel fillet height ≥ thickness of lead
- Solder fillet at toe is not required

Acceptable

- Solder heel fillet height ≥ ½ thickness of lead
- Solder fillet at toe is not required

Unacceptable

- Solder heel fillet height < ½ thickness of lead
- Solder fillet at toe is not required

MIL-STD-2000A specifies the following:

- Flat leads shall exhibit a visible fillet rising from the land to a minimum of 50 percent up the side of the lead whenever the lead is over the land. (In contrast, ANSI/J-STD-001A does not offer a fillet requirement for rising up the side of the lead.)
- The solder shall extend the length of the lead termination. (ANSI/J-STD-001A calls for minimum joint length D being equal to lead width for class 3 products.)
- The outline of the lead must be discernible. (ANSI/J-STD-001A does not specify.)
- The heel fillet shall be continuous between the heel of the lead and the circuit land. The heel fillet shall extend to the midpoint of the lower bend radius. (ANSI/J-STD-001A specifies that minimum heel fillet height F for class 3 products is the sum of lead thickness W and joint thickness G.)

18.3.3 Round or flattened coined leads

Table 18-4 sets the solder fillet criteria for round or flattened leads according to ANSI/J-STD-001A. Solder fillet configuration and parameters are displayed in Fig. 18-5.

Table 18-4. Solder Fillet Criteria for Round or Flattened Leads (ANSI/J-STD-001A)

Feature	Dimension	Class 1	Class 2	Class 3
Maximum side overhang	*A*	½*W*	½*W*	¼*W*
Maximum toe overhang	*B*	*	*	*
Minimum end joint width	*C*	†	†	*W*—*A*
Minimum side joint length	*D*	*W*	*W*	1½*W*
Maximum heel fillet height‡	*E*	‡	‡	‡
Minimum heel fillet height	*F*	†	*G* + ½*T*	*G* + *T*
Minimum thickness	*G*	†	†	†
Minimum side joint height	*Q*	†	*G* + ½*T* or 0.5, whichever is less	*G* + ½*T* or 0.5, whichever is less

**Shall not* violate minimum design conductor spacing.

†Properly wetted fillet evident.

‡Solder fillet may extend through the top bend. Solder *must not* touch package body or end seal, except for low profile SMD devices, e.g., SOICs and SOTs. Solder should not extend under the body of low-profile surface mount components whose leads are made of Alloy 42 or similar metals.

NOTE: *W* = lead width, *T* = lead thickness. All dimensions are in millimeters.

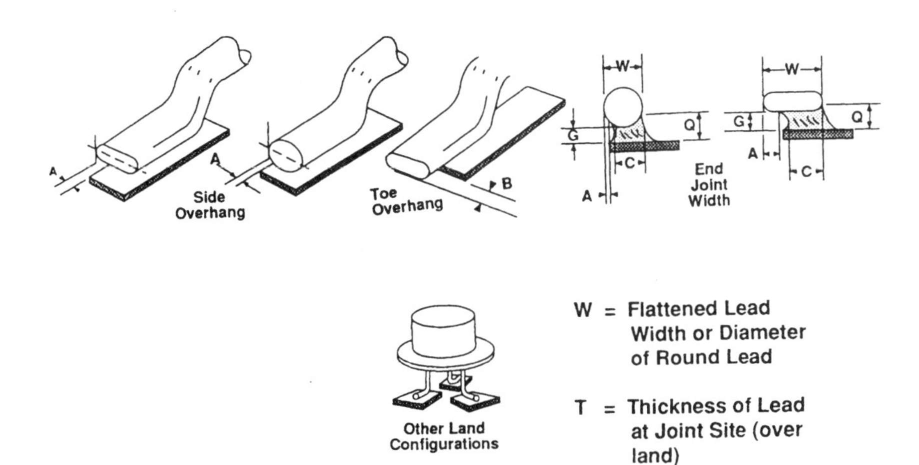

Figure 18-5. Schematic of round or flattened coined lead parameters.

MIL-STD-2000A specifies

- The heel fillet shall extend beyond the full bend radius. (ANSI/J-STD-001A offers minimum heel fillet height *F* being equal to the sum of lead thickness *T* and joint thickness *G*.)
- Minimum solder fillet height shall be 25 percent of the original lead diameter. (In contrast, ANSI/J-STD-001A class 3 calls for one half of lead thickness or lead diameter or 0.5 mm whichever is less.)
- The solder shall extend the length of the lead termination. (ANSI/J-STD-001A class 3 calls for $1\frac{1}{2}$ lead width or lead diameter.)
- The solder shall not overhang the land; and the outline of the lead must be discernible in the solder. (ANSI/J-STD-001A specifies solder fillet requirements.)

18.3.4 Rectangular or square end components

ANSI/STD-001 specifies solder fillet requirements as shown in Table 18-5 and corresponding parameters are illustrated in Fig. 18-6.

It should be noted that IEC intends to specify the minimum end overlap J as $\frac{2}{3}T$ for all three levels of assemblies.

The document proposed by the Netherlands specifies the following conditions. Preferred conditions are

1. When the height of termination is < 1.2 mm (0.048 in), the solder fillet height (equivalent to F of Fig. 18-6) is $\geq \frac{1}{3}$ height of termination.
2. When the height of termination is > 1.2 mm (0.048 in), the solder fillet height is > 0.4 mm (0.016 in).

Acceptable conditions are

1. When the height of termination is < 1.2 mm (0.048 in), the solder fillet height on the outside $= \frac{1}{3}$ height of termination and the solder fillet height on the other side is convex.
2. When the height of termination is > 1.2 mm (0.048 in), the solder fillet height on one side $= 0.4$ mm (0.016 in) and the solder fillet height on the other side is convex.

Unacceptable conditions are

1. When the height of termination is < 1.2 mm (0.048 in), the solder fillet height is $< \frac{1}{3}$ height of termination.

Table 18-5. Solder Fillet Criteria for Rectangular or Square End Components (ANSI/J-STD-001A)

Feature	Dimension	Class 1	Class 2	Class 3
Maximum side overhang	A	$\frac{1}{2}W$ or $\frac{1}{2}P$ or 1.5, whichever is less	$\frac{1}{2}W$ or $\frac{1}{2}P$ or 1.5, whichever is less	$\frac{1}{4}W$ or $\frac{1}{4}P$ or 1.5, whichever is less
Minimum end overhang	B	Not permitted	Not permitted	Not permitted
Minimum end joint width	C	$\frac{1}{2}W$ or $\frac{1}{2}P$, whichever is less	$\frac{1}{2}W$ or $\frac{1}{2}P$, whichever is less	$\frac{3}{4}W$ or $\frac{3}{4}P$, whichever is less
Minimum side joint length	D	Not required	Not required	Not required
Maximum fillet height*	E	*	*	*
Minimum fillet height	F	†	$G + \frac{1}{4}H$ or 0.5, whichever is less	$G + \frac{1}{4}H$ or 0.5, whichever is less
Minimum thickness‡	G	†	†	0.2‡
Minimum end overlap	J	Required	Required	Required

*The maximum fillet may overhang the land or extend onto the top of the end cap metallization; however, the solder *shall not* extend further onto the component body.

†Properly wetted fillet evident.

‡Unless satisfactory cleaning can be demonstrated with reduced clearance.

NOTE: W = width of termination area, T = length of termination, H = height of termination, P = width of land. All dimensions are in millimeters.

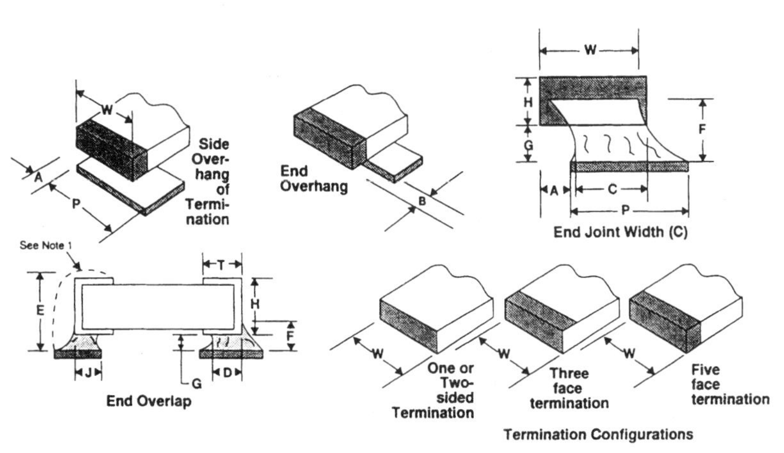

Figure 18-6. Schematic of rectangular or square end components.

2. When the height of termination is > 1.2 mm (0.048 in), the solder fillet height is < 0.4 mm (0.016 in).

MIL-STD-2000A proposes the following:

- The solder fillet shall extend at least 25 percent up the side of the device.
- As a minimum, 80 percent of the total end metallization shall have a solder fillet.
- A side fillet is not required.
- Solder may be present on the top of the end cap; however, the end cap shall not be fully encapsulated by solder.
- Wetting angles greater than 90°, which result from surface tension associated with large solder quantities are permissible provided good wetting is demonstrated.
- The difference between the thickness of the solder under each end of the part shall be less than 0.4 mm (0.015 in).

18.3.5 Cylindrical end cap terminations (MELF)

Figure 18-7 illustrates the key solder fillet parameters for MELF (metal electrode leadless face), and their recommended criteria are summarized in Table 18-6.

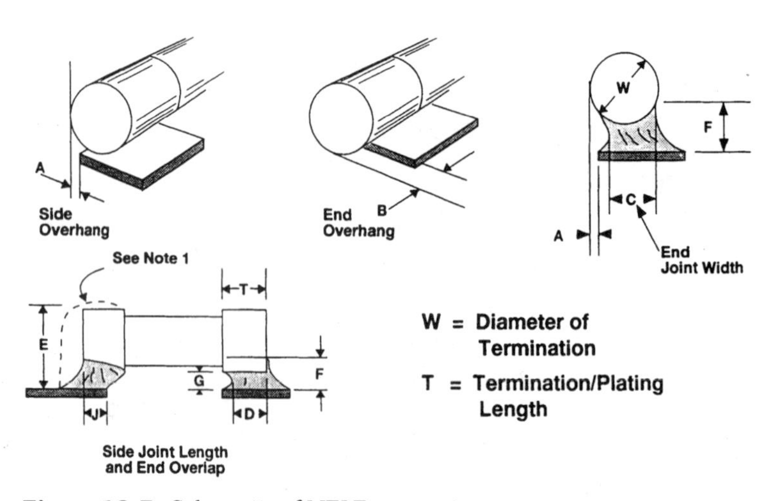

Figure 18-7. Schematic of MELF parameters.

Table 18-6. Solder Fillet Criteria for Cylindrical End Cap Terminations (ANSI/J-STD-001A)

Feature	Dimension	Class 1	Class 2	Class 3
Maximum side overhang	A	$\frac{1}{4}W$	$\frac{1}{4}W$	$\frac{1}{4}W$
Maximum end overhang	B	Not permitted	Not permitted	Not permitted
Minimum end joint width	C	*	$\frac{1}{2}W$	$\frac{1}{2}W$
Minimum side joint length†	D	*, †	$\frac{1}{2}T$†	$\frac{3}{4}T$†
Maximum fillet height‡	E	‡	‡	‡
Minimum fillet height (end and side)	F	*	*	$G + \frac{1}{4}W$ or 1.0,whichever is less
Minimum thickness	G	*	*	*
Minimum end overlap	J	Required	Required	Required

*Properly wetted fillet evident.

†Does not apply to components with end-only terminations.

‡The maximum fillet may overhang the land or extend onto the top of the end cap metallization; however, the solder *shall not* extend further onto the component body.

NOTE: W = lead width, T = lead thickness. All dimensions are in millimeters.

In contrast to Table 18-6, the IEC committee suggests that one-third of the termination diameter be the maximum side overhang A for levels A and B. It also recommends that the minimum end overlap J be two-thirds of the termination length for level A and B products and be equal to the termination length for level C products.

The Netherlands proposed the following criteria for the IEC's working group review:

Preferred: Solder fillet height (equivalent to F of Fig. 18-7) = component height and solder fillet is concave.

Acceptable: Solder fillet height $\geq$ 0.4 mm (0.016 in) and solder fillet is concave.

Unacceptable: Solder fillet height $<$ 0.4 mm (0.016 in) and solder fillet is convex.

MIL-STD-2000A requires

- The solder shall form a fillet extending at least 0.1 mm (0.004 in) or 25 percent of the part height up the side of the MELF.
- After soldering, the part shall be spaced between 0.05 and 0.4 mm

(0.002 and 0.015 in) off of the board. The solder connection shall form a fillet extending at least 0.1 mm (0.004 in) or 25 percent of the part height up the side of the MELF. (No requirement is specified for minimum thickness G, and $G + \frac{1}{4}W$ or 1.0 mm whichever is less is specified for minimum fillet height F in ANSI/J-STD-001A class 3 products.)

- At least 80 percent of the length of the metallized end shall extend over the land.

18.3.6 Bottom-only terminations of discrete chips

ANSI/J-STD-001A offers the solder fillet criteria for bottom-only termination discrete chips in Table 18-7 for parameters illustrated in Fig. 18-8.

18.3.7 Leadless chip carriers with castellation terminations

Solder fillet requirements for leadless chip carriers with castellation terminations are summarized in Table 18-8 for parameters displayed in Fig. 18-9.

MIL-STD-2000A proposes

- The solder fillet for leadless chip carriers shall be in the form of a horizontal fillet length equal to the vertical fillet rise.
- When the solder fillet extends beyond the castellation, the solder fillet shall be no closer than 0.25 mm (0.010 in) or the minimum electrical spacing, whichever is greater.
- There shall be evidence of good wetting; the wetting angle of the solder to the part and to the land shall be less than 90° except when the quantity of solder results in a rounded contour which extends over the edge of the land.
- When the leadless chip carrier has bottom-only terminations, the minimum solder connection height shall be 0.2 mm (0.008 in).

18.3.8 Butt joints

According to ANSI/J-STD-001A, butt joints are applicable only to two classes of products as featured in Table 18-9. Figure 18-10 illustrates the key parameters.

Table 18-7. Solder Fillet Criteria for Bottom-only Termination Chips (ANSI/J-STD-001A)

Feature	Dimension	Class 1	Class 2	Class 3
Maximum side overhang	*A*	*	*	*
Maximum end overhang	*B*	Not permitted	Not permitted	Not permitted
Minimum end joint width	*C*	$\frac{1}{2}W$ or $\frac{1}{2}P$, whichever is less	$\frac{1}{2}W$ or $\frac{1}{2}P$, whichever is less	$\frac{3}{4}W$ or $\frac{3}{4}P$, whichever is less
Minimum side joint length	*D*	*	*	*
Maximum fillet height	*E*	*	*	*
Minimum fillet height	*F*	*	*	*
Minimum thickness	*G*	†	†	0.2‡

*Unspecified parameter.

†Properly wetted fillet evident.

‡Unless satisfactory cleaning can be demonstrated with reduced clearance.

NOTE: W = termination width, P = land width. All dimensions are in millimeters.

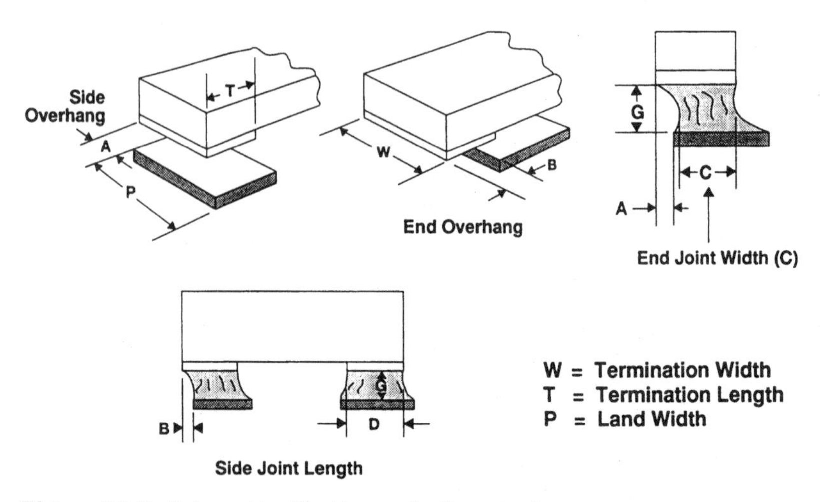

Figure 18-8. Schematic of bottom-only discrete chip parameters.

The document under IEC's review offers three levels of products with solder fillet as featured in Table 18-10.

18.4 Quality Assurance

Quality is a universal target. One way to define *quality* is the total conformance to the specified requirements for a specific use including performance, cost, and timeliness. Some of Genichi Taquichi's well-recognized quality ideas reflect that

- Continuous quality improvement and cost reduction are necessary for staying in business in a competitive economy.
- The final quality of a manufactured product is determined to a large extend by the engineering design of the product and its manufacturing process. This is basically to say, "do it right the first step."

Four fundamental elements to a well-planned quality product are design, material, equipment, and process control. With the sound design and right selection of materials and equipment, the establishment of the process and process control determines the manufacturing throughput and quality. ANSI/J-STD-001A section 12.2 offers the following:

The primary goal of process control is to continually reduce varia-

Table 18-8. Solder Fillet Criteria for Leadless Chip Carriers with Castellation Terminations (ANSI/J-STD-001A)

Feature	Dimension	Class 1	Class 2	Class 3
Maximum side overhang	A	$\frac{1}{2}W$	$\frac{1}{2}W$	$\frac{1}{4}W$
Maximum end overhang	B	Not permitted	Not permitted	Not permitted
Minimum end joint width	C	$\frac{1}{2}W$	$\frac{1}{2}W$	$\frac{3}{4}W$
Minimum side joint length†	D	*	$\frac{1}{2}F$ or P, whichever is less	$\frac{1}{2}F$ or P, whichever is less
Maximum fillet height	E	Not applicable	Not applicable	Not applicable
Minimum fillet height	F	*	$G + \frac{1}{4}H$	$G + \frac{1}{4}H$
Minimum thickness	G	*	*	0.2‡

*Properly wetted fillet evident.

†Length D is dependent upon fillet height F and is referenced to end of package.

‡Unless satisfactory cleaning can be demonstrated with reduced clearance.

NOTE: W = castellation width, H = castellation height, P = land length external to package. All dimensions are in millimeters.

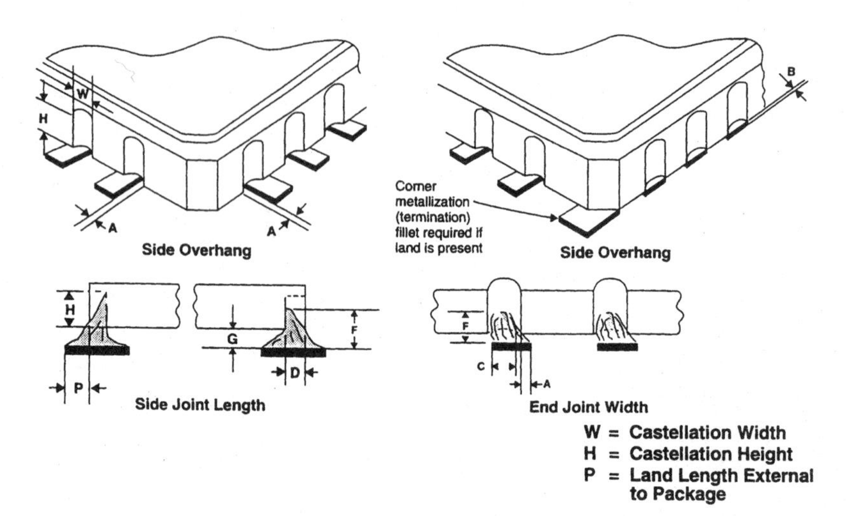

Figure 18-9. Schematic of leadless chip carrier parameters.

Table 18-9. Solder Fillet Criteria for Butt Joints (ANSI/J-STD-001A)

Feature	Dimension	Class 1	Class 2
Maximum side overhang	*A*	¼*W*	Not permitted
Maximum toe overhang	*B*	Not permitted	Not permitted
Minimum end joint width	*C*	¾*W*	¾*W*
Minimum side joint length	*D*	*	*
Maximum fillet height‡	*E*	†	‡
Minimum fillet height	*F*	0.5	0.5
Minimum thickness	*G*	†	†

*Unspecified parameter.

†Properly wetted fillet evident.

‡Maximum fillet may extend into the bend radius. Solder *must not* touch package body, except for low-profile SMD devices, e.g., SOICs and SOTs. Solder should not extend under the body of low-profile surface mount components whose leads are made of Alloy 42 or similar metals.

NOTE: 1. *W* = lead width, *T* = lead thickness. All dimensions are in millimeters. 2. Class 3 is not applicable.

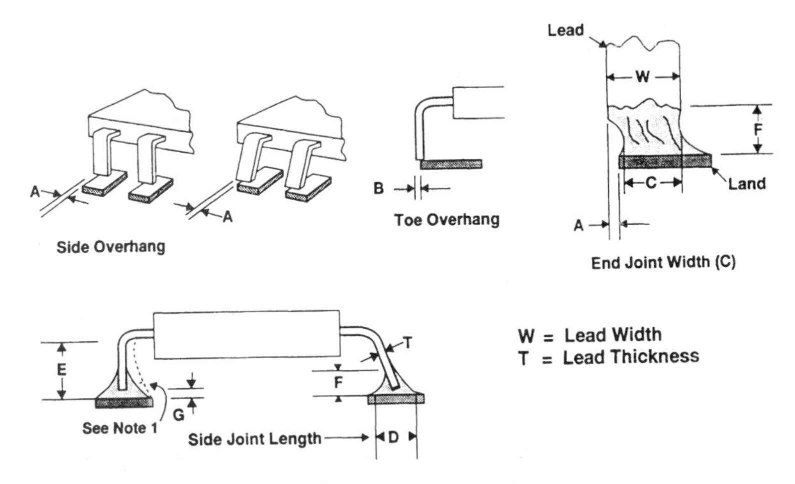

Figure 18-10. Schematic of butt joint parameters.

Table 18-10. Solder Joint Fillet Criteria for Butt Joints (IEC)

Feature	Dimension	Level A	Level B	Level C
Maximum side overhang	*A*	¼*W*	Not permitted	Not permitted
Maximum end overhang	*B*	Not permitted	Not permitted	Not permitted
Minimum end joint width	*C*	¾*W*	¾*W*	¾*W*
Minimum side joint length‡	*D*	*	*	*
Maximum fillet height	*E*	†	‡	‡
Minimum fillet height	*F*	0.5	0.5	*G* + ½*W*
Minimum thickness	*G*	†	†	†

*Unspecified parameter.

†Properly wetted fillet evident.

‡Maximum fillet may extend into the bend radius. Solder *must not* touch package body, except for low-profile SMD devices, e.g., SOICs and SOTs. Solder should not extend under the body of low-profile surface mount components whose leads are made of Alloy 42 or similar metals.

NOTE: *W* = lead width. All dimensions are in millimeters.

tions in the processes, products, or services to provide products or processes meeting or exceeding customer requirements. The process control system must include the following elements as a minimum.

1. Training must be provided to personnel with assigned responsibilities in the development, implementation, and utilization of process control and statistical methods that are commensurate with their responsibilities.
2. Quantitative methodologies and evidence must be maintained to demonstrate that the process is capable and in control.
3. Improvement strategies to define initial process control limits and methodologies leading to a reduction in the occurrence of process indicators in order to achieve continuous process improvement must be implemented.
4. Criteria for switching to sample-based inspection must be defined. When processes exceed control limits or demonstrate an adverse trend or run, the criteria for reversion to higher levels of inspection (up to 100 percent) must also be defined.
5. When defect(s) are identified in the lot sample, the entire lot must be 100 percent inspected for the occurrence(s) of the defect(s) observed.
6. A system must be in place to initiate corrective action for the occurrence of process indicators, out-of-control process(es), and/or discrepant assemblies.
7. A documented audit plan is defined to monitor process characteristics and/or output at a prescribed frequency.

Objective evidence of process control may be in the form of control charts or other tools and techniques of statistical process control derived from the application of process parameter and/or product parameter data. This data can be acquired from sources such as inspection, nondestructive evaluation, machine operation data, or periodic testing of production samples. For attribute data, the key is understanding and controlling parameters in the process that influence the response in question and establishing controls at that point. Attribute data, measured in parts per million of nonconforming product, can generally be correlated to a C_{pk} generated using variable data.

As the complexity of the board increases, often reflected by the increasing number of solder joints, the achievable production yield tends to decrease according to the concept of fundamental probability. At a given defect rate, Fig. 18.11 illustrates that the production yield

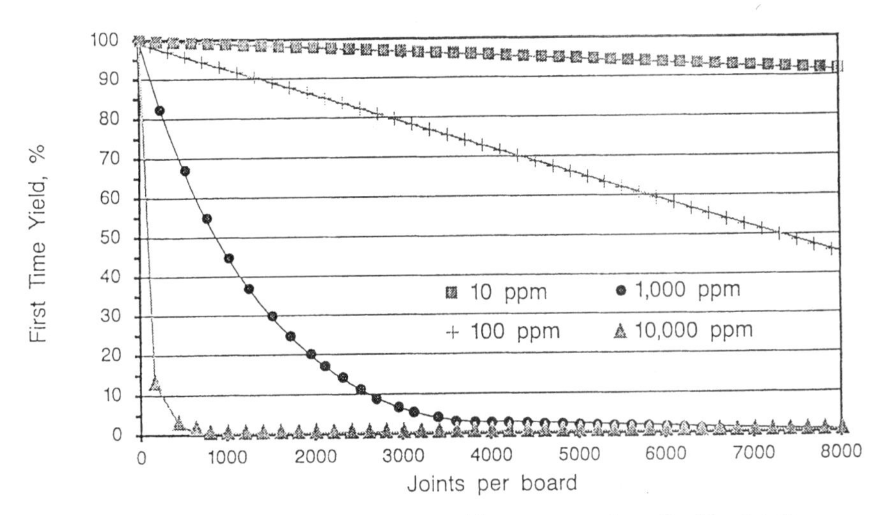

Figure 18-11. Quality and production yield versus number of solder joints.

decreases with increasing number of solder joints. For example, a defect rate of 100 ppm may maintain a 98 percent production yield for though-hole boards containing less than 300 solder joints. But the same defect rate (100 ppm) renders the production yield of a high-density surface mount board containing over 2000 solder joints below 87 percent.

19

Solder Joint Inspection, Rework, Repair

To achieve defect-free production is the common goal shared by all manufacturers, yet in reality, most production floors require some extent of inspection and rework. The main purpose of an inspection is to verify the quality and performance of the product for compliance with the specified requirements. The purpose of rework and repair is to change an unacceptable product to an acceptable level in accordance with the designated functional requirements.

In the past, the acceptance or rejection criteria for solder interconnections relied on visual inspection to detect the presence of any designated defects. Visual inspection with or without optical aids has been successful in identifying defects of through-hole solder joints for decades. The quality of the peripheral solder interconnections has largely been judged by the appearance of the wetting and how the solder joints meet the fillet requirements as discussed in Chap. 18. In addition, automated optical examination, laser, and x-ray inspection have also been used for inspection purposes.[1–15]

Array solder joints are essentially hidden. Visual inspection is rendered inapplicable. Inspection based on x-ray transmission is needed. The lack of access to the defective solder joint also makes the repair work a major task. In order to repair one solder joint, the whole BGA module has to be removed, cleaned, redressed, and remounted.

19.1 Inspection

The instruments which offer nondestructive evaluation are very much limited. The techniques to verify the quality of solder joints are grouped into three areas:

1. Examine the final solder joints by means of x-ray images
2. Examine the solder paste deposits prior to reflow
3. Continuously improve and establish a process system to assure the quality of solder joints produced, including the implementation of concurrent engineering

19.1.1 Examination of solder joint

For peripheral solder joints, inspection can be carried out by

- Manual inspection with or without a visual aid
- Laser
- X-ray image

Manual optical inspection detects exterior solder defects such as bridging, solder balls, misalignment, insufficient or open solder joints, and wetting quality. It can also identify solder joint fillet requirements for various solder joint configurations as illustrated in Figs. 18-3 through 18-10 and Tables 18-1 through 18-10.

Automated optical inspection using three-dimensional laser or x-ray imaging contributes to the improvement of process control. The three-dimensional laser analyzes the surface profile of solder joints to determine if their shape meets a specified criterion. X-ray imaging can detect interior (hidden) defects, such as voids, in addition to bridging, solder balls, misalignment, and insufficient or open solder joints.[5] As examples, two x-ray laminographic images of 0.5-mm-(0.020-in-) pitch QFP solder joints are shown in Figs. 19-1 and 19-2.[9] The inset profile of solder thickness exhibits the heel and toe fillet height and average solder thickness across the pad. The measurements are shown in Tables 19-1 and 19-2, respectively. The image and measurements exhibited in Fig. 19-2 and Table 19-2 indicate insufficient solder as compared with those in Fig. 19-1 and Table 19-1.

For array interconnections, their relatively large number of solder joints and thicker solder joints coupled with lack of accessibility make the inspection work more of a challenge. Conventional x-ray instru-

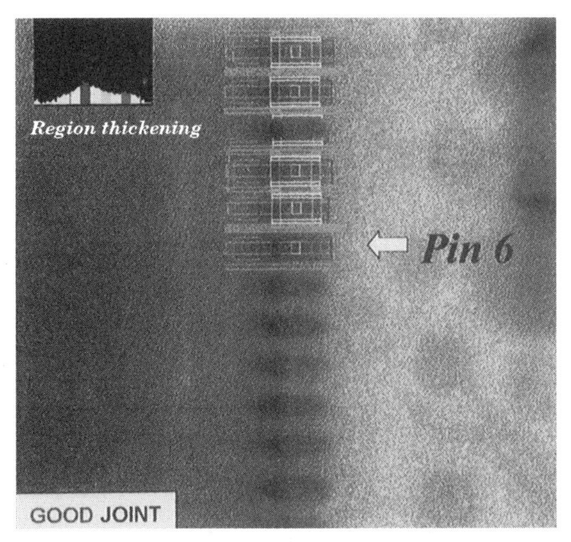

Figure 19-1. X-ray laminographic image of good QFP solder joints. (*Courtesy of Four Pi Systems.*)

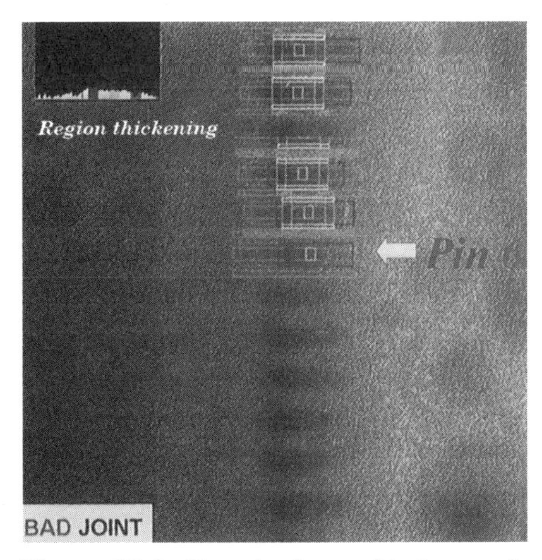

Figure 19-2. X-ray laminographic image of defective QFP solder joints. (*Courtesy of Four Pi Systems.*)

Table 19-1. Solder Joint Thickness for Good QFP Solder Joints

Reference designator	Inspection point	Thickness, 0.001 in
U^2 pin 5	Pad	1.65
	Heel	3.59
	Toe	1.95
U^2 pin 6	Pad	1.78
	Heel	4.15
	Toe	2.08
U^2 pin 7	Pad	1.80
	Heel	4.34
	Toe	2.03

Table 19-2. Solder Joint Thickness for Poor QFP Solder Joints

Reference designator	Inspection point	Thickness, 0.001 in
U^2 pin 5	Pad	0.94
	Heel	1.99
	Toe	0.99
U^2 pin 6	Pad	0.95
	Heel	2.19
	Toe	0.72
U]^2 pin 7	Pad	1.20
	Heel	3.07
	Toe	1.10

ments with radiation going perpendicularly through the board are found to be ineffective for detecting defects such as starved solder joints. Changing the design of solder pad dimensions on the board in relation to those on the BGA carrier may ease inspection using x-ray images. More objective inspection has to resort to cross-sectional x-ray imaging, such as laminography or scanned-beam laminography.

The scanned beam provides the image for each cross section at a specific location. Four typical sections, as shown in Fig. 19-3, are considered in order to reveal the most common defects.[1] The sections consist of

- Carrier slice
- Ball slice

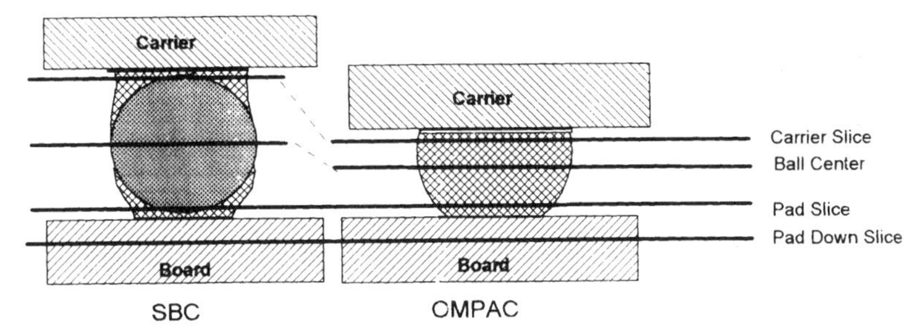

Figure 19-3. Locations of four typical cross sections. (*Courtesy of Four Pi Systems.*)

- Pad slice
- Pad down (board) slice

The x-ray images of three BGA solder interconnections are shown in Fig. 19-4, exhibiting the composite images (top row), and images of carrier slice, outer slice, and pad slice, respectively. The dark areas correspond to solder, and the light-gray shadows correspond to out-of-plane features of the interconnections.

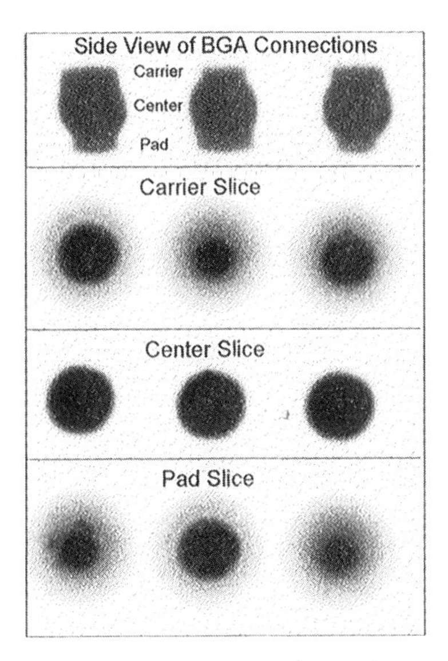

Figure 19-4. X-ray images of composite side view (top) and carrier, center, and pad cross sections of BGA solder interconnections. (*Courtesy of Four Pi Systems.*)

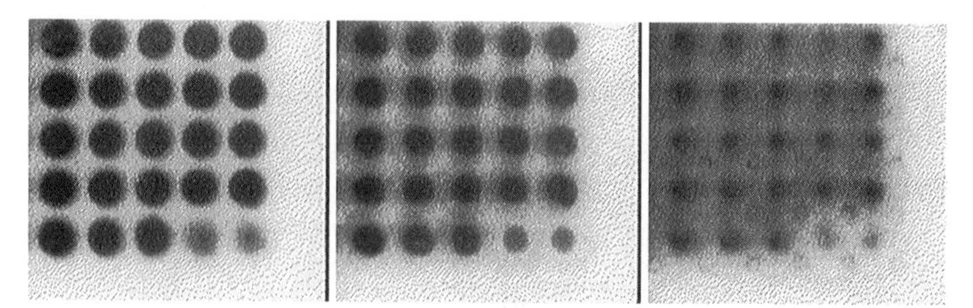

Figure 19-5. X-ray image showing two missing balls: Ball center (left), pad (center), pad down (right). (*Courtesy of Four Pi Systems.*)

The relevant number of slices needed depends on the level of process establishment and the confidence in data interpretation. Figure 19-5 offers one example: X-ray images indicate that two joints in the lower right corner have missing balls, resulting in open joints. In this case, the open joints are not due to solder paste deposition, since solder paste was deposited on the pad properly. Another example, shown in Fig. 19-6 reveals bridging between balls at the ball center slice and at the pad slice; its schematic side view is shown in Fig. 19-7. The open joint as a result of board warpage was detected based on the lighter

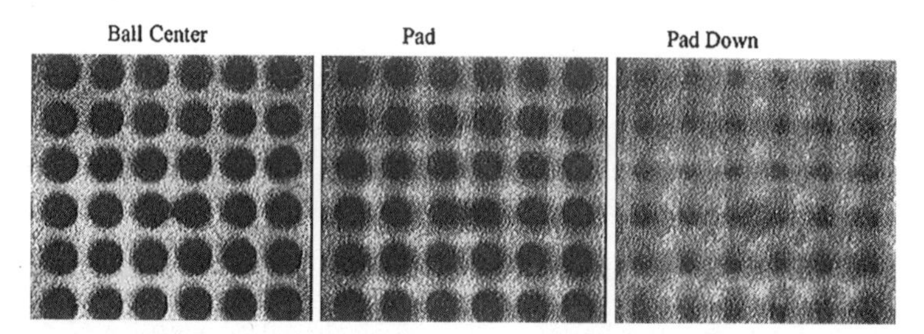

Figure 19-6. X-ray image showing bridging between balls at the ball center slice and at the pad slice.

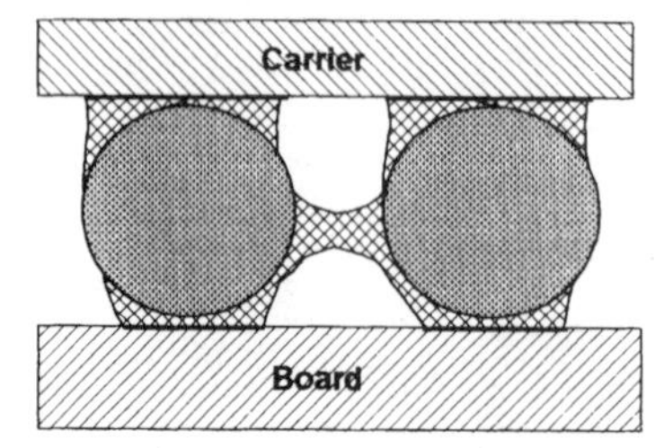

Figure 19-7. Schematic side view of bridging between solder balls. (*Courtesy of Four Pi Systems.*)

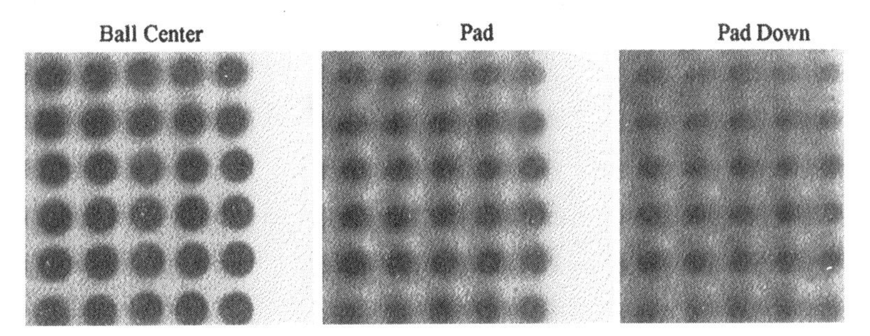

Figure 19-8. X-ray image of open joint as a result of board warpage. (*Courtesy of Four Pi Systems.*)

and smaller image area as viewed from the bottom to the top as shown in Fig. 19-8. Its schematic side view is represented by Fig. 19-9. The lighter image of the two top rows is due to the fact that the board is tilted away from the carrier which creates a longer distance from the ball bottom to the pad. The images of good solder joints as shown in Fig. 19-10 are compared with Fig. 19-11 for the defective (open) solder joints, as detected by x-ray laminography. The key measurements in joint radius at the pad and joint circularity for both good and defective solder joints are listed in Tables 19-3 and Table 19-4, respectively.

Defects such as open joints or bridged solder joints can also be

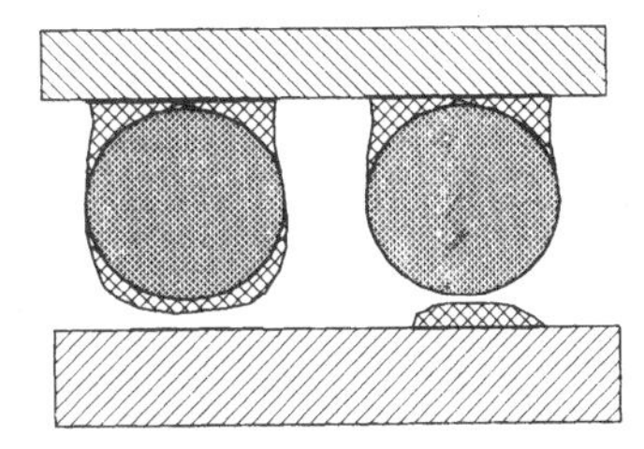

Figure 19-9. Schematic side view of an open joint.

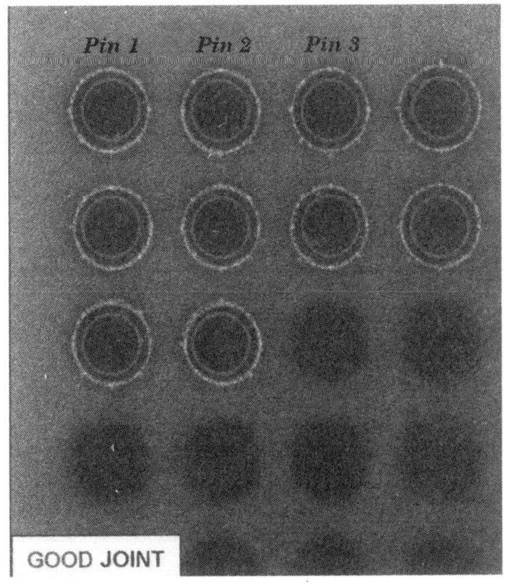

Figure 19-10. X-ray laminography of good BGA solder joint. (*Courtesy of Four Pi Systems.*)

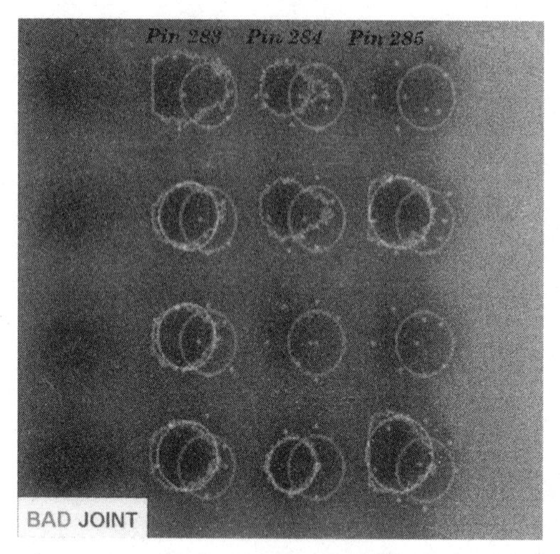

Figure 19-11. X-ray laminograph of defective BGA solder joint. (*Courtesy of Four Pi Systems.*)

Table 19-3. Radius and Circularity of Good BGA Solder Joints

Reference designator	Inspection point	Radius and circularity, 0.001 in
BGA pin 1	Ball joint radius at pad interface Circularity	16.92 0.31
BGA pin 2	Ball joint radius at pad interface Circularity	16.89 0.29
BGA pin 3	Ball joint radius at pad interface Circularity	16.94 0.34

Table 19-4. Radius and Circularity of Poor BGA Solder Joints

Reference designator	Inspection point	Radius and circularity, 0.001 in
BGA pin 283	Ball joint radius at pad Interface circularity	13.83 0.75
BGA pin 284	Ball joint radius at pad Interface circularity	13.37 0.58
BGA pin 285	Ball joint radius at pad Interface circularity	13.62 0.51

detected electrically. The board is designed with a daisy-chain pattern for each package which has two chains adjacent to each other starting from the center and spiraling to the outside of the array. Open solder joints are detected by probing the edge connector of the board, and bridged solder joints are detected as shorts between the two chains of each package.[2]

A versatile real-time x-ray inspection system is capable of revealing defects as small as 0.001 in. The system incorporated with patented x-ray camera technology[21] produces high-resolution imaging. Figure 19-12 is the x-ray image showing a missing ball; Fig. 19-13 reveals the

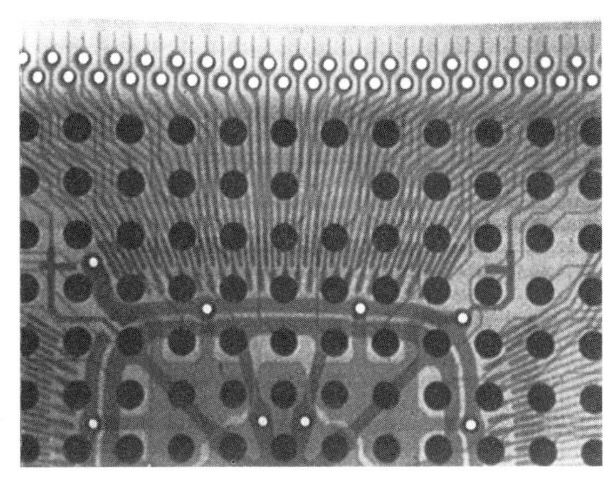

Figure 19-12. X-ray inspection—missing BGA balls. (*Courtesy of Glenbrook Technologies, Inc.*)

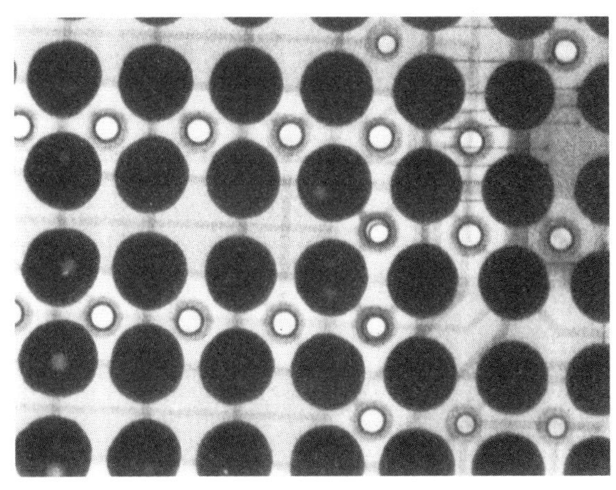

Figure 19-13. X-ray inspection—voids in solder joint. (*Courtesy of Glenbrook Technologies, Inc.*)

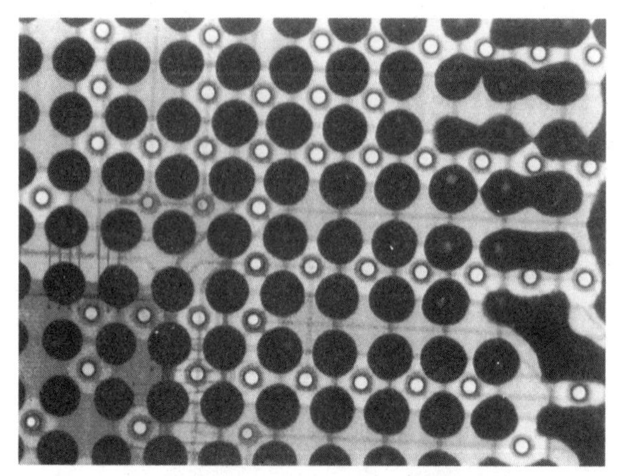

Figure 19-14. X-ray inspection—bridging between BGA balls. (*Courtesy of Glenbrook Technologies, Inc.*)

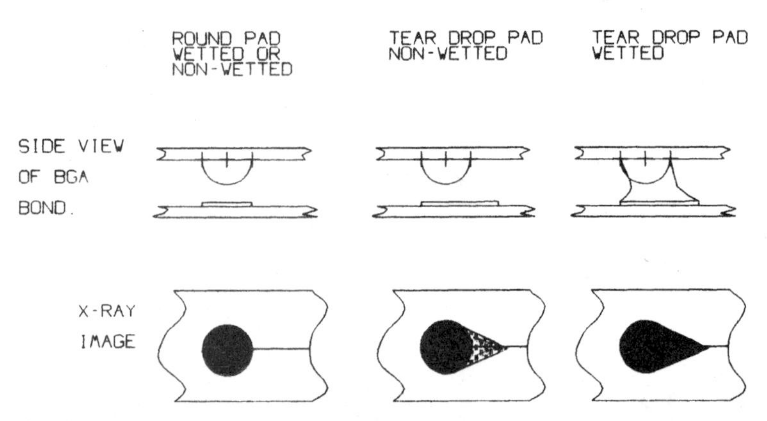

Figure 19-15. X-ray inspection—wetted or nonwetted on teardrop pad. (*Courtesy of Glenbrook Technologies, Inc.*)

voids in the solder joints; and Fig. 19-14 reveals bridging between balls. To detect the presence or absence of wetting, the shape of solder pads are designed to differ from round circles. Figure 19-15 illustrates the discernibility in x-ray images between wetted (right) and nonwetted (middle) pads by using teardrop-shaped pads. The system is compact and portable, capable of both real-time x-ray viewing and x-ray film imaging as shown in Fig. 19-16.

Figure 19-16. Desktop real-time x-ray inspection system. (*Courtesy of Glenbrook Technologies, Inc.*)

19.1.2 Examination prior to solder joint formation

The effect of defects on the cost of production intensifies as the production proceeds toward the final product. The detection of defects and their correction at an early stage of the process incur the least cost. For example, detection and correction at the solder paste deposition level costs less than that at the component placement level, which in turn costs less than at the solder joint level. Their relative cost effect is shown in Fig. 19-17.[4] Prior to the formation of solder joints, assurance of the following two areas is important to the quality of solder joints:

- Solder paste deposition
- Component placement

Examining solder paste deposits is an in-process step to prevent defective solder joints resulting from poor solder paste deposits. Techniques for examining the paste deposits have been introduced and adopted to

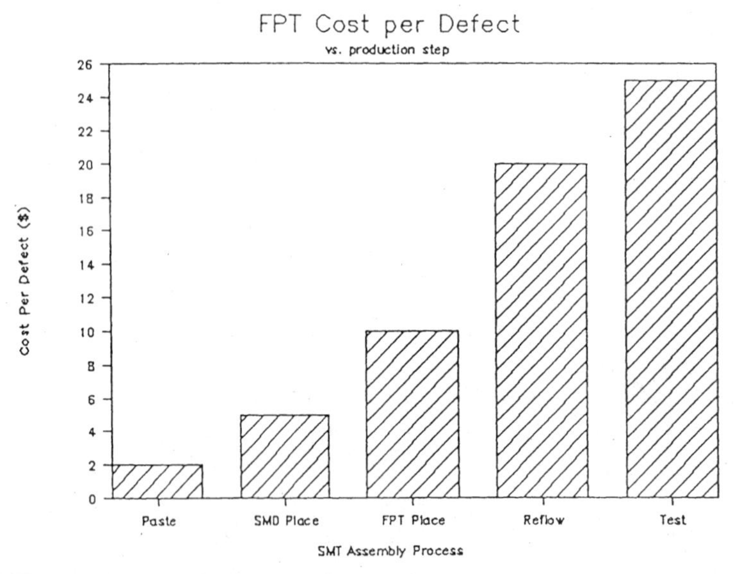

Figure 19-17. Relative cost effect versus production step detected.

different extents and include visual inspection, two-dimensional optical inspection, automated optical inspection (AO1), laser systems, and three-dimensional laser scanning. Three-dimensional examination of solder paste topography is achieved by rapid pulsing of the laser beam across the area to collect height data which are then analyzed by high-speed image processors.[6]

The recently introduced three-dimensional laser scanning technique[3] is designed to provide accurate volume measurement by offering accuracy in *x*, *y*, and *z* directions. Its capability of collecting a large number of data points from the top surface of the paste deposit and its ability to deliver the measurement of the paste deposit base (zero height) are considered primary features for obtaining accurate measurements. It is reported that 100 percent volumetric measurements on 912 solder paste deposits were achieved in a 30-s cycle time. The rate of data collection can reach 1.2 million samples per second. The in-line inspection system assists in sorting out boards with unacceptable paste deposits and systematically improves the printing process to achieve the minimal defect rate. With assurance of solder paste deposition, component placement in terms of accuracy and completeness is another area contributing to the quality of solder joints. This is especially important for finer-pitch components. As the pitch of components becomes smaller, the demands on placement accuracy increases. Three-dimensional laser scanning imaging has been an available technique which detects defects ranging from the presence or absence of components to the precise *x*, *y*, and theta (rotational) positions.

19.2 Rework and Repair

When conducting rework and repair of IC components on circuit boards, the following areas should be considered as general requirements:

- To remove the QFP component without damaging its leads and interior function and structure
- To remove the BGA component without damaging its interior function and structure
- To prevent any damage of land pattern and substrate
- To avoid excessive heat exposure of adjacent components and solder joints
- To consider the factors affecting the resulting solder joint properties
- To select the proper heating source including heat transfer efficiency and preheating needs

The key steps involve the removal of the fine-pitch component, preparation of the component (if not to be replaced) and the board, and placement of the component.[10–20] Common techniques for removing the components include

- Manual tool and soldering iron
- Hot gas
- Focused infrared
- Thermode
- Hot bar
- Hot bar plus hot gas
- Laser (CO_2 or YAG)

When using a soldering iron, the key parameters are recommended as summarized in Table 19-5.

To facilitate heat transfer, an external aid such as flux or solder wire can be used. The key performance to be achieved is to have all leads

Table 19-5. Key Parameters of Manual Repair Using a Soldering Iron

Parameter	Criterion
Tip temperature	260–371°C
Maximum temperature of component die	150°C or manufacturer's recommendation
Heating rate	2–5 °C/s
Maximum temperature for adjacent component solder joints	150°C
Time for component removal	1–3 s
Solder joint formation	2–4 s at 220°C

heated at one time. Another major factor affecting heat transfer is the size and configuration of the soldering tip.[12] A larger surface area of the tip provides better heat transfer. For example, a chisel with a ⅛-in tip delivers better heat transfer than one with a $^1\!/_{16}$-in tip, and a chisel tip is more efficient than a conical tip.

The preparation of leads and substrate involves the removal of residual solder and the reapplication of solder and flux. Residual solder can be applied by dispensing solder paste on the pads or using solder ribbon or tinning the leads. When using noncontact heating (hot gas, focused infrared, laser), thermal separation (isolation) of the component to be repaired from the rest is important to avoid the exposure of adjacent components to excess heat.[13] The maximum allowable temperature should not exceed the temperature specified by the manufacturer. It is generally accepted that adjacent devices should not be exposed to a temperature above 150°C, and that the heating rate should not exceed 5°C/s.

For mounting components, contact heating (thermode, hot bar, hot bar plus hot gas) is favored for its capability to provide lead-pad contact. Four-sided thermodes conform to the outline of the package leads on the pads and provide a heat source via electrically resistive material. The hot-bar method uses hot gas to heat the blades. For the hot-bar plus hot-gas system, the hot gas delivers direct heat to the leads and pads after heating the blades. This system features the capability of heating the leads and reflowing the pads before placing the component, thus reducing the lead skid problem.

For array packages, the techniques of rework and repair are still evolving. Nonetheless, the techniques developed for flip chips can be adopted.[19] Generally, the rework of array packages is not considered to be warranted for a single-chip package. Thus, it may be more cost-effective to use a new package. Three steps for repairing flip chips are involved:

- Package removal
- Solder cleanup
- Regular reflow to make new solder interconnections.

The removal process may include mechanical means using torque or ultrasound or a thermally enhanced process by preheating the component site and applying a separate heat source onto the top side of the component. The heat sources may involve conduction heating, inert gas, focused infrared with shield, and laser. Mechanical shaving, heated nitrogen flow to propel molten solder into the gas stream with a critical velocity, or slug-absorbing molten solder with sintered solderable metal are the techniques available for cleaning up residual solder.[19]

A device with dual hot-gas heating heads using nitrogen to enhance wetting has been developed.[20] A top-side head heats the component, and a second head applies heat to the general rework area from the bottom side providing a faster heating rate. It is recommended that, for PCMCIA cards, the bottom heat should be spread over the entire card to avoid card warpage caused by localized heat.

19.3 References

1. John A. Adams, "Using Cross-sectional X-Ray Techniques for Testing Ball Grid Array Connections and Improving Process Quality," *Proceedings NEPCON-West,* 1994, p. 1257.
2. David Gerke, "Ceramic Solder-Grid-Array Interconnection Reliability Over a Wide Temperature Range," *Proceedings NEPCON-West,* 1994, p. 1087.
3. Michael R. Beck and R. Joel Weldon, "3-D Solder Paste Inspections for High Yield Ball Grid Array (BGA) Assembly," *Proceedings NEPCON-West,* 1994, p. 686.
4. Michael Beck and Doreen Tarr, "Maximizing Yields Through Process Control Inspection," *Proceedings SMI,* 1991, p. 1147.
5. Michael Kral, "X-Ray Inspections: A Nicety or Necessity?" *Proceedings SMI,* 1991, p. 1133.
6. Russell J. Nieves, "Fine Pitch Print Inspection," *Proceedings NEPCON-West,* 1992, p. 181.
7. Alan Traug and Richard Alpen, "Laser/Infrared Technology for

Inspecting SMD Solder Joint Quality," *Proceedings NEPCON-West*, 1987, p. 669.

8. Dr. Riccardo Vanzetti, Vanzetti System, Private Communication, 1994.
9. Bruce Bolliger, Four Pi System, Private Communication, 1994.
10. Louis Abbagnaro, "A Low Temperature Method for Safe Manual Installations and Removal of Thru-Hole and Surface Mount Components," *Proceedings NEPCON-West*, 1993, p. 1119.
11. John A. Crawford, "Cost Effective Fine Pitch Rework," *Proceedings NEPCON-West*, 1993, p. 1119.
12. Louis Abbagnaro, "Manual Repair: New Techniques and Directors," *Surface Mount Technology*, November 1993, p. 27.
13. Douglas J. Peck, "Rework and Repair of TAB and FPT Devices," *Circuits Assembly*, May 1993, p. 43.
14. Manthos Economou, "Rework System Selection," *Proceedings NEPCON-West*, 1993, p. 1101.
15. James Bausell, "Refining Re-Assembly," *Circuits Assembly*, May 1993, p. 36.
16. Nick Hunn, "Getting Results from Hot-Gas Rework," *Circuits Assembly*, November 1990, p. 21.
17. Lino D. Andreti, "Essentials of Fine-Pitch Repair," *Circuits Assembly*, November 1990, p. 32.
18. Howard Markstein, "Reworking Assembled Printed Circuit Boards," *Electronic Packaging and Production*, April 1990, p. 75.
19. Karl J. Puttliz, "An Overview of Flip Chip Replacement Technology on MLC Multichip Modules," *Journal of Microcircuits and Electronic Packaging*, vol. 15, no. 3, 1992, p. 113.
20. Brian C. Devall, Jim McKaig, and Paul Min, "IBM Developed Attach and Removal Tool," *Proceedings NEPCON-West*, 1993, p. 763.
21. Product Brochure and private communication, Glenbrook Technologies, Inc., Morris Plains, New Jersey, 1995.

19.4 Suggested Readings

Anderson, James R., "An Innovative Approach to Fine Pitch Lead Alignment and Solder Delivery During Rework," *Proceedings NEPCON-West*, 1990.

Beauchesne, Robert C., "Solder Inspection: The Testing Challenge," *Printed Circuit Assembly*, December 1989, p. 16.

Caswell, Greg, and Steve Stach, "Repairing MCMs," *Advanced Packaging*, Fall 1993, p. 26.

Czaplicki, Brian P., "Verified BGA Repair," *Surface Mount Technology*, November 1994, p. 26.

Hendricks, Ryan, "On-Line Inspection Enables 6 Sigma Quality," *Circuits Assembly,* December 1990, p. 23.
Mani, B. S., et al., "A Process Control Plan for Fine Pitch SMTA Assemblies," *Proceedings SMI,* 1991, p. 1139.
Peck, Douglas J., "Thermal Considerations in Rework and Repair of TAB and Fine Pitch Devices," *Proceedings NEPCON-West,* 1992, p. 1459.
Romano, John A., "`Where' of the Rework Process," *Proceedings NEPCON-West,* 1991, p. 1965.
Rooks, Steve, Masa Okimuya, and Dick Urban, "Inspection of Very-Fine-Pitch Connections on PCMCIA Cards," *Proceedings NEPCON-West,* 1993, p. 752.
Salditt, Phil, "Deleting Open Solder Joints on Digital SMT Boards," *Electronics Engineering,* June 1993, p. 110.
Spigarelli, Donald J., "Thermal Separation in Surface Mount Attachment and Rework," *Proceedings NEPCON-West,* 1989, p. 89.
Trail, David K., Mark M. Mills, and Joyce A. Betts, "Solder Paste Application: Automated Inspection's Role in Process Control," *Surface Mount Technology,* April 1989, p. 23.

20
Future Trends

In the semiconductor die level, the IC continues to boast its density and functionality; the IC packaging proceeds from new and different packaging concepts to direct chip attachment, and accordingly module and board level assemblies are being developed with further advanced surface mounting techniques and processes.

20.1 Semiconductor Segment

With respect to technologies and products, semiconductors are progressing at a stunning pace since the IC was invented. It is revealing to compare the developments between the two decades—1980s and 1990s. Comparisons in key areas are summarized in Table 20-1.

The worldwide semiconductor market reached the $100 billion mark in 1994 with North America still holding the lead. Table 20-2 compares the respective market share of different regions in 1994 and 1995.

Microprocessors and memory chips continue to be the technology drivers. It appears that Moore's Law continues to predict well in IC development. For the next 15 years, the technology characteristics demanded by the marketplace[1] have been anticipated and are summarized in Table 20-3. As an example of the current technology (July 1995), the leading microprocessor (Intel Pentium) has achieved the following features:

133 MHz

8-in (200-mm) wafer

300 dies per inch wafer

0.35-μm process

Table 20-1. Comparison of Semiconductors in the 1980s and 1990s

	1980s	1990s
IC market	US $12–25 billion	US $50–200 billion
Wafer size	3–4 in	8–10 in
Circuit line	5 μm	0.1–0.8 μm
Pin count	40–80	300–800
Clean room	Class 1000	Class 0.1
Fabricators	Wet processing	Optical lithography (shorter wavelength, ultraviolet), x-ray lithography
Analytical tool	Microanalysis	Nanoanalysis
Packaging	Dual in-line package (DIP) surface mount technology (50 mil)	Surface mount technology —peripheral fine pitch ■ Array packages ■ MCM ■ Chip scale packaging ■ Chip scale assembly ■ Direct chip attach

Table 20-2. Worldwide Semiconductor Market[2]

	1994, US $billion	1995, US $ billion
North America	33.3	40.05 (+20.2%)
Japan	29.64	33.76 (+13.9%)
Western Europe	19.71	22.73 (+15.3%)
Rest of world	19.04	22.27 (+16.7%)
Total	101.72	118.81 (+16.6%)

Today, the personal computer industry is the biggest user of ICs. According to VIDA Research (Phoenix, Arizona), the personal computer industry will continue showing signs of growth. Its projected growth, as displayed in Fig. 20-1, shows personal computer shipments reaching 70 million units in 1997 worldwide, and its growth mirrors the IC market projection.

Table 20-3. Semiconductor Technology Characteristics

Minimum feature	1995	1998	2001	2004	2007	2010
First year DRAM shipment, μm	0.35	0.25	0.18	0.13	0.10	0.07
Memory, bits/chip	64M	256M	1G	4G	16G	64G
Memory cost/bit, millicent	0.017	0.007	0.003	0.001	0.0005	0.0002
Logic high-voltage microprocessor, transistor/cm^2	4M	7M	13M	25M	50M	90M
Logic low-voltage ASIC	2M	4M	7M	12M	25M	40M
Number of package pins						
Microprocessor	512	512	512	512	800	1024
ASIC	750	1100	1700	2200	3000	4000
Package cost, cents/pin	1.4	1.3	1.1	1.0	0.9	0.8
Chip frequency, on-chip clock, MHz	150	200	300	400	500	625
Power supply voltage, V	3.3	2.5	1.8	1.5	1.2	0.9
Battery voltage, V	2.5	1.8–2.5	0.9–1.8	0.9	0.9	0.9
Maximum power						
High performance with heat sink, W	80	100	120	140	160	180
Logic without heat sink, W/cm^2	5	7	10	10	10	10
Battery, W	2.5	2.5	3.0	3.5	4.0	4.5

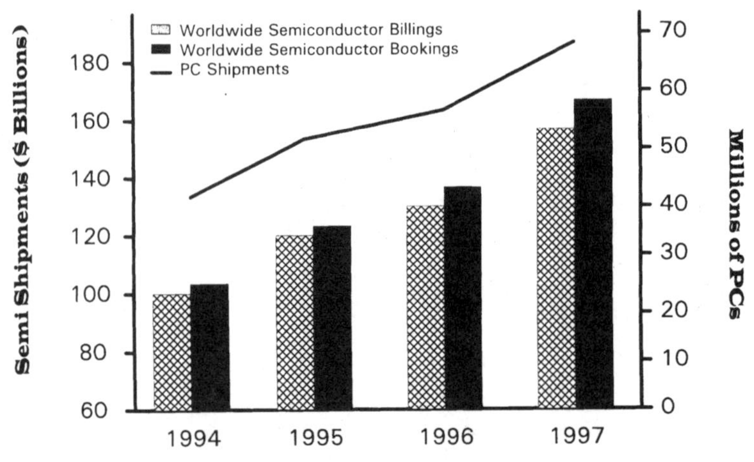

Figure 20-1. IC market and PC shipments.

20.2 IC Packages

Integrated circuit packages incorporating a high pin count and new package designs are expected to proliferate. Nonetheless, the biggest market share today goes to SOIC (including TSOP), and this is expected to continue in the foreseeable future. Table 20-4 depicts the worldwide market share of various packages.[3]

Table 20-4. Worldwide Merchant IC Package Market Share

Package type	1993	1998	1995–1998 CAGR
Plastic DIP	19,000 (47%)	9,625 (17%)	−13
Cerdip	1,300 (3%)	725 (1%)	−11
Ceramic PGA/BGA	80 (<1%)	200 (<1%)	27
Plastic PGA/BGA	50 (<1%)	400 (1%)	52
SOIC/TSOP	10,365 (26%)	21,275 (37%)	15
SOJ	2,300 (6%)	10,350 (18%)	35
PLCC	4,175 (10%)	6,900 (12%)	11
PQFP	1,725 (4%)	4,600 (8%)	22
LLCC/LDCC	375 (1%)	515 (1%)	7
Other (TAB, COB)	1,000 (2%)	2,875 (5%)	24
Total	40,370	57,465	

NOTE: Values are in millions.

The general trend of IC package evolution can be represented by the following chart:

Through-hole packages	DIP PGA	
	↓	
SMT packages	PLCC SOIC QFP	
	↓	
Advanced SMT packages	Fine-pitch QFP Fine-pitch SOIC Array package (BGA) Array-peripheral package Fine-pitch BGA	
	↙	↘
Chip-scale packages assembly	COB Wire bonding Bare chip	SMT(DCA) Flip chip Chip-scale packages

With further increases of I/O count, C-4 technology may leverage its strength to offer another level of electrical performance and power distribution. Figure 20-2 illustrates the minimum chip size versus I/O count of C-4 and ultrafine-pitch packages for I/O counts above 500.[4]

As I/O density increases, the system's thermal management becomes more demanding. The most efficient heat dissipation comes from the shortest distance for heat transfer in conjunction with most conductive media that make up the heat path. Ball grid array interconnections are expected to provide more efficient heat dissipation due to the shorter path of heat transfer compared with peripheral QFPs. Interconnections that are made of a metallic matrix (solder) are expected to surpass the polymeric matrix (adhesive) in conducting heat flow.

It is projected that 1994 worldwide BGA consumption will be around 78 million units out of 10 billion surface mount IC packages, constituting approximately a 0.8 percent share.[5] The demand for BGAs will continue to increase. By the year 2003, the use of BGAs will grow to 750 million units in comparison with 15 billion units of QFPs.

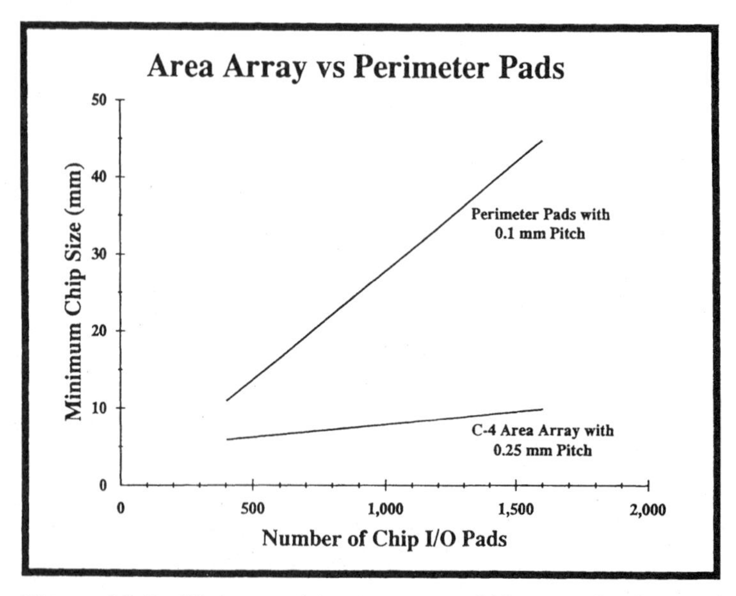

Figure 20-2. Minimum chip size versus I/O count for C-4 and ultrafine-pitch perimeter pads.

20.3 Trend of Soldering and Solder Paste

Today's state-of-the-art technology and mainstream manufacture include

- SMT manufacturing with fine pitch 20-mil (0.50-mm) QFP
- Mass printing with solder paste
- Paste with printing capability down to 12-mil (0.30-mm) pitch
- Manufacturability decreases with decreasing pitch
- Water-soluble and no-clean (air reflow) system
- Surface mount mass reflow (air or nitrogen)
- Wave soldering with no-clean flux (air or nitrogen) or water-soluble
- Surface mount mass reflow (Nitrogen) for minimal no-clean residue

Future technology and mainstream manufacturing are expected to involve

- Surface mount manufacturing using both peripheral fine-pitch QFPs and BGAs. The pitch of 20 mil (0.50 mm) may serve as a good separating point between the choice of QFP and BGA for a given number of I/O pins occupying a specific area.

- Advanced surface mount techniques may be needed to conduct chip scale assembly.
- Environment-friendly materials and processes

  Lead-free solders
  VOC-free chemistry
  Polymeric adhesives for some compatible applications
  Water-soluble and no-clean flux chemistry

- Enhanced reliability and quality in the manufacturing process and resulting solder joint, minimizing production-level inspection/repair.

20.4 Breakthrough Technologies

In addition to the innovations and improvements in IC circuitry miniaturization, IC packaging design, IC interconnection technologies, and surface mount board assembly techniques, key breakthroughs in fundamental material performance are warranted. This is particularly in demand for super-high-speed signal transmission (>300 MHz) applications.

- Conductor with lower resistance while maintaining other desirable properties
- Insulator with lower dielectric constant while maintaining other desirable properties
- Low-cost substrate with lower CTE, higher dimensional stability, and lower moisture susceptibility

By accomplishing the above, the transmission speed through the electronic mechanism, which is limited by on-chip connections (not by transistors), can be maximized.

20.5 Concluding Remarks

In order to serve the global market and to maintain a competitive edge, the manufacturing sector must constantly innovate and cut costs. The electronics industry has demonstrated the high spirit of innovation in its design of semiconductor chips, chip packaging, modules, and board assemblies. New technologies will continue to evolve leading to end-use products with better performance and lower cost that we can all enjoy.

As the global information era develops, it is a most fulfilling, rewarding, and challenging experience to be involved in and contribute to the backbone of the dynamic electronics industry. This is characterized by environmentally friendly surface mount manufacturing which put the advanced chips into real world uses, be it household appliances, entertainment gadgets, computer systems, or transportation and communication tools.

20.6 References

1. Semiconductor Industry Association.
2. World Semiconductor Trade Statistics and Cahners Economics (1995 forecast).
3. Integrated Circuit Engineering, Scottsdale, Arizona.
4. IBM Corporation.
5. BPA Technology and Management, New York.

Index

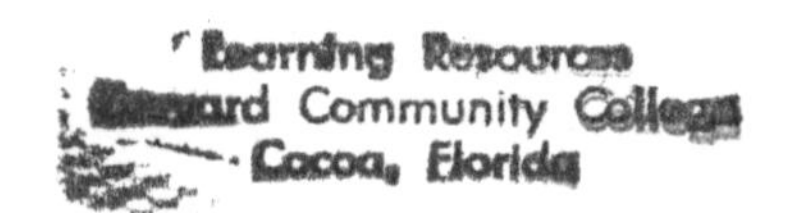